AF812783

Climate Smart Agriculture for Future Food Security

Sajid Faiz • Umair Ashraf
Kotb A. Attia • Rai Muhammad Amir
Editors

Climate Smart Agriculture for Future Food Security

Springer

Editors
Sajid Faiz
Institute of Molecular Biology and Biotechnology
The University of Lahore
Haripur, Pakistan

Umair Ashraf
Department of Botany, Division of Science and Technology
University of Education
Lahore, Punjab, Pakistan

Kotb A. Attia
Center of Excellence in Biotechnology Research
King Saud University
Riyadh, Saudi Arabia

Rai Muhammad Amir
Institute of Food and Nutritional Sciences
PMAS-Arid Agriculture University
Rawalpindi, Pakistan

ISBN 978-981-96-4498-8 ISBN 978-981-96-4499-5 (eBook)
https://doi.org/10.1007/978-981-96-4499-5

© The Editor(s) (if applicable) and The Author(s), under exclusive license to Springer Nature Singapore Pte Ltd. 2025

This work is subject to copyright. All rights are solely and exclusively licensed by the Publisher, whether the whole or part of the material is concerned, specifically the rights of translation, reprinting, reuse of illustrations, recitation, broadcasting, reproduction on microfilms or in any other physical way, and transmission or information storage and retrieval, electronic adaptation, computer software, or by similar or dissimilar methodology now known or hereafter developed.
The use of general descriptive names, registered names, trademarks, service marks, etc. in this publication does not imply, even in the absence of a specific statement, that such names are exempt from the relevant protective laws and regulations and therefore free for general use.
The publisher, the authors and the editors are safe to assume that the advice and information in this book are believed to be true and accurate at the date of publication. Neither the publisher nor the authors or the editors give a warranty, expressed or implied, with respect to the material contained herein or for any errors or omissions that may have been made. The publisher remains neutral with regard to jurisdictional claims in published maps and institutional affiliations.

This Springer imprint is published by the registered company Springer Nature Singapore Pte Ltd.
The registered company address is: 152 Beach Road, #21-01/04 Gateway East, Singapore 189721, Singapore

If disposing of this product, please recycle the paper.

Contents

About the Authors

Sajid Faiz, PhD, is an assistant professor working at the Institute of Molecular Biology and Biotechnology, The University of Lahore, Pakistan, and serving as adjunct scholar with Henan Institute of Science and Technology, Xinxiang, China. Dr. Fiaz received his PhD in crop genetics and breeding from the Chinese Academy of Agricultural Sciences. Dr. Fiaz's research interests include marker-assisted selection for agronomic traits, mutation breeding for biotic and abiotic stress resistance, quantitative trait loci (QTL) mapping, and genome editing for yield and quality traits in cereals. Currently, his lab group is working to identify genetic factors controlling drought stress tolerance in exotic and local landraces of rice. Dr. Fiaz is actively supervising undergraduate and postgraduate students. He has published more than approximately 170 research and review papers and around 12 book chapters with prestigious scientific magazines. He has been working as an academic editor for several journals. Dr. Fiaz has edited books with Springer and Wiley publishers on food security, genome editing, and omega mega information complex (OMICs)-based technology.

Umair Ashraf, PhD, is an assistant professor at the Department of Botany, University of Education, Lahore (Faisalabad Campus). He completed his PhD in "Crop Cultivation and Farming Systems" from South China Agricultural University (SCAU), Guangzhou, China. During his PhD, he was declared a "Distinguished International Young Scholar" by the China Scholarship Council. His research mainly focuses on the morpho-physiological, biochemical, and molecular basis of crop responses to various abiotic stresses and their management. He devised promising strategies for improving crop performance under sub-optimum conditions and suggested various novel indicators and effective management strategies augmenting stress tolerance in crop plants. He has published many research articles in SCI journals, and various book chapters with renowned publishers. He serves as a reviewer of several refereed journals and acts as an academic editor of several journals, also an active member of national and international scientific societies.

Kotb A. Attia is working as an associate professor at Center of Excellence in Biotechnology Research, King Saud University, Riyadh, Saudi Arabia. He completed his PhD in, "Genetic Engineering and Functional Genomics" from Fudan University, China, and pursued his postdoctoral fellowship from Niigata University,

Japan; Melbourne University, Australia; Toronto University, Canada. He has good experience with genetics from genes to genomes, genetic engineering, gene cloning and DNA analysis, functional genomics, microbial genetics, and bioinformatics. His research mainly focuses on physiological, morphological, and biochemical attributes of different plant species for achieving high yield and quality, stress breeding, genome editing, genome-wide association analysis, and improvement of crop physiological traits under climate change conditions. He has published several articles in scientific magazines and served as a reviewer, guest editor, and editorial board member. He is also actively engaged with various national and international scientific societies.

Muhammad Rai Amir is working as an associate professor at the Institute of Food and Nutritional Sciences, Pir Mehr Ali Shah (PMAS)-Arid Agriculture University Rawalpindi, Pakistan, and an additional director of industrial linkage at the same University. He is a seasoned academician and researcher with vast experience in the field of food science and nutrition. Dr. Rai has a PhD in food sciences and has made significant contributions to the advancement of knowledge in this field through his cutting-edge research and innovative teaching methods. He has a proven track record of excellence in teaching and has been instrumental in the development of several graduate programs in food science and nutrition and has published high-quality research articles that have been widely cited and recognized. Dr. Rai Muhammad Amir is also a highly regarded editor of several research international journals and has played a pivotal role in shaping the direction of research in food science and nutrition.

Novel Approaches and Practices for Sustainable Agriculture

1

Tanveer Ahmed, Huma Naz, Basharat Ali, Asma Shahbaz, Isha Akram, Ayesha Ilyas, Muhammad Hamza Babar, Noor Fatima, Javed Iqbal, Yasir Niaz, and Habib Ali

Abstract

Sustainable agriculture practices are essential for the entire world as they help in the security of food, checkclimate change, and come up with environmentally friendly ways of farming. This book chapter attempts to cover all the relevant aspects from the historical development of the concept to the innovations of the contemporary world. It describes sustainable agriculture, including the principles of the concept and the relevance of the practice in agriculture. It is focused on organic farming, permaculture, precision agriculture, conservation agriculture, agroforestry, urban agriculture, and climate-smart agriculture. Every approach is analyzed, including the theoretical foundations upon which it is based, the practices that support it, its advantages, its disadvantages, and examples of its usage. It also describes subsequent technologies and strategies, such as biotechnology, advanced technologies, and blockchain, and how they might assist in reinforcing sustainability. Further, it is associated with socio-economic topics, namely, policy actions, promotion strategies, and best practices. The last section of the chapter calls for future prospects and challenges that are the major debate of the future, as well as the need for further research and cooperation in the field of sustainability challenges. This book chapter's primary goal is to serve

T. Ahmed (✉) · A. Shahbaz · I. Akram · A. Ilyas · M. H. Babar · N. Fatima
Department of Life Sciences, Khwaja Fareed University of Engineering and Information Technology, Rahim Yar Khan, Pakistan

H. Naz
Department of Zoology, Cholistan Institute of Biological Sciences, Cholistan University of Veterinary and Animal Sciences, Bahawalpur, Pakistan

B. Ali (✉) · J. Iqbal · Y. Niaz · H. Ali
Department of Agricultural Engineering, Khwaja Fareed University of Engineering and Information Technology, Rahim Yar Khan, Pakistan
e-mail: dr.basharat@kfueit.edu.pk

© The Author(s), under exclusive license to Springer Nature Singapore Pte Ltd. 2025

S. Faiz et al. (eds.), *Climate Smart Agriculture for Future Food Security*,
https://doi.org/10.1007/978-981-96-4499-5_1

an important source for stakeholders, investigators, policymakers, or practitioners in the area of environmentally friendly farming.

Keywords

Sustainable agriculture · Organic farming · Precision farming · Agroforestry and biotechnology

1.1 Sustainable Agriculture

Sustainable farming can reliably supply meals as well as additional supplies to a world. An increasing population is necessary for human survival and, consequently, for all personal endeavors. Climate change, an elevated level of disappearing biodiversity, land degradation brought on by pollution, salinization, compaction, and soil erosion, the loss or contamination concerning water resources, higher costs of output, a continuous decrease in the number of farms, and, associated with that, impoverishment and a drop in rural population are just handful of the problems endangering agriculture's capacity to provide for human needs in the present and the future (Beus and Dunlap 1990; Wang et al. 2024). Not only does agriculture need to deal with these challenges, but the way it has been conducted in recent years has contributed significantly to all of these challenges (Goodland 1997). Together considering the more general idea of environmentally friendly growing acceptance, the idea of sustainable agriculture has gained popularity in response to these difficulties since the Brundtland Report's publication in 1987 (Tait and Morris 2000). However, the definition of ecological agriculture is unclear, much like the idea of environmentally friendly development altogether (Culleton et al. 1994). To address this issue and provide a more concrete definition of sustainable agriculture, many attempts have been made. Some definitional collections are as follows:

The integrated system of plant and animal production practices with a site-specific application that will, over time, "(a) satiate human needs for food and fiber, (b) improve environmental quality is the definition of sustainable agriculture, (c) efficiently utilize on-farm and nonrenewable resources assets then incorporate suitable organic processes or regulate sustain agriculture's financial sustainability, (d) and enhance farmers' and society's overall quality of life" (US Congress 1990). A sustainable farm tries to rely as much as feasible on resources that can be renewed and derived from the farm itself and advantageous natural processes rather than on products that must be purchased, such as fertilizers (Reganold et al. 1990). In order to prevent problems, we need to reduce trash and its effects on the environment, preserve every resource, or encourage agricultural systems' resilience, autonomy, and development. Sustainable agriculture refers to "management of the land, and sustained production for the nourishment and fulfillment of all methods which are compatible along with organic procedures" (MacRae et al. 1989). There is currently a need for criteria to be applied internationally in order to evaluate how important land use is in maintaining the cultural landscape. Whatever the case may be the emergence of an elevated degree of species richness in Central Europe was made possible by increased land use 1000 years ago; present land use tactics are blamed for the decline in biodiversity (Piorr 2003). The significance of creating agri-environmental indicators was emphasized by the Council of Europe at talks that took place in Cardiff in June 1998 and Vienna in December 1998.

Thus, what do we currently mean when we talk about agricultural sustainability? There are now numerous ways to suggest increased sustainability in some systems of agriculture (both preindustrial and industrialized) over dominant ones. Among them are biological, community-based, eco-agriculture, ecological, ecologically conscious, large-scale, fresh from the land, free-range, low feedback, natural, ecological design, environmentally friendly, and prudent usage (McNeely and Scherr 2003; Cox et al. 2004; Gliessman 2005; Pretty 1995; Conway and Pretty 1991. The question if systems of farming that use some of these concepts may be considered sustainable is still being hotly debated.

The following are the fundamentals of sustainability (Fig. 1.1):

- Introduce organic and as well as ecological mechanisms such as allopathy, nitrogen fixation, soil regeneration, competition, predation, and parasitism incorporated into the processes involved in producing food.
- Reduce the amount of nonrenewable inputs that harm the environment or the health of farmers and consumers.
- Make efficient use of farmers' expertise to help them become more self-sufficient and to replace costly outside inputs with human capital.
- Make good use of people's ability to work together to solve problems related to common agricultural and natural resource issues, such as pests, irrigation, watershed, forest, and credit control.

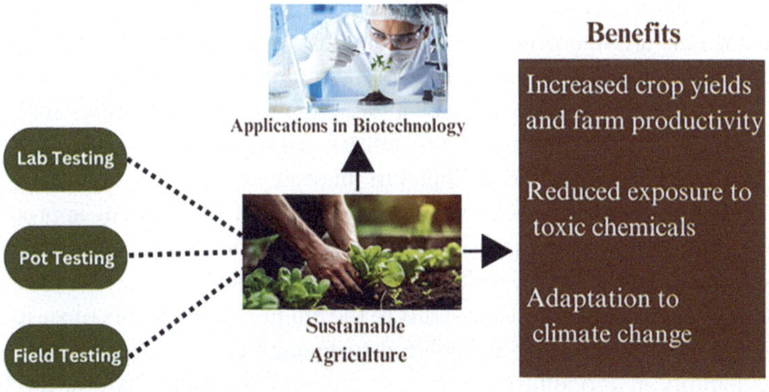

Fig. 1.1 Fundamentals of sustainable agriculture

Fig. 1.2 Importance of sustainable agriculture

However, the idea of agricultural sustainability does not suggest the denial of any techniques for technologies because of ideological disagreements. Modern technology is likely in order to maintain some sustainability benefits should it help farmers become more productive without negatively impacting the surroundings (Fig. 1.2). Agricultural methods that prioritize these ideas also frequently serve many purposes in economies and ecosystems (Dobbs and Pretty 2004; MEA 2005). Together, they create food and other products for farmers and markets, but they also support a number of important public goods, including groundwater recharge, wildlife and habitat preservation, clean water, carbon sequestration, flood protection, landscape amenity value, and leisure and tourism. This makes it possible to view sustainability as either case- or relative-specific and to infer a balance between various agricultural and environmental products and services. In order to maximize the benefits of nature's resources, more sustainable agriculture must tailor its technologies and methods to the specific needs of the local environment. These are most likely to result from new arrangements of social capital, which include trust relationships embodied in new social organizations, new institutional partnerships, both horizontal and vertical, and human resource capital, which includes qualities such as creativity, leadership, and management abilities. High social and human resource agricultural systems are better equipped to innovate in the face of unpredictability. This means that there are probably going to be a lot of different approaches to achieving agricultural sustainability. It also means that there is not likely to be a single combination of inputs, technology, and ecological management that works well for everyone. Fitting these elements to the unique requirements of various agricultural systems is implied by agricultural sustainability.

Sustainability in agriculture is sometimes mistakenly understood to imply a net reduction in input consumption, which would make such systems essentially expansive (requiring more acreage to produce the same amount of food). According to recent empirical research, the factors influencing agricultural production such as switching from fertilizers to nitrogen-fixing legumes, focusing on natural enemies instead of chemical pesticides, and plowing to zero-tillage are what lead to the success of agricultural sustainability initiatives and projects. An idea that focuses on resource intensification that better utilizes already-existing resources (such as land, water, and biodiversity) as well as technologies is preferable to one that emphasizes extensiveness (Conway and Pretty 1991). The "type of intensification" is the essential question. "Sustainable intensification" refers to intensification that minimizes or completely eliminates environmental impact while utilizing the greatest technology and inputs (best genotypes and best ecological management). This is accomplished by combining the use of natural, social, and cultural assets.

1.2 How Do We Achieve Sustainable Agriculture?

Based on environmental care and equitable economics, producers and other agricultural experts have developed a solid set of guiding principles for sustainability. Agro-ecology systems are becoming more productive and efficient at a rate that producers and researchers are accelerating every year. Sustainable techniques are receiving more support from agricultural institutions and Cooperative Extension offices. Additionally, more farmers recognize the benefits of these solutions on a human and financial level each year. Sustainable agriculture is slowly taking hold, with just one crop, a single plot, and one family at a time (Table 1.1). Away from the farm, farmers' policies and local markets that promote sustainable practices are being developed by grassroots activists and consumers. They are trying to increase customer knowledge of the processes involved in growing and processing food, including how plants, animals, soil, and water are handled. Additionally, they are attempting to strengthen the ties that will eventually lay the groundwork for both regional and locally independent food systems.

Table 1.1 Criteria of sustainable agriculture practices

Criteria	Example of practices
Weeds killer	Intercrops and crop rotations
Crop production	Planting trenches System of rice enhancement Intercrops
Soil fertility intensification	Composts and manure Green manure and biomass transfer
Soil and water administration	Water control structures such as terraces Help prevent soil erosion Contour planting Living barriers and hedgerows Conservation tillage

Know your markets, protect your profits, and add value to your products:

- Diversify enterprises.
- Market outside the commodity supply chains and corporate vertical integrators.
- Emphasize direct marketing and premium specialty markets.
- Consider forming a cooperative with other farmers.
- Add value through on-farm processing.

1.3 Historical Background and Evolution of Sustainable Agriculture

A belief about what ought to be, also referred to as the model, the objective, or even the ideology, has always directed agricultural evolution. A growth gap separates that objective from current agricultural practices. Similarly, the model's inclusion or breadth varies with time. We could spend a lot of time delving into the philosophical underpinnings of development, but we'll stick to a few main ideas that seem to be most directly connected to current sustainability theories. Some analysts use a Newtonian worldview to trace the origins of the current conflict. Four key components a mechanical conception of nature, a strict division between nature and civilization, trust in progress, and a consumerist ethic are used by Rifkin (1980) to characterize that viewpoint. However, the method is more important to us here than philosophical content. The industrialization of American agriculture was just getting started at the turn of the twentieth century. There was already tension between the perceived drastic changes brought about by industrialization and the urban "agrarian" lifestyle (Danborn 1979). The differences that existed between funders and the expanding group of "land-grant" scientists, however, were more significant. Eighty years later, these divisions are back to haunt us. Farmers' popular attitude at the beginning of the twentieth century led to the rejection of the Jeffersonian idea that individualism was the highest ideal. Politically, this resulted in the founding of groups like the Grange. Farmers believed it was their responsibility to advance and disseminate technological knowledge among themselves. The information came from two places. The new industry served as a model for the "systematic agriculturists." The second group, known as "scientific agriculturists," sought to rationalize and formalize their experiences as "natural historians" by using nature as their model (Marcus 1985). Land-grant scientists were starting to make an impression at the same time (Rossiter 1975). We can trace the origins of much of the present discussion on sustainable agriculture to these ideologies and to these groupings of farmer-scientists at the turn of the century. In the early 1900s, there was a significant expansionist trend in American agriculture. In the early 1930s, there were 6.8 million farms at its highest point (Hardin 1988). The rapid adoption of mechanization was fueled by rising costs and labor shortages brought on by both regional growth and conflicting needs from industry. Crop hybrids were developed and widely used as an example of how quickly technologies were advancing. A key factor in the formulation of the development paradigm was the land-grant system. The ideas of conservation changed in the early twentieth century, first emphasizing the preservation of natural regions.

The early twentieth-century progressive conservation movement laid the theoretical groundwork for later conservation initiatives (Batie et al. 1985). These shared origins are reflected in a number of conservative initiatives, such as the Food Security Act of 1985, the Soil Bank Programmed of 1956, and the Agricultural Adjustment Act of 1933 (Phipps and Crosson 1986). Through the use of land reserves or set-aside acres, these programs addressed the issues of increasing excess production and conserving soil (Hardin 1988). The "chemical" technologies spread in tandem with these advancements, new agricultural technology, and rapidly increasing mechanization. Following World War II, the use of industrially produced fertilizers and pesticides expanded quickly, resulting in what Rifkin refers to as "the age of alchemy" (Rifkin 1983).

1.4 Soil Management Technique in Organic Agriculture

1. **Crop Rotation**

   Crop rotation is the practice of planting several crops in the same spot during successive seasons. It aids in controlling diseases and pests, decreasing soil erosion, and maintaining soil fertility (Altieri 1995).
2. **Cover Cropping**

   The main purpose of growing cover crops such as clovers, grasses, and legumes is to preserve and improve the health of the soil. Their purpose is to cover the land, not to be harvested (Clark 2007).
3. **Composting**

   Compost is a nutrient-rich soil additive that is made by breaking down organic materials (such as plant leftovers and manure) (Rynk 1992).
4. **Green Manuring**

   To increase soil fertility, green manuring is cultivating and then adding green plant materials to the soil. Because leguminous plants can fix nitrogen, they are often employed (Sarrantonio 1992).
5. **Reduced Tillage**

   Minimizing soil disturbance through reduced tillage or no-till farming helps preserve soil structure and lowers erosion (Lal 1991).

1.5 Certification and Regulation of Organic Farming

Organic farming appears to be a significant example of environmental self-regulation with a solid foundation in values, with the goal of creating sustainable farming systems that are able to regulate themselves. Public certification was introduced as part of the political acceptance of organic farming, although this could interfere with self-regulation. The experiences of Denmark, which include significant rises in organic farming and domestic organic food markets, are used to explore this topic.

The organic agricultural movement was closely partnered with the establishment of public certification of organic products. However, this partnership was gradually weakened by the introduction of EU certification, and as a result, the organic social movement no longer influences production standards. Conversely, Danish organic farmers appear ready to create and implement new forms of self-regulation.

1.6 Permaculture: Designing Sustainable Systems

It became evident in the late nineteenth century that a more productive agricultural system was required to feed the world's population, particularly in the Global South (Krebs and Bach 2018). The green revolution of the 1960s appeared to be the answer to hunger and the avoidance of conflicts over food supplies with the development of high-yield cultivars, synthetic fertilizers and pesticides, and sophisticated technology (Beddington et al. 2012). These technological advancements brought a clear improvement in terms of increased production per unit of land, decreased workload, and enhanced food safety (Foley et al. 2011). Unfortunately, there was an unanticipated cost associated with these benefits. Modern industrial agriculture, which is predicated on the large-scale production of monocultures employing heavy machinery and copious amounts of agricultural chemicals, such as synthetic fertilizers and pesticides, is linked to many of the environmental issues facing humanity today (Power 2010). Biodiversity, soil organic matter, water reservoirs, greenhouse gases, the nitrogen cycle, the phosphorus cycle, and greenhouse gases are all significantly impacted by agriculture. According to Krebs and Bach (2018) each of these elements carries the potential to undermine the ecosystems' essential processes and jeopardize the availability of food for humans. Consequently, new designs for agricultural systems are required. As an alternative to the way land uses are designed systems. The scientific literature only briefly discusses permaculture, another strategy that offers long-term solutions for the human supply of resources and things (Ferguson and Lovell 2014).

Combining the terms "permanent" and "agriculture," the term "permaculture" refers to a design system and best practices framework for the development and maintenance of resilient and sustainable agroecosystems (Cordell et al. 2009). "Consciously designed landscapes, which mimic the patterns and relationships found in nature, while yielding an abundance of food, fiber, and energy for provision of local needs" is how co-founder David Holmgren describes permaculture. When planning agroecosystems, the most crucial elements of permaculture are (a) site characteristics; (b) the interaction between individual elements on multiple levels, ranging from mixed cultures at the field level to the diversity of land use at the agro-ecosystem level; and (c) the spatial arrangement of individual elements in Australia, permaculture first gained traction as an alternative agriculture movement in the late 1970s. The practical concept of permaculture teaches us to work with nature, not against it. Earth care, people care, and fair share—the guiding ethical principles of permaculture acknowledge the interconnectedness of agriculture with the rest of the planet and between humans and nonhuman animals. It takes more than simply sustainable food production to practice permaculture. A system known as permaculture integrates the various roles that would otherwise be performed by distinct components of human settlements, such as nutrients, species, farms, villages, industries, and sectors (Suh 2022).

1.7 Designing Resilient and Self-Sustaining Farming Systems

The goal of permaculture is to develop robust living systems that are modeled after the structures, functions, and patterns seen in nature. There are now design principles that are employed as a foundation for creating intricate agroecosystems. Depending on the nature of their work, certain permaculture designers have created their own sets of principles. The principles are succinct declarations that provide guidance and a range of options for action while working with complicated systems. The last six principles can be viewed from the viewpoint of a top-down designer, whereas the first six take a bottom-up approach. Additionally, as a result, there are certain areas where the concepts overlap (Krebs and Bach 2018). Observing and interacting empower one to be able to respond creatively to change, and striving to generate no waste and implementing self-regulation will lead to integration rather than segregation of aspects (Holmgren 2002). Designing industrial product and service systems with a systems approach is necessary to achieve sustainable development. Despite the fact that many businesses have set sustainability goals, developing sustainable systems is still difficult due to the wide range of social, environmental, and financial aspects it need to be considered throughout the system's life cycle (Fiksel 2003). Conventional systems engineering methods may be susceptible to unanticipated events despite their best efforts to anticipate and withstand disturbances. Alternatively, systems can be designed to have intrinsic "resilience" by leveraging basic characteristics such as variety, efficiency, adaptability, and cohesiveness. The majority of prior research on sustainable design has been on increasing ecological efficiency. Companies, for example, have discovered that reducing material and energy intensity and transforming waste into valuable secondary products benefit both shareholders and society as a whole (Altieri et al. 2015a, b). To promote systems thinking, a design protocol is presented that includes identifying system functions and boundaries, establishing requirements, selecting appropriate technologies, developing a system design, evaluating performance, and devising a practical deployment method. The method promotes explicit consideration of resilience in both engineered systems and the larger systems into which they are integrated. In order to encourage the development of climate-friendly food systems there is a need for other CCFS SMA techniques such as conservation agriculture (Sgroi 2022), agroforestry (Barrios et al. 2020), and soil water conservation.

1.8 Examples of Permaculture Design in Agriculture

In general, it is worth stating that the focus of the climate-resilient food systems research has been given to farming as the way toward sustainability and climate resilience in harsh environmental conditions. Among the undertakings required for the realization of this goal are the introduction of sustainable land management techniques, the diversification of the agricultural sector, engagement in cattle and poultry production, homestead gardening, and provision of access to technology, markets, and finance. In agriculture, permaculture enhances the physical properties of the soil by enhancing the amounts of organic matter within it, the structure of the soil, and the ability of the roots to penetrate it. Korže (2018) identifies that the capability of soil moisture content, improved infiltration rate, and the minimization of compaction, erosion, and landslides can guard the soil as well as its nutrient reservoir. It helps in accumulating the humus on the soil as well as reducing the velocity of rain droplets when they fall to the ground (Hathaway 2016). Reducing soil loss, increasing infiltration, and accumulating excess leached water can lead to future water reservoirs. These circumstances also promote the growth of microorganisms and plants in the soil (Pandey 2007). In an organic fertilizing study, the interaction of Compo Guano and lime was unclear. However, in the long run, it was found to be the best solution due to its positive effects on soil pH and nitrogen availability, as well as preserving microbial biomass and labile organic matter (Mursec 2011). The focus was on sustainable agriculture through permaculture, ensuring soil nourishment, food and economic security, and efficient use of local resources. There are limitations in the study and literature on permaculture, yet it is possible to compile several literatures. Permaculture improves soil quality by increasing organic matter, leading to better water retention, infiltration, and nutrient availability. Improve microbial activity and reduce soil and nutrient erosion (Yadav et al. 2023). It enhances the physical, chemical, and biological qualities of soil. It ensures food and economic security by offering a varied range of crops, increasing agricultural yields and income, and utilizing low-cost inputs to maximize output. It maximizes resource efficiency by utilizing local renewable resources and connecting them in a web-like structure.

1.9 Precision Agriculture: Enhance Efficiency and Sustainability

1.9.1 Fundamentals of Precision Agriculture

Precision agriculture also known as site-specific management is a current development. This permits corrections, in case of disparity within the field, with regard to such factors as fertility of the soil, soil moisture, weed incidence, and insect-pest density. The technology has the capability to decrease production costs through better utilization of crop inputs. Furthermore, it minimizes environmental deterioration because farmers are in a better position to use appropriate measures of agricultural inputs based on areas where these are required. Spatial, temporal, and predictive elements of the processes controlling soil and crop variability are the components of precision agriculture. It entails surveying, mapping, and planning of selected portions or "units" of a field in appreciation of spatiotemporal heterogeneity with regard to soil fertility, moisture supply, plant traits, or pest-insect populations. Variability of such a large scale has always been associated with varying cropping practices from one region to the other. Site-specific management referred to as precision agriculture allows for adjustments in view of within-field differences in factors such as weed density, soil moisture, fertility, and insect-pest populations (Adamchuk et al. 2004; Zeb et al. 2024). Since crop inputs are directly applied in a more effective and efficient, manner by the system it seeks to promote, the costs of production can be reduced (Basso et al. 2004).

The contingency factors of agricultural precision are the protective, temporal, and spatial dimensions of crop and soil heterogeneity. It involves samples, mapping, analysis, and management of certain fields in relation to the study of the geographical and temporal fluctuations in crop characters, soil fertility, moisture, and insect-pest population. Conventional cultivation practices have been known to relate to broad-scale variability across geographic diversity (Carrara et al. 2004). Precision agriculture is defined as the application of gear and software to assist farmers in making intelligent and precise decisions on a variety of tasks involved in farming including planting, application of fertilizers and pesticides, and even harvesting (Dutta et al. 2021; Farooq et al. 2024). From the given literature, Liaghat and Balasundrsm (2010) identified that precision agriculture enhances the efficiency of resource management and general farm management. Based on the kind of activities, production technology, and regions of agriculture, the distribution of energy resources is distinct. This leads to the growth of imported energy sources for the lubricants and nutrients hence posing the following issues on the agricultural sector (Dhaya and Kanthavel 2022). Through effective resource utilization and variability management, precision agriculture (PA) offers a paradigm shift toward a safe, sustainable approach. It offers numerous benefits in terms of profitability, productivity, sustainability, crop quality, environmental protection, and rural economic development (Liaghat and Balasundrsm 2010). Precision agriculture is a key component of sustainable soil and crop management in the modern period because of its capacity to use and analyze geographical and temporal data to measure, manage, and evaluate crop productivity (Pierce and Nowak 1999).

1.10 Precision Agriculture for Sustainable Soil and Crop Management

Today's agricultural practices which primarily use PA are primarily motivated by efficiency, economic, and environmental factors. In agricultural field settings, PA provides solutions for crop production through the selection and implementation of BMPs (Oshunsanya and Aliku 2016). The adoption of sustainable soil and crop management techniques will be fueled by the technological expertise and knowledge inherent in PA technologies. Soil sampling, geostatistics and GIS, farming by soil, site-specific farming, management zones, GPS, yield mapping, variable rate-nutrients, rate-herbicides, rate-irrigation, remote sensing, automated tractor navigation and robotics, proximal sensing of soils and crops, profitability, and adoption of precision farming are some of the important components of PA that are in line with the sustainability of soil and crop management (Mulla and Khosla 2015).

1.11 Importance of Precision Agriculture

PA is a method of using IT to manage farms in a way that increases agricultural output, profitability, and efficiency. In order to produce healthier crops and higher crop yields, PA can enhance crop management, minimize input consumption, and maximize land use (Lowenberg-DeBoer et al. 2021). Agriculture is changing as a result of technological advancements since the farmers profit greatly from their creativity. Farming efficiency is increasing thanks to technologies such as GIS, sensors, software, autonomous tractors, GNSS guidance, and agricultural drones (Adrian et al. 2005). In order to grow better crops, generate more food, and conserve resources, farmers can now accurately determine which inputs to use, how much to apply, and where to apply them thanks to technological advancements. Robert hinted that PA is an information revolution enabled by new technologies leading to more advanced and accurate agricultural management systems, rather than "just the introduction..." of fresh technology."

1.12 Technology and Tools Used in Precision Farming

1.12.1 Global Positioning System (GPS)

With an accuracy of between 100 and 0.01 m, a navigation system called GPS uses a satellite network designed to record location data, including positioning, elevation, and longitude (Lang 1992). Farmers can use GPS to pinpoint the precise location of field data, including the kind of soil, the presence of pests, weed invasion, water holes, boundaries, and obstacles. With an antenna, receiver, and light or sound guiding panel (DGPS), there is an automatic controlling system. GPS receivers can determine their position thanks to signals that GPS satellites send. Based on performance standards and past input applications, the system enables farmers to accurately pinpoint field locations so that inputs (seeds, fertilizers, insecticides, herbicides, and irrigation water) can be applied to a specific area (Batte and VanBuren 1999).

1.12.2 Sensing Technologies

Sensing technologies are essential to precision agriculture because they make it easier to gather and analyze data so that decisions about many elements of crop management can be made with knowledge (Lee et al. 2010). Advances in information and communication systems in conjunction with state-of-the-art sensing and actuation technologies offer great promise for progress in agricultural operations (Bochtis et al. 2014). Sensing technologies cover a wide range of technologies, including ground-based sensors such as weather stations, soil moisture sensors, and nutrient sensors, as well as remote sensing methods such as satellite images, aerial photography, and drones (Sishodia et al. 2020). In addition, the application of geographic information system (GIS) and global positioning system (GPS) applications in precision agriculture improves data collection and analysis for precise location tracking, mapping, and informed crop management decision-making, as well as resource optimization.

1.12.3 Remote Sensing

The science and skill of gathering data about the earth's surface without physically touching it is known as remote sensing. The process involves capturing energy that is either reflected or radiated from the surface of the planet. Following processing and analysis, the recorded data are utilized to create a prescription map that may be applied at a variable pace. There are several levels of information on crop and soil conditions that can be obtained using this technology. It enables the identification and/or description of an object, group of objects, or terrain without the sensor coming into direct touch with the ground. In order to detect agricultural stressors, injuries, or pest infestation, it uses aerial or satellite photography. Too far, the primary applications of satellite pictures have been in mapping and yielding potential prediction across wide areas. Mapping the crop's relative reflectance on a small scale is feasible with improved resolution, near-infrared, and red sensing (Adamchuk et al. 2003).

1.12.4 Geographic Information System (GIS)

In order to facilitate the collection, storage, feature extraction, retrieval, and analysis characteristics and geographic information to produce maps, this system is composed of hardware, software, and methods. GIS connects data in a single location so that it can be expanded upon as needed. Unlike traditional maps, computerized GIS maps are more detailed and include multiple layers of information (such as pests, crops, rainfall, soil nutrient levels, yield, and maps from soil surveys) (Berntsen et al. 2006).

1.12.5 Variable-Rate Technology (VRT)

PA is frequently misunderstood as a sophisticated technological intervention in developed-world agriculture. However, evidence suggests that it is profitable in less-developed parts of the globe. For instance, microdosing nutrient-starved soils in Africa with micronutrients increased cereal grain yields. A number of case studies showed how Chinese farmers profited by controlling in-field variability. To assess a farm's soil qualities' geographical distribution, geostatistical techniques are employed (Wang et al. 2006). A thorough comprehension of the spatial distribution of soil attributes makes it easier to use the SSM to maintain soil and agricultural yield while lowering expenses and having a smaller negative impact on the environment.

1.13 Challenges of Precision Agriculture

Farmers may find precision farming to be a useful management tool, and numerous manufacturers and research groups have invested significant resources in this technology. Many farmers remain hesitant despite the significant effort and numerous assurances from manufacturers for the following reasons: PFT has the potential to increase production efficiency, but because there is a lot of information to absorb, it complicates decision-making. Because of this, farmers may need assistance in incorporating the vast quantity of information that they have access to through various PFTs into their real management plans (Suryady et al. 2011). Farmers are generally concerned about the agronomic decision support system, which has not been able to handle the vast amount of agronomic data that is currently available. Most decision support systems' decision algorithms rely on one or a small number of data sources. Nonetheless, in order to manage the intricacy of the agronomic process and perform accurate and site-specific fertilizer delivery, numerous supplementary information sources must be used. Consequently, there are not many agronomic models available to farmers who use site-specific instruments (such as GPS, yield mapping, and sensors) to collect data in order to analyze the spatial information and modify their field decisions.

1.14 Conservation Agriculture: Practices for Soil and Water Conservation

1.14.1 Principles of Conservation Agriculture

Over the last 7 years, the term "Conservation Agriculture" has been used to distinguish between sustainable agriculture and the strictly defined "conservation agriculture". CA is built on three principles: tillage practices, crop residues, and crop rotation. According to about 96 million hectares of zero-ill CA were employed in rainfed production systems in five nations (USA, Brazil, Argentina, Canada, and Australia). The rest of the planet accounts for less than 10% of the zero CA area.

1.14.1.1 Tillage Practices

The development and establishment of germinated weed seeds are impacted by tillage methods, which lessen resistance to root and seedling penetration. A better chance of a successfully germinated seed emerging results from the lower resistance found in tilled soils. In addition, compared to untilled soils, tilled soils enable seedlings to emerge from deeper in the soil. Tillage affects not only seedling emergence but also seedling establishment. The radical of weed seeds that have germinated may find it difficult to penetrate NT soil surfaces if it is placed directly on the surface of untilled soils, which might lead to fatal germination. According to two studies, weed seeds dispersed 2–3 m in the direction of tillage but very little in untilled soils. The first research revealed that cultivation after harvest greatly boosted weed seed dispersal. Therefore, reducing tillage can lessen the amount of weed seed that spreads inside and across fields.

1.14.1.2 Crop Residues

Under either the CT or NT systems (CT + Res and NT + Res, respectively), crop residue may be maintained in the field. The residue in CT + Res is mixed into the soil to varying degrees, depending on the type of tillage used. According to Liebman and Mohler (2001), the impact of integrated residue on weeds through changed nutrient dynamics will vary greatly depending on the type of tillage utilized, the residue's carbon to nitrogen ratio, the type of soil, and the surrounding conditions. Through allelopathy, surface residues alter the weed seed's chemical environment. The phenomenon known as allelopathy occurs when a plant creates biochemicals that have an impact on the growth of other living things or on the plant itself. Live plants and decomposing leftovers might emit all pathogenic chemicals. Small seeds are often more sensitive to the allelopathic effects of agricultural residue. Although there might be a number of reasons behind this, in general, large-seeded crops exhibit preferred weed growth suppression. Crop residue can also absorb herbicide, but this does not always mean that weed control will be lessened. Research has demonstrated that decreased pesticide contact with weeds is more than made up for by the weed control offered by surface residue.

1.14.1.3 Crop Rotation

Arguably, crop rotations are the best method for weed and pest management (Fig. 1.3). A distinct set of biotic and abiotic limitations are imposed by each crop, which encourages the growth of certain weeds while suppressing the growth of others. According to, every particular crop may be viewed as a filter that only permits specific weeds to pass through its management regime. Three primary methods by which rotations modify selection pressures are (a) changes in management (e.g., the timing of field operations, herbicides), (b) different patterns of resource competition, and (c) allelopathy.

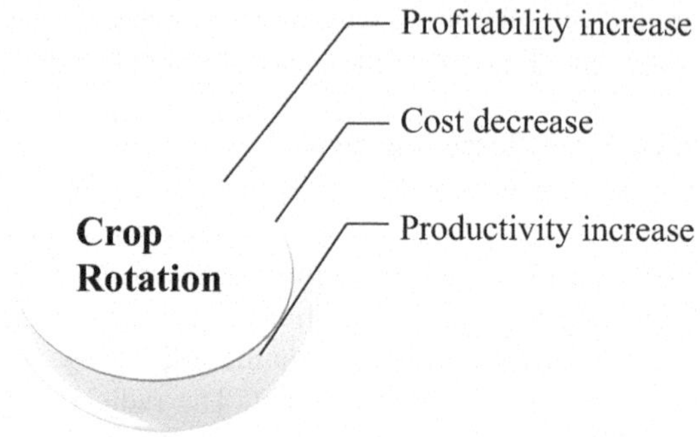

Fig. 1.3 Benefits of crop rotation

1.14.2 No Till and Minimum Tillage Practice

Varying loads will cause varying soil stresses and deformations depending on the tillage strategy used. The effects of contrasting tillage strategies on soil structure will vary depending on the physical, chemical, and biological characteristics of the soil as well as the temperature and moisture regime (Alaoui and Schwilch 2016). It is believed that implementing the minimum tillage (MT) and no-tillage (NT) systems will affect soil temperature and moisture content, as well as cause the soil to be more resistant to penetration than it would be with the conventional tillage system (CS). The volume and intensity of rainfall, the temperature of the water and soil, the slope and landform, the hydro-physical characteristics, the soil texture, and the compaction all affect or are influenced by rainwater penetration in soil and the increase of water storage within the soil profile. The tillage system affects all of these soil properties, which are highly dependent on one another (Sarauskis et al. 2009). Temperature, moisture, bulk density, porosity, penetration resistance, and soil structural state are all affected by the soil tillage system and its intensity (CS and MT), both directly and indirectly. The emergence of these features is more impacted by the intrinsic attributes of the soil, soil profile layers, weather, and management history in the NT system, where the soil is not mobilized (Guş et al. 2008). The results obtained by national and international research highly depend on soil type, climatic conditions of the area, and the presence of mulch in conservative systems (CT). Many authors have found that the bulk density of CT is higher in 319 as compared to CS; others have defined lower values of bulk density in conservative systems that have mulch on the soil surface or have not identified any discernible differences (Rusu et al. 2009).

1.15 Cover Cropping

The fundamental purpose of growing cover crops (CC) is to preserve the health of the soil and water while increasing the production of succeeding crops (Dabney et al. 2001). Generally speaking, CC can be categorized into three groups: grass (such as ryegrass and barley), legume (such as alfalfa, clover, cowpea, field pea, vetches, and lentils), and others (such as brassicas, forage, radish, turnip, rape, mustard and nonlegume broadleaves, spinach, and flax). It is possible to include several kinds of CC to satisfy the study requirement. Legume CC can enrich the soil with organic matter and fix atmospheric N. Legume CC has a higher turnover rate and a lower C: N ratio than grass CC, which makes its residue more readily available for the following crop (Sulas et al. 2017). It is unique from other Cucurbitaceae species that grass CC is capable of acquiring nutrients especially N from the remnants left in the soil after the previous harvest. It can also increase the rate of applying organic matter to the soil, affect the microbial processes in the soil, and produce large amounts of residue biomass C (Brandan et al. 2017). Grass CC has a comparatively greater C: N ratio which means that its residue will cover a more extended period and it will be beneficial when storing carbon. In order to incorporate the positive attributes of multiple CC types such as N scavenging, biomass generation, provision of simple C, C storage, enhancement of the biological diversification as well as enhancement of crop yields, cover crops mixtures can be sown into agricultural systems (Elhakeem et al. 2019). Thus it minimizes the chance for crop yield loss in several fields and weather conditions which in return improves CC establishment rate. To minimize fertilizer use and protect the water quality, cover crops can be used to smother weeds, disrupt diseases, provide nutrients, and avoid nutrient leaching (Singer et al. 2011). Forage radish and turnip, with their taproots may enhance the process of soil aeration, porosity, aggregation, water infiltration, C input after the root decomposition, and improvement of soil structure. They can also reduce the effects of compaction on the soil. Not only does it improve soil organic matter and microbial biomass, and avoid the production of bare soil ground, but the utilization of cover crops also reduces soil erosion by maintaining living indexes in the soil and on the surface of the soil. This means that to adopt CC there is a need to put into consideration the CC management since it has implications on the moisture content of the soil, the residues, and nutrients that are released to the ground. If the CC management plan is followed by the letter, then it means that there is a possibility of sustaining the production of crops and improving the soil at the same time. CC application in cropping systems has been perceived as a plausible approach toward altering crop residues both in quantity and quality, soil moisture regime, and the microbial biomass in the soil which are critical to soil fertility and conservation (Bechara et al. 2018).

1.16 Water Conservation Techniques in Agriculture

1.16.1 Weather Forecasting and Climate Modeling

Climate modeling and weather forecasting are input tools that form part of the Climate-Smart Agriculture approach. In this regard, by providing farmers with the necessary decision-making processes on issues such as planting, irrigation as well as harvesting, these technologies assist in reducing the risks hinged on climatic fluctuations. Farmers in India can get the weather data as well as the instructions regarding farming from the Indian Meteorological Department (IMD). These projections are now more refined and targeted because of such technologies and assist farmers in better organizing operations in the sector (Matteoli et al. 2020).

1.16.2 Resilient Crop Varieties

Looking at this in light of climatic change, the development and use of stable crop varieties are vital in order to feed the human population. These types are designed to withstand conditions such as heat, salinity, dryness, and even waterlogging. For instance, they are cultivating salt-resistant crops near the coastal areas besides wheat and types of rice that have been developed (Paik et al. 2020; Uzair et al. 2022).

1.17 Agroforestry

The term "agroforestry" was first used in 1977 as part of early international initiatives to research integrated production systems involving crops and trees. Potential agroforestry residues are produced through the integration of farming practices with woody perennials, or agroforestry. Despite being modest, agroforestry research investments over the past three decades have produced important advances in our understanding of the role that trees play in farmlands and the benefits that integrated farming systems offer both ecologically and economically. Until the Middle Ages, it was common in Europe to clear-fell abandoned forests, burn the slash, grow food crops on the cleared areas for varying periods, and plant or sow different kinds of trees before, concurrently with, or following the sowing of the agricultural crop. Although this "farming system" has lost its appeal in Europe, it was still widely practiced in a few places in Germany and Finland until the end of the twentieth century, as late as the 1920s.

1.18 Major Ecosystem Services and Environmental Benefits

Adding trees, crops, and/or animals to an agroforestry system can increase soil fertility, reduce erosion, enhance water quality, increase biodiversity, enhance aesthetics, and sequester carbon (Garrity et al. 2010).

1. **Agroforestry for Carbon Sequestration**

   Through physical or biological processes, carbon can be removed from the atmosphere and stored in carbon sinks, such as soils, vegetation, and oceans. When agroforestry systems are combined with trees or shrubs, the amount of carbon sequestered can be higher than in a field of crop plants or pasture.
2. **Agroforestry for Soil Enrichment**

   It is commonly known that agroforestry plays a significant role in promoting and preserving soil sustainability and long-term productivity. Tropical agroforestry systems frequently include crops and trees with the capacity to fix nitrogen through biological means. In agroforestry systems, nonfixing trees can also improve the physical, chemical, and biological qualities of the soil by releasing and recycling nutrients and contributing a sizable amount of organic matter both above and below ground. The impact of agroforestry on tropical soils has been investigated in many published papers as either original research papers or review papers (for example; Shrestha et al. 2018; Schroth and Sinclair 2003).

3. **Agroforestry for Biodiversity Conservation**

   Basic life-supporting organisms and the habitats in which they live are also rapidly being extinguished. As a result, there is a growing awareness across the globe of the necessity of taking prompt action to develop strategies that will effectively conserve biodiversity. The role that agroforestry plays in preserving biological diversity in both tropical and temperate regions of the world is becoming more and more recognized by scientists and policymakers. Numerous writers have looked into the ways that agroforestry systems contribute to biodiversity (Schroth et al. 2004).

4. **Agroforestry for Improved Air and Water Quality**

   Farming practices such as windbreaks and shelter belts are regarded as having many advantages. Some of such benefits are as follows: shielding of buildings and roads from drifting snows, reduced costs in livestock production due to moderation of wind chills, protection of crops, provision of wildlife habitat, removal of atmospheric CO_2, and addition of O_2, reduction in wind velocity and thus reduction of wind erosion or particulate matter in the air, reduction of noise pollution, control of odor from concentrated animal feeding operations, among others. More recently there has been increased concern toward the use of shelterbelts as a potential approach to managing livestock odor (Tyndall and Colletti 2007). Most of the foul-smelling molecules and bouquets are adsorbed on aerosols, which are particulates. Vegetative buffers have an added ability to clean airstreams of particulates by removing dust, gas, and microbial components. In the sphere of this group of issues with special reference to swine odor, Tyndall and Colletti (2007) depicted that with proper arrangement of the shelterbelts planting, it can successfully decrease the odor impact in a socio-economically effective manner. The provision of clean water is also another benefit that comes with agroforestry practices. Some of the problems associated with the current conventional intensive farming system include livestock fecal waste containing 68% N and 68% phosphorus which is filtered into the surrounding water sources, while only 40% of the applied N and phosphorus fertilizer is used by crops.

1.19 Agroforestry Practices for Enhancing Sustainability

Agroforestry, as described by is a versatile land-use system that involves cultivating woody perennials and agricultural crops within the same land management unit. This approach encompasses a diverse range of land use systems where animals and/or agricultural crops are intentionally intermixed with woody perennials in various spatial or temporal arrangements. Soil conservation is often cited as one of the primary benefits of agroforestry. In agroforestry systems with woody perennials, several biophysical and biochemical processes that evaluate soil substrate health may be influenced. The relatively noncontroversial impacts of trees on soil encompass a range of beneficial effects. These include the reduction of erosion, primarily achieved through the presence of understory vegetation and surface litter cover; the preservation or augmentation of organic matter and diversity; the improvement of physical characteristics such as moisture retention, porosity, and soil structure facilitated by the vast canopy cover and root system and additionally the improvement of nutrition consumption efficiency, attributed to the root system of the tree ability to recycle and intercept nutrients in the soil, thereby preventing their loss through leaching (Sobola et al. 2015). Nigeria possesses vast land and abundant vegetation, yet underscore the prevalent misconception among rural dwellers regarding the inexhaustible nature of forests, only recently awakening to the looming threat of extinction due to unregulated exploitation. Attribute the depletion of Nigeria's forest reserves to the rising need for agricultural land for shifting cultivation, urban expansion, and food production. Furthermore, the paragraph recognizes forestry, agriculture, and urban development as the predominant land usage practices in Nigeria.

1.20 Case Studies of Successful Agroforestry Projects

The *Faidherbia albida* (*Acacia albida*) in Niger: This project will demonstrate the advantage of using *Faidherbia albida* tree in the agricultural production process. It has benefited in enhancing crop productivity, improving soil productivity, and adding more value to farmer's cash earned. The Pekalongan Agroforestry Project in Indonesia: This project provides an example of the effective adoption of agroforestry practices in Indonesia where such enterprise has seen an increase in land productivity, enhanced status of biological diversity, and enhanced capacity to withstand adverse effects of climate change (Kuswanda and Herawati 2017). The Belize Agroforestry Project: This project aims at agroforestry of trees with crop and animal production in Belize thereby improving soil quality, and controlling erosion and income of farmers (Soto-Pinto et al. 2010).

1.21 Urban Agriculture; Sustainable Food Production in Cities

There are just a few examples; many more successes are recorded in the field of agroforestry as far as the promotion of sustainable agriculture and the conservation of the environment are concerned. Urban agriculture is based on rooftop gardens and indoor farming, through this; they get pure vegetables, fruits, and meat at their doorstep (Ackerman et al. 2014). So Urban agriculture plays an important role in sustainable food production in cities. According to some records, 200 million people are associated with this field. Urban agriculture also provides jobs and increases food security. It also helps to improve air quality and reduces water pollution. Urban agriculture is an important part of sustainable living (Ackerman et al. 2014). Additionally, urban agriculture can help reduce greenhouse gases and conserve natural resources. It also encourages a sense of community and encourages participation in local food production.

1.22 Importance and Challenges of Urban Agriculture

Urban agriculture may have a role in removing food insecurity problems in developing regions. This system can also help to reduce poverty and create jobs, as well as help to reduce environmental degradation. Urban agriculture can also help to reduce inequality and create a more sustainable food system. Urban agriculture represents many different things in different places for different people. It can make beautiful scenery and delightful scenes for all the people. It improves water infiltration, produces beautiful odors, and contaminates water (Zezza and Tasciotti 2010). This system has much more importance because it has worked according to policies and terms of form works. So, the importance of urban agriculture is increasing day by day due to its positive aspects. Urban agriculture helps to reduce food insecurity, create jobs, and support the local economy. It also helps to improve soil quality and reduce air pollution. Furthermore, urban agriculture can provide valuable educational opportunities for urban residents (Hallett et al. 2016). Urban agriculture also requires land, which is not always available in urban areas. Additionally, urban agricultural producers often lack the resources and knowledge to properly manage their farms. Facilitating and using network resources are very rare in these areas so they do not have enough information for the improvement of their field. Finally, urban agricultural operations are often small-scale, making it difficult to access the resources needed to succeed (Oberholtzer et al. 2014).

1.23 Urban Rooftop Gardening and Vertical Forming

> Green roofs are building roofs covered with a growth substrate and plants such as trees, shrubs, bushes, and grass. They are also referred to as rooftop gardens, living roofs, and eco-roofs.

A rooftop garden may transform abandoned space into a productive space that provides both ecological and commercial benefits, which is a fantastic approach to improving buildings in metropolitan settings. Rooftop gardens are common in Europe, for instance (Chowdhury et al. 2020). This system seeks the minimization of harm to the environment occasioned by conventional farming, the production of more food to satisfy the world's population, and the promotion of equitable rations of resources in societies. Urban agriculture is the farming of food in cities. This can be described as a method where people, mainly in the urban areas, cultivate plants and animals with a view of producing meals. This unique idea has come from North America where 70% of the projects are currently located in open air forms. New methods and technologies are used for urban rooftop gardening, these are very helpful and increase the production of food (Buehler and Junge 2016). It is the best solution to get year-round production with limited sources of water and space (Debangshi and Mondal 2021).

1.24 Community Gardens and Urban Agriculture Initiatives

Community gardens refer to the open spaces that are managed and operated by members of the local community in which food and flowers are produced. However, urban areas differ from backyard gardens that are privately managed by a family in a specific way. Community gardens are not only a source of food but also provide other benefits such as community building, education, and promoting health. These community gardens show beautiful views of buildings because urban areas are surrounded by huge marketplaces and industries. So community gardens are a source of relief and freshness for the urban population and also a source of food (Guitart et al. 2012). They also offer opportunities for education in terms of teaching communities about the technique for food production and why it is important to support local produce. Some of the benefits include: As can be seen from the above discussion, community gardens are one of the significant value additions to many cities as they assist in making access to fresh and healthy food products affordable for all the citizens, especially those who are in the humblest settings. The Urban Agriculture Initiatives are also policies and Strategies that are applied for food production and at the same time work as a greening factor in the urban society. Kanosvamhira and Tevera (2019) argued that the organization of urban farmers for resource and advocacy is pivotal in enhancing the success of urban agriculture, and social capital as a means of community in building public policies and programs. It is therefore necessary to identify and harness social capital for the achievement of commitment from farmers in the formulation process of policies; as well as for the sustenance and functionality of formulated policies.

1.25 Innovations in Sustainable Agriculture: Emerging Technologies and Practices

Sustainable agriculture is one of the most innovative solutions in agriculture causing a shift in how food is being produced, taking regard to environmental management, economic profitability, and social justice. This short chapter explains the novel innovative systems and methods in the sphere of agriculture in the future. From precision farming and regenerative agriculture to biotechnology and digital solutions, these innovations may afford a more resilient and sustainable food system. Through this advancement, it will be possible to offer solutions to calamities such as climate change, the decline in soil fertility, and food insecurity among others while conserving the environment and empowering society.

1.26 An Overview of Cutting-Edge Technologies Transforming Agriculture

Advanced and creative technologies are now playing a role in shaping the future of farming which has multiple solutions to global challenges of the food chain. Precision farming, crawler equipment, unmanned aerial vehicles, smart contracts, and genetic modification make farming efficient, productive, sustainable, and reactional to social-economic and environmental concerns. But, we must fully harness these technologies which means that coordinated work, adequate policies, and reinvestment in research, buildings, and capacities have to be made. Chosen stakeholders can pursue advanced agriculture that is richer, fairer, and less vulnerable by accepting the change and use of sustainable methods (Shamshiri et al. 2018).

1.26.1 Precision Farming

Precision farming uses different technologies such as satellite images, sensors, and machine learning techniques to support farm operations decisions. Precision farming involves soil variability, weather information, and crop yields, allowing farmers to make the right decision of seed type, amount of fertilizers to apply, time to water the crops, and when to apply pesticides. This increases resource productivity, decreases input costs, and raises crop returns with the least effect on the environment (Lowenberg-DeBoer 2018).

1.26.2 Autonomous Machinery

Self-driving tractors, inspiring robotic devices possessing ridges such as identification, environmental-mapping, and navigational systems, as well as artificial intelligence algorithms, are transforming farm work by performing chores previously done manually including planting, harvesting, and spraying. These machines are newer with better efficiency, precision, and safety than the conventional ways farmers can make the best use of their time and other resources. Self-driven tractors, UAVs, and mechanical harvesters are among the causes behind changed practices in farming efficiency and increased production and profitability (Dornbusch et al. 2018).

1.26.3 Drones

Unmanned aerial vehicles (UAVs) commonly referred to as drones are used in agriculture for mapping, surveillance, and crop monitoring. Fixed-wing UAVs with thermal imaging camera, multispectral sensors, and high resolution cameras are valuable for acquiring information concerning pest attacks, crop status, and field conditions. Due to the real-time data collected from drones, farmers are also able to manage plant health, irrigation, and nutrient management and thus produce maximum yield (Torres-Sánchez et al. 2018).

1.26.4 Blockchain

Blockchain technology provides solutions making the transactions in agriculture supply chains, from farm to fork, transparent, secure, and most importantly, traceable. Recognizing the transaction and all the communications that take place in a decentralized and unalterable manner, blockchain improves the supply chain value by increasing transparency, credibility, and responsibility of the members involved. Smart contracts can be used in making and receiving payments, ensuring product quality, and adherence to standards, and no more scams and foodborne illnesses (Nakamura et al. 2018).

1.26.5 Gene Editing

Some of the new-generation techniques in plant breeding that have attracted a lot of attention in recent years include CRISPR-Cas9 system genome editing. The trait release found in genetic engineering allows alterations in those genes in crops that are most likely to result in higher yield, disease resistance, and nutrients. It shortens the generation time, decreases the dependency on traditional breeding techniques, and increases the overall gene pool for the improvement of crop quality and productivity with increased resistance to biotic and abiotic stresses, enhanced nutritional value, and a better environmental adaptation (Khan et al. 2019).

1.26.6 Challenges and Future Directions

However, the use of new technologies in the production of food crops remains a challenge mainly due to the following challenges: cost factor, technology factor, policy factor, and social economic factor. Technology available must meet the needs of learners and practitioners for equal opportunity and access: the adoption of technology including innovations must be accessible for equal opportunities to everyone to enhance the creation of innovations in sustainable agriculture. The areas for further research and development should be associated with the mitigation of these challenges, as well as with the implementation of novel technologies in the formation of a more resistant, effective, and distributed food system.

1.27 Biotechnology and Genetic Engineering for Sustainable Farming

The application of biotechnology and genetic engineering for crop production and improvement has brought a revolution in agriculture to find options for high yield, improvement in crop resistance against pests and diseases, and reduction in the negative impacts on the environment. This chapter seeks to give an overview of the use of Biotechnology and Genetic Engineering in Sustainable Agriculture with emphasis on GMO, Gene editing tools, and biopharming. Relative to these technologies, it presents an assessment of their advantages, disadvantages, regulatory compliance, and future opportunities for sustainable agriculture.

1.27.1 Applications of Biotechnology and Genetic Engineering in Sustainable Farming

Because biotechnology deals with the integration and modification of biological systems in order to obtain the required outcomes, a number of devices and methods exist. Genetic engineering in agriculture, for example, recombinant DNA technology and gene editing are two techniques that enable the manipulation of the genome of crops to incorporate certain beneficial traits. The altruistic attributes that may be availed by genetically modified organisms (GMOs) include resistance to diseases, pests, and herbicides. Thus less chemistry is introduced and this leads to the enhancement of the use of environmentally friendly methods of farming. Gene editing tools like CRISPR-Cas9 make it possible and precise to introduce desired changes on selected loci within the genome of a crop, giving techniques for changing them more efficient for developing new traits. Biopharming involves the production of industrial enzymes, medicines, and vaccines that are produced using plants. It offers some applications in biome manufacturing and as a natural plant growth stimulator (Podevin et al. 2013).

1.27.2 Benefits of Biotechnology and Genetic Engineering for Sustainable Agriculture

The use of biotechnology and genetic engineering in agriculture has certain advantages toward the goal of sustainability and resistance. GM crops have shown higher productivity, low chemical use, and higher nutritional value, which is beneficial in food safety, and reduction of poverty. Gene editing tools allow for faster trait creation and adaptation since the number of generations has to be minimized when using these technologies. Biopharming is useful in generating high-value bioproducts in plants as a green approach to synthesis. Also, biotechnology makes crops resistant to climate change; pests, and diseases, thus enabling farmers to improve production and manage risks that are inherent in farming (James 2018).

1.27.3 Challenges and Regulatory Considerations

However, the introduction of biotechnology and genetic engineering services in the domain of agriculture has a few sorts of obstacles such as public perception toward it, regulations, intellectual property rights issues, and biosafety issues. Global consumers' attitudes toward GMOs and gene-edited crops thus differ depending on their risks, ethical concerns, and socioeconomic effects. International policies that govern the production and commercialization of biotech crops are still a major challenge since they differ from country to country and are likely to delay biotech crops' market access. Risk assessment, monitoring, and regulation of biotech products are important in order to evaluate the safety, effectiveness, and compatibility with the environment (Molla et al. 2020).

1.27.4 Future Directions

In conclusion, the biotechnology and genetic engineering of farming have a bright future as tools aimed at solving all the multifaceted issues of agriculture. Currently, major emphasis is given to improving the gene editing tools, widening the gene pool required for the development of new plant varieties, and increasing the knowledge of plant–microbe interactions. Multi-disciplinary, inter-sectoral, and societal engagement are crucial for the societal dialogue of ethico-policy issues in the use of biotechnology in agricultural production. Examining the evidence regarding the individual, political, and economic risk and reward factors associated with the commercial deployment of biotechnology and genetic engineering, stakeholders can work together to construct a better post-2015 food system for future generations.

1.28 Nanotechnology and Its Applications in Agriculture

At present, prospects of nanotechnology in scientific agriculture are certain significant opportunities offered by nanotechnology to overcome some of the complicated issues related to agriculture of the twenty-first century such as increase of crop production and efficiency of resource usage and reduction of adverse effects on the environment. Applying the properties of nanomaterials opens doors to using scientific inventions for the improvement of agriculture and the creation of durable and productive methods for the work of researchers, policymakers, and industry members. Despite these opportunities for nanotechnology in agriculture, it is mandatory that safety, ethical, and regulatory factors be considered to get the most out of nanotechnology's benefits to society (Rasheed et al. 2023).

1.28.1 Nanotechnology Applications in Crop Production

Nano-fertilizers, nano-pesticides, and nanosensors are the three foremost applications of nanotechnology that are changing the scenario of crop production by improving nutrient uptake, pest control, and real-time observation of the crop status. Since nano-fertilizers give the nutrients directly, there is little wastage and thus little pollution. Nano-pesticides kill pests precisely and leniently affecting the environment compared to normally used pesticides. We can therefore apply it in precision agriculture, for we get data in terms of moisture content of the soil, nutrient balance, and pest attacks among other aspects (de Rosa et al. 2010).

1.28.2 Nanotechnology Applications in Soil Remediation

Soils that have been contaminated present several difficulties to agricultural production and the environment. Nano-adsorbents, nano-sorbents, and nano-bioremediation agents are some of the most effective and eco-friendly technologies for soil remediation. These nanomaterials can adsorb, degrade or immobilize contaminants, and rejuvenate the soil fertility and the ecosystem in general without secondary pollution (Kah and Hofmann 2014).

1.28.3 Challenges and Regulatory Considerations

However, there are some issues that have to be solved while using nanotechnology in agriculture including the toxicity of the nanoparticles, their behavior in the environment, and the lack of legal regulation of nanotechnology use. The triumvirate of risk assessment, standard testing methodology, and the right regulatory measures are needed to develop nanomaterials for safe and responsible use. These challenges require stakeholder engagement, interprofessional collaboration, and an increased awareness of, and responsible use of nanotechnology in agriculture (Khodakovskaya et al. 2011).

1.28.4 Future Perspectives

This research proposes a great potential for nanotechnology in agriculture in the future because the current research is still on the application of new nanomaterials, delivery systems, and uses. The innovation in nanoscale science, biotechnology, and data analytics will shape the future nano-enabled agri-technology solutions. For the nanotechnology research findings to be applied in agriculture, interdisciplinary studies, collaboration, and technology transfer are key (Sarathy et al. 2016).

1.29 Blockchain and IoT in Supply Chain Management for Sustainable Agriculture

The combination of blockchain and IoT technologies in the supply chain can provide the best solutions to increase the transparency, accountability, and sustainability of the agricultural sector. This chapter gives a brief understanding of the use of blockchain and IoT in the agriculture supply chain specifically in food safety and ethical practices, and reduction of wastage. It explains the prospects and developments of using these technologies for sustainable agriculture with a focus on the opportunities and risks involved.

1.29.1 Applications of Blockchain and IoT in Agriculture Supply Chains

The use of the decentralized technology known as blockchain allows for creating an immutable record of transactions and data transfers throughout the supply chain. Together with sensors and smart tags, blockchain makes it possible to track agricultural products from farming right up to consumption. Smart contracts provide a way of executing contracts and transactions automatically by enforcing rules and standards without the risk of compromise, thereby encouraging supply (Zheng et al. 2018).

1.29.2 Benefits of Blockchain and IoT for Sustainable Agriculture

The integration of blockchain and IoT in supply chain management offers several benefits for promoting sustainability in agriculture. By providing transparent and verifiable information about the origin, handling, and quality of agricultural products, these technologies enhance food safety and quality assurance. Real-time monitoring of environmental conditions and supply chain activities enables timely interventions to prevent food spoilage, reduce waste, and optimize resource utilization. Additionally, blockchain-based systems promote ethical practices such as fair trade, organic certification, and labor rights by ensuring visibility and accountability throughout the supply chain (Kamilaris et al. 2019).

1.29.3 Challenges and Considerations

Nonetheless, the use of blockchain and IoT in the agriculture supply chain has some challenges that would need to be addressed regarding the adoption of the technology, which include technical limitations, compatibility, data privacy, and legal implications. Integration with legacy systems and existing structures and frameworks poses a big challenge; this will, therefore, call for adequate resource commitment. However, sustaining and maintaining data security, data accuracy, and regulatory compliance is the critical success factor in gaining stakeholders' trust. These challenges make it important for blockchain and IoT professionals to work with end-users, other stakeholders, and include the application of broad industrial best practices in order to mainstream the technologies in sustainable agriculture (Khan et al. 2019).

1.29.4 Future Directions

Augmenting the agriculture supply chains by adopting blockchain and IoT will open more pathways in the near future for sustainability, transparency, and improved resilience. Current research activities aim at improving the ability of these technologies to interconnect, grow in size, and ease of use. Further, integration with new technologies such as artificial intelligence, remote sensing, and additionally distributive computing will also advance the functioning of agricultural chain of supplies systems. Regardless of applying the concept of blockchain or IoT, the stakeholders can unite to develop a better food system to sustain for the next generations (Liang et al. 2017).

1.30 Future Challenges to Sustainable Agriculture

1.30.1 Climate Change

The effects of climate change on agriculture are dramatic through the extreme weather conditions that begin to happen frequently, and alterations in rainfall and temperature. As indicated above, these changes affect crop production, livestock production, and food security of the farmers (Porter et al. 2014).

1.30.2 Soil Degradation

For sustainable agriculture, one of the most significant and complex issues affecting the soil resource is degradation, through erosion, loss of fertility, and pollution. Low-quality soils are unfavorable to production, rich in unwanted inputs, and result in desertification and poor agricultural suitability (Lal 2015).

1.30.3 Water Scarcity

As mentioned earlier, water shortage reduces the volumes of water that can be used in irrigation hence limiting farming. This aspect puts in danger food security and the future of agricultural practices in general, but more especially where land is a major limiting factor such as arid and semi-arid lands (FAO 2011).

Table 1.2 Policies for sustainable agriculture

Agriculture potential based on biophysical and environmental factors	Low	Areas in arid and semi-arid regions: less favored agricultural areas	Less favored isolated and vulnerable locations: low and degraded soils, steep slope, scarcity of water
	High	Unfavorable agricultural regions: remote places with a rich biotic environment and a high level of plant biodiversity	Agriculturally friendly areas

1.30.4 Economic and Policy Barriers

Forces at the industry and government levels, including the high cost of markets, capital, and favorable policies prohibit the adoption of sustainable techniques (Table 1.2). The following barriers have to be unmasked in order to examine why smallholder and resource-poor farmers are unable to implement sustainable agricultural systems (Pretty et al. 2011).

Author Contribution Statement Each author contributed significantly to the development of this chapter. Tanveer Ahmed led the overall conceptualization and frameworks of the chapter, supervised the writing process, and coordinated the manuscript development. Huma Naz conducted primary research, contributed to data analysis, and wrote sections on the methodology. Basharat Ali provided expertise on case studies and played a key role in the conclusion. Asma Shahbaz co-wrote the introduction, discussed the implications of the findings, and assisted with revisions. Isha Akram reviewed related work and contributed to the literature review. Ayesha Ilyas assisted with data visualization, reviewed the drafts, and ensured clarity. Muhammad Hamza Babar coordinated data collection and contributed to the final proofreading. Noor Fatima assisted with data collection and reviewed drafts for accuracy. Javed Iqbal provided expertise in statistical analysis and data interpretation. Yasir Niaz contributed to the theoretical discussions and case study analysis. Habib Ali coordinated the structure and flow of the chapter and ensured consistency across sections, providing final editorial support. Together, the authors collaborated closely to ensure the chapter's depth and quality.

References

Ackerman K, Conard M, Culligan P, Plunz R, Sutto MP, Whittinghill L (2014) Sustainable food systems for future cities: the potential of urban agriculture. Econ Soc Rev 45(2):189–206

Adamchuk VI, Perk RL, Schepers JS (2003) Application of remote sensing in site-specific management. University of Nebraska, Cooperative Extension Institute of Agriculture and Natural Resources, Lincoln

Adamchuk VI, Hummel JW, Morgan MT, Upadhyaya SK (2004) On-the-go soil sensors for precision agriculture. Comput Electron Agric 44:71–91

Adrian AM, Norwood SH, Mask PL (2005) Producers' perceptions and attitudes toward precision agriculture technologies. Comput Electron Agric 48(3):256–271

Alaoui A, Schwilch G (2016) Soil quality and agricultural management practices inventory at case study sites, manual iSQAPER. Version, 1(26.07)

Altieri MA (1995) Agroecology: the science of sustainable agriculture. CRC Press, Boca Raton

Altieri MA, Funes-Monzote FR, Petersen P (2015a) Agroecologically efficient agricultural systems for smallholder farmers: contributions to food sovereignty. Agron Sustain Dev 32(1):1–13

Altieri MA, Nicholls CI, Henao A, Lana MA (2015b) Agroecology and the design of climate change-resilient farming systems. Agron Sustain Dev 35(3):869–890

Barrios E, Gemmill-Herren B, Bicksler A, Siliprandi E, Brathwaite R, Moller S et al (2020) The 10 elements of agroecology: enabling transitions towards sustainable agriculture and food systems through visual narratives. Ecosyst People 16(1):230–247

Basso B, Cammarano D, Vita PD (2004) Remotely sensed vegetation indices: theory and applications for crop management. Rivista Italiana di Agrometeorologia 1:36–53

Batie SS, Shabman LA, Kramer RA (1985) U.S. agriculture and natural resource policy. In: The dilemmas of choice. Resources for the Future, Washington, DC

Batte MT, VanBuren F (1999) Precision farming—factors influencing profitability. In: Northern Ohio crops day meeting, Wood County, Ohio, vol 21

Bechara E, Papafilippaki A, Doupis G, Sofo A, Koubouris G (2018) Nutrient dynamics, soil properties and microbiological aspects in an irrigated olive orchard managed with five different management systems involving soil tillage, cover crops and compost. J Water Clim Change 9:736–747

Beddington JR, Asaduzzaman M, Clark ME, Fernández Bremauntz A, Guillou MD, Howlett DJB et al (2012) What next for agriculture after Durban? Science 335(6066):289–290

Berntsen J, Thomsen A, Schelde K, Hansen OM, Knudsen L, Broge N et al (2006) Algorithms for sensor-based redistribution of nitrogen fertilizer in winter wheat. Precis Agric 7:65–83

Beus CE, Dunlap RE (1990) Conventional versus alternative agriculture: the paradigmatic roots of the debate. Rural Sociol 55(4):590–616

Bochtis DD, Sorensen CG, Busato P (2014) Advances in agricultural machinery management: a review. Biosyst Eng 126:69–81

Brandan CP, Chavarría D, Huidobro J, Meriles JM, Brandan CP, Gil SV (2017) Influence of a tropical grass (*Brachiaria brizantha* cv. Mulato) as cover crop on soil biochemical properties in a degraded agricultural soil. Eur J Soil Biol 83:84–90

Buehler D, Junge R (2016) Global trends and current status of commercial urban rooftop farming. Sustainability 8(11):1108

Carrara M, Comparetti A, Febo P, Orlando S (2004) Spatially variable rate herbicide application on durum wheat in Sicily. Biosyst Eng 87(4):387–392

Chowdhury MH, Eashat MFS, Sarkar C, Purba NH, Habib MA, Sarkar P, Shill LC (2020) Rooftop gardening to improve food security in Dhaka city: a review of the present practices. Int Multidiscip Res J 10:17–21

Clark A (2007) Managing cover crops profitably, 3rd edn. Sustainable Agriculture Research and Education (SARE)

Cordell D, Drangert JO, White S (2009) The story of phosphorus: global food security and food for thought. Glob Environ Chang 19(2):292–305

Cox TS, Picone C, Jackson W (2004) Research prioritiesin natural systems agriculture. In New dimensions inagroecology (eds D. Clements & A. Shrestha). Binghamp-ton, NY: Food Products Press

Conway GR, Pretty JN (1991) Unwelcome harvest: Agriculture and pollution. London, UK: Earthscan

Culleton N, Tunney H, Coulter B (1994) Sustainability in Irish agriculture. Irish Geogr 27(1):36–47. https://doi.org/10.1080/00750779409478697

Dabney S, Delgado J, Reeves D (2001) Using winter cover crops to improve soil and water quality. Commun Soil Sci Plant Anal 32:1221–1250

Danborn DB (1979) The resisted revolution: urban America and the industrialization of agriculture, 1900–1930. University of Iowa Press, Iowa City, p 195

Debangshi U, Mondal R (2021) Rooftop farming—an overview. Chronicle of bioresource management, vol 5, pp 63–68

Dhaya R, Kanthavel R (2022) Energy efficient resource allocation algorithm for agriculture IoT. Wirel Pers Commun 125:1361–1383

Dornbusch T, Marconi S, Sartori L (2018) Robotics and automation in agriculture: trends and future perspectives. Biosyst Eng 173:111–126. https://doi.org/10.1016/j.biosystemseng.2018.06.019

Dutta A, Roy S, Kreidl O, Boloni L (2021) Multi-robot information gathering for precision agriculture: current state, scope, and challenges. IEEE Access 9:161416–161430

Elhakeem A, van der Werf W, Ajal J, Lucà D, Claus S, Vico RA, Bastiaans L (2019) Cover crop mixtures result in a positive net biodiversity effect irrespective of seeding configuration, agriculture. Ecosyst Environ 285:106627

FAO (2011) The state of the world's land and water resources for Food and Agriculture (SOLAW) Managing Systems at risk, Food and Agriculture Organization of the United Nations. FAO, Rome

Farooq MS, Majeed A, Ghazy A et al (2024) Partial replacement of inorganic fertilizer with organic inputs for enhanced nitrogen use efficiency, grain yield, and decreased nitrogen losses under rice-based systems of mid-latitudes. BMC Plant Biol 24:919. https://doi.org/10.1186/s12870-024-05629-w

Ferguson RS, Lovell ST (2014) Permaculture for agroecology: design, movement, practice, and worldview, a review. Agron Sustain Dev 34:251–274

Fiksel J (2003) Designing resilient, sustainable systems. Environ Sci Technol 37(23):5330–5339

Foley JA, Ramankutty N, Brauman KA, Cassidy ES, Gerber JS, Johnston M et al (2011) Solutions for a cultivated planet. Nature 478(7369):337–342

Garrity DP, Akinnifesi FK, Ajayi OC, Weldesemayat SG, Mowo JG, Kalinganire A et al (2010) Evergreen agriculture: a robust approach to sustainable food security in Africa. Food Secur 2:197–214

Goodland R (1997) Environmental sustainability in agriculture: diet matters. Ecol Econ 23(3):189–200

Gliessman SR (2005) Agroecology and agroecosystems. In The earthscan reader in sustainable agriculture (ed. J. Pretty). London, UK: Earthscan

Guitart D, Pickering C, Byrne J (2012) Past results and future directions in urban community gardens research. Urban For Urban Green 11(4):364–373

Guş P, Rusu T, Bogdan I (2008) Factors which impose completing preserving effects of minimum soil tillage systems on arable fields situated on slopes. In: 5th international symposium soil minimum tillage system, pp 155–161. Ed. Risoprint Cluj-Napoca

Hallett S, Hoagland L, Toner E (2016) Urban agriculture: environmental, economic, and social perspectives. Hortic Rev 44(44):65–120

Hardin LS (1988) Thirty years of agriculture: a review of North America. e-Spania 30(3):98–101

Hathaway MD (2016) Agroecology and permaculture: addressing key ecological problems by rethinking and redesigning agricultural systems. J Environ Stud Sci 6:239–250

Holmgren D (2002) Permaculture. Principles and pathways beyond sustainability. Holmgren Design Services, Hepburn

James C (2018) Global status of commercialized biotech/GM crops in 2018: biotech crops continue to help meet the challenges of increased population and climate change, ISAAA brief no. 54. https://www.isaaa.org/resources/publications/briefs/54/

Kah M, Hofmann T (2014) Nanopesticides research: current trends and future priorities. Environ Int 63:224–235. https://doi.org/10.1016/j.envint.2013.11.021

Kamilaris A, Fonts A, Prenafeta-Boldú FX (2019) The rise of blockchain technology in agriculture and food supply chains. Trends Food Sci Technol 91:640–652. https://doi.org/10.1016/j.tifs.2019.08.014

Kanosvamhira TP, Tevera D (2019) Urban agriculture as a source of social capital in the cape flats of Cape Town. Afr Geogr Rev 39(2):175–187

Khan MA, Noshaba S, Mehmood Z (2019) Blockchain technology in agriculture and food supply chains: a systematic literature review. Comput Electron Agric 167:105074. https://doi.org/10.1016/j.compag.2019.105074

Khodakovskaya MV, de Silva K, Nedosekin DA, Dervishi E, Biris AS, Shashkov EV et al (2011) Complex genetic, photothermal, and photoacoustic analysis of nanoparticle-plant interactions. Proc Natl Acad Sci 108(3):1028–1033. https://doi.org/10.1073/pnas.1008856108

Korže AV (2018) Permaculture for sustainable life style 2:34–39. https://doi.org/10.5923/j.ijire.20180202.02
Krebs J, Bach S (2018) Permaculture scientific evidence of principles for the agroecological design of farming systems. Sustainability 10(9):3218
Kuswanda T, Herawati T (2017) Agroforestry as an adaptive strategy to climate change: a review. IOP Conf Ser Earth Environ Sci 101(1):012026
Lal R (1991) Tillage and agricultural sustainability. Soil Tillage Res 20(2–4):133–146
Lal R (2015) Restoring soil quality to mitigate soil degradation. Sustainability 7(5):5875–5895
Lang L (1992) GPS, GIS, remote sensing: an overview. Earth Observation Magazine
Lee W, Alchanatis V, Yang C, Hirafuji M, Moshou D, Li C (2010) Sensing technologies for precision specialty crop production. Comput Electron Agric 74(1):2–33
Liaghat S, Balasundrsm S (2010) A review: the role of remote sensing in precision agriculture. Am J Agric Biol Sci 5(1):50–55
Liang X, Zhao J, Shetty S, Liu J (2017) Integrating blockchain for data sharing and collaboration in mobile healthcare applications. In: IEEE 28th annual international symposium on personal, indoor, and mobile radio communications (PIMRC). IEEE, Piscataway, pp 1–5. https://doi.org/10.1109/PIMRC.2017.8292598
Lowenberg-DeBoer J (2018) Precision agriculture: a step change in agricultural innovation. Outlook Agric 47(4):259–264. https://doi.org/10.1177/0030727018804141
Lowenberg-DeBoer J, Franklin K, Behrendt K, Godwin R (2021) Economics of autonomous equipment for arable farms. Precis Agric 22:1992–2006
MacRae RJ, Hill SB, Henning J, Mehuys GR (1989) Agricultural science and sustainable agriculture: a review of the existing scientific barriers to sustainable food production and potential solutions. Biol Agric Horticult 6(3):173–219
Marcus AJ (1985) Agricultural science and the quest for legitimacy. Iowa State University Press, Ames
Matteoli F, Schnetzer J, Jacobs H (2020) Climate-smart agriculture (CSA): an integrated approach for climate change management in the agriculture sector. In: Handbook of climate change management: research, leadership, transformation. Springer, Berlin, pp 1–29
Molla KA, Karmakar S, Molla J, Sarkar D (2020) CRISPR-Cas: a revolutionary tool for crop improvement. Int J Mol Sci 21(4):1630. https://doi.org/10.3390/ijms21041630
Mulla D, Khosla R (2015) Historical evolution and recent advances in precision farming. In: Lal R, Stewart BA (eds) Soil-specific farming, 1st edn. CRC Press, Boca Raton, pp 16–51
McNeely JA, Scherr SJ (2003) Ecoagriculture. Washington, DC: Island Press.
Oberholtzer L, Dimitri C, Pressman A (2014) Urban agriculture in the United States: characteristics, challenges, and technical assistance needs. J Ext 52(6):28
Oshunsanya S, Aliku O (2016) GIS applications in agronomy. In: Geospatial technology-environmental and social applications. Intech Open, Rome, pp 217–234
Paik S, Le DTP, Nhu LT, Mills BF (2020) Salt-tolerant rice variety adoption in the Mekong River Delta: farmer adaptation to sea-level rise. PLoS One 15(3):e0229464
Pandey DN (2007) Multifunctional agroforestry systems in India. Curr Sci 92:455–463
Phipps TT, Crosson PR (1986) In: Phipps TT, Crosson PR, Price KA (eds) Agriculture and the environment: an overview in agriculture and the environment. Resources for the Future, Washington, DC, pp 3–29
Pierce FJ, Nowak P (1999) Aspects of precision agriculture. Adv Agron 67:1–85
Piorr HP (2003) Environmental policy, agri-environmental indicators and landscape indicators. Agric Ecosyst Environ 98:17–33
Podevin N, Davies HV, Hartung F, Nogue F, Casacuberta JM (2013) Site-directed nucleases: a paradigm shift in predictable, knowledge-based plant breeding. Trends Biotechnol 31(6):375–383. https://doi.org/10.1016/j.tibtech.2013.03.004
Porter JR, Xie L, Challinor AJ, Cochrane K, Howden SM, Iqbal MM et al (2014) Food security and food production systems. In: Climate change 2014: impacts, adaptation, and vulnerability. Cambridge University Press, Cambridge, pp 485–533

Power AG (2010) Ecosystem services and agriculture: tradeoffs and synergies. Philos Trans R Soc B Biol Sci 365(1554):2959–2971

Pretty J (1995) Regenerating agriculture: policies and practice for sustainability and self-reliance. Earthscan/National Academy Press, London/Washington, DC, p 320

Pretty J, Toulmin C, Williams S (2011) Sustainable intensification in African agriculture. Int J Agric Sustain 9(1):5–24. https://doi.org/10.3763/ijas.2010.0583

Rasheed A, Hussain S, Mushtaq W, et al (2023) Application of silver nanoparticles synthesized through varying biogenic and chemical methods for wastewater treatment and health aspects. Environ Saf Pollut Res. https://doi.org/10.1007/s11356-022-24761-4

Reganold JP, Papendick RI, Parr JF (1990) Sustainable agriculture. Sci Am 262(6):112–121

Rifkin J (1980) Entrophy: a new world view. Bantam, New York

Rifkin J (1983) Algeny. Viking Press, New York. 298 pp

Rossiter M (1975) The emergence of agricultural science: Justus Liebig and the Americans, 1840–1880. Yale University Press, New Haven

Rusu T, Gus P, Bogdan I, Moraru PI, Pop AI, Clapa D et al (2009) Implications of minimum tillage systems on sustainability of agricultural production and soil conservation. J Food Agric Environ 7(2):335–338

Rynk R (1992) On-farm composting handbook. Natural Resource, Agriculture, and Engineering Service (NRAES), Ithaca

Sarathy V, Chang C, Mercer K (2016) Agricultural nanotechnology: research progress and prospects. Nanotechnol Sci Appl 9:25–33. https://doi.org/10.2147/NSA.S78365

Sarauskis E, Romaneckas K, Buragiene S (2009) Impact of conventional and sustainable soil tillage and sowing technologies on physical-mechanical soil properties. Environ Res Eng Manag 49(3):36–43

Sarrantonio M (1992) Northeast cover crop handbook. Rodale Institute, Emmaus

Schroth G, Sinclair F (2003) Trees crops and soil fertility: concepts and research methods. CABI, Wallingford, p 464

Schroth G, da Fonseca GAB, Harvey CA, Gascon C, Vasconcelos H, Izac AN (2004) Agroforestry and biodiversity conservation in tropical landscapes. Island Press, Washington, DC

Sgroi F (2022) The circular economy for resilience of the agricultural landscape and promotion of the sustainable agriculture and food systems. J Agric Food Res 8:100307

Shamshiri RR, Weltzien C, Hameed IA, Yule IJ, Grift TE, Balasundram SK et al (2018) Research and development in agricultural robotics: a perspective of digital farming. Int J Agric Biol Eng 11(4):1–14

Shrestha BM, Chang SX, Bork EW, Carlyle CN (2018) Enrichment planting and soil amendments enhance carbon sequestration and reduce greenhouse gas emissions in agroforestry systems: a review. Forests 9(6):369

Singer J, Malone R, Jaynes D, Ma L (2011) Cover crop effects on nitrogen load in tile drainage from Walnut Creek Iowa using root zone water quality (RZWQ) model. Agric Water Manag 98:1622–1628

Sishodia RP, Ray RL, Singh SK (2020) Applications of remote sensing in precision agriculture: a review. Remote Sens 12(19):3136

Sobola OO, Amadi DC, Jamala GY (2015) The role of agroforestry in environmental sustainability. IOSR J Agric Vet Sci 8(5):20–25

Soto-Pinto L, Anzueto M, Mendoza J, Ferrer GJ, de Jong B (2010) Carbon sequestration through agroforestry in indigenous communities of Chiapas, Mexico. Agrofor Syst 78:39–51

Suh J (2022) Permaculture principles, practices, and environmentalism. In: Sustainable agriculture reviews 58: phosphorus use efficiency for sustainable agriculture. Springer International Publishing, Cham, pp 1–23

Sulas L, Mercenaro L, Campesi G, Nieddu G (2017) Different cover crops affect nitrogen fluxes in Mediterranean vineyard. Agron J 109:2579–2585

Suryady Z, Shaharil MHM, Bakar KA, Khoshdelniat R, Sinniah GR, Sarwar U (2011) Performance evaluation of 6LoWPAN-based precision agriculture. In: The international conference on information networking 2011 (ICOIN2011). IEEE, Piscataway, pp 171–176

Tait J, Morris D (2000) Sustainable development of agricultural systems: competing objectives and critical limits. Futures 32(3–4):247–260

Torres-Sánchez J, López-Granados F, Peña JM (2018) An automatic object-based method for optimal thresholding in UAV images: application for vegetation detection in herbaceous crops. Comput Electron Agric 151:84–96. https://doi.org/10.1016/j.compag.2018.05.032

Tyndall J, Colletti J (2007) Mitigating swine odor with strategically designed shelterbelt systems: a review. Agrofor Syst 69:45–65

US Congress (1990) Food, Agriculture, Conservation, and Trade Act of 1990, Public Law 101–624: U.S. Farm Bill, 28 November 1990

Uzair M, Ali M, Fiaz S, Attia K, Khan N, Al-Doss AA, Khan MR, Ali Z (2022) The characterization of wheat genotypes for salinity tolerance using morpho-physiological indices under hydroponic conditions. Saudi J Biol Sci 29:103299

Wang N, Zhang N, Wang M (2006) Wireless sensors in agriculture and food industry—recent development and future perspective. Comput Electron Agric 50(1):1–14

Wang FF, Qin P, Zhang MK, Xiong ZE, Cuthbertson AGS, Alharbi SA, Fiaz S et al (2024) Genome-wide analysis of developmental stage-specific transcriptome in Spodoptera litura for the identification of effective control method. J Crop Health 76:1039. https://doi.org/10.1007/s10343-024-01026-3

Yadav SPS, Lahutiya V, Ghimire NP, Yadav B, Paudel P (2023) Exploring innovation for sustainable agriculture: a systematic case study of permaculture in Nepal. Heliyon 9(5):e15899

Zeb U, Rahim F, Azizullah A et al (2024) Effects of copper sulphate stress on the morphological and biochemical characteristics of Spinacia oleracea and Avena sativa. BMC Plant Biol 24:889. https://doi.org/10.1186/s12870-024-05566-8

Zezza A, Tasciotti L (2010) Urban agriculture, poverty, and food security: empirical evidence from a sample of developing countries. Food Policy 35(4):265–273

Zheng Z, Xie S, Dai H, Chen X, Wang H (2018) An overview of blockchain technology: architecture, consensus, and future trends. In: IEEE international congress on Big Data (BigData Congress). IEEE, Piscataway, pp 557–564. https://doi.org/10.1109/BigDataCongress.2018.00100

Carbon Sequestration and Sustainable Soil Management

2

Abid Khan, Mushtaq Ahmad Khan, Muhammad Adnan, Ikram Ullah, Raheem Shahzad, Hafiza Samra Younas, Monsif Ur Rehman, Hongbing Zheng, Tian Shilin, and Mehmood Ul Hassan

Abstract

Carbon sequestration is crucial for mitigating climate change by storing atmospheric carbon dioxide (CO_2) in soil and vegetation. This approach is an underlying principle of sustainable soil management, a goal that involves maintaining and enhancing overall soil health for the purposes of agricultural productivity

A. Khan (✉) · R. Shahzad
Department of Horticulture, The University of Haripur, Haripur, Pakistan
e-mail: khanabid@uoh.edu.pk; raheem.shahzad@uoh.edu.pk

M. A. Khan · M. Adnan
Department of Agriculture, University of Swabi,
Anbar Swabi, Khyber Pakhtunkhwa, Pakistan
e-mail: mushtaq@uoswabi.edu.pk; madnan@uoswabi.edu.pk

I. Ullah
School of Biological Sciences, Universiti Sains Malaysia, Penang, Malaysia

H. S. Younas
Department of Soil Science, University of Layyah, Layyah, Pakistan
e-mail: samrayounas@ul.edu.pk

M. U. Rehman · M. U. Hassan
Department of Agriculture, Hazara University Mansehra,
Mansehra, Khyber Pakhtunkhwa, Pakistan
e-mail: drmonsif@hu.edu.pk

H. Zheng
Collage of Agronomy, Jilin Agricultural University, Changchun, China
e-mail: hongbingzheng@126.com

T. Shilin
College of Biological and Food Engineering, Huanghuai University,
Zhumadian, Henan, China
e-mail: 20070945@huanghuai.edu.cn

© The Author(s), under exclusive license to Springer Nature Singapore Pte Ltd. 2025

S. Faiz et al. (eds.), *Climate Smart Agriculture for Future Food Security*,
https://doi.org/10.1007/978-981-96-4499-5_2

and acting as a storehouse for CO_2. Soils are the most important land-based ecosystems that store approximately two-thirds of terrestrial soil organic carbon (SOC). The global SOC pool is estimated to be three times as large as the atmospheric carbon pool and four times larger than the biotic carbon. Implementing specific soil management practices, such as reforestation or afforestation and sustainable agricultural techniques, can increase C sequestration in the soil, leading to greater SOC levels. These farming practices not only capture carbon but also improve soil structure/fertility/water retention while increasing ecosystem resilience overall. On the other hand, human land use changes like deforestation and agricultural expansions have caused a big loss of carbon from soils. Conversion of natural vegetation into croplands is a major driver of SOC loss. However, the implementation of sustainable management and reforestation can multiply SOC to a greater extent. Practices like no-till farming, cover cropping, crop rotation, and the use of organic amendments are all a part of sustainable soil management. These practices have in common that they increase the degree of carbon sequestration by encouraging organic matter stocks to build up and, on the other side, decreasing soil disturbance. Furthermore, reforestation and afforestation activities also increase carbon sequestration in biomass and soil so that these projects have a long-term impact on climate change. There are important environmental co-benefits to carbon sequestration in addition to its mitigation potential of climate change. By reducing atmospheric CO_2 levels, it mitigates ocean acidification, protecting marine ecosystems. Moreover, increased SOC benefits improve soil health, biodiversity, and agricultural performance. In conclusion, soil carbon sequestration through sustainable land management is an important and effective strategy to fight climate change and environmental degradation. By adopting effective land management practices, we can improve SOC content, boost soil health, and add to a more resilient and sustainable ecosystem.

Keywords

Carbon sequestration · Carbon dioxide · Soil organic carbon · Afforestation · Climate change · Crop rotation

2.1 Carbon Cycle

One essential component of the Earth's nutrition cycle is the carbon cycle. Carbon is exchanged and circulated via photosynthesis, respiration, decomposition, mineralization, and other activities between the ocean layers, atmosphere, and land. In the air, carbon is present as gaseous CO_2, which assembles a lesser but substantial percentage of total terrestrial carbon. Living organisms and decomposing organic matter are two stages of the cycle on land. Minerals contain a major portion of carbon, especially in the form of calcium and magnesium carbonates ($CaCO_3$). Some of this carbon has been sequestered beneath the surface of the Earth by biogeochemical processes such as liquid petroleum, solid coal, and in the form of natural gas. The foundation process on Earth for life is photosynthesis. Primitive organisms evolved the ability to utilize the water and CO_2 from the environment to synthesize organic compounds necessary for their growth and development (Sha et al. 2022). The majority of photosynthesis occurs on land, and the tropical belt forest provides the largest contribution. Plants absorb carbon in the form of CO_2, which is released back into the atmosphere as CO_2 or transferred into the sea in the form of organic matter (Bhadra et al. 2022). This process happens by two pathways: through the process of respiration and through the micro-organisms' action, which breaks down the decaying organic material and finally puts it back in its mineral form. Both respiration and photosynthesis are responsible for the daily alteration of carbon dioxide in the atmospheric pool (Govani et al. 2022).

Carbon may be found in the water/ocean as molecules of CO_2, dissolved HCO_3, or as microscopic plants called phytoplankton. The carbon cycle varies in marine environments, where phytoplankton is the predominant photosynthetic organism. CO_2 captured in the upper water layers triggers the carbon to flow lower. Decomposers on the ocean floor utilize organic materials, generating CO_2, which is partially absorbed in the deep ocean and partially liberated into the air. The oceanic reservoir takes in CO_2 from both the Earth and the atmosphere via precipitation and surface absorption (Iversen 2023).

Approximately 38,000–40,000 Gt of carbon is held deep in the ocean, whereas 2000 Gt is found in soil and on land, the atmosphere contains 750 Gt, while a surface layer of the ocean contains 800 Gt (Pulselli 2008). Living organisms, the atmosphere, soil, and the surface layer of the ocean are interconnected, with continuous and comparatively much quicker exchanges of carbon occurring among them. Transfer of carbon between this rapidly interacting system and the deep ocean takes a considerably longer time. Essentially, the exchange with the deep ocean restricts CO_2 absorption; hence, the overall effect is upon the atmosphere's CO_2 build-up. This implies that the depths of the ocean cannot aid in reducing CO_2 accumulation (Farooqi et al. 2021).

Although these exchanges, primarily powered by solar energy, are roughly balanced annually, imbalances can occur, and feedback mechanisms may significantly impact atmospheric CO_2 levels over timescales ranging from years to centuries. The primary causes of this unpredictability are changes in land and ocean absorption by climatic events like El Niño and human activities. It might be challenging to differentiate changes caused by human actions from those caused by natural fluctuation (Govani et al. 2022). The two main ways that human activity leads to CO_2 accumulation are through the burning of fossil fuels and deforestation. Fossil fuels and biomass are combusted to satisfy humanity's increasing energy needs, causing the atmospheric release of CO_2. The biosphere's delicate equilibrium is seriously disturbed by this input, particularly since fossil fuels are consumed much more quickly than the eons it took for sluggish sedimentary processes to create this resource (Li et al. 2022).

2.2 Carbon Sequestration and Its Environmental Benefits

Carbon sequestration, mostly in relation to climate change mitigation, has numerous key environmental benefits. Carbon sequestration is vital for combating climate change, lowering the concentration of greenhouse gases in the atmosphere, and reducing global warming (Prajapati et al. 2023). It not only aims to halt climate change, in general, but also helps the protection of habitats and ecosystem services directly by stabilizing (and increasing the resilience) biodiversity conservation (Alam et al. 2023). It strengthens the soil's structure, improving its organic matter content, promoting microbial activity and nutrient cycling, improving water retention, buffering pH levels, encouraging root development, lowering erosion, and boosting climatic resistance, which improves soil fertility. Hence, better crop yields, healthier soils, and more environmentally friendly farming methods are the results of these advancements (Pandao et al. 2023). Nayak et al. (2022) noted that carbon sequestration helps to reduce ocean acidification by improving biological carbon uptake, safeguarding and reviving blue carbon ecosystems, raising ocean alkalinity, absorbing and storing CO_2 emissions, creating marine protected areas, and encouraging kelp and seaweed farming. These actions collectively help to stabilize ocean pH levels and support healthier marine ecosystems. Kabir et al. (2023) pointed out that carbon sequestration helps to fight against climate change by reducing CO_2 levels; it also helps air quality directly and indirectly through the removal of pollutants from mechanisms such as natural filtration systems like trees, and others, and promoting healthier ecosystems. Pandao et al. (2023) reported that carbon sequestration supports resilience to climate change by stabilizing climate patterns, enhancing ecosystem resilience, improving soil health and agricultural productivity, regulating water cycles, conserving biodiversity, promoting human health and well-being, and ensuring long-term climate stability. These multiple benefits also help to increase the ability of both natural and human systems to either resist or adapt to climate change impacts. Additionally, Prajapati et al. (2023) also reported that the carbon sequestration process plays a substantial role in capturing CO_2 for long-term storage through biological, oceanic, geological, and technological methods. These are all strategies that store carbon safely for the long term, ultimately decreasing greenhouse gas concentrations in the atmosphere and mitigating climate change.

2.3 Ways to Improve Soil Carbon Sequestration

2.3.1 Organic Farming

The term organic farming involves a range of practices that collectively define the "organic" farming system. Organic farming aims to produce high-quality food in a sustainable manner while safeguarding the environment. When it comes to soil health, organic farming's key aspects include varied crop rotations, the cultivation of legumes to provide nitrogen, minimal external nutrient inputs, and the avoidance of mineral fertilizers and synthetic chemicals, which are commonly employed for crop production in traditional agriculture. These guidelines and restrictions are included in national and international regulations on organic farming (IFOAM 2006). Due to significantly fewer exterior inputs of nutrients, for example, nitrogen, than traditional agriculture, mostly the farms that use organics operate as mixed systems that include forage production, livestock, and arable crops, with FYM from livestock being incorporated in the soil. The requirement for the symbiotic fixation of nitrogen and the production of forage generally need rotations of crops that are typically more diversified than those in traditional farming (Sher et al. 2021).

Traditional farming practices can cause disastrous impacts on soil health because they worsen the soil structure and water retention, prevent soil carbon (C) sequestration, and cause runoff and erosion (Hussain et al. 2021). Since most farming takes place in rural regions, poor soil health is likely to impact these communities the most, thereby disproportionately harming the economies of the rural (Stringer et al. 2020). Enhancing the health of the soil is now considered a key component in numerous climate change and agricultural initiatives at the state level (Syed et al. 2022), national, and global levels (FAO 2015). Presently, the investment is from private (e.g., General Mills, Danone, Indigo Ag) and public sources (Basche et al. 2020) to encourage farming methods that boost soil fertility by no/reduce tillage, strengthening crop diversification, keeping the soil covered, and incorporating crops with livestock (Tully and McAskill 2020). Hence, it is imperative that we have data-driven recommendations (Pullin and Knight 2009) on the expected advantages that particular methods will provide for soil health.

It has been demonstrated that organic farming improves SOC levels. Hence, it is recommended as a step to enhance the total amount of greenhouse gas equilibrium in agriculture in comparison to traditional agriculture (Niggli et al. 2009). This opinion is not universally accepted, though as certain research has indicated either no impact or an inconsistent impact on SOC (Atoloye et al. 2022). Although neither rotations of crops nor fertilization with organic materials is specifically organic farming, they are both recognized for their significant influence on SOC in any farming system. A notable distinction is the higher level of specialization in conventional farming, resulting in a more pronounced spatial segregation of livestock husbandry and cash crop cultivation. In terms of soil carbon, this means that comparing management practices fairly should start with comparable crop rotations and rates of organic fertilization. Alternatively, if farm type and rotation vary, the differences in system boundaries should be factored into the data interpretation. Under comparable conditions concerning crop rotation and organic fertilization, specific organic farming practices include the cessation of external inputs such as artificial chemicals and mineral fertilizers for the control of weeds and pests. These factors may have an impact on SOC by changing the number of plant residues available for SOC accumulation.

2.3.2 Cover Cropping

The cover crops, also known as green manures, catch crops, intercrops, and service crops, are typically cultivated for purposes other than harvest (Van Eerd et al. 2023). It has been suggested that in crop rotations, the addition of cover crops is a viable strategy for enhancing SOC stocks in the soil (Van Eerd et al. 2023) and also safeguarding the erosion losses of SOC (Jastrow and Miller 2018). Yet, there are several factors, including soil, climate, and agricultural management techniques, which can influence the SOC's potential sequestration by cover crops.

The quantity of SOC sequestered through the cultivation of cover crops relies upon the quality and the quantity of the residue and how it breaks down in the soil (Ghimire et al. 2017). Generally, the cover crop biomass is primarily influenced by the species of the cover crop cultivated, its management by agronomic practices, and climatic conditions (Quintarelli et al. 2022), along with soil fertility (Van Eerd et al. 2018). For instance, in contrast to the climate of the semiarid temperate, the humid temperate climate exhibited greater cover crop productivity (Ruis et al. 2019). There has also been evidence of a favorable relationship between the yearly temperature, precipitation, and sequestered SOC with cover crops (Jian et al. 2020). The quality of the residue affects how rapidly it decomposes. Though, consistent data about the impact of excellent cover crops in SOC buildup and stabilization have not been recorded. Residues having high ratios of C/N are expected to aid the stabilization of the soil organic matter (SOM), as they contain a significant amount of lignin and phenols that decompose slowly due to their physiochemical properties (Castellano et al. 2015). Certain studies have suggested that residues having low C/N ratios benefit from the stabilization of SOM, which is formed from the byproducts of the microbes (Cotrufo and Lavallee 2022). While, the texture of the soil, particularly clay content, is an additional intrinsic feature influencing SOM stabilization (Six and Paustian 2014). Additionally, the distribution of soil particle size can affect the productivity of the plant because of its significant impact on water retention and availability of nutrients (Kern and Muth 1995). The pH of the soil can also influence the decomposition of organic matter by affecting SOC solubility, microbial growth, and activity (Wang et al. 2022). Further soil factors, like levels of initial SOC, can influence the rate and amount of SOC accumulation in the cover cropping system. For instance, low SOC stock soils might sequester SOC more rapidly whenever the cover crops have been integrated into the system (Jian et al. 2020).

Furthermore, tillage and crop rotation are two further agronomic practices that impact SOC accumulation under cover cropping (McClelland et al. 2021). For instance, reduce tillage minimizes disruption of soil aggregation from frequent disturbances, hence physically shielding the decomposition of SOM (Ahmad and Wang 2023). Similarly, crop rotation not only varies carbon inputs but also tends to improve the stability of soil aggregates and the activity of microbes (Van Eerd et al. 2018). This could improve the residues of the crop and inputs of the root exudate, while decreasing losses of the SOC (Chamberlain et al. 2020). Furthermore, rotation of the crop can influence the cover crops growing season, therefore having an indirect impact on the quantity of biomass from cover crops that are returned to the soil (Van Eerd et al. 2018).

A recent meta-analysis showed that as compared to no cover crop, the cover crops increased SOC globally by 16% (Jian et al. 2020), with mean C sequestration rates of 167 kg C ha^{-1} $year^{-1}$ (Poeplau and Don 2015). In contrast, the sequestration rate of C relies on the species of the cover crop: higher rates in nonlegumes than for legumes because of larger C/N ratios (Utomo et al. 1990). Additionally, the SOC in mixtures of cover crops showed higher values compared to single species (Sainju et al. 2018). These mechanisms range from selective preservation of plants to increasing blockage inside aggregates and stronger chemical associations with the particles of minerals, respectively (Zhang et al. 2022). C sequestration in most of the studies was quantified in terms of aggregate protected C (Yao et al. 2019).

2.3.3 Conservation/Zero Tillage

Sustainability in crop production can be ensured by using alternative methods of sustainable crop production like conservation agriculture (CA). This approach is widely recognized as a strategy to enhance soil quality and improve the contents of organic C in ecosystems related to agriculture while also promoting infiltration rates and reducing erosion issues (Kahlon et al. 2013). Prasad et al. (2016) found that in rainfed agriculture CA practice reduces soil degradation problems. These practices include zero tillage or no-till farming and preservation of subsoil with crop mulches, adapted in combination on the basis of appropriate cropping systems (Somasundaram et al. 2020). It is also defined by FAO as an "approach to farm management that combines the best of environmental quality with economic profitability and social responsibilities" (Lal 2018). In contrast, Dumanski et al. (2006) argued that traditional agriculture operations primarily attempt to improve production at the expense of the environment. Using advanced agricultural technologies to enhance crop production and protecting the health and integrity of the ecosystem are the main outcomes of conservation agriculture (Dumanski et al. 2006). The CA system is important in order to sustainably manage agricultural land and maximize crop production with no or little environmental damage and mitigate the impacts of global warming (Pimentel and Burgess 2013). Over the past few years, CA has gained much importance because of its multiple benefits such as a significant improvement in soil properties (physicochemical and biological properties) as well; as sufficient nutrient availability and status of water for crops under it throughout besides long-term sustainable yield productivity also offering opportunities to ameliorate climate change (Busari et al. 2015).

According to Ella et al. (2016), conservation agricultural practices during the production of upland crop systems can increase the nutrient and water status of the soil along with its residual water content, rate of soil infiltration, and organic carbon content. As opposed to conventional systems, minimum soil disturbance through no-till practices, regular soil covers or mulching, and using crop rotations and intercropping systems are the three pillars on which CA is built (FAO 2015). Some authors included the inclusion of woody crops for increasing production on infertile soils without degrading the environment as a pillar of conservation agriculture (Vicente-Vicente et al. 2016). Lal (2015) on the basis of conceptual models, suggested that integrated nutrient management, soil degradation processes, and natural ecosystem function through the protection of ecosystem services to improve soil carbon sequestration with a concomitant reduction in GHGs (like CO_2) emissions are required for global warming abatement. Woody crops possess a range of features that may make them strong C sequesters in the atmosphere (Gonzalez-Sanchez et al. 2019). Because of their lengthy life cycle, they are able to store carbon in permanent structures (trunk, branches, and roots), as well as in the medium (soil) through processes such as rhizodeposition. Furthermore, such perennial woody crops through their extensive and deeply ingrained root systems promote the direct transfer of soil organic carbon (SOC) into the subsoil, which reduces the susceptibility of SOC to mineralization. A cropping system consisting of cover crops (CC) is also considered as a management tactic to boost the rate of soil organic carbon (SOC) sequestration (Ruibérriz et al. 2012). Generally, farming practices consisting of adopting cover crops increase SOC due to the rise in organic materials from plant residues and the reduction in the rate of mineralization brought about by using conservation tillage practices (Chahal et al. 2020). Apart from adding carbon materials, cover crops provide various agroecosystem benefits, such as reducing N losses through leaching (Ketterings et al. 2015), ensuring the production of crops in a sustainable manner (Chahal and Van Eerd 2018), and enhancing the quality of soil (de Torres et al. 2021).

Reports from previous research also demonstrated that soil carbon content has been increased due to CT in the tropics. This technique has the potential to significantly increase the amount of organic matter and carbon in the soil, as well as to significantly increase the sequestration of SOC (Ella et al. 2016). Mathew et al. (2012) reported that CT practices have been found to have a positive impact on both the biological and physicochemical aspects of the soil. Enhancement in the soil respiration rates and microbial biomass production were also observed by using conservation tillage techniques (García-Orenes et al. 2013). Das et al. (2022) reported that soil bulk density has significantly decreased by CT practices, thereby increasing the OC content, microbial biomass C of soil, and maintaining the activity of the soil organisms. Additionally, studies on conservation tillage practices showed that combining crop cover and reduced or no tillage improves the diversity of macro-fauna (Mutema et al. 2013). Positive effects of CT approach include decreased risk of soil erosion, improved organic matter content along with rate of water infiltration and storage capacity, populations of flora and fauna, agroecosystem's stability, energy use efficiency, and improvement in crop yields (Iqbal et al. 2005). In a no-till or minimal tillage system, a significant amount of carbon is added to the soil when crop residues are left on the soil surface, which also reduces soil disturbance. Consequently, the mineralization rates of organic matter reduced because the incorporation of residues and its vulnerability to physical disruption decreased (Mikha and Rice 2004). Likewise, in a reduced tillage system, retaining crop residues including stubbles and root biomass improves SOC, which establishes a physical barrier between substrates and soil microbes. This barrier promotes the formation of stable microaggregates and macroaggregates that safeguard microbial decomposition (Tripathi et al. 2014).

It is commonly known that soil microbial biomass C is generally higher in the ZT system compared to the CT system. In the same way, CO_2 emissions were considerably lower under the ZT system when compared to the CT system. Under the ZT system, the input of biomass carbon increases and the losses of labile C decrease, which suggests improving C sequestration in the soil within the same tillage system. Adopting practices of conservation agriculture like the NT system, the effectiveness of the agricultural system can be improved by increasing water infiltration rates along with minimizing soil erosion, improving aggregate stability, moderating soil conditions, and decreasing weed populations. Adopting this approach can save time, resources, and cost of production with high net profits. Further, it improves crop yields through timely sowing, minimizes pest infestations by supporting soil biological communities, and cuts greenhouse gas emissions (Hobbs 2007). Therefore, this system supports both the sequestration of more C and the improvement in SOM concentration.

2.3.4 Afforestation

Soils and forests are the main places responsible for carbon sequestration. The amount of CO_2 accumulating in the atmosphere is directly correlated with variations in the amount of carbon absorbed by vegetation and soils. Therefore, measuring changes in carbon sequestration capacity in forest ecosystems is urgently needed (Tolunay 2019). Various studies reported (Mobley et al. 2015; Wang et al. 2018) that afforestation can enhance the resilience of the ecosystem and increase soil carbon sequestration under arid and semi-arid climates. Though, it can also lead to significant changes in land use, and hence may also affect soil water and the carbon pool. It is crucial to carefully examine carbon dynamics and soil water balance as the area is continuously re-vegetating.

The soil carbon pools can be characterized into soil inorganic carbon (SIC) pools and soil organic carbon (SOC) pools. Following afforestation, the dynamics of soil carbon pools undergo changes. Several studies have indicated that the age of forest stands significantly affects soil carbon stores (Deng et al. 2014). A study by Bradford and Kastendick (2010) reported that in forest ecosystems, the age of a forest stand is a key indicator of carbon reservoir in soil. According to a study by Mao et al. (2010), soil carbon storage increases with the age of the forest stand. On the other hand, a reduction in soil carbon storage was observed by Smal and Olszewska (2008), whereas no major change in soil carbon storage was observed by Sartori et al. (2007) in relation to forest age. Soils are important terrestrial ecosystems that have the capability to store considerable amounts of organic carbon. Lal (2002) reported four times larger sequestration of global SOC in the biotic C pool and three times larger under the atmospheric pool. Localized changes in land use and land cover, especially in grasslands, agriculture, and forestry activities, may lead to a decrease in the global CO_2 emissions to the atmosphere (Wang et al. 2024). Deforestation and erosion contribute to the atmosphere to a loss of roughly 1.7 and 0.1 $PgCyear^{-1}$, respectively (Bruce et al. 1999). One of the key reasons for the decline in soil organic carbon (SOC) is the removal of natural vegetation for agricultural purposes (Li et al. 2022). However, if management techniques are modified (Lal 2002) or afforestation efforts are taken (Nave et al. 2024), the amount of SOC could be significantly increased.

2.3.5 Pasture Management

Soil organic carbon (SOC) accumulation is influenced by the organic matter (OM) that is added to soil both above and below ground, which is determined by the management of pasture (species of the pasture, management of grazing, and soil nutrition) (McDonald et al. 2023). In pasture systems, grazing management may have a positive impact on SOC accumulation by encouraging perennial plants with deep roots (Stanley et al. 2024), increasing the growth of the roots and plant (Hao et al. 2024), encouraging the input of legumes nitrogen (N), and promoting the turnover of root (Chen et al. 2015) and trash resurfaces on the soil. Furthermore, preserving ground cover can lessen soil erosion, compaction, and disturbance, which can improve OM conservation in soils (García-Palacios et al. 2018). To enhance sequestration and prevent more SOC losses in grazing lands, it is critical to comprehend the connections among the management of grazing and the major factors influencing SOC sequestration (McDonald et al. 2023).

Reviews from the entire globe have shown that SOC stocks are negatively impacted by moderate to high stocking intensities, but not by mild intensities of the stocking (Byrnes et al. 2018; Jiang et al. 2020) in contrast to the exclusion of grazing. Less research has been done on the effects of stocking techniques, such as grazing frequency, timing, or duration. Nonetheless, a worldwide meta-analysis discovered that rotational grazing increased SOC stocks more than continuous grazing did (Byrnes et al. 2018). Likewise, Conant et al. (2017) found that "improved grazing" and other management techniques, such as enhancing soil fertility, irrigation, and spreading grass and legume species, raised SOC.

The scientific literature reports clear differences in the impact of different grazing types on carbon sequestration, but one pattern stands out clearly: perennial forage systems retain more carbon than annual field crops do. Further increase was observed in the accumulation of this C for a longer time in the rotation of field crops with pasture than only pasture in the fields (Lin et al. 2020). This does not mean that the impacts of all grazing systems are equal. Rather, when pastures are under poor management characterized by overgrazing then pasture will create more atmospheric carbon instead of working as a source of sink. Overgrazing removes the top growth of plants thereby reducing green tissues for photosynthesis and putting the plants under energy deficit status. Consequently, plants transfer the carbon from the rhizosphere to upper plant parts to be used for leaf growth. This pulling of C makes the root system shrink, and the pastures become more vulnerable to both, moisture deficits and the removal of root exudates used for organic matter. Hence, bare patches are created in the pasture as some plants die out in stressed conditions and the pasture becomes susceptible to erosion. Such pastures might also upset animals' performance, including their possible mortality, as the lack of feed from these pastures compels the animals to feed on some unusual poisonous plants, especially at the end of the growing season. These soils become equal or even fall below in their carbon content to those that are under annual crops. Still, when brought under improved management, such stressed, overgrazed, or degraded pasture fields will be more beneficial than those that are already under good management (Klumpp and Fornara 2018).

Differences were comparable among the effects of continuously grazed and rotationally grazed pastures. The effects may range from slight or no decrease to substantial increases in soil C content of rotationally grazed compared to continuously grazed pasture fields (Sarkar et al. 2020). Although such differences may seem unclear, studies on these effects found a common element that carbon sequestration was better or equal in the continuously grazed fields, which are under closely managed grazing systems. Such management systems were characterized by vigilant monitoring of grass growth in relation to grazing density and providing supplement feed in case grazing exceeds grass growth. On the other hand, the "laissezfaire" approach to grazing is characterized by using pastures continuously for livestock grazing from spring to fall annually. These studies were also in regions with consistent pasture production around the crop season with no significant "summer slump." The studies would have different results if the research areas consisted of paddocks that had been over-grazed because over-grazing has a negative impact on soil C content (Zhou et al. 2020). Grazing systems that are under good management often pose challenges to study because the differences among them are small which are difficult to measure. In conclusion, however, intensive rotational grazing with good management appears to be more beneficial to retain soil carbon as shown in an intensive study by Bouman et al. (2018) in Nova Scotia.

There is certainly evidence that compared to cool-season grasses and legumes (C3 plants), warm-season grasses using the C4 photosynthetic pathway can support more storage of soil carbon and growth during summer even though their spring growth may not start quickly. Plants with a deeper root system have the ability to move carbon from root exudates into the subsoil. The minerals in the soil can easily bind with carbon to stabilize. Promoting a diverse mix of plant species in long-term pastures enables best-adopting plant species to dominate under variable conditions. This in turn will improve the overall performance of the mixed pasture by increasing the quantity of carbon translocated into the soil. In addition, limited evidence also suggests that broadleaf nonlegume plants (forbs) in pasture with mixed plants promote microbial activity thereby increasing carbon sequestration (Jordon et al. 2022).

2.4 Carbon Sequestration and Soil Health

2.4.1 Role in Nutrient Availability

At the moment, the SOC content and SCseq investigation at the national or regional levels are highly intriguing, especially in agricultural soils. This is partly because SOC plays a significant role in CC and environmental quality and partly because it is the most important factor influencing the fertility and quality of the soil, as well as the ultimate productivity of the crop and its sustainability (Dad et al. 2021). In order to improve SOC, agricultural management should increase the inputs of SOM, reduce SOM oxidation and breakdown, or do both (Yadav et al. 2019). An increase in the SOM stock is beneficial for the fertility of the soil because it may allow organic matter to mineralize and release nutrients for plants. To maximize soil carbon storage, however, localized agriculture techniques must be implemented that will increase soil C inputs with steady or declining outputs (Dignac et al. 2017).

According to Grover et al. (2015) findings, soil C input is the main factor influencing SCseq during the development of the plant, which is influenced by the availability of N and nutrients needed to sustain N_2 fixation. Artificial fertilizers have an impact on soil health and carry a C charge even though they do not add organic matter. Therefore, increasing the ratio of nutrient solubilizers/mobilizers to biological N2 fixers may be a sustainable and economical way to improve SCseq in agroecosystems (Grover et al. 2015). Incorporating green manure crops into traditional cropping systems via multiple crops (growing various crops in the same area during different seasons) and crop rotation (growing different crops in the same area during different years) is one of the options that exist for the soil C increment, thereby enhancing the agricultural yield and the soil quality. Because crop rotation and multiple crops reduce soil erosion and increase soil structure and permeability, they are well known for being efficient methods of improving the physical qualities of soil. Because it increases the content of SOM and soil microbial activity, this approach also improves the fertility of the soil. The rotational green manure crops would be beneficial for several soil-related techniques. Legumes are the most widely grown among green manure crops because of their special capacity to fix atmospheric N2 through their root nodules. In many parts of the world, legumes particularly in a cropping system can (1) increase biological diversity; (2) provide net supplements of the C and N in soil; and (3) preserve the fertility of the soil (Sánchez-Navarro et al. 2018).

In various soil conditions, crop production performance and the dynamics of plant nutrients are significantly influenced by SOC and N (Meena et al. 2019). Thus, their soil dynamics are interlinked. Because of their special relationship, there are comparable overall trends in the accumulation, degradation, and subsurface transportation of total C and OC. As a result, their patterns are consistent throughout (Elbasiouny and Elbehiry 2019). The majority of agricultural soils had minimal N in the 1950s. During the production and turnover of belowground biomass, careful N fertilizer application can improve crop quality and yield, which can be used to counteract rising atmospheric CO_2 emissions and restore the absorbed CO_2 to the SOC stock (Shah et al. 2021). However, excessive and inefficient use of N fertilizer not only wastes money but also contaminates groundwater and surface water, acidifies the soil, and increases emissions of greenhouse gases (Niu et al. 2016). The addition of the chemical nitrogen lowers the ratio of the C:N and encourages the growth of labile C-containing non-N2-fixing bacteria. As a result, this low C:N ratio promotes plant development, which raises carbon sequestration through rhizodeposition and plant biomass. However, over time, the lack of other nutrients may become a limitation. Furthermore, the manufacturing of chemical N costs C and yields no additional OM (Grover et al. 2015). According to Singh et al. (2017), microbial biomass drives efforts at turnover and establishes a connection between soils and these efforts. The soil processes are regulated by the crop residues/SOM C:N ratio. For soil microbial biomass (MB), humus, and stable forms of SOC, the typical C:N ratio is approximately 10–12, 10, and 7, respectively. The MB's nutrients and C are constantly replenished in the ground and are linked to persistent long-term changes in SOC. The net result of the respiration rate, MB turnover, and mineralization attempts is the total SOC at any given time. These procedures have a close connection to soil microbiological metabolism.

Lal et al. (2007) noted that soil organic matter (SOM) is constructed from plant and microbe residues, which require a variety of components, while soil carbon sequestration necessitates other nutrients (Table 2.1 and Fig. 2.1). The weight-based C:N ratio of SOM is 12:1, the C:P ratio is 50:1, and the C:S ratio is 70:1. For one compendium of C sequestration, one must have 83, 20, and 14 kg of N, P, and S, respectively (Lal et al. 2007). Additionally, Richardson et al. (2014) suggested that managing inorganic sulfur (S), phosphorus (P), and nutrients (N) is a crucial viewpoint for methods of constructing the C of soil from residues of the crops high in C. Furthermore, Dignac et al. (2017) stressed the need to not undervalue the interactions between practices and more cycles of the nutrient (S, P, N, etc.) in SOM while considering them from the perspective of C cycling and soil C storage enhancement. An extremely prevalent stoichiometric ratio of the C to SOM in crop residues has significant consequences for the rate, whereby C is converted from crop residue to soil C. For instance, the amount of N, P, and S in wheat stubbles is roughly 17, 2, and 3 parts per 1000 units of C, while the amount in soil humus, or the fine fraction of SOM, is roughly 90, 19, and 14 parts per 1000 units. All soils have this robust nutrient coupling in SOM, independent of the type of soil, location, or history of management. Similar to the ratios of nutrients in the microorganisms of the soil (i.e., 250, 49, 26, and 103, 11, 9 parts N, P, and S per 1000 units of C in fungus and bacteria, correspondingly), the SOM's C, N, P, and y stoichiometry is comparable to that of the residues of the crop. In a variety of soils and ecosystems, this seems to be highly controlled and mostly stable. Long-term carbon storage in sediment and marine ecosystems has also been suggested to be driven by the bio-sequestration of carbon through the deposition of microbes (Richardson et al. 2014) (Table 2.2 and Fig. 2.2).

Table 2.1 Integrated effect of biochar and nitrogen sources on soil organic carbon (g kg^{-1})

Biochar (t ha^{-1})	Years 2015–2016	2016–2017	Mean	% increase
0	0.494 d	0.510 d	0.502 d	
10	0.605 c	0.676 c	0.641 c	27.62
20	0.707 b	0.777 b	0.742 b	47.86
30	1.077 a	1.145 a	1.111 a	121.38
LSD$_{(0.05)}$	0.016	0.017	0.011	
Nitrogen management (kg ha^{-1})				
Control	0.500 g	0.539 h	0.520 g	
90 N urea	0.611 f	0.642 g	0.626 f	20.45
120 N urea	0.653 e	0.692 f	0.673 e	29.34
150 N urea	0.689 d	0.727 e	0.708 d	36.16
90 N FYM	0.688 d	0.760 d	0.724 d	39.23
120 N FYM	0.729 c	0.797 c	0.763 c	46.74
150 N FYM	0.806 b	0.874 b	0.840 b	61.56
90 N PM	0.820 b	0.887 b	0.853 b	64.11
120 N PM	0.824 b	0.892 b	0.858 b	64.91
150 N PM	0.892 a	0.961 a	0.926 a	78.16
LSD$_{(0.05)}$	0.025	0.026	0.018	
Mean	0.721 b	0.777 a		
Interactions				
Y × BC	ns	**Y × N**	ns	
BC × N	*	**Y × BC × N**	ns	

N, FYM, and PM stands for nitrogen, farmyard manure, and poultry manure, respectively

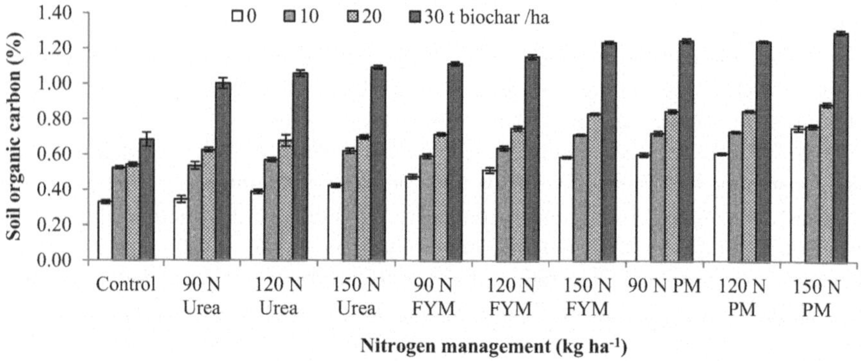

Fig. 2.1 Interactive effect of biochar and N sources on soil organic carbon content. The small bars at the top of each column represent the standard error of the mean (Khan et al. 2022)

Table 2.2 Integrated effect of biochar and nitrogen sources on soil organic matter (%)

Biochar (t ha^{-1})	Years 2015–2016	2016–2017	Mean	% increase
0	0.850 d	0.877 d	0.864 d	
10	1.041 c	1.163 c	1.102 c	27.54
20	1.217 b	1.337 b	1.277 b	47.76
30	1.924 a	1.970 a	1.911 a	121.24
LSD$_{(0.05)}$	0.027	0.028	0.019	
Nitrogen management (kg ha^{-1})				
Control	0.860 g	0.928 h	0.894 g	
90 N urea	1.050 f	1.104 g	1.077 f	20.51
120 N urea	1.123 e	1.191 f	1.157 e	29.40
150 N urea	1.86 d	1.250 e	1.218 d	36.22
90 N FYM	1.183 d	1.307 d	1.245 d	39.29
120 N FYM	1.254 c	1.371 c	1.312 c	46.81
150 N FYM	1.386 b	1.504 b	1.445 b	61.63
90 N PM	1.411 b	1.525 b	1.468 b	64.19
120 N PM	1.416 b	1.534 b	1.475 b	64.99
150 N PM	1.534 a	1.652 a	1.593 a	78.24
LSD$_{(0.05)}$	0.043	0.045	0.031	
Mean	1.240 b	1.337 a		
Interactions				
Y × BC	ns	**Y × N**	ns	
BC × N	*	**Y × BC × N**	ns	

N, FYM, and PM stand for nitrogen, farmyard manure, and poultry manure, respectively

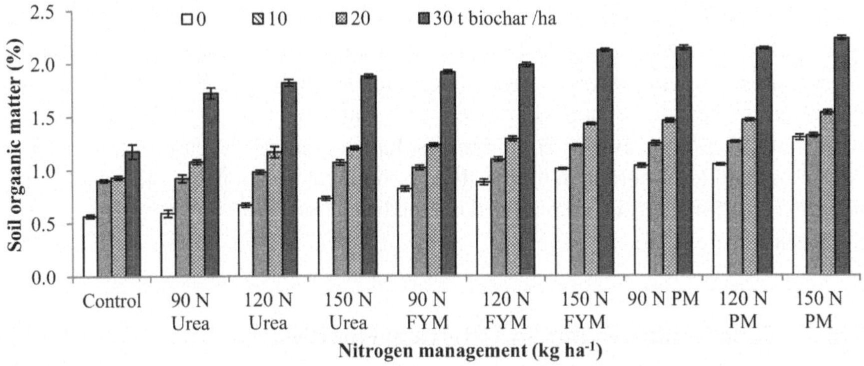

Fig. 2.2 Interactive effect of biochar and N sources on soil organic matter content. The small bars at the top of each column represent the standard error of the mean (Khan et al. 2022)

2.4.2 Role in Improving Soil Chemistry

Changes in the chemical and biological characteristics of soil are closely correlated with variations in SOC concentration. The primary component and roughly 58% of the soil organic matter is soil organic C. More research has been done on the impact of SOC on soil chemical characteristics than on microbial processes and behaviors. Variations in SOC content have an impact on soil respiration, microbial biomass, and community structure. By raising the concentration of SOC, management techniques like no-till might encourage the activity of the microbes in the topmost layer. Dick (1984) discovered a strong correlation ($r = 0.91$) between SOC and enzyme (invertase) activity during the course of prolonged no-till and traditional tilling throughout the whole soil profile. According to Grigera et al. (2007), soil microbial biomass varies between fields in correlation with SOC. Additionally, they discovered that the concentrations of soil actinomycete and bacteria biomarkers were extra strongly associated with fine in contrast to coarse organic matter, although the concentrations of fungal and mycorrhizal biomarkers were exclusively correlated with coarse organic matter. Changes in microbial activity carried on by organic C have an impact on the availability and transformation of nutrients, the turnover of the organic matter, and the overall resilience and health of the soil.

The SOC and other soil chemical processes and characteristics are mutually dependent. For instance, by providing substrate and habitat (i.e., stable aggregates), SOC influences microbial biomass and activity (Joshi and Garkoti 2023). The microbial processes also control SOC turnover. Furthermore, it is commonly known that SOC contributes positively to soil pH and cation exchange capacity. A rise in the concentration of SOC boosts the cation exchange capacity and buffers pH (Catania et al. 2022). Additionally, runoff contaminants can be absorbed and filtered by soil organic C, lowering pollution from nonpoint sources and enhancing the quality of the water. Cations and organic pollutants (such as herbicides and pesticides) can react with organic materials due to the high specific surface area and common net negative charge on organic materials. Because organic matter has a higher capacity for adsorbing contaminants than clay minerals, cations from pollutants are drawn to it more than to clay. Every chemical reaction and microbiological activities that impact soil behavior and the quality of water are influenced by organic C (Mishra et al. 2023).

2.4.3 Role in Improving Soil Physical Health

Soil physical properties are affected by variations in SOC content. An increase in SOC lowers the danger of soil compaction, encourages macro-aggregation, and improves water retention. The production of binding agents by organic compounds causes soil particles to clump together to form aggregates. It was shown that in numerous tillage and cropping system studies carried out in the central Great Plains and eastern United States, variations in SOC content accounted for 40% of the variation in the size of water-stable aggregates in the 0–5 cm (0–2 in.) soil level. These findings imply that for every 1% increase in SOC, the aggregate size might increase by 0.06 in., or 1.5 mm (Blanco-Canqui and Benjamin 2013).

The enhancement of soil aggregate stability can have an impact on near-surface soil processes that control water infiltration and runoff, including splash detachment, surface sealing, and crusting. Increased water runoff is caused by surface sealing and aggregate dissociation, which restricts soil pores. Soil aggregate slaking and surface seal development are less common in high SOC soils. Stable aggregates can preserve the consistency of macropores connecting to the surface for improved infiltration of water, precipitation collection, and decreased runoff when they combine with an abundance of surface residue cover. If soils could absorb water more quickly during heavy rainstorms and hold water more efficiently during droughts, some issues with runoff, floods, and droughts might be resolved (Blanco-Canqui and Benjamin 2013).

In certain situations, a rise in SOC lowers the compatibility of soil. The Proctor test is commonly employed to estimate the soil's relative compaction and calculate the maximum compatibility. For instance, adding cover crops to a no-till winter wheat–grain sorghum rotation in south-central Kansas for 15 years raised the surface's SOC concentration at 0–7.5 cm (0–3 in.) depth, while this rise in SOC reduced the maximum bulk density (maximum compatibility) found by the Proctor test in a linear fashion. The maximum soil compatibility can be decreased by around 0.1 Mg m^{-3} (0.003 tn ft^{-3}) by increasing SOC from 10 g kg^{-1} (0.16-ounce lb^{-1}) to 20 g kg^{-1} (0.32 oz. lb^{-1}) using cover cropping. Additionally, organic C raises the critical content of water in the soil, which is the point at which maximum compatibility occurs. This implies that, in comparison to soils with higher SOC concentrations, soils with lower SOC concentrations can be compacted at lower water contents and under the same compaction force (Blanco-Canqui and Benjamin 2013).

By improving strong physicochemical connections among soil particles, lowering the bulk density of the soil, inducing certain elastic qualities, and increasing soil aggregate stability, an increase in SOC may diminish maximum compatibility (Blanco-Canqui and Benjamin 2013). Forty percent of inorganic particle density is found in soil organic materials. Consequently, raising the existing levels of SOC through better management techniques might be one way to lessen the dangers associated with excessive compaction. Increasing SOC could potentially mitigate excessive compaction, especially in the vicinity of the surface layers where the majority of SOC is found.

The ability of SOC to retain water is another advantage. Water absorption and retention in the soil matrix are both increased by organic C. One gram (0.0022 lb) of C can retain 1–10 g (0.0022 and 0.022 lb) of plant-available water, depending on the textural class of the soil (Emerson 1995). This indicates that for medium-textured soils, an increase of 1% in SOC may result in an increase in accessible water of 12.5 mm (0.49 in.) on average in the upper 20 cm (8 in.) depth of the soil profile, with a bulk density of 1.25 Mg m^{-3} (0.039 tn ft^{-3}). Because organic C increases the area of the particular surface per mass of soil, it helps retain water in the soil. Soil with a greater specific surface area absorbs more water at the particle surface and releases it more slowly than the same amount of soil with a lower specific surface area. During dry spells, the gradual release of water helps lessen excessive water loss. Hence, one strategy to encourage soil water storage and increase plant-available water is to increase SOC (Blanco-Canqui and Benjamin 2013).

2.4.4 Role in Soil Biology/Biodiversity

Even though soil biological activity is necessary for the stabilization of soil organic matter (Lehmann and Kleber 2015), current assessments demonstrate that biological markers of the quality of soil are even inadequately represented in scientific publications from the previous few decades in comparison to the physical or chemical indicators (Lehmann et al. 2020). In addition to being valuable in and of itself, high biological soil quality promotes other biological processes that help in crop production, such as nutrient mineralization, symbiont colonization of roots, and disease control in plants (Van Der Heijden et al. 2016). Microbial soil quality indicators that are most frequently utilized are those that measure carbon and nitrogen (Cmic and Nmic) (Bünemann et al. 2018). Another commonly examined biological soil quality indicator is soil basal respiration, which is the constant respiration rate resulting from the mineralization of organic materials (van Eekeren et al. 2008). The metabolic quotient (qCO2), which represents the metabolic efficiency of a particular soil microbial community, is obtained from the ratio between soil respiration and Cmic (Anderson and Domsch 1993). The biotic pathways of phosphorus, which are essential for plant growth, are governed by the activity of phosphatase enzymes (Caldwell 2005). While the pH of the soil is a key indicator of soil quality and drives the variety of soil bacterial communities, the pH appears to have less of an impact on the composition of fungus communities (Rousk et al. 2010).

Despite the fact that numerous studies have demonstrated that organic farming methods enhance soil quality in comparison to conventional methods (Lori et al. 2017), discussions over the modifications and variations in SOC contents are frequently controversial (Leifeld and Fuhrer 2010). A recent meta-analysis found that under organic management, SOC contents were higher than under conventional management (Gattinger et al. 2012). Nonetheless, 21 years of SOC content monitoring has been conducted in the DOK study, the most old system of farming comparison trial in the world, which is situated in the mild environment of northwest Switzerland (Therwil), documented declining or, at most, stable SOC contents in both conventional and organic systems (Fliessbach et al. 2007).

Additionally, it acts as a food source for soil microorganisms and is a significant metabolite for bacteria. It has also been acknowledged as a tool that could minimize the impact of climate change through sequestration in agricultural soils. Furthermore, over the past few years, there has been a rising focus within the scientific community on the biological attributes of soil. These attributes are noted to exhibit a higher sensitivity to even minor changes compared to chemical or physical properties (Dindar et al. 2015). Microorganisms, among the native biological constituents of soil, hold a pivotal position in numerous vital biochemical processes, which include the cycling of elements like carbon, nitrogen, phosphorus, and sulfur, as well as facilitating energy transfers within the soil environment (Martinez et al. 2018). In addition to that, enzyme activity also stands out as a highly valuable parameter for evaluating soil quality. Soil enzymes predominantly originate from microorganisms, but they can also be produced by plant roots and soil-dwelling organisms (Borowik et al. 2020). In the opinion of several researchers, enzyme activity is the most sensitive measure of the biological as well as chemical status of soil because it participates in all microbiological reactions, including nutrient cycles in the soil, and it responds fast to changes in soil brought about by human-induced or natural factors (Antonious et al. 2020).

2.5 Conclusion

Carbon sequestration and sustainable soil management are integral components in the fight against climate change and the promotion of environmental sustainability. Through effective practices such as reforestation, afforestation, and improved agricultural techniques, it is possible to significantly enhance soil organic carbon (SOC) levels, thereby reducing atmospheric CO_2 concentrations. These practices not only mitigate climate change but also improve soil structure, fertility, and water retention, contributing to overall ecosystem health and resilience. However, current land use changes, particularly deforestation and agricultural expansion, pose significant challenges by depleting SOC levels. Thus, adopting sustainable management practices is essential for reversing this trend and achieving long-term environmental benefits.

2.6 Future Perspectives

The future of carbon sequestration and sustainable soil management lies in the widespread adoption and refinement of effective land management practices. Areas we still need to explore: By selecting innovative agricultural practices, including precision agriculture, agroforestry, and incorporation of biochar that enhance soil carbon sequestration while improving crop productivity. Policies and incentives for sustainable land management should be adopted by governments and international bodies. These would include subsidies for farmers, no-till planting, cover crops, and other practices that sequester carbon. More in-depth studies are required to grasp the intricate dynamics of soil carbon sequestration under different land uses and management practices. Monitoring and evaluating the long-term influences of these practices will provide valuable insights and monitor future efforts. Educating the farming community, land managers, and the general public to know why carbon sequestration is so important with their soils and what they can do to change this. It is useful to have strong outreach programs that can help transfer knowledge and promote the use of best practices. Technological developments like remote sensing or soil health monitoring tools provide us with the exact data on how much carbon is in our soils and what practices really work to sequester more. Ensuring global access to know-how is key to collaboratively strategize carbon sequestration and sustainable soil management across diverse regions and ecosystems. By keeping an eye on these areas, we can improve the ability of the soils to sequester carbon, boost soil health, and maintain a more sustainable and resilient global environment. To fully realize the potential of carbon sequestration in reducing climate change, legislative frameworks, scientific research, and practical land management must be integrated.

Authors Contribution

Conceptualization Abid Khan, Mushtaq Ahmad Khan, Muhammad Adnan.

Original Drafting Abid Khan, Mushtaq Ahmad Khan, Muhammad Adnan, Raheem Shahzad.

Software Ikram Ullah, Raheem Shahzad, Hafiza Samra Younas, Monsif Ur Rahman.

Reviewing and Editing Abid Khan, Mushtaq Ahmad Khan, Hongbing Zheng, Tian Shilin, Ikram Ullah, Hafiza Samra Younas, Mehmood Ul Hassan.All the authors read and approved the final manuscript.

References

Ahmad KW, Wang G (2023) Conservation tillage: a sustainable approach for carbon sequestration and soil preservation. A review. J Agric Sustain Environ 2:1–24. https://doi.org/10.56556/jase.v2i1.770

Alam SA, Kivinen S, Kujala H, Tanhuanpää T, Forsius M (2023) Integrating carbon sequestration and biodiversity impacts in forested ecosystems: concepts, cases, and policies. Ambio 52(11):1687–1696. https://doi.org/10.1007/s13280-023-01931-3

Anderson TH, Domsch AK (1993) The metabolic quotient for CO2 (q CO2) as a specific activity parameter to assess the effects of environmental conditions, such as pH, on the microbial biomass of forest soils. Soil Biol Biochem 25(3):393–395. https://doi.org/10.1016/0038-0717(93)90140-7

Antonious GF, Turley ET, Dawood MH (2020) Monitoring soil enzymes activity before and after animal manure application. Agriculture 10:166. https://doi.org/10.3390/agriculture10050166

Atoloye IA, Jacobson AR, Creech JE, Reeve JR (2022) Soil organic carbon pools and soil quality indicators 3 and 24 years after a one-time compost application in organic dryland wheat systems. Soil Tillage Res 224:105503. https://doi.org/10.1016/j.still.2022.105503

Basche A, Tully K, Álvarez-Berríos NL, Reyes J, Lengnick L, Brown T et al (2020) Evaluating the untapped potential of US conservation investments to improve soil and environmental health. Front Sustain Food Syst 4:547876. https://doi.org/10.3389/fsufs.2020.547876

Bhadra P, Maitra S, Shankar T, Hossain A, Praharaj S, Aftab T (2022) Climate change impact on plants: plant responses and adaptations. In: Plant perspectives to global climate changes. Academic Press, New York, pp 1–24. https://doi.org/10.1016/b978-0-323-85665-2.00004-2

Blanco-Canqui H, Benjamin JG (2013) Impacts of soil organic carbon on soil physical behavior. In: Quantifying and modeling soil structure dynamics, vol 3. Wiley, New York, pp 11–40. https://doi.org/10.2134/advagricsystmodel3.c2

Borowik A, Wyszkowska J, Kucharski M, Kucharski J (2020) The role of dactylis glomerata and diesel oil in the formation of microbiome and soil enzyme activity. Sensors 20:3362. https://doi.org/10.3390/s20123362

Bouman OT, Fredeen AH, Mazzocca AM (2018) Soil organic matter storage in a perennial rotation pasture in north-central Nova Scotia from 2007 to 2015. Can J Soil Sci 98(2):385–388. https://doi.org/10.1139/cjss-2017-0122

Bradford JB, Kastendick DN (2010) Age-related patterns of forest complexity and carbon storage in pine and aspen–birch ecosystems of northern Minnesota, USA. Can J For Res 40(3):401–409. https://doi.org/10.1139/x10-002

Bruce JP, Frome M, Haites E, Janzen H, Lal R, Paustian K (1999) Carbon sequestration in soils. J Soil Water Conserv 54(1):382–389

Bünemann EK, Bongiorno G, Bai Z, Creamer RE, De Deyn G, De Goede R et al (2018) Soil quality—a critical review. Soil Biol Biochem 120:105–125. https://doi.org/10.1016/j.soilbio.2018.01.030

Busari MA, Kukal SS, Kaur A, Bhatt R, Dulazi AA (2015) Conservation tillage impacts on soil, crop and the environment. Int Soil Water Conserv Res 3:119–129. https://doi.org/10.1016/j.iswcr.2015.05.002

Byrnes RC, Eastburn DJ, Tate KW, Roche LM (2018) A global meta-analysis of grazing impacts on soil health indicators. J Environ Qual 47(4):758–765. https://doi.org/10.2134/jeq2017.08.0313

Castellano MJ, Mueller KE, Olk DC, Sawyer JE, Six J (2015) Integrating plant litter quality, soil organic matter stabilization, and the carbon saturation concept. Glob Chang Biol 21(9):3200–3209. https://doi.org/10.1111/gcb.12982

Caldwell BA (2005) Enzyme activities as a component of soil biodiversity: a review. Pedobiologia 49(6):637–644. https://doi.org/10.1016/j.pedobi.2005.06.003

Catania V, Bueno RS, Alduina R, Grilli E, La Mantia T, Castaldi S (2022) Soil microbial biomass and bacterial diversity in southern European regions vulnerable to desertification. Ecol Indic 145:109725. https://doi.org/10.1016/j.ecolind.2022.109725

Chahal I, Van Eerd L (2018) Evaluation of commercial soil health tests using a medium-term cover crop experiment in a humid, temperate climate. Plant Soil 427:351–367. https://doi.org/10.1007/s11104-018-3653-2

Chahal I, Vyn RJ, Mayers D, Van Eerd LL (2020) Cumulative impact of cover crops on soil carbon sequestration and profitability in a temperate humid climate. Sci Rep 10:1–11. https://doi.org/10.1038/s41598-020-70224-6

Chamberlain LA, Bolton ML, Cox MS, Suen G, Conley SP, Ane JM (2020) Crop rotation, but not cover crops, influenced soil bacterial community composition in a corn-soybean system in southern Wisconsin. Appl Soil Ecol 154:103603. https://doi.org/10.1016/j.apsoil.2020.103603

Chen J, Carrillo Y, Pendall E, Dijkstra FA, Dave Evans R, Morgan JA et al (2015) Soil microbes compete strongly with plants for soil inorganic and amino acid nitrogen in a semiarid grassland exposed to elevated CO 2 and warming. Ecosystems 18:867–880. https://doi.org/10.1007/s10021-015-9868-7

Conant RT, Cerri CE, Osborne BB, Paustian K (2017) Grassland management impacts on soil carbon stocks: a new synthesis. Ecol Appl 27(2):662–668. https://doi.org/10.1002/eap.1473

Cotrufo MF, Lavallee JM (2022) Soil organic matter formation, persistence, and functioning: a synthesis of current understanding to inform its conservation and regeneration. Adv Agron 172:1–66. https://doi.org/10.1016/bs.agron.2021.11.002

Dad JM, Dand SA, Pala NA (2021) The effect of bi-culture cover crops on soil quality, carbon sequestration, and growth characteristics in apple orchards of North Western Himalayas. Agrofor Syst 95:1745–1758. https://doi.org/10.1007/s10457-021-00687-7

Das BS, Wani SP, Benbi DK, Muddu S, Bhattacharyya T, Mandal B et al (2022) Soil health and its relationship with food security and human health to meet the sustainable development goals in India. Soil Secur 8:100071. https://doi.org/10.1016/j.soisec.2022.100071

de Torres MARR, Carbonell-Bojollo RM, Moreno-García M, Ordóñez-Fernández R, Rodríquez-Lizana A (2021) Soil organic matter and nutrient improvement through cover crops in a Mediterranean olive orchard. Soil Tillage Res 210:104977. https://doi.org/10.1016/j.still.2021.104977

Deng L, Liu GB, Shangguan ZP (2014) Land-use conversion and changing soil carbon stocks in China's 'Grain-for-Green' Program: a synthesis. Glob Change Biol 20(11):3544–3556. https://doi.org/10.1111/gcb.12508

Dick WA (1984) Influence of long-term tillage and crop rotation combinations on soil enzyme activities. Soil Sci Soc Am J 48(3):569–574. https://doi.org/10.2136/sssaj1984.03615995004800030020x

Dignac MF, Derrien D, Barre P, Barot S, Cécillon L, Chenu C et al (2017) Increasing soil carbon storage: mechanisms, effects of agricultural practices and proxies. A review. Agron Sustain Dev 37:14. https://doi.org/10.1007/s13593-017-0421-2

Dindar E, Sagban FOT, Baskaya HS (2015) Evaluation of soil enzyme activities as soil quality indicators in sludge-amended soils. J Environ Biol 2015(36):919–926. https://doi.org/10.30638/eemj.2013.292

Dumanski J, Peiretti R, Benites J, McGarry D, Pieri C (2006) The paradigm of conservation agriculture. Proc World Assoc Soil Water Conserv 1:58–64

Elbasiouny H, Elbehiry F (2019) Soil carbon and nitrogen stocks and fractions for improving soil quality and mitigating climate change. Egypt J Soil Sci 59:131–144. https://doi.org/10.21608/ejss.2019.9984.1251

Ella VB, Reyes MR, Mercado A, Adrian A, Padre R (2016) Conservation agriculture increases soil organic carbon and residual water content in upland crop production systems. Eurasian J Soil Sci 5:24–29. https://doi.org/10.18393/ejss.2016.1.024-029

Emerson WW (1995) Water-retention, organic-C and soil texture. Soil Res 33(2):241–251

FAO (2015) Global forest resources assessment, how are the world's forests changing? Food and Agriculture Organization of the United Nations, Rome

Farooqi ZUR, Hussain MM, Qadeer A, Ayub MA (2021) Role of carbon cycle in soil productivity and carbon fluxes under changing climate. In: Frontiers in plant-soil interaction. Academic Press, New York, pp 29–48

García-Orenes F, Morugán-Coronado A, Zornoza R, Scow K (2013) Changes in soil microbial community structure influenced by agricultural management practices in a Mediterranean agro-ecosystem. PLoS One 8:e80522. https://doi.org/10.1371/journal.pone.0080522

García-Palacios P, Gattinger A, Bracht-Jørgensen H, Brussaard L, Carvalho F, Castro H et al (2018) Crop traits drive soil carbon sequestration under organic farming. J Appl Ecol 55(5):2496–2505. https://doi.org/10.1111/1365-2664.13113

Gattinger A, Muller A, Haeni M, Skinner C, Fliessbach A, Buchmann N et al (2012) Enhanced top soil carbon stocks under organic farming. Proc Natl Acad Sci 109(44):18226–18231. https://doi.org/10.1073/pnas.1209429109

Ghimire B, Ghimire R, VanLeeuwen D, Mesbah A (2017) Cover crop residue amount and quality effects on soil organic carbon mineralization. Sustainability 9(12):2316

Gonzalez-Sanchez EJ, Veroz-Gonzalez O, Conway G, Moreno-Garcia M, Kassam A, Mkomwa S et al (2019) Meta-analysis on carbon sequestration through conservation agriculture in Africa. Soi Tillage Res 190:22–30. https://doi.org/10.1016/j.still.2019.02.020

Govani S, Kundu S, Bhattacharya N, Vidyadharani G, Angel SJ, Sugumar S (2022) Introduction to the impact of microbes on climate: an ocean of opportunities. In: Microbiome under changing climate. Woodhead Publishing, Sawston, pp 393–414

Grigera MS, Drijber RA, Wienhold BJ (2007) Redistribution of crop residues during row cultivation creates a biologically enhanced environment for soil microorganisms. Soil Tillage Res 94(2):550–554. https://doi.org/10.1016/j.still.2006.08.016

Grover M, Maheswari M, Desai S, Gopinath KA, Venkateswarlu B (2015) Elevated CO2: plant associated microorganisms and carbon sequestration. Appl Soil Ecol 95:73–85. https://doi.org/10.1016/j.apsoil.2015.05.006

Hao X, Yang J, Dong S, He F, Zhang Y (2024) The influence of grazing intensity on soil organic carbon storage in grassland of China: a meta-analysis. Sci Total Environ 924:171439. https://doi.org/10.1016/j.scitotenv.2024.171439

Hobbs PR (2007) Conservation agriculture: what is it and why is it important for future sustainable food production? J Agric SciCamb 145:127

Hussain S, Hussain S, Guo R, Sarwar M, Ren X, Krstic D et al (2021) Carbon sequestration to avoid soil degradation: a review on the role of conservation tillage. Plants 10(10):2001. https://doi.org/10.3390/plants10102001

Iqbal M, Ali A, Rizwanullah M (2005) Residual effect of tillage and farm manure on some soil physical properties and growth of wheat (*Triticum aestivum* L.). Int J Agric Biol 7:54–57

Iversen MH (2023) Carbon export in the ocean: a biologist's perspective. Annu Rev Mar Sci 1(1):357–381. https://doi.org/10.1146/annurev-marine-032122-035153

IFOAM (2006) The IFOAM Norms for Organic Production and Processing. Version 2005. International Federation of Organic Agriculture Movements. Germany

Jastrow JD, Miller RM (2018) Soil aggregate stabilization and carbon sequestration: feedbacks through organomineral associations. In: Soil processes and the carbon cycle. CRC Press, Boca Raton, pp 207–223

Jian J, Du X, Reiter MS, Stewart RD (2020) A meta-analysis of global cropland soil carbon changes due to cover cropping. Soil Biol Biochem 143:107735

Jiang ZY, Hu ZM, Lai DY, Han DR, Wang M, Liu M et al (2020) Light grazing facilitates carbon accumulation in subsoil in Chinese grasslands: a meta-analysis. Glob Change Biol 26(12):7186–7197. https://doi.org/10.1111/gcb.15326

Joshi RK, Garkoti SC (2023) Influence of vegetation types on soil physical and chemical properties, microbial biomass and stoichiometry in the central Himalaya. Catena 222:106835. https://doi.org/10.1016/j.catena.2022.106835

Jordon MW, Smith P, Long PR, Bürkner PC, Petrokofsky G, Willis KJ (2022) Can Regenerative Agriculture increase national soil carbon stocks? Simulated country-scale adoption of reduced tillage, cover cropping, and ley-arable integration using RothC. Sci Total Environ 825:153955. https://doi.org/10.1016/j.scitotenv.2022.153955

Kabir M, Habiba UE, Khan W, Shah A, Rahim S, Patricio R et al (2023) Climate change due to increasing concentration of carbon dioxide and its impacts on environment in 21st century; a mini review. J King Saud Univ Sci 35(5):102693

Kahlon MS, Lal R, Ann-Varughese M (2013) Twenty two years of tillage and mulching impacts on soil physical characteristics and carbon sequestration in Central Ohio. Soil Tillage Res. 126:151–158. https://doi.org/10.1016/j.still.2012.08.001

Kern PV, Muth TM (1995) Multifactor productivity in the metal stampings industry. Monthly Lab Rev 118:48

Ketterings QM, Swink SN, Duiker SW, Czymmek KJ, Beegle DB, Cox WJ (2015) Integrating cover crops for nitrogen management in corn systems on northeastern US dairies. Agron J 107:1365–1376

Khan MA, Basir A, Fahad S, Adnan M, Saleem MH, Iqbal A et al (2022) Biochar optimizes wheat quality, yield, and nitrogen acquisition in low fertile calcareous soil treated with organic and mineral nitrogen fertilizers. Front Plant Sci 13:879788

Klumpp K, Fornara DA (2018) The carbon sequestration of grassland soils–climate change and mitigation strategies. Sustainable meat and milk production from grasslands, p 509

Lal R (2002) Soil carbon dynamics in cropland and rangeland. Environ Pollut 116(3):353–362. https://doi.org/10.1016/s0269-7491(01)00211-1

Lal R (2018) Digging deeper: a holistic perspective of factors affecting soil organic carbon sequestration in agroecosystems. Glob Chang Biol 24:3285–3301

Lal R, Follett RF, Stewart BA, Kimble JM (2007) Soil carbon sequestration to mitigate climate change and advance food security. Soil Sci 172:943–956

Lehmann J, Kleber M (2015) The contentious nature of soil organic matter. Nature 528(7580):60–68. https://doi.org/10.1038/nature16069

Lehmann J, Bossio DA, Kögel-Knabner I, Rillig MC (2020) The concept and future prospects of soil health. Nat Rev Earth Environ 1(10):544–553

Leifeld J, Fuhrer J (2010) Organic farming and soil carbon sequestration: what do we really know about the benefits? Ambio 39(8):585–599

Li S, Li X, Ho SH (2022) How to enhance carbon capture by evolution of microalgal photosynthesis? Sep Puri Technol 291:120951. https://doi.org/10.1016/j.seppur.2022.120951

Lin D, McCulley RL, Nelson JA, Jacobsen KL, Zhang D (2020) Time in pasture rotation alters soil microbial community composition and function and increases carbon sequestration potential in a temperate agroecosystem. Sci Total Environ 698:134233

Lori M, Symnaczik S, Mäder P, De Deyn G, Gattinger A (2017) Organic farming enhances soil microbial abundance and activity—a meta-analysis and meta-regression. PLoS One 12(7):e0180442. https://doi.org/10.1371/journal.pone.0180442

Mao R, Zeng DH, Hu YL, Li LJ, Yang D (2010) Soil organic carbon and nitrogen stocks in an age-sequence of poplar stands planted on marginal agricultural land in Northeast China. Plant Soil 332:277–287. https://doi.org/10.1007/s11104-010-0292-7

Martinez AV, Cano A, Johnson J (2018) Simultaneous determination of multiple soil enzyme activities for soil health biogeochemical indices. Appl Soil Ecol 126:121–128

Mathew RP, Feng Y, Githinji L, Ankumah R, Balkcom KS (2012) Impact of no-tillage and conventional tillage systems on soil microbial communities. Appl Environ Soil Sci 2012:1–10

McClelland SC, Paustian K, Schipanski ME (2021) Management of cover crops in temperate climates influences soil organic carbon stocks: a meta-analysis. Ecol Appl 31(3):e02278. https://doi.org/10.1002/eap.2278

McDonald SE, Badgery W, Clarendon S, Orgill S, Sinclair K, Meyer R et al (2023) Grazing management for soil carbon in Australia: a review. J Environ Manag 347:119146. https://doi.org/10.1016/j.jenvman.2023.119146

Meena AL, Jha P, Dotaniya ML, Kumar B, Meena BP, Jat RL (2019) Carbon, nitrogen and phosphorus mineralization as influenced by type of organic residues and soil contact variation in vertisol of Central India. Agric Res 9:232–240

Mishra RK, Mentha SS, Misra Y, Dwivedi N (2023) Emerging pollutants of severe environmental concern in water and wastewater: a comprehensive review on current developments and future research. Water-Energy Nexus 6:74

Mikha MM, Rice CW (2004) Tillage and manure effects on soil and aggregate-associated carbon and nitrogen. Soil Sci Soc Am J 68(3):809–816. https://doi.org/10.2136/sssaj2004.8090

Mobley ML, Lajtha K, Kramer MG, Bacon AR, Heine PR, Richter DD (2015) Surficial gains and subsoil losses of soil carbon and nitrogen during secondary forest development. Glob Change Biol. 21(2):986–996

Mutema M, Mafongoya P, Nyagumbo I, Chikukura L (2013) Effects of crop residues and reduced tillage on macrofauna abundance. J Org Syst 8:5–16

Nave LE, DeLyser K, Domke GM, Holub SM, Janowiak MK, Keller AB et al (2024) Land use change and forest management effects on soil carbon stocks in the northeast US. Carbon Bal Manag 19(1):5. https://doi.org/10.1186/s13021-024-00251-7

Nayak N, Mehrotra R, Mehrotra S (2022) Carbon biosequestration strategies: a review. Carbon Capt Sci Technol 4:100065. https://doi.org/10.1016/j.ccst.2022.100065

Niggli U, Fließbach A, Hepperly P, Scialabba N (2009) Low greenhouse gas agriculture: mitigation and adaptation potential of sustainable farming systems. Ökologie Landbau 141:32–33

Niu L, Manxia C, Xiumei G, Xiaohua L, Hongbo S, Zhaopu L et al (2016) Carbon sequestration and Jerusalem artichoke biomass under nitrogen applications in coastal saline zone in the northern region of Jiangsu, China. Sci Total Environ 568:885–890. https://doi.org/10.1016/j.scitotenv.2016.06.074

Pandao MR, Jejal AD, Shukla EP, Prabhumitrareddy S, Rout S, Baral K et al (2023) Unlocking the benefits of carbon sequestration for enhancing soil health. Int J Environ Clim Change 13(12):1349–1359. https://doi.org/10.9734/ijecc/2023/v13i123803

Pimentel D, Burgess M (2013) Soil erosion threatens food production. Agriculture 3:443–463. https://doi.org/10.3390/agriculture3030443

Poeplau C, Don A (2015) Carbon sequestration in agricultural soils via cultivation of cover crops—a meta-analysis. Agric Ecosyst Environ 200:33–41

Prajapati SK, Choudhary S, Kumar V, Dayal P, Srivastava R, Gairola A et al (2023) Carbon sequestration: a key strategy for climate change mitigation towards a sustainable future. Emerg Trends Clim Change 2(2):1–14

Prasad J, Rao CS, Srinivas K, Jyothi CN, Venkateswarlu B, Ramachandrappa B et al (2016) Effect of ten years of reduced tillage and recycling of organic matter on crop yields, soil organic carbon and its fractions in alfisols of semi arid tropics of southern India. Soil Tillage Res 156:131–139. https://doi.org/10.1016/j.still.2015.10.013

Pulselli FED (2008) Global warming potential and the net carbon balance. In: Encyclopedia of ecology, vol 3. Elsevier BV, Amsterdam, pp 1741–1746

Pullin AS, Knight TM (2009) Doing more good than harm–Building an evidence-base for conservation and environmental management. Biol Conserv 142(5):931–934. https://doi.org/10.1016/j.biocon.2009.01.010

Quintarelli V, Radicetti E, Allevato E, Stazi SR, Haider G, Abideen Z et al (2022) Cover crops for sustainable cropping systems: a review. Agriculture 12(12):2076

Richardson AE, Kirkby CA, Banerjee S, Kirkegaard J (2014) The inorganic nutrient cost of building soil carbon. Carbon Manag 5:265–268. https://doi.org/10.1080/17583004.2014.923226

Rousk J, Brookes PC, Bååth E (2010) The microbial PLFA composition as affected by pH in an arable soil. Soil Biol Biochem 42(3):516–520

Ruis SJ, Blanco-Canqui H, Creech CF, Koehler-Cole K, Elmore RW, Francis CA (2019) Cover crop biomass production in temperate agroecozones. Agron J 111(4):1535–1551. https://doi.org/10.2134/agronj2018.08.0535

Ruibérriz MAR, Bojollo RMC, Braña CA, Lizana AR, Fernández RMO (2012) Carbon sequestration potential of residues of different types of cover crops in olive groves under Mediterranean climate. Span J Agric Res 10(3):649–661. https://doi.org/10.5424/sjar/2012103-562-11

Sartori F, Markewitz D, Borders BE (2007) Soil carbon storage and nitrogen and phosphorous availability in loblolly pine plantations over 4 to16 years of herbicide and fertilizer treatments. Biogeochemistry 84:13–30. https://doi.org/10.1007/s10533-007-9072-8

Sainju UM, Singh HP, Singh BP, Whitehead WF, Chiluwal A, Paudel R (2018) Cover crop and nitrogen fertilization influence soil carbon and nitrogen under bioenergy sweet sorghum. Agron J 110(2):463–471. https://doi.org/10.2134/agronj2017.05.0253

Sánchez Navarro V (2018) Legumes crop for a sustainable agriculture: study of soil fertility, greenhouse gas emission, carbon sequestration and nutritional status of the crops

Sánchez-Navarro V, Zornoza R, Faz A, Fernández JA (2018) Comparing legumes for use in multiple cropping to enhance soil organic carbon, soil fertility, aggregates stability and vegetables yields under semiarid conditions. Sci Hortic 246:835–841. https://doi.org/10.1016/j.scienta.2018.11.065

Sarkar R, Corriher-Olson V, Long C, Somenahally A (2020) Challenges and potentials for soil organic carbon sequestration in forage and grazing systems. Rangel Ecol Manag 73(6):786–795. https://doi.org/10.1016/j.rama.2020.04.002

Sha Z, Bai Y, Li R, Lan H, Zhang X, Li J et al (2022) The global carbon sink potential of terrestrial vegetation can be increased substantially by optimal land management. Commun Earth Environ 3(1):8

Shah J, Wang X, Khan SU, Khan S, Fiaz S, Qayyum A (2021) Optical sensor based nitrogen management in oat for yield enhancement. Sustainability (Basel) 13(12):6955. https://doi.org/10.3390/su13126955

Sher A, Wang X, Sattar A, Ijaz M, Manaf A, Fiaz S (2021) Exogenous applied thiourea for improving the productivity and nutritional quality of bread wheat (Triticum aestivum L.). Agronomy (Basel) 11(7):1432

Singh SB, Saha S, Dutta SK, Singh AR, Boopathi T (2017) Impact of secondary forest fallow period on soil microbial biomass carbon and enzyme activity dynamics under shifting cultivation in North Eastern Hill region, India. Catena 156:10–17

Six J, Paustian K (2014) Aggregate-associated soil organic matter as an ecosystem property and a measurement tool. Soil Biol Biochem 68:A4–A9

Smal H, Olszewska M (2008) The effect of afforestation with Scots pine (Pinus silvestris L.) of sandy post-arable soils on their selected properties. II. Reaction, carbon, nitrogen and phosphorus. Plant Soil 305:171–187. https://doi.org/10.1007/s11104-008-9538-z

Somasundaram J, Sinha N, Dalal RC, Lal R, Mohanty M, Naorem A et al (2020) No-till farming and conservation agriculture in South Asia—issues, challenges, prospects and benefits. Crit Rev Plant Sci 39:236–279. https://doi.org/10.1080/07352689.2020.1782069

Stanley PL, Wilson C, Patterson E, Machmuller MB, Cotrufo MF (2024) Ruminating on soil carbon: applying current understanding to inform grazing management. Glob Change Biol 30(3):e17223. https://doi.org/10.1111/gcb.17223

Stringer LC, Fraser ED, Harris D, Lyon C, Pereira L, Ward CF et al (2020) Adaptation and development pathways for different types of farmers. Environ Sci Pol 104:174–189. https://doi.org/10.1016/j.envsci.2019.10.007

Syed A, Raza T, Bhatti TT, Eash NS (2022) Climate impacts on the agricultural sector of Pakistan: risks and solutions. Environ Chall 6:100433

Tolunay D (2019) Biomass factors used to calculate carbon storage of Turkish forests. Forestist 69(2):145–155. https://doi.org/10.26650/forestist.2019.110719

Tripathi R, Nayak A, Bhattacharyya P, Shukla A, Shahid M, Raja R et al (2014) Soil aggregation and distribution of carbon and nitrogen in different fractions after 41 years long-term fertilizer experiment in tropical rice–rice system. Geoderma 213:280–286

Tully KL, McAskill C (2020) Promoting soil health in organically managed systems: a review. Org Agric 10(3):339–358. https://doi.org/10.1007/s13165-019-00275-1

Utomo M, Frye WW, Blevins RL (1990) Sustaining soil nitrogen for corn using hairy vetch cover crop. Agron J 82(5):979–983. https://doi.org/10.2134/agronj1990.00021962008200050028x

Van Der Heijden MG, De Bruin S, Luckerhoff L, Van Logtestijn RS, Schlaeppi K (2016) A widespread plant-fungal-bacterial symbiosis promotes plant biodiversity, plant nutrition and seedling recruitment. ISME J 10(2):389–399

Van Eekeren N, Bommelé L, Bloem J, Schouten T, Rutgers M, de Goede R et al (2008) Soil biological quality after 36 years of ley-arable cropping, permanent grassland and permanent arable cropping. Appl Soil Ecol 40(3):432–446. https://doi.org/10.1016/j.apsoil.2008.06.010

Van Eerd LL, DeBruyn AH, Ouellette L, Hooker DC, Robinson DE (2018) Quantitative and qualitative comparison of three wet aggregate stability methods using a long-term tillage system and crop rotation experiment. Can J Soil Sci 98(4):738–742

Van Eerd LL, Chahal I, Peng Y, Awrey JC (2023) Influence of cover crops at the four spheres: A review of ecosystem services, potential barriers, and future directions for North America. Sci Total Environ 858:159990. https://doi.org/10.1016/j.scitotenv.2022.159990

Vicente-Vicente JL, García-Ruiz R, Francaviglia R, Aguilera E, Smith P (2016) Soil carbon sequestration rates under Mediterranean woody crops using recommended management practices: a meta-analysis. Agric Ecosyst Environ 235:204–214

Wang S, Adhikari K, Wang Q, Jin X, Li H (2018) Role of environmental variables in the spatial distribution of soil carbon (C), nitrogen (N), and C: N ratio from the northeastern coastal agroecosystems in China. Ecol Indic 84:263–272

Wang S, Wang Z, Heinonsalo J, Zhang Y, Liu G (2022) Soil organic carbon stocks and dynamics in a mollisol region: a 1980s–2010s study. Sci Total Environ 807:150910. https://doi.org/10.1016/j.scitotenv.2021.150910

Wang N, Chen X, Zhang Y, Pang J, Long Z, Chen Y et al (2024) Integrated effects of land use and land cover change on carbon metabolism: based on ecological network analysis. Environ Impact AssessRev 104:107320. https://doi.org/10.1016/j.eiar.2023.107320

Yadav GS, Lal R, Meena RS, Babu S, Das A, Bhowmik SN et al (2019) Conservation tillage and nutrient management effects on productivity and soil carbon sequestration under double cropping of rice in northeastern region of India. Ecol Indic 105:303–315. https://doi.org/10.1016/j.ecolind.2017.08.071

Yao Z, Zhang D, Liu N, Yao P, Zhao N, Li Y et al (2019) Dynamics and sequestration potential of soil organic carbon and total nitrogen stocks of leguminous green manure-based cropping systems on the Loess Plateau of China. Soil Tillage Res 191:108–116. https://doi.org/10.1016/j.still.2019.03.022

Zhang Y, Li P, Liu X, Xiao L (2022) Changes in soil aggregate fractions, stability, and associated organic carbon and nitrogen in different land use types in The Loess Plateau, China. Sustainability 14(7):3963. https://doi.org/10.3390/su14073963

Zhou G, Zhou L, Zhou X (2020) Effects of grazing intensity on belowground carbon and nitrogen cycling. In: Grasses and grassland aspects. IntechOpen, London

Cultivation of New Crops Under Changing Climate

3

Yunus Emre Arvas, Ali Devlet, and Yilmaz Kaya

Abstract

Global climate change poses serious threats to ecosystems and agriculture. It has been observed that crop growth and productivity are negatively affected due to climate change. Various methods have been developed in the past to address these issues. The Green Revolution aimed to increase agriculture by using high-yield crop varieties and chemical fertilizers, but these practices have only addressed the effects of climate change to a limited extent. Therefore, there is a need for sustainable and integrated agricultural approaches. Genetically modified plants are considered an effective strategy for adapting to climate change. These plants can enhance carbon sequestration capacities and provide climate-resilient food systems. In this context, genetically engineered "smart plants" have been developed to withstand climate variations. Genetically modified plants have the potential to optimize various agricultural traits. Transgenic technology is recognized as a rapid method for gene transfer. This technology allows the introduction of desired genes into plants, enabling the enhancement of specific traits. It can increase plants' resistance to climate change and other stress factors. However, developing high-efficiency protocols for genetic transformation is nec-

Y. E. Arvas (✉)
Faculty of Science, Department of Molecular Biology and Genetics, Van Yüzüncü Yıl University, Tusba, Van, Turkey

Faculty of Science, Department of Biology, Karadeniz Technical University, Ortahisar-Trabzon, Turkey

A. Devlet
Faculty of Agriculture and Natural Science, Department of Field Crops, Bilecik Seyh Edebali University, Bilecik, Turkey

Y. Kaya
Faculty of Agriculture, Department of Agricultural Biotechnology, Samsun Ondokuz Mayıs University, Samsun, Turkey

© The Author(s), under exclusive license to Springer Nature Singapore Pte Ltd. 2025

S. Faiz et al. (eds.), *Climate Smart Agriculture for Future Food Security*,
https://doi.org/10.1007/978-981-96-4499-5_3

essary. Transgenic plants play a significant role in reducing greenhouse gas emissions. These plants can lower emissions by reducing pesticide use and fossil fuel consumption. Additionally, developing plants with high carbon sequestration capacity aims to decrease the amount of CO_2 in the atmosphere. This summary provides an overview of the potential contributions of genetically modified plants and transgenic technology in combating climate change.

Keywords

Genetically modified plants (GMPs) · Transgenic technology · Climate resilience · Carbon sequestration · Emission reduction · Climate change · Sustainable agriculture

3.1 Introduction

Global warming and the resulting climate changes are among the most critical global issues of the twenty-first century and are a significant source of concern due to their impact on the world. Climate change is a global crisis that threatens not only humans but also entire ecosystems and biodiversity. The disruption of natural balance leads to a reduction in biodiversity and irreversible damage, also manifested through the depletion of natural resources. The Industrial Revolution is seen as one of the fundamental causes of this process (Kumar et al. 2024a; Scafetta 2024; Usman et al. 2024). Climate change has been triggered primarily by increased fossil fuel use with industrialization and rising concentrations of greenhouse gases in the atmosphere; this trend is expected to continue accelerating in the future (Bilgili et al. 2024; Leichenko and O'Brien 2024).

Both natural processes and human activities contribute to greenhouse gas emissions in the atmosphere. Major greenhouse gases include carbon dioxide (CO_2), methane (CH_4), nitrous oxide (N_2O), and water vapor; these gases create the greenhouse effect by absorbing infrared radiation in the atmosphere (Mahmood et al. 2022). Human activities increase the concentration of greenhouse gases, leading to global warming and disruptions in hydrological cycles (Filonchyk et al. 2024; Kumari 2024). Before the Industrial Revolution, the concentration of carbon dioxide in the atmosphere was around 280 ppm, whereas today it has risen to over 410 ppm, representing a 45% increase. This level is projected to reach between 730 and 1000 ppm by 2100 (Franta 2022; Jain 2022b). The accumulation of increased carbon dioxide and other greenhouse gases has raised average annual temperatures by approximately 0.8 °C as of 2017. According to the Intergovernmental Panel on Climate Change (IPCC), the increase in atmospheric CO_2 concentration is expected to raise global temperatures between 1.0 and 3.7 °C (IPCC 2014). Simultaneously, this process could disrupt precipitation patterns and increase the frequency of drought events, particularly in arid regions (Deng et al. 2024; Javaid et al. 2022).

All living organisms that are part of the ecosystem are exposed to the effects of climate change. Although climate change negatively affects all life on Earth, plants are more adversely affected than other organisms. Humans depend on plants for survival, as everything we consume is directly from plants or from animals that feed on plants. Additionally, plants, as fundamental components of natural ecosystems, absorb approximately 30% of the carbon dioxide released into the atmosphere by humans each year (Hubau et al. 2020; Terrer et al. 2021; Jain 2022a).

As the effects of climate change intensify, predicting the impacts of increasing atmospheric CO_2 levels and temperatures on plant populations becomes increasingly difficult. Plants' long life cycles limit their ability to adapt to changing environmental conditions and increase their sensitivity to climate change. Environmental changes resulting from climate change lead to significant declines in plant productivity and yield, with long-term negative impacts expected to include reduced vegetation cover, decreased nutrient levels, increased droughts, and the spread of pests and diseases (Zhao et al. 2020; Janni and Pieruschka 2022; Janni et al. 2024).

Increasing climate change is causing significant alterations in plant community structure and composition, biological productivity and diversity, and distribution patterns. According to the Intergovernmental Panel on Climate Change (IPCC), this process will further increase the extinction risk for many plant species (IPCC 2014). It is estimated that 3–21% of endemic plant species in Europe may face extinction risk by 2050. This situation will lead to major changes in ecosystem productivity and biodiversity. Increases in temperatures, significant changes in precipitation patterns, and more frequent and intense extreme weather events may alter plant distribution within ecosystems. To adapt to climate change, some plant species may need to migrate to higher elevations or face extinction (García Criado et al. 2020; Palmquist et al. 2021).

Certain plant species that successfully grow under specific climate conditions must adapt to new conditions or require suitable growing areas to survive. Otherwise, these species will face the risk of extinction. In this context, plant species that cannot adapt to climate change face increased extinction risk due to their inability to change habitats (Casu et al. 2024; Yadav et al. 2024).

Today, environmental stresses caused by global warming and climate change are creating significant challenges in agricultural production, severely limiting crop yields and threatening food security (Mohanty et al. 2024). Temperature fluctuations and changes in precipitation patterns are primary indicators of these environmental stresses. Abiotic stress factors such as extreme temperatures, drought, salinity, and heavy metal toxicity, combined with biotic stresses such as pathogens and weeds, can reduce the yields of essential crops by up to 50%. Therefore, enhancing plant resistance to abiotic and biotic stresses is crucial in combating climate change (Shabbir et al. 2022; Zandalinas and Mittler 2022).

Plants have developed various adaptive mechanisms that create specific cellular responses to mitigate the effects of environmental stresses (Islam et al. 2022). In addition, developing new agricultural methods to increase plant tolerance to abiotic and biotic stresses, applying different crop diversification strategies, and adopting both traditional and innovative approaches are crucial. Improving plant cultivation techniques will facilitate the development of climate change-resistant crops that can better adapt to stress conditions such as drought and high temperatures (El-Saadony et al. 2022). Genome-wide association studies (GWAS), advanced phenotyping methods, genomic selection (GS), and genotyping strategies are critical in identifying key genes for crop improvement under climate change. Moreover, developing abiotic and biotic stress-resistant transgenic plants using genetic engineering approaches is considered an important strategy against climate change (Bhat et al. 2015; He and Li 2020; Srivastava et al. 2020; Afzal et al. 2023).

The effects of climate change extend beyond plants to affect plant–soil interactions and soil microorganisms. Soil microbiota, like plants, are significantly impacted by the negative effects of global warming. Changes in plant species, rapid extinction of generations, and the emergence of new species lead to alterations in soil structure, which, in turn, changes the roles of soil microorganisms in ecosystem cycles. Understanding how plants respond to environmental changes is of utmost importance in this context. Laboratory and field studies have examined the effects of increased CO_2 levels and temperature fluctuations on plant germination, photosynthesis, transpiration, morphology, anatomy, physiology, and yield. However, these negative effects have further emphasized the importance of understanding how plants can adapt to climate change, and scientists continue to discover genes that could contribute to this process (Classen et al. 2015; Jansson and Hofmockel 2020; Morris and Blackwood 2024; Qiao et al. 2024; Roy et al. 2024).

This study will examine environmental stress factors caused by climate change, their effects on plants, changes in photosynthetic performance due to changing environmental conditions, modern plant cultivation techniques, and biotechnological strategies for combating global climate change. Additionally, the potential solutions that transgenic plants developed using genetic engineering techniques may offer for food security issues caused by climate change will be evaluated.

3.2 Effects of Climate Change on Plant Growth and Development

Plants are among the organisms most exposed to environmental changes and adverse conditions due to their lack of active mobility. Abiotic stress factors such as drought, salinity, excessive rainfall, and extreme temperatures that they encounter throughout their life cycle directly affect plant growth and development (Enebe and Babalola 2018; Oshunsanya et al. 2019). Plants adapt their growth and development strategies to minimize the impact of environmental changes, and when exposed to specific stress conditions for extended periods, they enhance their ability to adapt to these conditions (Basu et al. 2016; Seleiman et al. 2021). The distribution of plants of the same species across regions with different climate conditions is the strongest indicator of these adaptation abilities. Plants exposed to environmental stresses are expected to undergo significant changes not only physiologically but also metabolically (Bhargava and Sawant 2013; Ramírez et al. 2024). Depending on the severity and duration of the stress, plants can even restructure their metabolic processes to such an extent that it may alter their life cycle (Nick 2024; Qiao et al. 2024).

The functional differences between species that adapt to different environmental conditions are rooted in the diversity of their developmental processes (Farooq et al. 2023). Environmental stress factors create various effects in different plant tissues and organs; therefore, molecular, cellular, and morphological responses to stress vary among tissues and according to the plant's developmental stage (Raza et al. 2020; Dos Santos et al. 2022). The diversity in developmental processes between species and the ability of a plant of a specific species to dynamically modify these processes in response to environmental conditions are key to the plant's survival success in both natural and agricultural environments. Plants experience changes in the initiation and timing of developmental events as a result of their response to environmental stress, leading to differences at both the organ level and the whole plant level (Farkas and Dobránszki 2024; Shah et al. 2024). For example, the suppression of lateral root development in response to stress factors like water deficiency is an example of changes in developmental events. Changes in the timing of developmental processes can be observed in *Arabidopsis thaliana* transitioning from vegetative to reproductive development earlier under high temperatures. Such changes can be observed in plant organs or at the whole plant level. For instance, the formation of additional leaf nodes and larger leaves in soybean plants in response to increased CO_2 levels is a good example of such adaptation (Gray and Brady 2016; Huang et al. 2024a).

3.3 Effects of Temperature on Plant Development

Temperature is one of the key factors determining plant distribution, growth rate, and developmental processes in ecosystems. Each plant species grows within optimal temperature ranges (usually 25–30 °C), and the morphological and physiological functions of plants are shaped according to these conditions (Khalil et al. 2018; Ul Hassan et al. 2021). According to IPCC projections, global temperatures are expected to increase by 1.2–3.7 °C over the next 30–50 years (IPCC 2014). This increase will raise the likelihood of plants being exposed to temperature stress by increasing the frequency, duration, and intensity of extreme temperatures. Excessive temperatures above the optimal range can cause heat-related damage to plants (Hussain et al. 2021). Particularly, temperature fluctuations during the summer months can damage the molecular interactions necessary for normal growth, adversely affecting fruit set and development. Prolonged exposure to extreme temperatures can result in cellular disorganization, disruptions in metabolic activities, and growth retardation (Zhu et al. 2021; Bacelar et al. 2024).

Extreme temperatures produce excessive reactive oxygen species (ROS) by denaturing cellular structures, reducing protein stability, and damaging cell membranes, leading to severe oxidative stress. The growth stage most affected by such stress is often germination. High temperatures can reduce cell water content, adversely affecting germination rates, viability, and root growth in plants (Abhijith Shankar et al. 2024; Shahbaz 2024). For example, studies have shown a reduction in germination rates in crops such as wheat and peas due to increased temperatures.

Similarly, the germination rates of some plant seeds can drop to zero under high temperatures (Mohamed et al. 2021). This situation is associated with the direct effects of temperature stress on metabolism (Tarnawa et al. 2023; Abshire et al. 2024).

Additionally, high temperatures reduce the net assimilation rate (NAR) of plants, thereby decreasing growth rates. The negative effects of temperature stress not only impact metabolic processes but also vital physiological processes such as photosynthesis (Mathur et al. 2014; Kaushal et al. 2016). Photosystem II (PSII) is one of the most temperature-sensitive photosynthetic components, and the rate of photosynthesis can decrease significantly due to temperature stress (Parsons et al. 2011; Anjana Jajoo and Allakhverdiev 2017). This reduces the overall growth rate and yield. Under moderate temperature stress, there is a decrease in the activity of Rubisco, one of the main enzymes in photosynthesis, which can halt the photosynthesis process. High temperatures can also lead to reduced plant height and accelerated flowering times (Qu et al. 2021; Wijewardene et al. 2021).

Low temperatures can also negatively impact physiological and biochemical processes in plants. Sudden temperature drops can cause changes in plant cell membranes, reduce the synthesis of antioxidant enzymes, and delay carbon transfer (Nievola et al. 2017; Bhattacharya 2022). Especially temperatures below freezing can lead to frost damage in plants. Considering all these negative effects, it is known that breeding and genetic research are being conducted with the aim of developing plant species that are resistant to both high and low temperatures (Parsons et al. 2011; Driedonks et al. 2016; Wijewardene et al. 2021; Zafar et al. 2022).

3.4 Effects of Drought on Plant Development

Drought is one of the most significant issues triggered by global climate change. Drought is defined as the lack of adequate rainfall over a specific period from a meteorological perspective (Sun et al. 2021). Plants are exposed to drought stress when there is insufficient water in the soil, and this stress is exacerbated by water loss through evaporation and transpiration (Amgain and Lal 2024; Garg et al. 2024). From an agricultural standpoint, drought occurs when plants are unable to obtain the water they need from the soil for growth and development. This situation is one of the most critical environmental stress factors that seriously affect plant growth, biomass, and yield (Ahmad et al. 2022). In recent years, the increased frequency and severity of droughts have posed a threat to global food security. By 2050, drought could negatively impact plant growth on more than 50% of arable land (Haider et al. 2024; Mondal et al. 2024).

Drought leads to a significant decrease in agricultural productivity by causing rainfall to fall below expected levels and reducing water resources due to unplanned use. With consequences such as ecosystem degradation, changes in plant patterns, increased plant diseases, desertification, forest losses, and reduced biodiversity, drought is one of today's most pressing environmental issues (Amgain and Lal 2024; Haider et al. 2024).

3.5 Effects of Drought Stress on Plants

Drought stress disrupts the molecular, biochemical, physiological, and morphological functions of plants. Water deficiency negatively affects plant yield and quality, with the severity, duration of the stress, and the plant's growth stage determining the extent of these effects (Li et al. 2021). Many plant species are particularly sensitive to drought during the early stages of germination and seedling development. Water deficiency disrupts osmotic balance, adversely affecting cellular metabolism; it causes changes in DNA, RNA, and protein structures, leading to membrane damage in cells and excessive production of reactive oxygen species (ROS) (Sharma et al. 2012; Abhijith Shankar et al. 2024; Shahbaz 2024). As a result, seed germination and seedling viability decrease. One of the stages most affected by drought is the generative (reproductive) phase; during this phase, severe drought can cause yield losses of up to 95% in plants. Drought during the flowering stage, in particular, can lead to infertility in plants (Tarnawa et al. 2023; Deng et al. 2024; Qiao et al. 2024; Staniak et al. 2024).

When drought conditions persist for a long time, plants stop stem elongation and increase root development in an attempt to access more water (Ahmed et al. 2020; Hannan et al. 2020; Wang et al. 2021). However, severe and prolonged drought leads to the cessation of both stem and root growth, and a reduction in leaf area and leaf number. With a decrease in photosynthesis rate, cell division and expansion stop, and yellowing and shedding of leaves occur (Wang et al. 2021). The adverse effects of drought stress are manifested through physiological and morphological changes such as decreased leaf water potential, stomatal closure, senescence (aging), and leaf curling (Deng et al. 2024).

Drought affects important biochemical and physiological processes such as photosynthesis, chlorophyll synthesis, nutrient uptake, respiration, and carbohydrate metabolism, severely limiting plant growth. These negative effects of drought stress raise questions about how plants survive in environments with insufficient water. Interestingly, plant DNA contains genes that regulate defense strategies against drought. These defense strategies determine how plants respond to stress and their chances of survival (Jarin et al. 2024; Qiao et al. 2024).

3.6 Plant Responses and Defense Mechanisms to Drought Stress

Plants develop two main defense strategies against drought: stress avoidance and stress tolerance. Plants that avoid stress can survive under moderate drought conditions, while stress-tolerant plants activate protective mechanisms to sustain life under more severe drought conditions (Saddiq et al. 2021). For example, some resilient plant species can reduce their tissue water content to as low as 5% during periods of limited water and quickly rehydrate when water becomes available again. Other plants adapt to drought through structural changes; for instance, succulent plants have thick, fleshy leaves that can store water. Additionally, some plants minimize water loss by reducing leaf size or shedding their leaves entirely (Jarin et al. 2024; Qiao et al. 2024; Sharma et al. 2024).

Plants also exhibit internal strategies to tolerate drought, such as osmotic adjustment, antioxidant defense mechanisms, and stomatal closure. The regulation of gene expression and signaling pathways in these processes enhances the plants' resistance to stress (Tarnawa et al. 2023; Amgain and Lal 2024; Haider et al. 2024).

3.7 Drought in Turkey and Future Perspectives

Due to global climate change, a significant portion of Turkey is affected by semi-arid climate conditions. This situation positions Turkey as one of the countries most impacted by global warming. Therefore, developing drought-resistant plant species, increasing genetic research, and adopting sustainable agricultural practices to maintain agricultural productivity are of utmost importance (Erkol et al. 2024; Korkmaz and Kuriqi 2024).

3.8 Effects of Salinity on Plant Growth

The condition where mineral salts such as Na^+ and Cl^- damage plants is referred to as "salt stress." Salt stress is one of the most significant environmental factors limiting plant development, and it is expected to become more widespread due to global climate change and increasing irrigation practices (Siddiqui et al. 2021; Yan et al. 2021). Factors such as low rainfall, high evaporation, and rising temperatures cause arid and semi-arid areas to be at risk of salt stress. Worldwide, 50% of irrigated agricultural lands are affected by high salinity, and this rate is increasing by 10% annually. At this rate, by 2050, more than half of the arable land will be damaged by high salt levels. The expansion of saline soils poses a serious threat to global food security (Corwin 2021; Dos Santos et al. 2022; Junaid and Gokce 2024). Soil salinity is continuously accumulating in 6% of agricultural lands each year at a rate of 1–2%, leading to significant declines in the yield of staple crops such as corn, wheat, and rice (Elhag and Bahrawi 2017; Corwin 2021; Genua-Olmedo et al. 2022).

The use of saline irrigation water and high evaporation rates lead to the accumulation of salt on the soil surface. This situation also causes groundwater to become saline and increases the concentration of soluble ions such as Na^+ and Cl^-. These ions cause oxidative stress by accumulating reactive oxygen species (ROS) in plant cells, leading to disruptions in plant metabolism and consequently yield losses (Abhijith Shankar et al. 2024; Shahbaz 2024; Sharma et al. 2024).

3.9 Effects of Salt Stress on Plants

The negative effects of salt stress on plants vary depending on environmental conditions, plant species, and soil composition. Salt stress is primarily detected by plants through their roots and impedes plant growth both in the short and long term. In the short term, salt stress reduces water availability in plants, leading to osmotic stress. In the long term, plants are exposed to ion toxicity due to excessive salt accumulation. This stress negatively affects many physiological processes, from germination and growth to photosynthesis, water balance, oxidative stress, and reproductive processes (Akhtar 2019; Razzaq et al. 2020; Siddiqui et al. 2021; Amerian et al. 2024).

Plants develop various biochemical and molecular defense mechanisms against salt stress. To resist salt stress, plants use mechanisms such as ion accumulation or excretion, control of ion uptake by roots, and regulation of ion transport to shoots (Abdullah et al. 2022). Additionally, plants facilitate the accumulation of ions in specific cellular compartments and activate their antioxidant systems by synthesizing osmotic regulatory molecules. The activation of genes associated with salt stress, synthesis of transcription factors, and production of stress proteins are key strategies that enhance plant tolerance to salt stress (Yarra and Wei 2021; Aizaz et al. 2024; Bhat et al. 2024).

These biochemical and molecular strategies contribute to maintaining agricultural productivity by improving plants' survival capabilities under salt stress (Jhanani et al. 2024; Zhou et al. 2024).

3.10 Strategies Developed by Plants to Combat Climate Change

Global environmental changes have profound effects on plant growth and agricultural productivity. Extreme climate events seriously threaten food security, necessitating innovative solutions. However, it is known that current strategies for effectively combating climate change are inadequate. Since agricultural production is largely based on the selection of crop species suited to specific climate conditions, agriculture has always been considered a bioscience dependent on regional climate characteristics (Ranasinghe and Mapa 2024; Sabat and Tripathy 2024; Usman et al. 2024).

According to the latest report by the Food and Agriculture Organization (FAO) of the United Nations, if current anthropogenic greenhouse gas (GHG) emissions and climate change trends continue, significant reductions in the yield of many essential products critical for human nutrition are expected. Therefore, innovative and sustainable strategies must be developed to enhance plant production and crop resilience (Mbow et al. 2019; FAO 2022).

One of the strategies developed to adapt to climate change is to increase the tolerance capacity of plant varieties. Traditional plant breeding techniques aim to create resistant species by utilizing resistance traits from wild relatives of plants. Additionally, biotechnology offers a faster and more precise solution through modern genetic engineering techniques to produce genetically modified crops that are resistant to climate change (Chege et al. 2024; Janni et al. 2024; Shah et al. 2024; Usman et al. 2024).

Recent advances in genetic engineering represent a significant revolution in plant breeding and crop productivity. Genomic-assisted breeding, next-generation sequencing techniques, and genome editing methods are among the main innovative tools used in the development of climate-resilient, high-yield, and better-adapted crop varieties (Platani et al. 2023; Saha et al. 2023; Bharti et al. 2024; Kumar et al. 2024b).

3.11 Agricultural Practices

Recent studies have highlighted various beneficial agricultural approaches for coping with climate change and helping plants adapt. These approaches include improving irrigation techniques, crop rotation, optimizing planting and harvesting periods, selecting crops with short life cycles, and making changes in crop patterns (Shah and Wu 2019; Kalaiselvi et al. 2024). Altering planting times, cultivating stress-resistant species, and developing new crop varieties can help reduce the negative effects of environmental stress, allowing plants to adapt to new conditions and contributing to global food security (Zhang et al. 2018). Mylonas and colleagues reported that early planting of wheat in Southern Europe led to better development of the root system before winter, making the plants more resistant to low winter temperatures. Many studies have shown that delaying planting times negatively impacts wheat yield (Mäkinen et al. 2018; Beillouin et al. 2020; Huang et al. 2024b; Jarecki 2024; Pinke et al. 2024).

Abiotic stress factors, particularly drought and salinity, significantly limit crop production. One of the main factors affecting crop productivity on a global scale is limited water resources (Rubab et al. 2023). Implementing drought strategies will alleviate the impact of drought stress on crop plants and enhance crop development under these conditions. Mondal and colleagues reported that selecting early-maturing wheat varieties reduced stress effects by ensuring the completion of grain filling before drought stress. Seed treatments are another method that improves plant adaptation to drought environments by increasing water use efficiency and performance (Imran et al. 2021; Arvas et al. 2022; Shabbir et al. 2022; Tarnawa et al. 2023; Jarin et al. 2024; Oyebamiji et al. 2024).

Additionally, appropriate fertilization techniques are crucial for mitigating the effects of abiotic stress and enhancing plant adaptation capacity. Fertilizers can increase productivity by providing energy to plants while maintaining soil fertility. For example, nitrate fertilizers can improve the survival rate of waterlogged plants. Moreover, the application of bacteria that promote plant growth is an important factor in positively affecting crop stress tolerance (Mahmud et al. 2021; Arvas and Durak 2022; Barłóg et al. 2022).

3.12 Breeding Studies and Genomics

Plant breeding is the process of systematically altering and developing the genetic makeup of economically important plant species to meet the expectations of growers and consumers. The main objective of breeding is to use the variations obtained through genetic modifications to develop new plant varieties that are high-quality, efficient, disease-resistant, and capable of adapting to environmental stress conditions in the shortest possible time. In environmental stress conditions, plant breeding provides a dynamic tool in the crop development process (Anders et al. 2021; Hasan et al. 2021; Afzal et al. 2023; Aziz and Masmoudi 2025).

Raza and colleagues have highlighted the importance of genetic divergence analysis in breeding methods that use polymorphism, hybridization, and recombination techniques to obtain new plant varieties (Raza et al. 2021). Breeding studies employ strategies such as phenotypic screening of stress-tolerant plant species and collection of genetic material. Molecular markers are used for various purposes, including genetic mapping, QTL (quantitative trait locus) analysis, characterization of gene sources, and selection of parents for breeding programs. Additionally, molecular marker technology plays a crucial role in identifying stress-related genomic regions and providing information about their function (Jing et al. 2018; Prakash et al. 2019; Kulkarni et al. 2020; Najeeb et al. 2020).

Marker-assisted selection (MAS) and genome-wide association studies (GWAS) are fundamental approaches in molecular and integrated breeding techniques that provide stress tolerance. These technologies are effectively used in developing new crop varieties adapted to climate change with high yield potential. Integrated breeding methods with genomics play a significant role in producing plant species adapted to changing climate conditions (Das and Rao 2015; Nogoy et al. 2016; Nakano and Kobayashi 2020; Shimizu et al. 2020; Hasan et al. 2021; Sahito et al. 2024).

3.13 Transgenic Plant Technology

Global climate change continues to pose a serious threat to ecosystems, causing ongoing concern. Variables associated with climate change have negative impacts on plant growth and agricultural productivity. Various methods have been tried in the past to overcome these adverse effects (Bolan et al. 2023; Mishra 2023). The Green Revolution is one such initiative that aimed to promote the use of high-yielding plant varieties, chemical fertilizers, and pesticides. However, changing climate conditions have limited the effectiveness of these practices (Rebouh et al. 2023; Bajoria et al. 2024). Therefore, it is necessary to adopt sustainable and integrated approaches in agricultural production. Developing genetically modified (transgenic) plants can be considered an effective adaptation strategy in combating climate change. The development of transgenic plants, such as "smart plants" that can sequester more carbon and withstand climate crises, may play a critical role in ensuring food security. For instance, staple food sources such as wheat, corn, and rice are being made more resilient to climate conditions through the addition of different genes. Additionally, genes adapted to climate conditions can be identified from the natural genetic variation of existing plants and their wild relatives and integrated into breeding programs (Arvas and Kaya 2019; Arvas and Kocacaliskan 2020; Das et al. 2023; Khokharvoytas et al. 2023).

Transgenic technology is a rapidly advancing method in agriculture that enables gene transfer. Unlike traditional breeding methods, this technology allows the transfer of desired genes into the plant gene pool, thereby optimizing specific traits in plants. Recombinant DNA technology facilitates the precise transfer of targeted genes, preserving important economic traits. Moreover, the sources of resistance genes are not limited to closely related plant species. While traditional breeding and marker-assisted selection (MAS) are applicable, especially in annual and perennial plants, these techniques are limited in plants with long reproductive cycles. Through transgenic technology, many agriculturally important traits can be successfully integrated from different organisms into plants. However, the development of high-efficiency genetic transformation protocols for this technology is crucial (Glıck et al. 2010; Adetunji et al. 2022; Aziz and Masmoudi 2025; Moran et al. 2024).

Transgenic plants contribute to reducing greenhouse gas emissions, thus mitigating the effects of climate change (Delangiz et al. 2019). For example, transgenic agriculture practices since 1996 are estimated to have contributed to the sequestration of 133.639 million tons of CO_2 in the soil worldwide (Sutherland et al. 2021; Brookes 2022). Additionally, some transgenic plants, such as genetically modified poplar trees, have been reported to increase the rate of environmental pollutant metabolism, thereby facilitating the removal of harmful substances like hydrocarbons from the environment (James et al. 2024; Yeşilyurt et al. 2024).

Additionally, the development of "carbon-sequestering plants" is emerging as a solution to climate change. For example, integrating naturally occurring substances in plants, such as suberin, into agricultural products such as corn and canola through genetic engineering could enable deeper root growth, allowing carbon to be stored in the soil for extended periods (Coulson 2024; Tang et al. 2024). Gene editing technologies such as CRISPR are being used to enhance the carbon capture capabilities of plants and soil microorganisms. Moreover, it is reported that genetically modified plants could be made up to 40% more productive, optimizing photosynthesis and reducing atmospheric CO_2 levels (Khan et al. 2024; Nawaz et al. 2024).

While the ability of genetically modified plants to ensure a sustainable food supply in the future remains a topic of debate, the economic benefits and potential for increased productivity of transgenic crops are seen as significant opportunities (Qaim 2020; Ghimire et al. 2023; Husaini and Sohail 2023).

3.14 Conclusion and Recommendations

Climate change is currently recognized not only as an environmental issue but also as a serious threat to the sustainability of human life. The increasing frequency and intensity of natural disasters such as floods, fires, and droughts on a global scale pose significant risks to all ecosystems, particularly affecting agricultural productivity. The impacts of climate change on plants necessitate urgent measures for agricultural production and food security. Climate change exacerbates abiotic stress factors such as temperature increases and changes in water availability. Plants, having less mobility compared to other organisms in the ecosystem, are more significantly affected by climate change. Rising temperatures and decreasing precipitation can lead to yield losses and quality declines in plants. Additionally, these changes create suitable conditions for plant diseases and pests, negatively affecting crop quantity and quality.

A series of strategies have been developed to help plant varieties adapt to climate change. These strategies include modifications to irrigation techniques, crop rotations, changes in planting and harvesting times, cultivation of crops with shorter life cycles, and adjustments in crop patterns. Such agronomic approaches are crucial for helping plants adapt to changing climate conditions and ensuring global food security. Genetically modified (transgenic) plants are considered an effective adaptation strategy against climate change. Modern biotechnological methods enable the development of plant varieties with climate-resistant traits. Specifically, genetic engineering is used to create plants with high carbon sequestration capacity and resilient food systems.

Transgenic plants have significant potential to enhance agricultural productivity and reduce greenhouse gas emissions. To address abiotic stress factors in agriculture, irrigation and fertilization strategies should be optimized. For example, appropriate fertilization and irrigation methods can improve water use efficiency and plant adaptation in drought-prone environments. Additionally, the use of fertilizers that maintain soil fertility and promote plant growth can be effective in increasing crop productivity. Genetic variations and genome editing techniques are important tools in developing climate-resistant plant varieties. Molecular markers, genetic mapping, and genome editing methods contribute to the development of stress-resistant plants. Understanding how plant varieties respond to abiotic stress conditions and using this knowledge to create new, resilient varieties is crucial in breeding research. Investments in biotechnology and genetic engineering should be made to accelerate the development of plant varieties that can adapt to climate change. These investments can expedite the creation of plant varieties capable of coping with climate change.

Considering changes in temperature and water availability, agricultural management strategies should be updated. Irrigation and fertilization methods should be optimized to adapt to changing climate conditions. Plant growers, researchers, and agricultural experts should be informed about the effects of climate change and strategies to address these effects. This information can aid in the adoption of effective and sustainable agricultural practices. International collaboration and information sharing should be encouraged to enhance agricultural productivity on a global scale and address climate change. Knowledge and experiences gained from different regions can contribute to the development of applicable solutions in other regions.

In conclusion, effective and sustainable strategies must be developed to cope with climate change. Biotechnology, agricultural management strategies, and international collaboration can help successfully implement these strategies and play a crucial role in meeting future food supply needs.

References

Abdullah et al (2022) Physiological, biochemical and yield responses of linseed (Linum usitatissimum L.) in alpha tocopherol mediated alleviation of salinity stress. Front Plant Sci 13:867172

Abhijith Shankar P, Parida P, Bhardwaj R, Yadav A, Swapnil P, Seth CS, Meena M (2024) Deciphering molecular regulation of reactive oxygen species (ROS) and reactive nitrogen species (RNS) signalling networks in Oryza genus amid environmental stress. Plant Cell Rep 43:185

Abshire N, Hauck AL, Walia H, Obata T (2024) Tissue-and time-dependent metabolite profiles during early grain development under normal and high night-time temperature conditions. BMC Plant Biol 24:568

Adetunji CO, Ukhurebor KE, Olaniyan OT, Ubi BE, Oloke JK, Dauda WP, Hefft DI (2022) Recent advances in molecular techniques for the enhancement of crop production. In: Agricultural biotechnology, biodiversity and bioresources conservation and utilization. Routledge, London, pp 201–220

Afzal S, Mubeen M, Hussain S, Ali M, Javeed HMR, Al-Ashkar I, Soufan W, Pandey S, Islam MS, El Sabagh A (2023) Modern breeding approaches for climate change. In: Climate change impacts on agriculture: concepts, issues and policies for developing countries. Springer, Berlin, pp 299–313

Ahmad HM, Wang X, Mahmood-Ur-Rahman MU et al (2022) Morphological and physiological response of Helianthus annuus L. to drought stress and correlation of wax contents for drought tolerance traits. Arab J Sci Eng 47:6747–6761. https://doi.org/10.1007/s13369-021-06098-1

Ahmed MU, Arise RO, Sudi YI (2020) Evaluation of the antidiarrhoeal activity of aqueous root and stem bark extracts of Annona senegalensis. Int J Life Sci Biotechnol 3:1–17

Aizaz M, Lubna, Jan R, Asaf S, Bilal S, Kim K-M, Al-Harrasi A (2024) Regulatory dynamics of plant hormones and transcription factors under salt stress. Biology 13:673

Akhtar MS (2019) Salt stress, microbes, and plant interactions: mechanisms and molecular approaches. Springer Singapore, Singapore

Amerian M, Palangi A, Gohari G, Ntatsi G (2024) Enhancing salinity tolerance in cucumber through selenium biofortification and grafting. BMC Plant Biol 24:24

Amgain NR, Lal R (2024) 14.1 Drought. In: Managing soil drought. Routledge, London, p 400

Anders S, Cowling W, Pareek A, Gupta KJ, Singla-Pareek SL, Foyer CH (2021) Gaining acceptance of novel plant breeding technologies. Trends Plant Sci 26:575–587

Anjana Jajoo AJ, Allakhverdiev SI (2017) High-temperature stress in plants: consequences and strategies for protecting photosynthetic machinery. In: Plant stress physiology. CABI, Wallingford, pp 138–154

Arvas YE, Durak MZ (2022) Biofertilizers in agricultural production. In: Polat T, Okant M (eds) Designing the latest trends in agriculture. Iksad Publications, Ankara, pp 123–140

Arvas YE, Kaya Y (2019) Genetiği Değiştirilmiş Bitkilerin Biyolojik Çeşitliliğe Potansiyel Etkileri. Yüzüncü Yıl Üniversitesi Tarım Bilimleri Dergisi 29:168–177

Arvas YE, Kocacaliskan İ (2020) Genetiği değiştirilmiş bitkilerin biyogüvenlik riskleri. Türk doğa ve fen dergisi 9:201–210

Arvas YE, Yolci MS, Marakli S (2022) Investigation the effect of drought stress on tolerant and resistant rice (Oryza sativa L.) varieties in terms of morphologic and genetic characters. Genet Resour Crop Evol 70:427–436

Aziz MA, Masmoudi K (2025) Molecular breakthroughs in modern plant breeding techniques. Hortic Plant J 11:15–41

Bacelar E, Pinto T, Anjos R, Morais MC, Oliveira I, Vilela A, Cosme F (2024) Impacts of climate change and mitigation strategies for some abiotic and biotic constraints influencing fruit growth and quality. Plants 13:1942
Bajoria A, Kanpariya J, Bera A (2024) Greenhouse gases and global warming. In: Advances and technology development in greenhouse gases: emission, capture and conversion. Elsevier, Amsterdam, pp 121–135
Barłóg P, Grzebisz W, Łukowiak R (2022) Fertilizers and fertilization strategies mitigating soil factors constraining efficiency of nitrogen in plant production. Plants 11:1855
Basu S, Ramegowda V, Kumar A, Pereira A (2016) Plant adaptation to drought stress. F1000Research 5:F1000 Faculty Rev-1554
Beillouin D, Schauberger B, Bastos A, Ciais P, Makowski D (2020) Impact of extreme weather conditions on European crop production in 2018. Philos Trans R Soc B 375:20190510
Bhargava S, Sawant K (2013) Drought stress adaptation: metabolic adjustment and regulation of gene expression. Plant Breed 132:21–32
Bharti U, Gaur P, Kaur K, Singh M (2024) Tracking genetically modified (GM) rice ingredients in samples of packed rice and food products from the marketplace in India: a pilot study for regulatory compliance. J Environ Health Sci Eng 22:1–7
Bhat J, Salgotra R, Dar M (2015) Phenomics: a challenge for crop improvement in genomic era. Mol Plant Breed 6:1–11
Bhat BA, Mir RA, Mir WR, Hamdani SS, Mir MA (2024) Transcription factors-golden keys to modulate the plant metabolism to develop salinity tolerance. Plant Stress 11:100409
Bhattacharya A (2022) Physiological processes in plants under low temperature stress. Springer, Berlin
Bilgili M, Tumse S, Nar S (2024) Comprehensive overview on the present state and evolution of global warming, climate change, greenhouse gasses and renewable energy. Arab J Sci Eng 49:1–29
Bolan S, Padhye LP, Jasemizad T, Govarthanan M, Karmegam N, Wijesekara H, Amarasiri D, Hou D, Zhou P, Biswal BK (2023) Impacts of climate change on the fate of contaminants through extreme weather events. Sci Total Environ 909:168388
Brookes G (2022) Genetically modified (GM) crop use 1996–2020: environmental impacts associated with pesticide use change. GM Crops Food 13:262–289
Casu A, Camardo Leggieri M, Toscano P, Battilani P (2024) Changing climate, shifting mycotoxins: a comprehensive review of climate change impact on mycotoxin contamination. Compr Rev Food Sci Food Saf 23:e13323
Chege C, Onyango K, Bodjerenou S, Schmidt M, Heike H, Ostermann H (2024) Adapting green innovation centers to climate change: reducing post-harvest loss and improving processing for target groups
Classen AT, Sundqvist MK, Henning JA, Newman GS, Moore JA, Cregger MA, Moorhead LC, Patterson CM (2015) Direct and indirect effects of climate change on soil microbial and soil microbial-plant interactions: what lies ahead? Ecosphere 6:1–21
Corwin DL (2021) Climate change impacts on soil salinity in agricultural areas. Eur J Soil Sci 72:842–862
Coulson A (2024) The environmental impact of AI in the lab: a double-edged sword. BioTechniques 76:353
Das G, Rao G (2015) Molecular marker assisted gene stacking for biotic and abiotic stress resistance genes in an elite rice cultivar. Front Plant Sci 6:698
Das A, Mahanta M, Pramanik B, Purkayastha S (2023) Genetically modified crops and crop species adapted to global warming in dry regions. In: Enhancing resilience of dryland agriculture under changing climate: interdisciplinary and convergence approaches. Springer, Berlin, pp 385–409
Delangiz N, Varjovi MB, Lajayer BA, Ghorbanpour M (2019) The potential of biotechnology for mitigation of greenhouse gasses effects: solutions, challenges, and future perspectives. Arab J Geosci 12:1–14

Deng S, Zhao D, Chen Z, Liu L, Zhu Y, Wang K, Gao X, Wu H, Zheng D (2024) Global distribution and projected variations of compound drought-extreme precipitation events. Earths Future 12:e2024EF004809

Dos Santos TB, Ribas AF, De Souza SGH, Budzinski IGF, Domingues DS (2022) Physiological responses to drought, salinity, and heat stress in plants: a review. Stress 2:113–135

Driedonks N, Rieu I, Vriezen WH (2016) Breeding for plant heat tolerance at vegetative and reproductive stages. Plant Reprod 29:67–79

Elhag M, Bahrawi JA (2017) Soil salinity mapping and hydrological drought indices assessment in arid environments based on remote sensing techniques. Geosci Instrum Methods Data Syst 6:149–158

El-Saadony MT, Saad AM, Soliman SM, Salem HM, Desoky E-SM, Babalghith AO, El-Tahan AM, Ibrahim OM, Ebrahim AA, Abd El-Mageed TA (2022) Role of nanoparticles in enhancing crop tolerance to abiotic stress: a comprehensive review. Front Plant Sci 13:946717

Enebe MC, Babalola OO (2018) The influence of plant growth-promoting rhizobacteria in plant tolerance to abiotic stress: a survival strategy. Appl Microbiol Biotechnol 102:7821–7835

Erkol ZI, Yesilyurt SN, Dalkilic HY (2024) Impacts of climate change on spatial drought distribution in the Mediterranean Basin (Turkey): different climate models and downscaling methods. Theor Appl Climatol 155:4065–4087

FAO (2022) FAO strategy on climate change 2022–2031. FAO, Rome

Farkas D, Dobránszki J (2024) Vegetal memory through the lens of transcriptomic changes—recent progress and future practical prospects for exploiting plant transcriptional memory. Plant Signal Behav 19:2383515

Farooq et al (2023) Major challenges in widespread adaptation of aerobic rice system and potential opportunities for future sustainability. S Afr J Bot 159:231–251

Filonchyk M, Peterson MP, Zhang L, Hurynovich V, He Y (2024) Greenhouse gases emissions and global climate change: examining the influence of CO2, CH4, and N2O. Sci Total Environ 935:173359

Franta BA (2022) Big carbon's strategic response to global warming, 1950–2020. Stanford University, Stanford

García Criado M, Myers-Smith IH, Bjorkman AD, Lehmann CE, Stevens N (2020) Woody plant encroachment intensifies under climate change across tundra and savanna biomes. Glob Ecol Biogeogr 29:925–943

Garg KK, Anantha K, Jat M, Kumar S, Sawargaonkar G, Singh A, Akuraju V, Singh R, Ahmed MI, Rao CS (2024) 6 Drought management in soils. In: Managing soil drought. Routledge, London, p 161

Genua-Olmedo A, Temmerman S, Ibanez C, Alcaraz C (2022) Evaluating adaptation options to sea level rise and benefits to agriculture: the Ebro Delta showcase. Sci Total Environ 806:150624

Ghimire BK, Yu CY, Kim W-R, Moon H-S, Lee J, Kim SH, Chung IM (2023) Assessment of benefits and risk of genetically modified plants and products: current controversies and perspective. Sustainability 15:1722

Glıck BR, Pasternak JJ, Patten CL (2010) Recombinant DNA technology, molecular biotechnology: principles and applications of recombinant DNA. ASM Press, Washington, DC

Gray SB, Brady SM (2016) Plant developmental responses to climate change. Dev Biol 419:64–77

Haider S, Bibi K, Munyaneza V, Zhang H, Zhang W, Ali A, Ahmad IA, Muhammad M, Xu F, Yang C (2024) Drought-induced adaptive and ameliorative strategies in plants. Chemosphere 364:143134

Hannan A, Hassan L, Hoque MN, Tahjib-Ul-Arif M, Robin AHK (2020) Increasing new root length reflects survival mechanism of rice (Oryza sativa L.) genotypes under PEG-induced osmotic stress. Plant Breed Biotechnol 8:46–57

Hasan N, Choudhary S, Naaz N, Sharma N, Laskar RA (2021) Recent advancements in molecular marker-assisted selection and applications in plant breeding programmes. J Genet Eng Biotechnol 19:1–26

He T, Li C (2020) Harness the power of genomic selection and the potential of germplasm in crop breeding for global food security in the era with rapid climate change. Crop J 8:688–700

Huang PK, Schmitt J, Runcie DE (2024a) Exploring the molecular regulation of vernalization-induced flowering synchrony in Arabidopsis. New Phytol 242:947–959

Huang X, Zhou X, Liu X, Zhong W, Wang X, Ju Z, Yin Y, Xin Q, Liu N, Liu X (2024b) Structural and physicochemical effects on the starch quality of the high-quality wheat genotype caused by delayed sowing. Front Nutr 11:1389745

Hubau W, Lewis SL, Phillips OL, Affum-Baffoe K, Beeckman H, Cuní-Sanchez A, Daniels AK, Ewango CE, Fauset S, Mukinzi JM (2020) Asynchronous carbon sink saturation in African and Amazonian tropical forests. Nature 579:80–87

Husaini AM, Sohail M (2023) Robotics-assisted, organic agricultural-biotechnology based environment-friendly healthy food option: beyond the binary of GM versus organic crops. J Biotechnol 361:41–48

Hussain R, Ayyub CM, Shaheen MR, Rashid S, Nafees M, Ali S, Butt M, Ali M, Maqsood A, Fiaz S, Ahmar S, Mehmood T, Mora-Poblete F (2021) Regulation of osmotic balance and increased antioxidant activities under heat stress in Abelmoschus esculentus L. triggered by exogenous proline. Agronomy (Basel) 11:685. https://doi.org/10.3390/agronomy11040685

Imran QM, Falak N, Hussain A, Mun B-G, Yun B-W (2021) Abiotic stress in plants; stress perception to molecular response and role of biotechnological tools in stress resistance. Agronomy 11:1579

IPCC (2014) Climate change 2014: synthesis report. In: Pachauri RK, Meyer LA (eds) Contribution of working groups I, II and III to the fifth assessment report of the intergovernmental panel on climate change (Geneva, Switzerland)

Islam MR, Naveed SA, Li Z, Zhao X, Fiaz S, Zhang F, Wu Z, Hu Z, Fu B, Shi Y, Shah SM, Xu J, Wang W (2022) Identification of candidate genes for salinity and anaerobic tolerance at the germination stage in rice by genome-wide association analyses. Front Genet 13:822516

Jain AK (2022a) Global carbon budget 2022. Earth Syst Sci Data 14:4811–4900

Jain M (2022b) Increasing atmospheric extreme events and role of disaster risk management: dimensions and approaches. In: Extremes in atmospheric processes and phenomenon: assessment, impacts and mitigation. Springer, Berlin, pp 303–328

James A, Rene ER, Bilyaminu AM, Chellam PV (2024) Advances in amelioration of air pollution using plants and associated microbes: an outlook on phytoremediation and other plant-based technologies. Chemosphere 358:142182

Janni M, Pieruschka R (2022) Plant phenotyping for a sustainable future, vol 73. Oxford University Press, Oxford, p 5085

Janni M, Maestri E, Gullì M, Marmiroli M, Marmiroli N (2024) Plant responses to climate change, how global warming may impact on food security: a critical review. Front Plant Sci 14:1297569

Jansson JK, Hofmockel KS (2020) Soil microbiomes and climate change. Nat Rev Microbiol 18:35–46

Jarecki W (2024) Response of winter wheat to delayed sowing and varied nitrogen fertilization. Agriculture 14:121

Jarin AS, Islam MM, Rahat A, Ahmed S, Ghosh P, Murata Y (2024) Drought stress tolerance in rice: physiological and biochemical insights. Int J Plant Biol 15:692–718

Javaid MM, Wang X, Florentine SK, Ashraf M, Mahmood A, Li FM, Fiaz S (2022) Effects on photosynthetic response and biomass productivity of Acacia longifolia subspecies longifolia under elevated CO2 and water-limited regimes. Front Plant Sci 13:817730

Jhanani G, Govindasamy C, Raghavendra T (2024) Germination and biological adaptation approaches as salt-stress tolerance process in selected paddy cultivars under salinity stress. Environ Res 259:119566

Jing L, Rui X, Chunchao W, Lan Q, Xiaoming Z, Wensheng W, Yingbin D, Lizhen Z, Yanyan W, Yunlian C (2018) A heading date QTL, qHD7.2, from wild rice (Oryza rufipogon) delays flowering and shortens panicle length under long-day conditions. Sci Rep 8:1–11

Junaid M, Gokce A (2024) Global agricultural losses and their causes. Bull Biol Allied Sci Res 2024:66

Kalaiselvi K, Sujarani P, Sakthivel V (2024) Cropping pattern in farming. In: Advanced technologies for smart agriculture. River Publishers, Nordjylland, pp 283–300

Kaushal N, Bhandari K, Siddique KH, Nayyar H (2016) Food crops face rising temperatures: an overview of responses, adaptive mechanisms, and approaches to improve heat tolerance. Cogent Food Agric 2:1134380

Khalil U, Ali S, Rizwan M, Rahman KU, Ata-Ul-Karim ST, Najeeb U, Ahmad MN, Adrees M, Sarwar M, Hussain SM (2018) Role of mineral nutrients in plant growth under extreme temperatures. In: Plant nutrients and abiotic stress tolerance. Springer, Berlin, pp 499–524

Khan N, Choi S-H, Lee C-H, Qu M, Jeon J-S (2024) Photosynthesis: genetic strategies adopted to gain higher efficiency. Int J Mol Sci 25:8933

Khokharvoytas A, Shahbaz M, Maqsood MF, Zulfiqar U, Naz N, Iqbal UZ, Sara M, Aqeel M, Khalid N, Noman A (2023) Genetic modification strategies for enhancing plant resilience to abiotic stresses in the context of climate change. Funct Integr Genomics 23:283

Korkmaz M, Kuriqi A (2024) Regional climate change and drought dynamics in Tunceli, Turkey: insights from drought indices. Water Conserv Sci Eng 9:49

Kulkarni SR, Balachandran S, Ulaganathan K, Balakrishnan D, Praveen M, Prasad AH, Fiyaz R, Senguttuvel P, Sinha P, Kale RR (2020) Molecular mapping of QTLs for yield related traits in recombinant inbred line (RIL) population derived from the popular rice hybrid KRH-2 and their validation through SNP genotyping. Sci Rep 10:1–21

Kumar S, Chatterjee U, David Raj A, Sooryamol K (2024a) Global warming and climate crisis/extreme events. In: Climate crisis: adaptive approaches and sustainability. Springer, Berlin, pp 3–18

Kumar T, Wang J-G, Xu C-H, Lu X, Mao J, Lin X-Q, Kong C-Y, Li C-J, Li X-J, Tian C-Y (2024b) Genetic engineering for enhancing sugarcane tolerance to biotic and abiotic stresses. Plants 13:1739

Kumari S (2024) Consequences and mitigation of global warming. In: Environment conservation and sustainable development: proceedings of international conference. Blue Rose Publishers, Noida, p 314

Leichenko R, O'Brien K (2024) Climate and society: transforming the future. John Wiley & Sons, New York

Li W, Huai X, Li P, Raza A, Mubarik MS, Habib M, Faiz S, Zhang B, Pan J, Khan RSA (2021) Glutathione peroxidase (GPX) gene family in rapeseed (Brassica napus L.): genomics analysis and expression profiling against different stresses. Antioxidant 10(9):1481. https://doi.org/10.3390/antiox10091481

Mahmood A, Wang X, Ali A, Awan MI, Ali L, Fiaz S et al (2022) Bio-methane production from maize with varying nitrogen levels and harvesting times under semi-arid conditions of Pakistan. Pol J Environ Stud 31(3):1–9. https://doi.org/10.15244/pjoes/143506

Mahmud AA, Upadhyay SK, Srivastava AK, Bhojiya AA (2021) Biofertilizers: a nexus between soil fertility and crop productivity under abiotic stress. Curr Res Environ Sustain 3:100063

Mäkinen H, Kaseva J, Trnka M, Balek J, Kersebaum K, Nendel C, Gobin A, Olesen JE, Bindi M, Ferrise R (2018) Sensitivity of European wheat to extreme weather. Field Crop Res 222:209–217

Mathur S, Agrawal D, Jajoo A (2014) Photosynthesis: response to high temperature stress. J Photochem Photobiol B Biol 137:116–126

Mbow C, Rosenzweig C, Barioni LG, Benton TG, Herrero M, Krishnapillai M, Liwenga E, Pradhan P, Rivera-Ferre MG, Sapkota T, Tubiello FN, Xu Y (2019) Intergovernmental panel on climate change (IPCC). Food security. In: Shukla PR, Skea J, Calvo Buendia E, Masson-Delmotte V, Pörtner H-O, Roberts DC, Zhai P, Slade R, Connors S, Van Diemen R, Ferrat M, Haughey E, Luz S, Neogi S, Pathak M, Petzold J, Portugal Pereira J, Vyas P, Huntley E, Kissick K, Belkacemi M, Malley HDOJ (eds) Climate change and land: an IPCC special report on climate change, desertification, land degradation, sustainable land management, food security, and greenhouse gas fluxes in terrestrial ecosystems

Mishra RK (2023) Fresh water availability and its global challenge. Br J Multidiscip Adv Stud 4:1–78

Mohamed AAE, Fiaz S, Ali M, Abdallah AA, Ahmed SI (2021) Performance of some rice (Oryza sativa L.) cultivars under water shortage and high temperature stress. Sanis Malysia 50(3):617–628. http://journalarticle.ukm.my/16903/

Mohanty U, Sinha P, Nageswara Rao M, Swain DK, Singh K (2024) Climate variability and food security. In: Climate risk management in agriculture: monthly and seasonal forecast application. Springer, Berlin, pp 9–32

Mondal SK, Su B, Huang J, Zhai J, Wang G, Kundzewicz ZW, Wang Y, Jiang S, Jiang H, Zhou J (2024) Climate change will aggravate South Asian cropland exposure to drought by the middle of 21st century. Earths Future 12:e2023EF003872

Moran MN, Jones D, Jensen S, Marcoli R, Jerry D (2024) Optimising commercial traits through gene editing in aquaculture: strategies for accelerating genetic improvement. Rev Aquac 16:1127–1159

Morris SJ, Blackwood CB (2024) The ecology of soil biota and their function. In: Soil microbiology, ecology and biochemistry. Elsevier, Amsterdam, pp 275–302

Najeeb S, Ali J, Mahender A, Pang Y, Zilhas J, Murugaiyan V, Vemireddy LR, Li Z (2020) Identification of main-effect quantitative trait loci (QTLs) for low-temperature stress tolerance germination- and early seedling vigor-related traits in rice (Oryza sativa L.). Mol Breed 40:10

Nakano Y, Kobayashi Y (2020) Genome-wide association studies of agronomic traits consisting of field- and molecular-based phenotypes. Rev Agric Sci 8:28–45

Nawaz T, Gu L, Fahad S, Saud S, Harrison MT, Zhou R (2024) Sustainable protein production through genetic engineering of cyanobacteria and use of atmospheric N2 gas. Food Energy Secur 13:e536

Nick P (2024) Towards a grammar of plant stress: modular signalling conveys meaning. Theor Exp Plant Physiol 36:503–521

Nievola CC, Carvalho CP, Carvalho V, Rodrigues E (2017) Rapid responses of plants to temperature changes. Temperature 4:371–405

Nogoy FM, Song J-Y, Ouk S, Rahimi S, Kwon SW, Kang K-K, Cho Y-G (2016) Current applicable DNA markers for marker assisted breeding in abiotic and biotic stress tolerance in rice (Oryza sativa L.). Plant Breed Biotechnol 4:271–284

Oshunsanya SO, Nwosu NJ, Li Y (2019) Abiotic stress in agricultural crops under climatic conditions. In: Sustainable agriculture, forest and environmental management. Springer, Berlin, pp 71–100

Oyebamiji YO, Adigun BA, Shamsudin NAA, Ikmal AM, Salisu MA, Malike FA, Lateef AA (2024) Recent advancements in mitigating abiotic stresses in crops. Horticulturae 10:156

Palmquist KA, Schlaepfer DR, Renne RR, Torbit SC, Doherty KE, Remington TE, Watson G, Bradford JB, Lauenroth WK (2021) Divergent climate change effects on widespread dryland plant communities driven by climatic and ecohydrological gradients. Glob Chang Biol 27:5169–5185

Parsons A, Edwards G, Newton P, Chapman D, Caradus J, Rasmussen S, Rowarth J (2011) Past lessons and future prospects: plant breeding for yield and persistence in cool-temperate pastures. Grass Forage Sci 66:153–172

Pinke Z, Decsi B, Demeter G, Kalicz P, Kern Z, Acs T (2024) Continental lowlands face rising crop vulnerability: structural change in regional climate sensitivity of crop yields, Hungary (Central and Eastern Europe), 1921–2010. Reg Environ Chang 24:33

Platani M, Sokefun O, Bassil E, Apidianakis Y (2023) Genetic engineering and genome editing in plants, animals and humans: facts and myths. Gene 856:147141

Prakash C, Sevanthi AM, Shanmugavadivel P (2019) Use of QTLs in developing abiotic stress tolerance in rice. In: Advances in rice research for abiotic stress tolerance. Elsevier, Amsterdam, pp 869–893

Qaim M (2020) Role of new plant breeding technologies for food security and sustainable agricultural development. Appl Econ Perspect Policy 42:129–150

Qiao M, Hong C, Jiao Y, Hou S, Gao H (2024) Impacts of drought on photosynthesis in major food crops and the related mechanisms of plant responses to drought. Plants 13:1808

Qu Y, Sakoda K, Fukayama H, Kondo E, Suzuki Y, Makino A, Terashima I, Yamori W (2021) Overexpression of both rubisco and rubisco activase rescues rice photosynthesis and biomass under heat stress. Plant Cell Environ 44:2308–2320

Ramírez CF, Cavieres LA, Sanhueza C, Vallejos V, Gómez-Espinoza O, Bravo LA, Sáez PL (2024) Ecophysiology of Antarctic vascular plants: an update on the extreme environment resistance mechanisms and their importance in facing climate change. Plants 13:449

Ranasinghe P, Mapa MS (2024) A comprehensive approach to develop rice varieties to cope with climate change impacts. YSF Thematic Publication 2024, p 128

Raza A, Ashraf F, Zou X, Zhang X, Tosif H (2020) Plant adaptation and tolerance to environmental stresses: mechanisms and perspectives. In: Plant ecophysiology and adaptation under climate change: mechanisms and perspectives I: general consequences and plant responses. Springer, Berlin, pp 117–145

Raza A, Razzaq A, Mehmood SS, Hussain MA, Wei S, He H, Zaman QU, Xuekun Z, Yong C, Hasanuzzaman M (2021) Omics: the way forward to enhance abiotic stress tolerance in Brassica napus L. GM Crops Food 12:251–281

Razzaq A, Ali A, Safdar LB, Zafar MM, Rui Y, Shakeel A, Shaukat A, Ashraf M, Gong W, Yuan Y (2020) Salt stress induces physiochemical alterations in rice grain composition and quality. J Food Sci 85:14–20

Rebouh NY, Khugaev CV, Utkina AO, Isaev KV, Mohamed ES, Kucher DE (2023) Contribution of eco-friendly agricultural practices in improving and stabilizing wheat crop yield: a review. Agronomy 13:2400

Roy S, Kapoor R, Mathur P (2024) Revisiting changes in growth, physiology and stress responses of plants under the effect of enhanced CO2 and temperature. Plant Cell Physiol 65:4–19

Rubab M, Jannat S, Freeg H et al (2023) Evaluation of functional kompetitive allele-specific PCR (KASP) markers for selection of drought-tolerant wheat (Triticum aestivum) genotypes. Funct Plant Biol 51:NULL. https://doi.org/10.1071/FP23032

Sabat M, Tripathy A (2024) Genetically modified and gene-edited food crops: recent status and future prospects. In: Food production, diversity, and safety under climate change. Springer, Berlin, pp 211–222

Saddiq MS, Wang X, Iqbal S, Hafeez MB, Khan S, Raza A, Iqbal J, Maqbool MM, Faiz S, Qazi MA, Bakhsh A, Zaib J, Gulshan AB (2021) Effect of water stress on grain yield and physiology characters of quinoa (Chenopodium quinoa) genotypes. Agronomy (Basel) 11(10):1934. https://doi.org/10.3390/agronomy11101934

Saha PU, Ali A, Khan J (2023) Genetically engineered microorganisms. In: Genetic engineering. Apple Academic Press, Palm Bay, pp 125–157

Sahito JH, Zhang H, Gishkori ZGN, Ma C, Wang Z, Ding D, Zhang X, Tang J (2024) Advancements and prospects of genome-wide association studies (GWAS) in maize. Int J Mol Sci 25:1918

Scafetta N (2024) Impacts and risks of "realistic" global warming projections for the 21st century. Geosci Front 15:101774

Seleiman MF, Al-Suhaibani N, Ali N, Akmal M, Alotaibi M, Refay Y, Dindaroglu T, Abdul-Wajid HH, Battaglia ML (2021) Drought stress impacts on plants and different approaches to alleviate its adverse effects. Plants 10:259

Shabbir R, Singhal RK, Mishra UN, Chauhan J, Javed T, Hussain S, Kumar S, Anuragi H, Lal D, Chen P (2022) Combined abiotic stresses: challenges and potential for crop improvement. Agronomy 12:2795

Shah F, Wu W (2019) Soil and crop management strategies to ensure higher crop productivity within sustainable environments. Sustainability 11:1485

Shah MH, Pramanik K, Ajaharuddin S, Kundu S, Alam MA, Islam S, Pal A, Atta K, Pande CB (2024) Effects of climate change on plant growth and development. In: Climate-resilient agriculture. Apple Academic Press, Palm Bay, pp 9–79

Shahbaz M (2024) Heat and wheat: adaptation strategies with respect to heat shock proteins and antioxidant potential; an era of climate change. Int J Biol Macromol 256:128379

Sharma P, Jha AB, Dubey RS, Pessarakli M (2012) Reactive oxygen species, oxidative damage, and antioxidative defense mechanism in plants under stressful conditions. J Bot 2012:217037

Sharma A, Dheer P, Rautela I, Thapliyal P, Thapliyal P, Bajpai AB, Sharma MD (2024) A review on strategies for crop improvement against drought stress through molecular insights. 3 Biotech 14:173

Shimizu T, Kacar YA, Cristofani-Yaly M, Curtolo M, Machado MA (2020) Markers, maps, and marker-assisted selection. In: The citrus genome. Springer, Berlin, pp 107–139

Siddiqui ZS, Oh SD, Kim EJ, Jang YJ, Lee SK, Yun DW, Kwon TR, Wajid D, Ansari HH, Park SC, Cho JI (2021) Physiological and photochemical evaluation of pepper methionine sulfoxide reductase B2 (CaMsrB2) expressing transgenic rice in saline habitat. Plant Physiol Biochem 167:198–209

Srivastava RK, Singh RB, Pujarula VL, Bollam S, Pusuluri M, Chellapilla TS, Yadav RS, Gupta R (2020) Genome-wide association studies and genomic selection in pearl millet: advances and prospects. Front Genet 10:1389

Staniak M, Szpunar-Krok E, Wilczewski E, Kocira A, Podleśny J (2024) The function of macronutrients in helping soybeans to overcome the negative effects of drought stress. Agronomy 14:1744

Sun C, Ali K, Yan K, Fiaz S, Dormatey R, Bi Z, Bai J (2021) Exploration of epigenetics for improvement of drought and other stress resistance in crops: a review. Plants (Basel) 10(6):1226. https://doi.org/10.3390/plants10061226

Sutherland C, Gleim S, Smyth SJ (2021) Correlating genetically modified crops, glyphosate use and increased carbon sequestration. Sustainability 13:11679

Tang S, Qian J, Lu B, He Y, Liu Y, Xu K, Shen J (2024) Adsorption and uptake of functionalized nanoplastics (NPs) by wetland plant (sphagnum): a unique pathway for polystyrene-NPs reduction in non-vascular plants. Sci Total Environ 950:175367

Tarnawa Á, Kende Z, Sghaier AH, Kovács GP, Gyuricza C, Khaeim H (2023) Effect of abiotic stresses from drought, temperature, and density on germination and seedling growth of barley (Hordeum vulgare L.). Plants 12:1792

Terrer C, Phillips RP, Hungate BA, Rosende J, Pett-Ridge J, Craig ME, Van Groenigen KJ, Keenan TF, Sulman BN, Stocker BD (2021) A trade-off between plant and soil carbon storage under elevated CO2. Nature 591:599–603

Ul Hassan M, Rasool T, Iqbal C, Arshad A, Abrar M, Abrar MM, Habib-Ur-Rahman M, Noor MA, Sher A, Fahad S (2021) Linking plants functioning to adaptive responses under heat stress conditions: a mechanistic review. J Plant Growth Regul:1–18

Usman S, Jayeoba JO, Kundiri A (2024) Climate change at a global concept: impacts and adaptation measures. Int J Environ Clim Change 14:445–459

Wang X, Deng S, Zhou Y, Long J, Ding D, Du H, Lei M, Chen C, Tie BQ (2021) Application of different foliar iron fertilizers for enhancing the growth and antioxidant capacity of rice and minimizing cadmium accumulation. Environ Sci Pollut Res Int 28:7828–7839

Wijewardene I, Shen G, Zhang H (2021) Enhancing crop yield by using rubisco activase to improve photosynthesis under elevated temperatures. Stress Biol 1:2

Yadav V, Pal D, Poonia AK (2024) A study on genetically engineered foods: need, benefits, risk, and current knowledge. Cell Biochem Biophys 82:1–16

Yan L, Baoxiang W, Jingfang L, Zhiguang S, Ming C, Yungao X, Bo X, Bo Y, Jian L, Jinbo L, Tingmu C, Zhaowei F, Baiguan L, Dayong X, Bello BK (2021) A novel SAPK10-WRKY87-ABF1 biological pathway synergistically enhance abiotic stress tolerance in transgenic rice (Oryza sativa). Plant Physiol Biochem 168:252–262

Yarra R, Wei W (2021) The NAC-type transcription factor GmNAC20 improves cold, salinity tolerance, and lateral root formation in transgenic rice plants. Funct Integr Genomics 21:473–487

Yeşilyurt S, Gürgan M, Duban S (2024) Utilization of wastewater sludge and poplar trees for remediation. In: Recent trends in management and utilization of industrial sludge. Springer, Berlin, pp 255–274

Zafar MM, Zhang Y, Farooq MA, Ali A, Firdous H, Haseeb M, Fiaz S, Shakeel A, Razzaq A, Ren M (2022) Biochemical and associated agronomic traits in Gossypium hirsutum L. under high-temperature stress. Agronomy (Basel) 12(6):1310

Zandalinas SI, Mittler R (2022) Plant responses to multifactorial stress combination. New Phytol 234:1161–1167

Zhang H, Li Y, Zhu J-K (2018) Developing naturally stress-resistant crops for a sustainable agriculture. Nat Plants 4:989–996

Zhao J, Lu Z, Wang L, Jin B (2020) Plant responses to heat stress: physiology, transcription, non-coding RNAs, and epigenetics. Int J Mol Sci 22:117

Zhou H, Shi H, Yang Y, Feng X, Chen X, Xiao F, Lin H, Guo Y (2024) Insights into plant salt stress signaling and tolerance. J Genet Genomics 51:16–34

Zhu T, Fonseca De Lima CF, De Smet I (2021) The heat is on: how crop growth, development, and yield respond to high temperature. J Exp Bot 72:7359–7373

Role of Neglected Crops Under Multi-stress Conditions

4

Muhammad Zain ul Abideen, Muhammad Abu Bakar Zia, Ifrah Munawar, and Muntaha Sanaullah

Abstract

Neglected crops, which are generally not focused on the current and commercialized form of agriculture, have the potential to offer solutions to issues related to food security and malnutrition as well as sustainable farming. The study aims to identify the needs, opportunities, and potential hindrances of these crops based on the contribution they make in supporting efforts toward the improvement of biodiversity, soil health, efficient use of natural resources, and combating the effects of climate change. Neglected crops are important because of their ability to greatly contribute to genetic variation and crop differentiation, which are important considerations in maintaining crop sustainability and minimizing crop loss. They help to restore physical properties, nutrient dynamics, and levels of organic matter in the soil, which promotes healthier and sustainable soil. Moreover, these crops have low water demand and require other inputs, making them useful for producing crops in arid regions and leading to low inputs for the agriculture industry. Neglected crops, on the other hand, are associated with some constraints, such as a lack of investment in research and development, weak market structures, low consumer demand and awareness, and insufficient

M. Z. ul Abideen
Faculty of Agricultural Sciences, Department of Plant Breeding and Genetics, University of the Punjab, Lahore, Pakistan

M. A. B. Zia (✉)
Faculty of Agricultural Sciences and Technology, Department of Plant Breeding and Genetics, University of Layyah, Layyah, Pakistan
e-mail: abzia@ul.edu.pk

I. Munawar
Department of Plant Breeding and Genetics, University of Agriculture, Faisalabad, Pakistan

M. Sanaullah
Department of Botany, University of Agriculture, Faisalabad, Pakistan

© The Author(s), under exclusive license to Springer Nature Singapore Pte Ltd. 2025

S. Faiz et al. (eds.), *Climate Smart Agriculture for Future Food Security*,
https://doi.org/10.1007/978-981-96-4499-5_4

policy responses. Climate change is also a limitation that affects their output and ability to adapt to the new climatic conditions. Thus, there is a great potential for neglected crops to gain attention through increased research on such crops, a better marketing framework, improved policies, and consumer awareness. For future directions and recommendations, it is recommended to focus on the improvement of genetics through various breeding processes, better and improved farming technologies, markets, improved and better products, consumer awareness, and research and knowledge exchange all over the world. In this way, it will be possible to incorporate these overlooked crops back into the agricultural system, which will help diversify diets, provide better standards of living, and develop more sustainable food chains.

Keywords

Green Revolution · Agricultural biodiversity · Conventional crops · Genetic diversity · Neglected crops

4.1 Introduction to Neglected Crops

Neglected crops, often referred to as "underutilized" or "orphan" crops, represent a diverse group of plant species that have traditionally played significant roles in the diets, cultures, and agricultural systems of various regions but have been largely overlooked by modern agricultural practices and research (Dansi et al. 2012). These crops, which include grains, legumes, roots, tubers, vegetables, and fruits, are integral to the livelihoods of many communities, particularly in developing countries. However, the Green Revolution and subsequent agricultural advancements have largely focused on a few major staple crops such as wheat, rice, maize, and soybeans, leading to the marginalization of these valuable plant species (Nelson et al. 2019).

Neglected crops are of paramount importance for several reasons. They contribute significantly to agricultural biodiversity, possessing unique genetic traits crucial for breeding programs to develop new crop varieties capable of withstanding pests, diseases, and changing climate conditions (Baldermann et al. 2016). The genetic diversity within neglected crops is invaluable for ensuring food security and sustainability. Furthermore, many of these crops are well adapted to local climatic conditions and soil types, often being more resilient to extreme weather events, drought, and poor soil quality compared to major staple crops. This adaptability makes them crucial for farming systems in marginal areas, where conventional crops may fail (Sambo 2014).

Additionally, neglected crops are often nutritionally rich, offering a wide range of essential nutrients, vitamins, and minerals. They can play a vital role in combating malnutrition and promoting dietary diversity (Li et al. 2020). For instance, crops such as amaranth and moringa are known for their high protein content and rich nutrient profiles. Beyond their nutritional value, these crops are deeply embedded in the cultural and culinary traditions of many communities. They hold significant cultural value and are often associated with traditional agricultural practices and knowledge systems that have been passed down through generations (Naidoo 2023).

There are several examples of neglected crops that illustrate their diversity and potential. Amaranth, a versatile crop grown for its leaves and seeds, is high in protein, vitamins, and minerals. Teff, native to Ethiopia, is a tiny grain known for its high iron and calcium content and is the main ingredient in injera, a traditional Ethiopian flatbread. Fonio, a small-grained cereal from West Africa, is renowned for its resilience to poor soil conditions and rapid growth cycle. Quinoa, indigenous to the Andean region, is a highly nutritious grain that has gained global popularity due to its high protein content and balanced amino acid profile (Bhattarai 2018; Gebru et al. 2020; Vega-Gálvez et al. 2010; Zhu 2020). Other examples include sorghum, taro, Bambara groundnut, and moringa, each with unique properties and benefits.

Promoting neglected crops offers numerous benefits. Diversifying agricultural systems with these crops can reduce dependency on a limited number of staple crops, enhancing food security and resilience to food supply disruptions. Incorporating neglected crops into diets can address nutritional deficiencies and promote better health outcomes, particularly in regions where malnutrition is prevalent. Many neglected crops are more environmentally sustainable, requiring fewer inputs such as water, fertilizers, and pesticides. Their cultivation can enhance soil health and reduce agricultural footprints (Mabhaudhi et al. 2022). Additionally, promoting these crops can create new economic opportunities for farmers, particularly in marginalized areas. Developing value chains and market access for neglected crops can improve livelihoods and stimulate rural economies.

Despite their potential, promoting neglected crops faces several challenges. There is limited scientific research and investment in breeding programs for these crops, hindering their improvement and wider adoption. Poor market infrastructure and limited consumer awareness can make it difficult for farmers to sell neglected crops profitably. Consumers may be unfamiliar with these crops or perceive them as inferior to more commonly known crops, reducing their demand and market value (Fig. 4.1).

4.2 Understanding the Multi-stress Conditions of Neglected Crops

Neglected crops are frequently cultivated in regions that face multiple environmental stresses, which include extreme weather conditions, poor soil quality, water scarcity, and pest and disease pressures. These crops have developed unique adaptations that allow them to thrive under such challenging conditions, making them critical for sustainable agriculture, particularly in the face of climate change. Understanding the multi-stress conditions that these crops endure is essential for leveraging their potential to enhance food security, nutritional quality, and agricultural resilience.

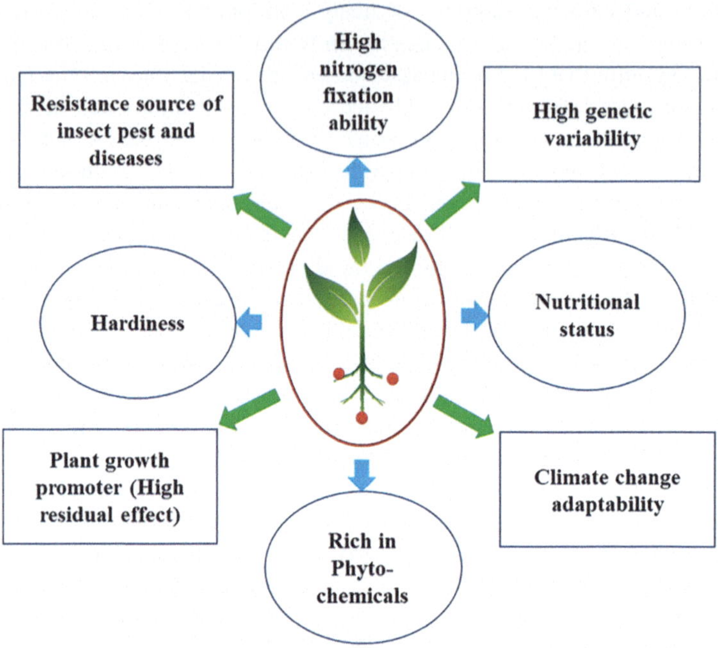

Fig. 4.1 Important features of neglected legumes (Samal et al. 2023)

4.2.1 Environmental Stress Factors

4.2.1.1 Drought and Water Scarcity

Many neglected crops, such as sorghum and fonio, have evolved to grow in arid and semi-arid regions where water is a limiting factor. These crops possess physiological and morphological adaptations such as deep root systems, reduced leaf area, and efficient water-use mechanisms that enable them to survive prolonged periods of drought (Gidi 2023; Rahman et al. 2024). Research into these adaptations can provide insights into developing drought-resistant varieties of other staple crops.

4.2.1.2 Poor Soil Quality

Neglected crops often grow in soils with low fertility, high salinity, or poor structure. For example, quinoa can tolerate high soil salinity, while amaranth thrives in nutrient-poor soils. These crops have developed mechanisms such as efficient nutrient uptake and usage, and the ability to fix nitrogen (in the case of legumes such as Bambara groundnut), which make them viable options for improving soil health and productivity in degraded lands (Talabi et al. 2022).

4.2.1.3 Extreme Temperatures

High-temperature tolerance is another characteristic of many neglected crops. Teff, for instance, can withstand both high daytime temperatures and cooler nights, a feature that allows it to be cultivated in diverse climatic conditions (Gelaw and Qureshi 2020). Understanding the genetic and physiological bases for this tolerance can aid in breeding heat-resistant varieties of other crops.

4.2.1.4 Pest and Disease Resistance

Neglected crops often exhibit resistance to local pests and diseases, a trait developed through centuries of natural selection and traditional breeding. For example, the genetic diversity in sorghum includes resistance to pests such as the sorghum midge and diseases such as anthracnose (Mengistu et al. 2018). This resilience reduces the need for chemical inputs, promoting more sustainable farming practices.

4.2.2 Socio-economic and Cultural Stress Factors

4.2.2.1 Market and Economic Stresses

Neglected crops typically face limited market access and lower economic incentives compared to major staple crops. Farmers growing these crops often lack the infrastructure and market linkages to sell their produce at profitable prices (Mayes et al. 2011). Additionally, there is a lack of value-added products and processing facilities for these crops, which hampers their economic viability.

4.2.2.2 Research and Development Gaps

There is a significant gap in research and development for neglected crops compared to major staples. This includes a lack of breeding programs, agronomic research, and extension services that could improve yields, pest resistance, and stress tolerance. Investment in R&D is crucial to unlock the full potential of these crops.

4.2.2.3 Cultural and Perception Challenges

In some regions, neglected crops are perceived as "poor people's food" or are associated with traditional, less modern agricultural practices (Cloete and Idsardi 2013). This cultural bias can hinder their adoption and promotion, despite their nutritional and environmental benefits. Efforts to change perceptions through education, awareness campaigns, and showcasing success stories are essential.

4.2.3 Strategies for Supporting Neglected Crops Under Multi-stress Conditions

4.2.3.1 Breeding and Genetic Research

Developing improved varieties of neglected crops through breeding programs that focus on stress tolerance, yield improvement, and nutritional enhancement is vital. Modern biotechnological tools, such as genomic selection and CRISPR gene editing, can accelerate this process (Rauf et al. 2024).

4.2.3.2 Agroecological Practices

Promoting sustainable agricultural practices that enhance the resilience of farming systems, such as intercropping, agroforestry, and conservation agriculture, can support the cultivation of neglected crops under multi-stress conditions. These practices also improve soil health, water retention, and biodiversity.

4.2.3.3 Policy and Market Support

Governments and international organizations need to create supportive policies and infrastructure to improve market access, provide incentives for growing neglected crops, and invest in value chain development. This includes establishing processing facilities, improving transportation, and creating market linkages.

4.2.3.4 Education and Awareness

Raising awareness about the benefits of neglected crops among consumers, farmers, and policymakers is crucial. Educational programs, culinary initiatives, and media campaigns can help change perceptions and increase demand for these crops.

Understanding and addressing the multi-stress conditions of neglected crops are key to harnessing their potential for sustainable agriculture. By leveraging their inherent resilience and promoting supportive measures, these crops can play a significant role in ensuring food security, enhancing nutrition, and building resilient agricultural systems in the face of global challenges.

4.3 Role of Neglected Crops in Mitigating Multi-stress Conditions

Neglected crops play a pivotal role in mitigating multi-stress conditions in agriculture, particularly in regions prone to environmental challenges such as drought, poor soil quality, extreme temperatures, and pest and disease pressures. These crops, through their unique adaptations and resilience, offer promising solutions for sustainable agriculture, enhancing food security, nutrition, and the resilience of farming systems. Their role in mitigating multi-stress conditions can be understood through several key aspects.

4.3.1 Enhancing Agricultural Resilience

Neglected crops are inherently resilient to a range of environmental stresses, making them crucial for enhancing the overall resilience of agricultural systems. For instance, crops such as sorghum and fonio are well adapted to drought-prone areas due to their deep root systems and efficient water-use mechanisms (Mbosso et al. 2020). By integrating these crops into farming systems, farmers can maintain productivity even under adverse climatic conditions, thereby reducing the risk of crop failure and ensuring a stable food supply.

4.3.2 Soil Health Improvement

Many neglected crops have the ability to thrive in poor soil conditions, including soils with low fertility, high salinity, or poor structure. Crops such as amaranth and quinoa can grow in nutrient-deficient soils and still produce significant yields. Additionally, leguminous crops such as the Bambara groundnut have nitrogen-fixing capabilities, which can improve soil fertility and reduce the need for chemical fertilizers (Tan et al. 2020). By cultivating these crops, farmers can enhance soil health, increase soil organic matter, and promote sustainable land use.

4.3.3 Diversifying Food Sources

Neglected crops contribute to dietary diversity and nutritional security by providing a variety of essential nutrients, vitamins, and minerals. For example, moringa leaves are rich in vitamins A and C, calcium, and protein, while teff is high in iron and calcium (Gahukar 2014). Incorporating these crops into local diets can help address nutritional deficiencies and improve overall health outcomes. Moreover, dietary diversification reduces reliance on a limited number of staple crops, spreading the risk and enhancing food security.

4.3.4 Pest and Disease Management

Neglected crops often exhibit natural resistance to local pests and diseases, reducing the need for chemical pesticides and enhancing the sustainability of agricultural practices. For instance, certain varieties of sorghum are resistant to pests such as the sorghum midge and diseases such as anthracnose (Mengistu et al. 2018). This resistance is particularly valuable in regions where access to agricultural inputs is limited, as it can lead to more stable yields and reduce the economic burden on farmers (Azmat et al. 2024).

4.3.5 Adaptation to Climate Change

Climate change poses significant challenges to global agriculture, with increased frequency of extreme weather events, altered precipitation patterns, and rising temperatures. Neglected crops, with their inherent resilience to environmental stresses, are well suited to adapt to these changes. For instance, teff's ability to grow under variable temperature conditions makes it a viable option for regions experiencing temperature fluctuations due to climate change (Cheng et al. 2015). By promoting the cultivation of these crops, agricultural systems can become more adaptive and resilient to climate-related challenges.

4.3.6 Socio-economic Benefits

The promotion and cultivation of neglected crops can also have significant socio-economic benefits, particularly for smallholder farmers in developing countries. These crops often require fewer inputs and can be grown on marginal lands, providing a source of income and food security for vulnerable communities. Additionally, developing markets for neglected crops and value-added products can create new economic opportunities, enhance livelihoods, and stimulate rural economies.

4.3.7 Sustainable Agricultural Practices

Neglected crops can be integral to sustainable agricultural practices such as intercropping, agroforestry, and conservation agriculture. These practices enhance biodiversity, improve ecosystem services, and promote sustainable land management. For example, intercropping sorghum with legumes can improve soil fertility, reduce pest and disease incidence, and increase overall productivity (Barasa et al. 2022). Such practices contribute to the sustainability and resilience of farming systems, particularly in regions facing multi-stress conditions.

4.4 Nutritional and Economic Benefits of Neglected Crops

Neglected crops offer substantial nutritional and economic benefits that are increasingly recognized in the context of global food security, health, and sustainable development. These crops, traditionally important in specific regions, are not only rich in essential nutrients but also hold significant potential for improving the livelihoods of smallholder farmers and stimulating local economies.

4.4.1 Nutritional Benefits

- **Rich Nutrient Profiles**: Neglected crops are often nutritionally dense, providing essential vitamins, minerals, and other nutrients. For example, moringa leaves are exceptionally rich in vitamins A and C, calcium, potassium, and protein. Quinoa is another nutrient-dense crop, containing all nine essential amino acids, making it a complete protein source, which is rare among plant foods (Dhakar et al. 2011; Pathan et al. 2019).
- **Dietary Diversity**: Incorporating neglected crops into diets promotes dietary diversity, which is crucial for balanced nutrition. Many of these crops offer unique nutritional benefits that complement those of more common staples. For instance, teff, a staple in Ethiopian cuisine, is high in iron and calcium, helping to prevent anemia and support bone health (Gebru et al. 2020).
- **Combating Malnutrition**: In regions where malnutrition is prevalent, neglected crops can play a vital role in addressing micronutrient deficiencies. For example,

amaranth is rich in iron, magnesium, and calcium, making it an excellent dietary component for populations at risk of these deficiencies. The high nutrient content of these crops can significantly improve public health outcomes (Joshi and Verma 2020).

- **Functional Foods and Health Benefits**: Many neglected crops possess bioactive compounds with health-promoting properties. For instance, fonio is not only a good source of essential amino acids but also has a low glycemic index, making it suitable for people with diabetes. The antioxidants and anti-inflammatory properties of various neglected crops contribute to their role as functional foods that support overall health and wellness (Zhu 2020).

4.4.2 Economic Benefits

- **Income Diversification**: Neglected crops can provide an additional source of income for smallholder farmers, particularly in regions where agricultural opportunities are limited. By cultivating these crops, farmers can diversify their income sources, reducing economic vulnerability. For example, farmers in West Africa growing fonio have found new markets for this ancient grain, improving their economic stability.
- **Market Opportunities**: The growing global interest in healthy, organic, and sustainable foods has created new market opportunities for neglected crops. These crops, often considered superfoods, are gaining popularity in international markets, providing higher income potential for farmers. For instance, quinoa has seen a significant rise in demand worldwide, leading to increased economic benefits for farmers in the Andean region (Alandia et al. 2020).
- **Value Addition**: Processing and adding value to neglected crops can enhance their economic benefits. Developing products such as flour, snacks, and beverages from these crops can create new business opportunities and boost local economies. For example, teff flour is now used in a variety of gluten-free products, expanding its market reach and economic value (Minarovičová et al. 2019).
- **Agro-Economic Stability**: Cultivating neglected crops can contribute to the stability of local agricultural economies. These crops are often well adapted to local environmental conditions, requiring fewer inputs and reducing dependency on expensive and environmentally damaging agricultural practices. This resilience can lead to more stable yields and reduced financial risk for farmers.
- **Employment Opportunities**: The production, processing, and marketing of neglected crops can create employment opportunities in rural areas. This can help reduce rural–urban migration by providing viable livelihoods within agricultural communities. The entire value chain, from cultivation to market, offers numerous job opportunities that can stimulate rural development.
- **Sustainable Development**: Promoting neglected crops aligns with sustainable development goals (SDGs), particularly those related to zero hunger, good health and well-being, and sustainable economic growth. By supporting the cultivation

and commercialization of these crops, policymakers can contribute to broader economic development and environmental sustainability.

4.5 Agroecological Benefits of Neglected Crops

Neglected crops offer numerous agroecological benefits that make them valuable for sustainable farming systems. These crops, which have been largely overlooked by modern agricultural practices, possess unique characteristics that enhance biodiversity, improve soil health, promote efficient resource use, and support resilient agricultural ecosystems. Understanding and leveraging the agroecological benefits of neglected crops is essential for developing sustainable and resilient food systems.

4.5.1 Enhancing Biodiversity

- **Genetic Diversity**: Neglected crops contribute significantly to genetic diversity within agricultural landscapes. This diversity is crucial for the resilience of ecosystems, providing a broader genetic pool that can help crops adapt to changing environmental conditions and resist pests and diseases. For example, the genetic variability in sorghum includes traits for drought tolerance and pest resistance, which are vital for breeding programs aimed at developing resilient crop varieties (Dansi et al. 2012).
- **Crop Diversification**: Incorporating neglected crops into farming systems promotes crop diversification, which is a fundamental principle of agroecology. Crop diversification can reduce the risk of total crop failure due to pests, diseases, or adverse weather conditions. For instance, intercropping amaranth with other crops can create a more balanced and resilient agroecosystem, reducing the reliance on a single crop and enhancing overall farm productivity (Innazent et al. 2024).

4.5.2 Soil Health Improvement

- **Nutrient Cycling**: Many neglected crops play a crucial role in nutrient cycling within the soil. Leguminous crops such as the Bambara groundnut fix atmospheric nitrogen into the soil, improving soil fertility and reducing the need for synthetic fertilizers (Hasan et al. 2018). This natural process enhances soil health and supports sustainable crop production.
- **Soil Structure and Erosion Control**: The root systems of neglected crops can improve soil structure and prevent erosion. For example, the deep root systems of crops such as quinoa and fonio help bind the soil, reducing erosion and promoting water infiltration (Kell 2011). This is particularly important in regions prone to soil degradation and erosion.

- **Organic Matter and Soil Microbiota**: Neglected crops often contribute to higher levels of organic matter in the soil. The residues from these crops can improve soil organic content, which in turn supports beneficial soil microbiota. A healthy soil microbiome enhances nutrient availability, soil structure, and overall soil fertility, creating a more productive and sustainable farming system.

4.5.3 Efficient Resource Use

- **Water Use Efficiency**: Many neglected crops are adapted to grow in arid and semi-arid regions, exhibiting high water use efficiency. Crops such as sorghum and teff require less water compared to conventional staple crops, making them suitable for cultivation in water-scarce areas (Davidson et al. 2020). This efficiency helps conserve water resources and makes agricultural systems more sustainable in regions facing water scarcity.
- **Low Input Requirements**: Neglected crops generally require fewer agricultural inputs, such as fertilizers and pesticides, due to their adaptation to local conditions and resistance to pests and diseases. This reduces the environmental impact of farming, lowers production costs for farmers, and supports sustainable agricultural practices.

4.5.4 Resilience to Climate Change

- **Climate Adaptation**: Neglected crops often possess traits that make them highly adaptable to changing climatic conditions. For example, fonio is known for its ability to grow in poor soils and withstand drought conditions, making it a resilient crop in the face of climate variability (Zhu 2020). The cultivation of such crops can enhance the resilience of farming systems to climate change.
- **Temperature Tolerance**: Some neglected crops can thrive under extreme temperature conditions. Teff can grow well in both hot and cool environments, offering a reliable crop option as global temperatures fluctuate (Habte et al. 2019). This adaptability is crucial for maintaining agricultural productivity in the face of global warming.

4.5.5 Supporting Agroecological Practices

- **Intercropping and Agroforestry**: Neglected crops can be effectively integrated into intercropping and agroforestry systems. These practices enhance biodiversity, improve soil health, and increase overall farm productivity. For example, intercropping moringa with other crops can provide shade, improve soil fertility, and offer additional sources of food and income (Dhakar et al. 2011).
- **Pest and Disease Management**: The natural resistance of many neglected crops to local pests and diseases can reduce the need for chemical pesticides. This

promotes a healthier agroecosystem and supports the principles of integrated pest management (IPM). For instance, the pest-resistant traits of certain sorghum varieties can minimize pest outbreaks and reduce crop losses.

4.6 Challenges Faced by Neglected Crops and Their Opportunities

Neglected crops are plant species that have traditionally been important in certain regions but have been overlooked by modern agricultural practices. While these crops offer numerous benefits, they also face significant challenges. However, with strategic interventions, there are considerable opportunities to enhance their role in global food systems, environmental sustainability, and economic development.

4.6.1 Challenges Faced by Neglected Crops

- **Limited Research and Development**: One of the primary challenges facing neglected crops is the lack of investment in research and development. Compared to major staple crops such as wheat, rice, and maize, there is relatively little scientific research focused on improving the yields, disease resistance, and stress tolerance of neglected crops. This limits their potential productivity and adaptation to changing environmental conditions.
- **Poor Market Infrastructure**: Neglected crops often suffer from underdeveloped market infrastructure. Farmers growing these crops may have limited access to markets, which restricts their ability to sell their produce at profitable prices. The lack of established value chains and processing facilities further hampers market access and economic viability.
- **Consumer Awareness and Perception**: In many regions, there is a lack of consumer awareness about the nutritional benefits and culinary potential of neglected crops. These crops are sometimes perceived as "poor people's food" or inferior to more widely recognized crops. Changing these perceptions is crucial for increasing demand and market value.
- **Policy and Institutional Support**: Neglected crops often receive limited support from agricultural policies and institutions. Government policies tend to favor major staple crops, providing subsidies, extension services, and research funding that neglect underutilized species. This lack of institutional support can hinder the promotion and development of neglected crops.
- **Climate Change and Environmental Stress**: While many neglected crops are resilient to environmental stresses, they still face challenges from climate change. Altered precipitation patterns, extreme weather events, and changing pest and disease dynamics can impact their productivity. Addressing these challenges requires targeted research and adaptation strategies.

4.6.2 Opportunities for Promoting Neglected Crops

- **Enhancing Research and Development**: Increased investment in research and development can unlock the potential of neglected crops. This includes breeding programs to improve yield, disease resistance, and climate adaptability, as well as agronomic research to optimize cultivation practices. Advances in biotechnology and genomics offer new tools to accelerate the improvement of these crops.
- **Developing Market Infrastructure**: Improving market infrastructure is essential for integrating neglected crops into mainstream markets. This involves establishing efficient value chains, processing facilities, and transportation networks. Creating market linkages between producers and consumers can enhance economic opportunities and profitability for farmers.
- **Raising Consumer Awareness**: Educating consumers about the nutritional and health benefits of neglected crops can boost their demand. Public awareness campaigns, culinary initiatives, and incorporating these crops into school feeding programs can help change perceptions and increase their popularity. Highlighting their role in sustainable diets and food security can also attract health-conscious and environmentally aware consumers.
- **Policy and Institutional Support**: Policymakers can play a crucial role in promoting neglected crops by providing subsidies, research funding, and extension services tailored to these crops. Integrating neglected crops into national agricultural policies and food security strategies can enhance their cultivation and utilization. Support from international organizations and NGOs can also drive their development.
- **Leveraging Climate Resilience**: The inherent resilience of many neglected crops to environmental stresses offers significant opportunities in the context of climate change. Promoting these crops can enhance the resilience of agricultural systems, particularly in regions prone to drought, poor soil conditions, and extreme weather events. Developing climate-resilient varieties through breeding programs can further strengthen their adaptability.
- **Promoting Agroecological Practices**: Neglected crops can be integrated into sustainable farming practices such as intercropping, agroforestry, and conservation agriculture. These practices enhance biodiversity, improve soil health, and promote sustainable land management. Encouraging the use of neglected crops in agroecological systems can provide environmental benefits and improve farm productivity.
- **Economic Diversification and Rural Development**: Supporting the cultivation and commercialization of neglected crops can stimulate rural economies and provide new income opportunities for smallholder farmers. Developing value-added products and niche markets can enhance economic returns. For example, quinoa has gained global popularity as a health food, providing substantial economic benefits to farmers in the Andean region.
- **International Collaboration and Knowledge Sharing**: International collaboration and knowledge sharing can drive the promotion and development of neglected crops. Sharing best practices, research findings, and successful case

studies across regions can accelerate their adoption and integration into global food systems. Collaborative initiatives can also attract funding and support from international donors and development agencies.

4.7 Future Directions and Recommendations for Neglected Crops in Research and Global Context

Neglected crops hold tremendous potential for contributing to global food security, nutrition, and sustainable agriculture. However, realizing this potential requires concerted efforts in research, policy, and practical implementation. Here are some future directions and recommendations for promoting neglected crops both in the research area and at the global level:

4.7.1 Research Directions

4.7.1.1 Genetic Improvement and Breeding Programs

- **Genomic Research**: Invest in genomic sequencing and genetic studies to identify traits such as drought tolerance, pest resistance, and nutritional quality. Utilize advanced breeding techniques, including CRISPR and marker-assisted selection, to develop improved crop varieties (Arora and Narula 2017).
- **Participatory Breeding**: Engage local farmers in participatory breeding programs to ensure that the developed varieties meet the specific needs and preferences of the communities where these crops are traditionally grown.

4.7.1.2 Agroecological and Agronomic Research

- **Sustainable Farming Practices**: Conduct research on integrating neglected crops into sustainable farming systems such as intercropping, agroforestry, and conservation agriculture. Focus on practices that enhance soil health, biodiversity, and resilience to climate change.
- **Crop Management**: Develop optimized agronomic practices for neglected crops, including planting techniques, pest and disease management, and efficient water use strategies.

4.7.1.3 Nutritional and Health Research

- **Nutrient Profiling**: Conduct comprehensive nutritional profiling of neglected crops to highlight their health benefits. Research the bioavailability of nutrients and the presence of bioactive compounds that promote health.
- **Dietary Interventions**: Investigate the role of neglected crops in improving diet quality and addressing malnutrition, particularly in vulnerable populations. Explore their potential in dietary diversification and food fortification programs.

4.7.1.4 Climate Resilience Studies

- **Climate Adaptation**: Study the climate resilience of neglected crops and develop strategies to enhance their adaptability to changing environmental conditions. Focus on understanding how these crops can contribute to climate change mitigation and adaptation.

4.7.2 Global Recommendations

4.7.2.1 Policy Support and Advocacy

- **Inclusion in Agricultural Policies**: Advocate for the inclusion of neglected crops in national and international agricultural policies. Governments should provide incentives, subsidies, and support for the cultivation and commercialization of these crops.
- **International Agreements**: Promote international agreements and collaborations focused on the conservation and sustainable use of neglected crops. Organizations like FAO and CGIAR can play a pivotal role in this regard.

4.7.2.2 Market Development and Value Chains

- **Market Infrastructure**: Develop market infrastructure to facilitate the commercialization of neglected crops. This includes improving transportation, storage, and processing facilities to reduce post-harvest losses and enhance product quality.
- **Value Addition**: Promote the development of value-added products from neglected crops. This can include processed foods, health products, and culinary innovations that appeal to both local and global markets.

4.7.2.3 Consumer Awareness and Education

- **Public Awareness Campaigns**: Launch public awareness campaigns to educate consumers about the nutritional benefits and culinary potential of neglected crops. Use media, social platforms, and community programs to change perceptions and increase demand.
- **Educational Programs**: Integrate information about neglected crops into school curricula and agricultural training programs. Highlight their role in sustainable agriculture and healthy diets.

4.7.2.4 Research Collaboration and Knowledge Sharing

- **Global Research Networks**: Establish global research networks and partnerships to share knowledge, research findings, and best practices related to neglected crops. Encourage collaboration between research institutions, universities, and international organizations.
- **Data Repositories**: Create and maintain open-access databases and repositories for genetic, agronomic, and nutritional information on neglected crops. This can facilitate research and development efforts worldwide.

4.7.2.5 Support for Smallholder Farmers

- **Capacity Building**: Provide training and capacity-building programs for smallholder farmers to enhance their knowledge and skills in cultivating and marketing neglected crops. Focus on sustainable farming practices and business development.
- **Access to Resources**: Ensure that smallholder farmers have access to the necessary resources, including seeds, tools, and financial services, to successfully grow and market neglected crops.

4.7.2.6 Conservation and Biodiversity

- **Ex Situ and In Situ Conservation**: Strengthen efforts to conserve the genetic diversity of neglected crops both in situ (on farms) and ex situ (in gene banks). Ensure that traditional knowledge related to these crops is preserved and respected.
- **Agrobiodiversity Promotion**: Promote the cultivation of a diverse range of crops, including neglected ones, to enhance agrobiodiversity and ecosystem resilience. Encourage policies and practices that support biodiversity-friendly farming.

4.8 Conclusion

Agricultural crops that are rarely addressed in current and advanced farming strategies directly relate to the improvement of food security and sustainable practices in farming systems. These attributes present great value in climate change and environmental stress since they are genetically diverse and flexible under low input conditions. Thus, by increasing soil fertility, contributing to biological diversity, and optimizing the usage of the resources, these crops help to establish a more sustainable and stable agricultural environment. Nonetheless, they are confronted with many challenges such as lack of research, poor market facilities, and absence of policy provision. Efforts for overcoming these threats include more investment in research, market development, policy support, and consumer awareness to fully harness underutilized crops improving people's diets, incomes, and nutrition in rural areas, and global food security. It is thus imperative to reclaim these crops as endemic to agroecological practices and climate change mitigation processes.

References

Alandia G, Rodriguez JP, Jacobsen S-E, Bazile D, Condori B (2020) Global expansion of quinoa and challenges for the Andean region. Glob Food Sec 26:100429. https://doi.org/10.1016/j.gfs.2020.100429

Arora L, Narula A (2017) Gene editing and crop improvement using CRISPR-Cas9 system. Front Plant Sci 8:1932. https://doi.org/10.3389/fpls.2017.01932

Azmat S et al (2024) Morphogenetic characterization of Stenotrophomonas maltophilia infecting white stripe disease of rice (Oryza sativa L.). Notulae Botanicae Horti Agrobotanici Cluj-Napoca 52(2):13547. https://doi.org/10.15835/nbha52213547

Baldermann S, Blagojević L, Frede K, Klopsch R, Neugart S, Neumann A, Ngwene B, Norkeweit J, Schröter D, Schröter A, Schweigert FJ, Wiesner M, Schreiner M (2016) Are neglected plants the food for the future? Crit Rev Plant Sci 35(2):106–119. https://doi.org/10.1080/07352689.2016.1201399

Barasa J, Ochuodho J, Kebeney S, Barasa AW (2022) The role of sorghum legume intercroping system in improving soil productivity on small holder farmers in Western Kenya. Int J Res Innov Soc Sci 6(7):1–8. https://doi.org/10.47772/ijriss.2022.6701

Bhattarai G (2018) Amaranth: a golden crop for future. Himal J Sci Technol 2:108–116. https://doi.org/10.3126/hijost.v2i0.25854

Cheng A, Mayes S, Dalle G, Demissew S, Massawe F (2015) Diversifying crops for food and nutrition security—a case of teff. Biol Rev 92(1):188–198. https://doi.org/10.1111/brv.12225

Cloete PC, Idsardi EF (2013) Consumption of indigenous and traditional food crops: perceptions and realities from South Africa. Agroecol Sustain Food Syst 37(8):902–914. https://doi.org/10.1080/21683565.2013.805179

Dansi A, Vodouhè R, Azokpota P, Yedomonhan H, Assogba P, Adjatin A, Loko YL, Dossou-Aminon I, Akpagana K (2012) Diversity of the neglected and underutilized crop species of importance in Benin. Sci World J 2012:1–19. https://doi.org/10.1100/2012/932947

Davidson JM, Aiken RM, Min DH, Kluitenberg GJ (2020) Water use and productivity of teff, a dairy quality forage crop. Kansas Agricultural Experiment Station. Res Rep 6(5). https://doi.org/10.4148/2378-5977.7928

Dhakar R, Pooniya B, Gupta M, Maurya S, Bairwa N (2011) Moringa: the herbal gold to combat malnutrition. Chron Young Sci 2(3):119. https://doi.org/10.4103/2229-5186.90887

Gahukar RT (2014) Potential of minor food crops and wild plants for nutritional security in the developing world. J Agric Food Inf 15(4):342–352. https://doi.org/10.1080/10496505.2014.952429

Gebru YA, Sbhatu DB, Kim K-P (2020) Nutritional composition and health benefits of teff (Eragrostis tef (Zucc.) Trotter). J Food Qual 2020:1–6. https://doi.org/10.1155/2020/9595086

Gelaw AM, Qureshi AS (2020) Tef (Eragrostis tef): a superfood grain from Ethiopia with great potential as an alternative crop for marginal environments. Environ Policy:265–278. https://doi.org/10.1007/978-3-319-90472-6_11

Gidi M (2023) Sorghum as a model crop for drought stress tolerance. Adv Biosci Bioeng 11(3):54–65. https://doi.org/10.11648/j.abb.20231103.14

Habte E, Muktar MS, Negawo AT, Lee S-H, Lee K-W, Jones CS (2019) An overview of teff (Eragrostis teff (Zuccagni) Trotter) as a potential summer forage crop in temperate systems. J Korean Soc Grassland Forage Sci 39(3):185–188. https://doi.org/10.5333/kgfs.2019.39.3.185

Hasan M, Uddin MK, Mohamed MTM, Zuan ATK (2018) Nitrogen and phosphorus management for Bambara groundnut (Vigna subterranea) production—a review. Legume Res 41(4):483–489. https://doi.org/10.18805/lr-379

Innazent A, Anitha S, Jacob D (2024) Yield potential and economics of chilli + amaranth intercropping system as influenced by plant geometry and fertigation. Agric Sci Digest (10). https://doi.org/10.18805/ag.d-5931

Joshi N, Verma KC (2020) A review on nutrition value of amaranth (Amaranthus caudatus L.): the crop of future. J Pharmacogn Phytochem 9:1111–1113

Kell DB (2011) Breeding crop plants with deep roots: their role in sustainable carbon, nutrient and water sequestration. Ann Bot 108(3):407–418. https://doi.org/10.1093/aob/mcr175

Li X, Yadav R, Siddique KHM (2020) Neglected and underutilized crop species: the key to improving dietary diversity and fighting hunger and malnutrition in Asia and the Pacific. Front Nutr 7:593711. https://doi.org/10.3389/fnut.2020.593711

Mabhaudhi T, Hlahla S, Chimonyo VGP, Henriksson R, Chibarabada TP, Murugani VG, Groncr VP, Tadele Z, Sobratee N, Slotow R, Modi AT, Baudron F, Chivenge P (2022) Diversity and diversification: ecosystem services derived from underutilized crops and their co-benefits for sustainable agricultural landscapes and resilient food systems in Africa. Front Agron 4:859223. https://doi.org/10.3389/fagro.2022.859223

Mayes S, Massawe FJ, Alderson PG, Roberts JA, Azam-Ali SN, Hermann M (2011) The potential for underutilized crops to improve security of food production. J Exp Bot 63(3):1075–1079. https://doi.org/10.1093/jxb/err396

Mbosso C, Boulay B, Padulosi S, Meldrum G, Mohamadou Y, Niang A, Coulibaly H, Koreissi Y, Sidibé A (2020) Fonio and Bambara groundnut value chains in Mali: issues, needs, and opportunities for their sustainable promotion. Sustainability 12(11):4766–4766. https://doi.org/10.3390/su12114766

Mengistu G, Shimelis H, Laing M, Lule D (2018) Breeding for anthracnose (Colletotrichum sublineolum Henn.) resistance in sorghum: challenges and opportunities. Aust J Crop Sci 12(12):1911–1920. https://doi.org/10.21475/ajcs.18.12.12.p1230

Minarovičová L, Lauková M, Karovičová J, Kohajdová Z, Kepičová V (2019) Gluten-free rice muffins enriched with teff flour. Potravinarstvo Slovak J Food Sci 13(1):187–193. https://doi.org/10.5219/1045

Naidoo P (2023) Greener on the other side: tracing stories of amaranth and moringa through indenture. Agenda 37(1):102–110. https://doi.org/10.1080/10130950.2023.2183138

Nelson ARLE, Ravichandran K, Antony U (2019) The impact of the green revolution on indigenous crops of India. J Ethnic Foods 6(1):8. https://doi.org/10.1186/s42779-019-0011-9

Pathan S, Eivazi F, Valliyodan B, Paul K, Ndunguru G, Clark K (2019) Nutritional composition of the green leaves of quinoa (Chenopodium quinoa Willd.). J Food Res 8(6):55. https://doi.org/10.5539/jfr.v8n6p55

Rahman S, Ikram AR, AlHusnain L et al (2024) Genome-wide profiling of bZIP transcription factors in Camelina sativa: implications for development and stress response. BMC Genom Data 25:88. https://doi.org/10.1186/s12863-024-01270-6

Rauf A, Sher MA, Farooq U et al (2024) An SNP based genotyping assay for genes associated with drought tolerance in bread wheat. Mol Biol Rep 51:527. https://doi.org/10.1007/s11033-024-09481-x

Samal I, Bhoi TK, Raj MN, Majhi PK, Murmu S, Pradhan AK, Kumar D, Paschapur AU, Joshi DC, Guru PN (2023) Underutilized legumes: nutrient status and advanced breeding approaches for qualitative and quantitative enhancement. Front Nutr 10:1110750. https://doi.org/10.3389/fnut.2023.1110750

Sambo BE (2014) Endangered, neglected, indigenous resilient crops: a potential against climate change impact for sustainable crop productivity and food security. IOSR J Agric Vet Sci 7(2):34–41. https://doi.org/10.9790/2380-07223441

Talabi AO, Vikram P, Thushar S, Rahman H, Ahmadzai H, Nhamo N, Shahid M, Singh RK (2022) Orphan crops: a best fit for dietary enrichment and diversification in highly deteriorated marginal environments. Front Plant Sci 13:839704. https://doi.org/10.3389/fpls.2022.839704

Tan XL, Azam-Ali S, Goh EV, Mustafa M, Chai HH, Ho WK, Mayes S, Mabhaudhi T, Azam-Ali S, Massawe F (2020) Bambara groundnut: an underutilized leguminous crop for global food security and nutrition. Front Nutr 7:601496. https://doi.org/10.3389/fnut.2020.601496

Vega-Gálvez A, Miranda M, Vergara J, Uribe E, Puente L, Martínez EA (2010) Nutrition facts and functional potential of quinoa (Chenopodium quinoa willd.), an ancient Andean grain: a review. J Sci Food Agric 90(15):2541–2547. https://doi.org/10.1002/jsfa.4158

Zhu F (2020) Fonio grains: physicochemical properties, nutritional potential, and food applications. Compr Rev Food Sci Food Saf 19:3365. https://doi.org/10.1111/1541-4337.12608

Crop Modeling Under Climate Change

5

Abdullah Ahmed Butt and Zahra Ahmed

Abstract

"Crop Modeling Under Climate Change" is the exploration of crop–climate coupled systems and their pertinence to relevant agricultural practices and the importance of an approach by integrating crop models. After a brief introduction to crop modeling and its relevance in the era of climate change, the basic structures and components of various types of crop models are examined. The chapter deals with how climate change affects agrarian production, tearing apart the resistance of various crops and strategies presented in the UN 2030 Agenda to address challenges. It covers climate data integration into crop models and shows how crops respond to different climatic elements such as temperature, precipitation, and CO_2 concentrations. The chapter also describes adaptation strategies and presents responses in the form of genetic enhancement, agronomic practices, and water management. It presents an overview of the key equipment and software needed to conduct crop modeling and guidance on how to validate and calibrate models. The climate impacts section enables the evaluation of future impacts using scenario analysis. The chapter concludes with the challenges and restrictions in existing crop models and discusses future directions that can help to prepare for climate change under current solutions consistent with the UN 2030 Agenda aiming at sustainable stable food security.

A. A. Butt (✉)
Department of Food Science and Technology, Government College University, Faisalabad, Pakistan

Z. Ahmed
Department of Pharmacology, King Edward Medical University, Lahore, Pakistan

© The Author(s), under exclusive license to Springer Nature Singapore Pte Ltd. 2025

S. Faiz et al. (eds.), *Climate Smart Agriculture for Future Food Security*,
https://doi.org/10.1007/978-981-96-4499-5_5

Keywords

Agricultural resilience · Climate adaptation · Sustainable farming · Food security · Environmental impact · Data integration

5.1 Introduction

This chapter zeroed in on the matter of agriculture concerning this new global climate. Given the rapidly increasing impact of climate variability on agricultural systems, there is a real necessity for higher level tools to asses them and predict their effects (Arora 2019). In this respect, crop modeling has been recognized as an invaluable tool that provides knowledge on how a wide range of crops can respond to different climatic scenarios—which is essential for fostering practices leading to improved resilience in agriculture (Corbeels et al. 2018). Crop models predict the growth and development of crops taking into account temperature, precipitation, CO_2 concentration, or extreme weather events (Zhao et al. 2019). Models managed well are informative and can give insightful analysis to predict future crop performance, assess the possible impacts of climate change, and help provide effort directed at developing an adaptation strategy (Rodriguez et al. 2018). This chapter mulls over the principles of crop modeling with a particular focus on the kinds and ingredients used in these models.

In this chapter, we also reveal the global consequences of climate change warming and its effect on agriculture/soil health, looking at what crops are more varied to these threats. It focuses on the incorporation of climate data into crop models, highlighting that good post-processing and quality control techniques are paramount. The adaptation measures such as genetic manipulations, agronomical practices water management, etc., discussed in the chapter are important tools to maintain the productivity of crops in a changing climate.

This chapter offers a clear analysis of the different dimensions of crop modeling. It does not simply mean showing how it is done using APSIM and STICS or discuss model validation, calibration problem statements, and scenarios, rather it provides an overview of crop modeling as an ecosystem so that this way to understand sustainable systems is beyond an isolated context. However, it also accounts for constraints and gaps and provides a direction to combat risk better by suggesting tips for making the agricultural system more resilient that will further help the global food security agenda and the sustainability practice of farming.

5.1.1 Overview of Crop Modeling

Crop modeling is a mathematical simulation of the growth, development, and production of crops under changing environmental conditions which provides an important tool for researchers, agronomists, and policymakers to understand the complex nature of the interaction between crops and the environment with the purpose to predict future performance and develop strategies aimed at attaining sustainable increased agricultural productivity (Ahmed and Ahmad 2020). The basic principle of crop modeling is usually rooted in several mathematical equations and algorithms describing the physiological process in crops that combine information regarding weather conditions with soil properties, genetics characteristics for genotype-by-environment interactions, and testing performance or yield under different loss factors due to planting practices (Velten et al. 2024). The models are important for understanding how climate change will affect agriculture and create strategies to adapt agricultural practices in a manner that ensures food security while supporting the sustainability of farming systems.

Types of Crop Models can be classified into three: Empirical models will have historical data and some sensible statistical approximations which allow for much simpler quick estimation to be performed (Shirley et al. 2020). They quickly become impotent when predicting conditions outside the areas of previously used input. Nor can experience-based mechanistic models (process-type models) simulate biological processes in great detail and hence require an enormous amount of data input to predict better the responses of crops or yield from different environmental factors such as temperature, water availability, and nutrient levels (Trivedi et al. 2013). These hybrid models combine the best attributes of both empirical and mechanistic approaches to produce highly accurate predictions that can be applied in a wide variety of contexts.

The objective of crop modeling is to predict agronomic consequences in response to changes in the external environment and provide assistance for resource management (e.g., water, nutrients) and supporting decision-making in agriculture (Challinor et al. 2018). Crop models use simulations of different management practices and climate scenarios to determine the most efficient methods for achieving yield while minimizing unwanted environmental effects (Jägermeyr et al. 2021). In the face of increasing climate concerns, such models function as an essential line of defense guiding what tropical threats might be forthcoming and predicting for rebuilding sustainable agricultural systems that can adapt to a changing world. Crop modeling is, therefore, a key element for sustainable agricultural development and long-term food security under global climate change conditions.

5.1.2 Importance of Crop Modeling in Climate Change Context

Crop modeling is a major determinant for understanding and controlling the effect of climate change on agriculture. Given the potential that untapped resources underneath soil may offer to make future scenarios for crop production a possible reality in a climate change-constrained world, where the warming trend is accelerating globally and patterned rainfall amount and frequency/occurrence are changing with increasing no-returning extreme events (Seneviratne et al. 2021). This method is being widely used in simulating and analyzing the crop growth process, which can help us understand how crops respond to different climatic conditions systematically, of great significance for not only food security but also promoting sustainable agriculture. However, the most important advantage in a climate change context is that crop modeling gives an insight into likely performance under current and future (climatic) conditions before we grow them (Brown et al. 2018). Such models often forecast changes rather crudely using weather data (e.g., temperature, precipitation, and CO_2 concentrations) of extreme climatic events affecting crop development and yield (Kukal and Irmak 2018). Such predictive power is necessary for scientists and policymakers to know what lies ahead on the horizon in agriculture.

It can additionally be beneficial for crop models to recognize vulnerabilities in weather alternate throughout one-of-a-kind sorts of vegetation (Challinor et al. 2018). Process-based crop growth models can be used to identify what climatic factors are the types causing stress that compromises crops' health and productivity (Chaudhry and Sidhu 2022). Such information is necessary for devising adaptive strategies, like selecting climate-smart crop varieties and timely planting dates to undertake an optimal irrigation schedule (Mulla et al. 2020). It can be in the form of adaptation to reduce the risk of damage, and it leads to stabilizing agricultural productivity. Crop modeling also assists in the assessment and advancement of climate-smart agricultural practices (Gupta et al. 2022). Crop models can simulate different management scenarios to assess conservation tillage, crop rotation, and integrated pest management practices under future climate conditions (Yang et al. 2018). This assessment will help to determine the best methods for maintaining or increasing crop yields with less environmental damage (Shah and Wu 2019). In this context, crop modeling plays an important role in achieving the objective of sustainable agriculture which is crucial for long-term food security.

Crop models also offer important data for policymakers in formulating and implementing appropriate climate adaptation and mitigation policies (Shelia et al. 2019). Crop modeling has important agricultural implications that can be valuable as inputs for national and international agricultural policies to channel subsidies, investments in research, and development of infrastructure including technology (Bellmann 2019). If we align agricultural policy to projections of climate change, states can more effectively support farmers and safeguard food systems in the context of global change (Rawe et al. 2019). And, crop modeling is also a cornerstone in scientific research into climate change and its effects on agriculture. It will help the community to merge heterogeneous datasets, from climate projections to soil characteristics and crop physiology that support a comprehensive assessment of dynamics between climatic variables and crop performance (Maestrini et al. 2022). By gaining insights into the nature of connections within complex agricultural systems, this research supports efforts to devise new approaches that address many climate challenges.

In a nutshell, crop modeling is an essential component in the climate change scenario to give predictions on the future performance of crops while identifying their vulnerabilities and providing supported information for coming up with possible solutions using best agricultural practices compatible with changing climates additionally driving research all around. The predictive capabilities of crop models will help us to anticipate and adjust to impacts from the changing climate, so food systems remain robust globally.

5.2 Fundamentals of Crop Modeling

5.2.1 Definition and Scope

Crop modeling is the use of mathematical and computational tools to simulate complex interactions between crop physiology-related parameters (yield, growth, development, etc.) with soil processes as well operations on site required for cropping such as sowing or fertilization. It is key for predicting the behavior of crops when exposed to different conditions and, hence appropriate decision-making in agri-management and policy formulations (Keating and Thorburn 2018). Crop modeling is much broader than only representing relationships for biological processes such as photosynthesis, respiration, nutrient uptake, and phenological development that are in turn influenced by environmental drivers such as temperature, light interception efficiency, available water content, or soil properties (Piao et al. 2019).

At its heart, the goal of crop modeling is to represent the complex system that drives plant growth and yield to allow agents virtual space within which one can guardedly explore how different agricultural management practices or climatic scenarios might impact upon it. Crop models are integrative to simulate potential outcomes of agricultural systems under hypothetically changed conditions and thus useful as a tool for proactive management by merging disparate data, i.e., weather-based prediction, soil analysis, and genetic trait information (Cooper et al. 2020). The predictive capacity of these models is a consequence of plant growth models that include dynamic simulation in the time domain and approach actual crop response toward its environment, which also disciplines uncertainties making them useful tools for resource efficiency optimization and sustainability enhancement (Akhavizadegan et al. 2021). It also allows us to take a deeper view than yield prediction only using crop modeling. It is an important tool in determining how future climate change may influence agricultural productivity and can therefore provide information to assist in responding measures. This means operating scenarios of enriched CO_2, and rainfall/temperature changes to gauge how it will change crop behavior (Pastore et al. 2020). These models are of great importance for selection in breeding programs and to support management strategies adapted to changes in global climate.

These characteristics make crop models useful for a wide range of applications and scales. They can be used either for point-wise or area-oriented growth assessments and short-term seasonal forecasts as long-term climatic scenarios. This is a kind of flex that can mean finding solutions right down on the farm (e.g., tweaking certain management practices) or it might be nation/regional-wide, more important agricultural policy changes in the way we grow food.

5.2.2 Types of Crop Models

The crop modeling arena is gaining enrichment with many such models, suitable for varied research objectives and application requirements which make these unique in terms of providing insights into the multiple facets governing dynamics within a crop system. Common are empirical models, which have their basis in statistical analyses of historical data to find correlations between environmental variables and crop responses (Delmar 2019). Such models are appreciated for their simplicity and speed which makes quick assessments or predictions possible (Yin et al. 2021). But because many tree-ring chronologies are built relying on historical data, they can be less helpful when the climate changes or something crazy happens in an ecosystem. At a more explicit and mechanistic level, considering process-based or mechanistic models makes it possible to account for the physiological and biochemical processes controlling crop growth (Bhatla and Lal 2023). These models characterize the crops and their environment by representing processes, such as photosynthesis, respiration, nutrient uptake, and water use that generate a mechanistic understanding of plant function. Mechanistic models are data- and calibration-hungry, a fact that they compensate chewing it all to expansive much precise insights about the crop–environment relationships (Wallach et al. 2021).

Hybrid models, combining the advantages of both empirical and mechanistic approaches, serve as a complement to these methods. These hybrid models are calibrated to real-world data to improve prediction and thus bridge the mechanistic-under determinism gap while remaining relatively simple (Gao et al. 2020). This pairing ensures that the simulations are highly data-driven as well and they bring with them a functional approach also which offers versatility to their applications. This representative is also an emerging frontier in crop-simulating functional-structural plant models (FSPMs). Integrating the plant architecture with its physiological functions, these models provide a mechanism for simulation at fine scales and connections between form (architecture) and function. They are especially useful to investigate plant plasticity and hence the effect of different environmental regimes or crop management practices, providing valuable information on how to design crops with an optimized aboveground architecture for increased productivity.

These crop models are designed to provide insight at various levels of matching (for individual plants, field plot scale, and the regional/eco-system level) as well under different time scales from short-term weather extremes through long-term climate change projections. An appropriate model will be selected depending on the research question, type of available data, and outcomes aimed to be achieved. Such models, with the adaptability and accuracy that they provide in an age where agricultural systems are facing rapid change due to environmental factors such as but not limited to climate conditions, turn global focus on sustainable agriculture needs. They not only have helped us to predict crop yield outcomes and optimize the community, e.g., thereby underlying mechanisms of trees and how they can work in synch with accurate predictions development for example through upregulation demography because leaves capture canopy light interception program. Therefore, various models representing different aspects of crops as a whole crop model has helped to have a better understanding of the agricultural systems that can be useful for innovations and resilient methods against changes.

5.2.2.1 Empirical Models

Empirical data-driven models represent a bedrock of agricultural research using historical information to create statistical associations between environmental variables and crop performances. They inherently involve the transfer of information, based upon extensive datasets describing past observations taken for estimating crop trace gas emissions or yield (crop yields, weather conditions, and actions toward the soil in irrigated rainfed systems) (Keating and Thorburn 2018). Empirical models reveal trends and correlations through regression analysis, machine learning, and time-series investigation to give insights into how crops perform in various conditions (Elavarasan et al. 2018). The main advantage of empirical methods lies in their capacity to provide quick and easy-to-use predictions, which is particularly salient for practical applications as a precision farming decision-support tool (Laurent et al. 2021). Empirical models, because of their simplicity and how fast they are to build have quite some limitations. The reliance on historical transactions means that they cannot properly predict results of emerging or extremely vital conditions that are not typically present in a training dataset. Especially considering the increasing prevalence of climate change, along with its uninhibited and seemingly endless weather patterns (Jin et al. 2018). Nevertheless, given the importance of considering interactions among processes within simulated datasets, empirical models continue to be an essential part of a crop model uniform and serve as proxies for better integration across more sophisticated large-scale simulation approaches that include pragmatic neighboring resistance/resilience analyses.

Crop yield prediction tools developed using empirical models help inform planting and harvesting decisions, and water allocation scheduling to fields according to specific needs (site-specific application), leading toward modern agriculture practices (Schauberger et al. 2020). The inputs for these models can span from basic weather data to more sophisticated indices such as the Normalized Difference Vegetation Index (NDVI), an indicator of plant health and vigor (Pasqualotto et al. 2019). While relying on historical data to predict the future remains useful in certain contexts, empirical models have an edge because they can adapt and improve with new information. Furthermore, the use of more advanced statistical and machine learning algorithms has greatly improved what could be done using empirical methods, leading to better-informed and differentiated analyses. More specifically, machine learning which is a subfield of artificial intelligence enables the processing of large amounts and/or complex datasets where it can find patterns that may be too subtle to detect using traditional statistical methods. The previous characteristics make them very good for anomaly detection, trend analysis, or scenario testing.

Beyond this, they are also frequently employed for validation and calibration of more sophisticated mechanistic models as a reference model to understand their precision and credibility (Basso and Liu 2019). Comparing model outputs with the data from examples in natural history allows researchers to identify where models may be inaccurate and which processes require more fidelity to reflect real-world conditions. The model was validated and calibrated in an iterative way which prevented the crop model from being outdated for both scientific research and practical use. Empirical approaches to crop modeling offer a flexible and robust methodology for studying, understanding, and predicting crop response. Although they are far from capturing the full complexity of crop–environment interactions, their simplicity lies in the widespread use and adaptability to be implemented with ease by practitioners worldwide make them an important asset for both daily farm management under a variety of conditions as well as strategic long-term planning within climate change adaptation scenarios.

5.2.2.2 Mechanistic Models

Mechanistic models, also called process-based crop growth simulation models, with a strong scientific foundation on the explicit understanding of important biological and physical processes contribute to plant dry matter accumulation potential (Basso and Liu 2019). While empirical models measure historical statistical correlations, mechanistic ones work with the mechanisms by which plant physiology is controlled (e.g., photosynthesis, respiration, nutrient and water uptake) (Legris et al. 2019). These models comprise rather explicit descriptions of physiological processes and plant–environment interactions in terms of temperature, light, carbon dioxide concentration, soil moisture, nutrient availability, etc. This means that mechanistic models can simulate these processes as closely as possible and in high detail, therefore being powerful tools for understanding how crops respond under a diverse set of environmental conditions and management practices (Peng et al. 2020).

Mechanistic models are strong at dynamic crop growth simulation that can estimate the multiple feedback loops relating to a plant system and interactions between plants with their biophysical environment. This makes it a highly versatile model that can simulate transitions in climate, soil characteristics, and farming practices providing an objective tool to quantify the impact of these changes on regional or global crop production (Sun et al. 2019). For instance, mechanistic models can be employed to realize the consequences of mounting atmospheric CO_2 on crop photosynthesis and growth which is an important input for securing climate change adaptation strategies (Jayawardena et al. 2019). The disadvantage of mechanistics is that they are complex. They involve comprehensive and specific data inputs such as meteorological datasets, soil properties, crop phenology of the production area or cropping practices), and physiological parameters (Pasquel et al. 2022). This makes them costly to build and maintain. Predictions made by mechanistic models are accurate predictions that heavily rely on well-calibrated input data (in quality and resolution) as well as the quality of the assumptions, and parameters underlying them (Triana et al. 2019). Although mechanistic models have inherent limitations, improvements in computational power and remote sensing techniques, combined with further availability of data at various scales, are increasing their usability.

Mechanistic models are incredibly useful in research, education, and policy-making. They are used to simulate the effects of management practices on soil and crop, such as irrigation and fertilization, which aims to help farmers make the optimal use of resources/inputs impacting sustainability (Yue et al. 2022). They also help breeding programs identify favorable traits that improve the resilience of crops to important environmental stresses, such as drought or nutrient deficiency. For studies in climate change, mechanistic models are an absolute necessity to design and assess both adaptation and mitigation strategies that provide accurate guidance for policymakers toward making the best-informed decisions securing food supply while considering environmental sustainability (Razavi et al. 2021).

In general mechanistic models are a very rich and flexible approach to crop modeling that gives the unique potential for appropriately detailed representation of biophysical mechanisms controlling crop growth, so the results have much deeper insights into all biological processes resulting in plant productivity. Their power to simulate intricate plant–environment interactions makes them indispensable for deepening and revising our knowledge of agricultural systems, thus contributing to technological advances meeting challenges that the climatic changes bring along with ever-growing global food demand. The mechanistic models will still be essential for many scientific and practical applications in those systems because they can still be refined, and combined with new tools such as lidar or drone imagery to map fields and crops at high spatial scales automatically without human labor.

5.2.3 Key Components of Crop Models

The major set of components included in crop models are the basic elements that form these complex tools capable of simulating integrated dynamics and functioning on growth, and phenology under different environmental conditions. At the core of CTLs are models that describe key processes controlling plant physiology such as photosynthesis, respiration, transpiration, and nutrient uptake (Lambers et al. 2019). These processes are powered by intricate sub-models of how plants use sunlight to make energy, exchange gases with the atmosphere, and extract water and nutrients from soil (Talbot and Monfet 2024). These physiological processes are modulated by some important environmental factors, and these interactions should be accurately modeled to predict crop performance such as air temperature, light intensity in the growth chamber, CO_2 concentration, and soil moisture.

Crop models also simulate phenological development stages that register along a crop-specific progression through attributes such as germination, flowering, and maturity (Potgieter et al. 2021). This component of modeling is essential for informing about the effect different environmental conditions and management practices have on the timing and duration of these growth stages that in turn modify whole season yield and quality. Another key ingredient is soil processes, including the physical properties of soils (soil water balance), nutrient cycling, and root–soil interactions. These sub-models describe soil water and nutrient transport in the skeletons, which can then be further stressed by plant roots to provide information on how different soil properties and management practices affect crop growth (Rawat et al. 2019). Crop models use climate and weather data input as a basis for simulations of the external conditions that drive processes in simulation (Boote et al. 2018). Such data consist of observations, e.g., temperature, precipitation, solar radiation, and wind speed made on a daily or hourly basis, etc. The conjuncture of high-resolution climate projections is important to consider when evaluating the effects on future crop productivity due to specific Weather Classification and Climate

Change scenarios. Management practices such as planting dates, irrigation schedules, fertilization, and pest control measures which can influence the growth environment and resource availability for crops are also important.

In addition, all crop models need to be tested for calibration and validation against measured data. Calibration is adjusting model parameters to comply with field experiments observations, and validation refers to model predictive capacity by means of independent datasets. The iterative and recursive nature of this development is a part of creating powerful models that accurately predict what the real world will look like.

5.3 Climate Change and Agriculture

Climate change not only disregards but also transcends administrative borders, nature, and agriculture that exist in a closer co-habitation to one another than we may think from the outset. As most climate variability affects the agricultural sector, differences in temperature increase trend (degrees Centigrade), precipitation pattern, and frequency of extreme weather events directly influence crop-production yields on one hand and also livestock productivity, whereas these factors endanger the operational sustainability level of farming systems (Shi et al. 2021). Higher global temperatures can contribute to more unstable weather, in turn increasing the food-security risks—a major headache for feeding an ever-growing populace its daily dietary requirements (El Mokhtar et al. 2022). Simultaneously, agriculture is a major contributor to environmental change through deforestation for farmland expansion as well as immediate emissions of methane from livestock and fugitive nitrous oxide gas emissions from fertilizers (Pendrill et al. 2019).

The global warming impacts on food production are region-specific and interconnected. In some areas, crops could benefit from more abundant water and longer growing seasons: in others they may suffer from heat stress almost out of the gate or lack of irrigation down the line, to grow at all. These types of natural disasters can easily decimate crop fields rendering them unusable as well as undermining entire food supply chains (Prince et al. 2023). On top of all this, climate change tends to exacerbate the aforementioned difficulties even more, e.g., pests or disease epidemics may be able to spread rapidly in warmer conditions which can damage yet other crops and yields (Velásquez et al. 2018). Confronting the multifaceted adaptive challenge of agriculture and climate change requires significant amounts of yet another type of adaptation. This includes the breeding and adoption of climate-resilient crops, new water management options, as well sustainable agriculture practices that restore soil health to reduce greenhouse gas emissions. Key research investments in cutting-edge technologies precision agriculture, and advanced crop modeling for example are what will provide farmers worldwide with the ability to adjust their operations on a real-time data-enabled basis. Important components of this mitigation action are the development-driving policy frameworks that facilitate a transition to more sustainable farming practices, as well as specifically those highlighting carbon sequestration from agriculture all supported with R&D efforts.

5.3.1 Understanding Climate Change

Climate change is a driver for this level of mitigation and also requires broader critiques similar in scope. Climate change can be broadly regarded as variations in average weather such as temperature, rain, or snow over some time often 30 years (Koutsoyiannis 2021). Global warming is one type of expression of climate change which refers to the long-term changes within Earth's climate induced primarily by human activities such as burning fossil fuel and deforestation among other industrial processes (Naz et al. 2022). These greenhouse gases include carbon dioxide, methane gas, and nitrous oxide driven into the atmosphere mainly by human activities geared toward an enhancement of this greenhouse effect causing global warming (Jones et al. 2023). This warming then initiates a set of climatic changes: rising sea levels, melting polar ice caps, and variations in weather (Kumar et al. 2021). Although some may still doubt this, science has shown through a long chain of repeated tests and increasingly intricate weather models that the changes have occurred faster than they should in geological time scales. For example, reconstructed temperatures go back over the last century and show a clear warming trend; satellite data confirm significant polar ice losses as well as changed precipitation patterns around the globe. An accurate understanding of these changes must be based on not only an analysis of historical climate data but also the projection into future scenarios by sophisticated models. They integrate the atmospheric composition, solar radiation, ocean currents, and land surface processes to project how climate might change under different greenhouse gas emission pathways. This knowledge is essential for implementing efficient climate change mitigation and adaptation policies in agriculture and other sectors. It is also important to acknowledge that the impacts of climate are different on every part (due at least in part to fine-scale differences); not only longer-lasting droughts and more frequent heatwaves here but increased rainfall and flooding there (Raza et al. 2019). Such an understanding could help to anticipate the impacts of climate change on agricultural systems, build resilience, and secure global food supplies in a warmer world.

5.3.2 Impacts of Climate Change on Agriculture

Climate change has complex and widespread consequences for agriculture which cause multiple challenges in terms of regions, crop type, etc. With a shorter growing season, crops could mature too quickly and produce lower yields of inferior grain (Hanson 2021). This is through, for example, heat stress during flowering which can have a devastating impact on pollination and seed set, leading to significant yield loss (Jagadish 2020). This new climate aggravates these issues by changing the pattern of precipitation and therefore causes longer drought periods to dry soil moisture or in a different aspect prolonged rain which leads to waterlogging and deteriorating sediment. These shifts in equilibrium cause hardships in determining when and where to provide water for proper crop growth, this is why farmers need to adapt water management practices (Ayyam et al. 2019). The question for these changes aside from weather and temperature involves climate change also bringing (s) an elevation within frequency—intensity concerning extreme events including hurricanes/floods/ droughts plague that destroy crops. For instance, stronger and more frequent storms can harm crops or the ability of arable land to produce anything (Hanley et al. 2020). Even the agricultural schedule is being controlled, which induces planting and harvesting in different periods than those habitually used for agriculture. Farmers will have to modify their crop system or even grow other crops more adapted to this situation.

Under pressure from a changing climate, in which more sun and less rain, thanks to warmer temperatures pests (including diseases) get bolder lawns to grow faster. The impact of these drivers is likely to exacerbate biotic stresses on crops with severe effects not only on wave health and productivity but also in increasing the requirement for more integrated methods of pest and disease management (Khan et al. 2019). The regime of soil temperature and moisture which influences nutrient availability is also changing because of anthropogenic impact, it promotes processes of erosion and degradation. Animal diseases are expected to be impacted both directly and indirectly by climate change (by increases in extreme events)—whereas other types of health risks might rise from key cassation leads as changes for example vector interactions (Baker et al. 2022; Caminade et al. 2019).

With the scientific tools and technologies we need underpin an entire palette of crop breeding techniques, and a variety of precision agriculture approach on a large scale worldwide, across different local conditions; these provide farmers with promising opportunities to develop climate-resilient yet productive farming practices. The policymakers should try to come up with measures and enhance the adaptability of the agricultural system so that both food security and sustainable development go hand in hand while facing climate change.

5.3.3 Vulnerability of Crops to Climate Change

The crop vulnerability to climate change in agricultural systems is a complex interplay of factors that reveals the intricate interdependencies within. Climate change can alter temperature, precipitation, and the frequency of extreme weather events substantially, and crops are especially affected (Clarke et al. 2022). High temperatures can hasten metabolic plant processes, shift the duration of growth phases against the optimal durations for production, and ultimately reduce both productivity and quality (Zhu et al. 2021). Heat stress affects critical growth stages such as flowering and may reduce pollination and fruit set, which in turn can lead to significant yield losses (Lohani et al. 2022). Global warming is already causing temperature fluctuations, which are particularly harmful to crops (e.g., wheat, maize, and rice), a staple food for large parts of the human population (Bisbis et al. 2018). Both extreme rainfall and long droughts can devastate the productivity of crops, as are patterns in precipitation (Ayanlade et al. 2018). When land becomes arid and cracked in the absence of water, drought conditions prevail; soil lacks moisture to support plants still growing their roots (Dietz et al. 2021). On the other end of the scale, too much rainwater can cause flooding and waterlogging which cut off plant roots from air to breathe by displacing oxygen present in soil pores, facilitating several diseases that occur due to fungi living in saturated soils (Pouyat et al. 2020; Nieder et al. 2018). Severe weather conditions such as hurricanes, floods, and heatwaves can also wreak immediate havoc on crops by leveling fields and infrastructure or throwing a wrench into agricultural calendars (Abbas et al. 2018).

Warmer, wet winters are predicted to spread pesticide-resistant pests, while climate change influences which weeds and diseases will flourish (El-Husseini 2018). Thereby, biotic stress can aggravate which further leads to intense crop losses resulting in low productivity and thus more use of chemical inputs with long-term environmental implications. An example of this kind of threat to food security is the spread and migration of pests and plant diseases in areas where they were unknown before, negatively affecting crop health. Similarly, such changes in temperature and precipitation may also affect the soil by altering its structure or impacting organic matter turnover and nutrient availability among others. All this could contribute to soil degradation, erosion, and loss of fertility making it even more difficult to continue with agricultural production. The types of crops, as well as the local climate and soil conditions and agricultural practices, all affect how much a crop will be vulnerable to various impacts from global warming. Strategies, such as the release of climate-resilient crop varieties and treatments that rely on them to address these vulnerabilities are equally important. The resilience of our crop species is vital to ensure food security in the years ahead and making use of innovative science approaches alongside new scientific tools provides an opportunity for strengthening crops from being negatively affected by climate change. Such a comprehensive assessment will be critical for designing the appropriate adaption and mitigation measures necessary in an atmosphere that is less predictable than it used to be.

5.3.4 Addressing Agricultural Problems Using the UN 2030 Agenda

Agriculture is the most sensitive sector to climate change and variability. The solution to the problem in agriculture with the UN 2030 Agenda is based on utilizing Sustainable Development Goals (SDGs) that pave the way for building a strong and robust agricultural sector that cannot be deposed by adverse weather conditions, especially changes resulting from climate change (Abraham and Pingali 2020). The UN 2030 Agenda provides a roadmap for addressing these challenges, taking into account all three dimensions—economic, social, and environmental—that are needed to cope with the multifaceted nature of those affecting agriculture (Carpentier and Braun 2020). Agricultural sustainability and resilience are directly linked to key goals (SDG 2: Zero Hunger, SDG 8) as well as Supportive Goals (Sustainable Development Goal indicators help fine tune the responsibilities of specific targets) (Arora and Mishra 2022). Goal 2 seeks to end hunger and secure food by encouraging sustainable agricultural practices, improving access to nutritious foods, and supporting smallholder farmers as an example (Monaco 2024). Through the wider adoption of climate-smart agricultural practices that increase productivity, decrease greenhouse gas emissions, and are more resilient to changes in temperature, utilizing a land-efficient approach (Aryal et al. 2018). This calls for swift measures to be taken on how best climate adaptation and mitigation actions can blend in directly with national policies, and practices of agriculture. This requires early warning systems, sustainable land management, and increased disaster risk reduction policies. SDG 15 is all about protecting, restoring, and promoting sustainable use of terrestrial ecosystems, combating desertification, and halting and reversing land degradation (Yao et al. 2021). Governments, international organizations, the private sector, and local communities will work together to assist all countries in their efforts but also absorption capacity among many others. It also encourages the application of technology and innovations such as precision agriculture and advanced crop modeling to increase efficiency in resource use and improve cultivar resilience for food security purposes. By embedding the principles of its New Environmentalism, the government should call for agricultural policies and practices to be fully integrated into initiatives aimed at addressing climate change in a comprehensive way that also helps deliver sustainable food production systems as well as progress toward poverty reduction. This systems approach not only prescribes solutions for direct agricultural problems but serves as a groundwork so that global agriculture can be economically resilient and environmentally sustainable in an unpredictable future climate.

5.4 Integration of Climate Data in Crop Models

This is very important since climate data are needed routinely inside crop models to be able to simulate and forecast agricultural performance for present and future climates. With the addition of specific climate factors including temperature, rainfall, solar radiation sunlight level, and CO_2 concentrations, crop models can make more accurate predictions on how crops grow, and yield as well as the ability to withstand different versions of climatic conditions. This coupling allows us to assess the effects of climate change on agricultural productivity, improve farming systems, and design responses toward global food security. As a result, they can complement the routine of computerized crop models in this context accentuating their power and establishing them as essential tools for modern agriculture under changing climate conditions.

5.4.1 Sources of Climate Data

The model requires such information to simulate agricultural systems under changing environmental conditions, and a multitude of sources are available for the simulation of crops. These data sources are the weather stations based on the ground that give very local as well accurate measurements of temperature, precipitation, humidity average, and highest wind speed during 5 minutes or an hour, and the intensity of solar radiation (Colston et al. 2018). The significant benefit of this network is that the stations are usually run by national meteorological agencies and agricultural research institutions which means these data have a high quality—exactly what our models need for calibrations and validation (Behailu et al. 2021). Furthermore, ubiquitous spatial coverage and the capability of satellite remote sensing to monitor large-scale climatic patterns and trends have had far-reaching effects on climate data collection (Kogan 2019). Data from satellites operated by NASA, ESA, and NOAA provide information about surface temperature, vegetation health indices (VHI), and soil moisture-related data (Climate Data Record of Soil Moisture) (Pei et al. 2018). They also provide data on atmospheric content such as the total precipitation in mm, contributing to a more comprehensive understanding of how agriculture is affected due climate factors (Huffman et al. 2018). Climate reanalysis datasets are produced by assimilating observations with climate models to give a coherent description of the observed system from available and often incomplete observational data. In addition, general and regional circulation models (GCMs; RCMs) provide climate change scenarios ranging from low to high levels of greenhouse gas emissions used by end-users for long-term planning in agriculture (Chokkavarapu and Mandla 2019). Such models are critical in the management of risks and evidence-based policymaking in climate change scenarios. Such integration of data from multiple sources leads to robust and reliable crop models that are better able to represent complexities stemming from climate variability, thereby enabling improved decision-making in agricultural planning and management.

5.4.2 Processing and Quality Control of Climate Data

Climate data preprocessing and data quality control are the most important proceedings in the accurate outcome of a crop simulation modeling analysis. Climate data in its raw form, collected from different platforms such as weather stations, models, or satellites, are often unreliable and several preprocessing steps are required to make this information uniform for use within the crop influence format. The process starts with data cleaning that includes finding and removing bad values, and variances in smoothed-out data, resolving missing or unreported readings via imputation methods (interpolation) to ensure continuity of the time series; and normalizing units so they are easier to work with (Mitra et al. 2023). The model predictions are processed using advanced statistical methodologies and algorithms to remove any possible outliers one may have encountered (Burdett and Wellen 2022). Quality control includes cross-validation using independent datasets to confirm the consistency and accuracy of the climate data (Lewis et al. 2021). These include the comparison of observational data from ground stations to satellite observations and the reanalysis products (Ling et al. 2021). Further modifications are needed to focus data at the scales in which crop models require temporal and spatial resolution changes for proper representation of local climatic conditions regarding crop growth (Pasquel et al. 2022). Climate model outputs are also bias-corrected to account for systematic deviations of the climate models from observed climate patterns.

Validating Climate Data by processing and quality control results; not only adds to crop model predictability but also encourages a higher level of confidence in predictions, both computing systems will lead toward more empowered decisions within agriculturally recognized management and policy (Filippi et al. 2019).

5.4.3 Techniques for Integrating Climate Data into Crop Models

Statistical downscaling methods have been widely utilized to generate finer-scale climate data from coarse resolution (Abdollahipour et al. 2022). GCMs are inputted into the crop model in a way that bridges the gap not covered by local differences (Yang and Beyer 2018). Dynamic downscaling, on the other hand, involves running regional climate models (RCM), which resolve some of the local-scale atmospheric processes directly and thus help to create higher-resolution fields (Tapiador et al. 2020). Bias correction methods are employed to address systematic errors so that the climate model outputs can be corrected and made consistent with observation data (Mendez et al. 2020). Data assimilation methods may also be exploited in which observational data are assimilated online with model predictions to update the estimate of the current state, thus increasing its forecasting capacity (Kivi et al. 2022). Temporal aggregation methods have cumulative monthly or seasonal summaries of daily weather data that match with the climate variable resolution output from phenological models and synoptic crop growth stages. This study also uses regionalized kriging and inverse distance weighting to interpolate measured climate data, under the assumption that these are spatially representative of six explanatory variables (climate metrics).

5.5 Modeling Crop Responses to Climate Variables

As climate change is the main concern nowadays, therefore understanding the crop responses to various climatic variables through numerical modeling and predicting their impact on agricultural productivity resulting from altered temperature/precipitation-associated components or increases in CO_2 levels are of paramount importance. These models predict physiological and essentially developmental advancement of crops when grown in diverse climatic circumstances but they do ensure extremely very important aspects concerning potential yields, growth responses, or stress resistance. In that way researchers as well as farmers can learn from these trials to manage in advance climate forces they will alternatively be faced with for decades or centuries by the seeding of atmosphere sustained while at the same time maximizing agricultural yields through entirely other more sustainable means. Such climate data need to be combined using crop models to better predict future agricultural scenarios and food security.

5.5.1 Temperature Effects and Precipitation Patterns

Knowledge of the possible effects of temperature and precipitation is imperative particularly as these directly relate to growth and productivity which are primary factors affecting how a crop responds to climate variables. The major physiological processes such as photosynthesis, respiration, and transpiration are directly affected by temperature in addition to the developmental stages of crops such as germination, flowering commencement, and maturing stretches (Dusenge et al. 2019; Bhattacharya 2022). Yields and quality can be drastically reduced by either heat stress or frost damage due to extreme temperatures (Sehgal et al. 2018). For example, during flowering warm temperatures can decrease pollen viability and fertilization and thus reduce grain set in wheat or maize (Wang et al. 2019). The availability of water on the other hand, essentially driven by precipitation patterns, plays a critical role in maintaining moisture under the soil as well as supporting functions that relate to plant metabolism (Walter 2018). Disparity in the rain, such as long periods of drought or prolonged rainfall can bring opportunities for either water stress or high watering conditions that affect root health and nutrient uptake. Unfortunately, any other simulation centered around accurate modeling of those climatic variables relies on downscaling affluent behavioral temperature also precipitation data among plant increase executes after mean how unique weather stage impacts agricultural improvement including harvest yield. This means you use historical climate records to study what has happened and future projections for a glimpse at possible changes. These models can also help design adaptive strategies, like identifying heat-tolerant and drought-resistant crop varieties as well as developing irrigation practices and planting schedules that would buffer against the impacts of climate variability.

5.5.2 CO_2 Concentrations

This is because CO_2 concentrations artificially regulate plant responses to climate variables (Dusenge et al. 2019). Excess carbon dioxide can have beneficial and adverse impacts on plant growth from an industry perspective. When all else is constant, higher CO_2 often stimulates photosynthesis—especially in plants such as wheat, soybeans, and rice—as shown in unique simulation of climate forcing (Sharma et al. 2022). This can potentially increase the biomass of crops, along with water use efficiency (that is higher levels of CO_2 lead to a reduction in stomatal opening which reduces transpiration and therefore loss of its source) (Driesen et al. 2020). Nevertheless, the benefits of increased CO_2 can also be conditioned by other abiotic factors such as temperature, water supply, and nutrient availability (Rienth et al. 2021). Limitations in nutrients like nitrogen—or even just heat stress—could counteract the positive effects of higher CO_2 and result in decreased yield improvements (Hassan et al. 2021). Moreover, the modulation of CO_2 levels causes changes in plant physiology which are limited to growth rates as well but include nutritional quality too (e.g., reduced protein and concentrations of micronutrients on the consumable part) (Fowler et al. 2019). To model the CO_2 effects correctly, detailed physiological data have to be incorporated into crop models simulating how changing levels of atmospheric CO_2 interact with other climatic factors and management practices. This consists of the experimentation and field trials that provide calibration or validation, to predict reliably how future elevated atmospheric CO_2 will affect crop performances. Using this knowledge, researchers and farmers can devise strategies to improve crop yield in high CO_2 environments by breeding the crops that benefit most from CO_2 fertilization or changing fertilizer management practices to avoid nutrient dilution.

5.5.3 Extreme Weather Events

However, one cannot ignore extreme weather events since they dramatically change crop responses to climate changes; which describes large fluctuations in agricultural productivity. Monsoon-based extreme events such as heat waves, cold snaps coupled with heavy rainfalls, and floods can have even more devastating impacts on crops at times while storms across different seasons will produce varying results (Goyal et al. 2018). Heat waves during flowering can inhibit pollination which means reduced yields, downpours cause waterlogging or root issues just about anywhere and diseases pop up left, right, and center (Misra and Ansari 2022). On the other side of the spectrum, long droughts completely negate any water that falls because it is dry for so long, and plant growth is stunted if at all present (GaneshamurthyA et al. 2020). The modeling of these events requires historical climate data, combined with predictive climate models to simulate the frequency intensity, and the duration of extreme weather conditions. This will help researchers to identify the potential hazards and develop measures for adaptation of problems as well. Models need to account for the size and time of these, as well as crop-specific environmental stresses influencing crops (Webber et al. 2020). For instance, modeling the effect of late-season frost for a certain crop likely requires another set of parameters than those used in case of mid-season drought (Raymundo et al. 2021). Techniques such as stochastic weather generators and high-resolution climate models also assist in increasing the reality of extreme scenarios, thereby making less uncertain predictions. Integrating this climate-related risk into crop models provides a key element for preparing policies and practices that could help farmers face current vulnerabilities due to weather extremes as well the future challenges for agriculture in light of these pessimistic scenarios.

5.6 Adaptation Strategies in Crop Modeling

Climate change and variability impact on major crop productivity can be mitigated through adaptation strategies in crop modeling. Stress, drought stress, and disease will impact global agriculture in increasingly unpredictable ways as the impacts of climate change are realized around the world. Adaptation strategies include the implementation of sophisticated crop modeling tools combined with agronomic approaches which could help in making crops and farming systems more resilient (Brouziyne et al. 2018). Their strategies are to better use resources, including water and fertilizer; practice improvements in how crops are managed during all stages of the growing season and diversify crop plants so they can tolerate extreme range-topping temperatures (Wiréhn 2018). The assessment of the potential impacts on crop growth, yield, and quality from various climate change scenarios can help design effective possible adaptation strategies by using simulations provided by different types of crop models (Huang et al. 2019). Bringing this level of readiness to farmers, not only helps them predict and manage risks but also promotes the kind of sustainable agricultural practices that can survive in a changing climate. In the end, adaptation practices in crop modeling are necessary to maintain global food security, the livelihoods of rural people, and essentially environmental sustainability in agriculture.

5.6.1 Genetic Improvements and Crop Breeding

Plant modeling in crops requires genetic improvements and crop breeding as essential components of adaptation strategies to develop resilient varieties that will survive under different climatic conditions (Swarup et al. 2021). With the use of modern breeding programs, scientists can select and genotype plants that better reflect traits (e.g., drought tolerance, heat resistance) or that are not favorable for pest/disease attack in a specific environment (Langridge and Reynolds 2021). A range of advanced genomic tools such as molecular markers, marker-assisted selection (MAS), and quantitative trait loci (QTL) which now include high-density SNP array maps and GBS are available to help in the detection as well as introgression of favorable genes/QTL from diverse sources or landrace genetic into a specific base, which can be deployed for developing latest breeding materials with better rates, faster predictability, and higher accuracies (Paudel et al. 2020; Mwamahonje et al. 2021). Models of crop growth are an important tool in this context since they can predict the behavior of hypothetical breeding lines across a series of environments, predicting their yield potential and evaluating them for adaptation to specific climatic stresses (Jarquin et al. 2021). It combines genetic data with environmental conditions such as temperature, precipitation, and soil to develop breeding strategies for the development of crop cultivars that are high in resistance along with a better yield. Such an iterative methodology not only promotes faster development of climate-resilient crops but also aids sustainable agricultural practices by decreasing the need for agrochemical inputs and reducing environmental impacts. Thus, proactive genetic- and breeding-based improvements represent a science-driven tool to increase food security and provide resilient crops when dealing with variations in climate.

5.6.2 Agronomic Practices and Water Management Techniques

Agronomic practices and water management strategies are important components of any adaptation strategy in the context of crop modeling aimed at improving resource use while increasing resilience to changing climatic conditions (Rahn et al. 2018). Agronomic practices include the strategies of conservation tillage, crop rotation, and intercropping which are ways to increase soil health by retaining the availability of nutrients (Chahal et al. 2021). These practices help control soil erosion and nutrient runoff and improve the efficiency of water and nutrient use in crop production (Zhu et al. 2020). Moreover, by way of exactitude farming technologies like GPS elements and remote sensing help farmers track soil wetness levels, crop growth stages, and plague infestations (Zhang et al. 2019). These agronomic practices can be embedded into crop models and used to simulate how the same biophysical changes would improve simulation of their effects on crop productivity and resilience, across environmental scenarios, thus enabling farmers to put in place sustainable and adaptive farming.

Appropriate water management is vital in reducing the negative effects of low and variable rainfalls on cropland production (Lu et al. 2019). Methods for optimum use of water such as drip irrigation, rainwater harvesting, and moisture-conserving soil techniques should be adopted to ensure that the efficiency in utilization of this important asset is high at values required by crops needing copious amounts during critical growth stages (Zhang et al. 2021). In addition, crop modeling could also be an effective tool in evaluating suppression strategies enabling estimation of benefits and trade-offs at various hydro-meteorological conditions for different stakeholders involved in identifying optimal management interventions to improve agricultural water scarcity resilience across/arising from droughts/ erratic seasonal patterns. They not only contribute to increasing the profitability of farms with knowledge-based management but also promote sustainability and climate change adaptability in agriculture at local scales globally relevant for food security.

5.6.3 Integrated Pest Management

IPM is one such means of integrated farming, conspicuously in crop modeling, and implies sound practices in pest and disease control types (Bagheri et al. 2019). Moreover, IPM is a holistic management strategy that combines biological control and cultural methods such as resistant crop strains and application rates that will kill the pest but allow more to live because they are bio-controls (Ambethgar and Rameshkumar 2024). Crop models are essential components of IPM when applied to simulate pest dynamics, life cycles, growth, and the role of the surrounding environment, such as temperature and humidity, and phenology of the host crop factors of the pest population (Savary et al. 2018). Within these units, the model input data, focusing on the pest's life cycle, biological characteristics, and environmental factors occurring concurrently with the unit when the potential area's data are known in advance, allow predicting when pests cause an outbreak. The Calibrator Unit due to the genetic algorithm implemented with the role of seeking proximity numbers for coefficients that make an even curve between two separate years evaluates different types of control methods (Ahmed et al. 2020a). With the high accuracy of the pest risk model, farmers could take preventive measures through simple methods, visual scouting, and relief, to reduce the reliance on the chemical means needed for control and the potential for chemical resistance (Sharifzadeh et al. 2018). Furthermore, IPM control measures achieve the sustainability requirements in farming by lessening the harm caused to the environment, maintaining the beneficial insects, and removing pests (Romeh 2019). Incorporating IPM in crop modeling is not only profitable to healthy crops and enhanced productivity but it also ensures that the farming system would withstand new or altering weather and other factors prevailing across the globe.

5.7 Tools and Software for Crop Modeling

Tools and software for crop modeling were critical components that enabled us to merge complex computational ideas with detailed agricultural understanding, thus broadening our knowledge of processes related to crops experiencing multiple environmental drivers. Here is how we use this toolkit—with a range of advanced mathematical models and simulation frameworks as well as data-driven analytics software or visualization tools. As a result of using these techniques, researchers and agronomists can model intricate crop–environment interactions that involve soil characteristics, climatic factors, or management practices. This ability to explore a range of scenarios, including the impacts of climate change and options for enhancing higher productivity and sustainable agricultural operations. These tools can also be used to generate crop management plans that inform policy and are compatible with the massive data sets—historical climate records, satellite images, and the genetic information they allow into their programs. To tackle increasing agricultural challenges related to climate variability and expected global demand for food, advanced modeling tools are pivotal in promoting innovative agronomic practices on one hand and enhancing the resilience of our agriculture systems worldwide on the other.

5.7.1 DSSAT (Decision Support System for Agrotechnology Transfer)

One of the main crop modeling tools, the DSSAT (Decision Support System for Agrotechnology Transfer) framework, is supported by rich models that allow good representations of agricultural systems under varying environments. The DSSAT is the combined product of a collective group of researchers, who have attempted to unite with science different crop models for crops as well integrated soil databases and weather generators along with many management modules that voluntarily share substantial features (Gao et al. 2022). It is made to provide a flexible input framework with detailed soil, crop management, and physiology that allows for comprehensive analysis of how crops will respond at scale to climate variability nutrient availability or water-management scenarios (Hoogenboom et al. 2019). The DSSAT employs strong analytical algorithms for model calibration and validation against field data, ensuring the predictive accuracy of crop performance across agro-ecological zones (Feleke et al. 2021). This in turn provides answers to questions like when is the best time of year for planting, how much and what type of fertilizer should I be using, or even irrigation scheduling which can all help aid decision-making leading toward more sustainable agriculture as well as food security (Anar et al. 2019). Additionally, the properties in DSSAT come with documentation and support and will be accessible to researchers "on the ground"—field workers as well as extension agents or policymakers; thus it has an enormous potential for influence over agricultural research activities worldwide.

5.7.2 APSIM (Agricultural Production Systems Simulator)

APSIM is a complex, crop modeling tool with the ability to simulate many agricultural systems in an agile environment (Holzworth et al. 2018). APSIM is a widely used modeling system developed by the Commonwealth Scientific and Industrial Research Organisation (CSIRO) in collaboration with other collaborators (Keating 2024). It aids in simulating biophysical processes, soil water dynamics, nutrient cycling, and plant physiology (Ahmed et al. 2020b). The software has a modular shape which in turn enables users to tailor simulations using information on soil properties, climate variables, crop genetics, and management practice (Cichota et al. 2021). The power of APSIM lies in its ability to predict behavior, and to be able to apply them across scales where all potential interactions by the crop with their environment can be simulated ranging from climate change impacts assessment (long-term), to short-term analyses simulating different management practices at farm or agro-ecosystem scale. Robust validation of the simulator against experimental data guarantees it to be a reliable source for estimating crop responses in response to certain scenarios, hence supporting devices that help farmers make decisions (Kang and Özdoğan 2019). APSIM is open source and has a user-friendly interface that promotes the coupling of researchers, farmers, and policymakers alike to innovate adaptation strategies for performance-based farming practices worldwide. APSIM benefits from its versatility and substantial modeling capabilities, fortifying it as an indispensable tool for assessing options to support the tripling of food production sustainably in changing climate conditions.

5.7.3 STICS (Simulateur Multi Disciplinaire Pour les Cultures Standard)

STICS can be seen as a universal model in crop modeling that is capable of simulating a large number, 15 so far with some calibration required sometimes values omitted for crops that are not or very seldom grown, even at the international level, and include cereals, sugar beet, sunflower, and standard crops under various environmental conditions. Developed by INRAE (French National Research Institute for Agriculture, Food and Environment) (Tribouillois et al. 2018), STICS has benefitted from the synergy between disciplines such as crop physiology, soil physics, agronomy, and climate science. This model has a complete framework that allows detailed simulations of crop growth, development, and yield under stress conditions such as soil properties, weather data (past or future changes) also the effects of pest interactions (Courault et al. 2021). STICS uses sophisticated mathematical models and algorithms to reconstruct complex processes that govern photosynthesis, nutrient uptake, and water dynamics all along the phenology of a crop plant to anticipate possible responses or impacts due to climate change as well as for optimizing fertilization and considering potentially new agricultural practices (Evers et al. 2018). Its modular architecture, combined with a wide parameterization makes it possible to adapt the simulator for different crops and agricultural systems promoting its use in research as well as agronomic decision-making and policy development. Its coupling with GIS (Geographic Information Systems), its user-friendly interface, and online access help make it highly applicable by scientists, agronomists, or stakeholders in customizing crop management under different farming contexts on a global scale optimizing production inputs efficiency, resilience to uncertain climate/weather conditions; and environmental sustainability.

5.7.4 Features and Applications

Crop model software includes numerous features and functionality to facilitate continued progress in the field of agricultural research. These tools use complex algorithms, coupled with extensive databases containing a wide range of information on soil properties, weather data, and patterns in the climate system (Manivasagam and Rozenstein 2020). They also use germplasm characteristics and management practices to predict crop growth processes (e.g., leaf area index) and status at each time point during the season (e.g., photosynthesis and transpiration); but also yield under different set(s) of environment variables which can depend both upon current year conditions or be influenced (Stöckle and Kemanian 2020). They provide an opportunity for researchers worldwide to model intricate crop–environment interactions, including expected climate variability impacts on outputs of crops as well under nutrient management regimes and pest attacks. Furthermore, these modeling frameworks enable scenario analysis for future crop productivity and resilience (Peterson et al. 2020). These tools extend beyond research, helping farmers to use their inputs most efficiently (for example, how much irrigation or fertilizer is being used) and minimizing environmental footprint while improving overall resource utilization (Liu and Song 2020). Given the ease of use and strength of validation against field data, these tools are particularly helpful for guiding agri-environmental decision-making (from the farm management level to policy development targeted at sustainable food production) addressing global challenges related to food security.

5.8 Challenges and Limitations of Crop Modeling

Although it has made significant advances over the years, crop modeling faces several major challenges as well as intrinsic limitations that demand particular care in agricultural practice and research. Genetic factors, (micro)climate and response traits, and the complexity of the biology inputs include intricate interplay among genetic entities all modulating effect size interacting with both physiological processes, driving biomass production, yield components linked to harvestable regenerative tissue development against a backdrop of environmental/management conditions (Tian et al. 2021). Adequate and reliable field data on soil properties, climate variables, and management practices are needed to represent these dynamics realistically which is challenging due to the wide range of agro-ecological conditions involved. In addition, there is a lot of variability in climate patterns and extreme weather events due to the absence of accuracy with which current models can provide forecasts on how crops will respond under future climate scenarios because it introduces many uncertainties. In addition, several assumptions and simplifications in the design of crop simulation models can call into question the reliability of including them to quantify yield variability. This can lead to models relying on only empirical relationships and generalizations with no underlying mechanisms which will not accurately predict the complex responses of crops over time. Calibration and validation based on empirical data are key, but may also be difficult because of temporal as well as spatial variability under field conditions. Secondly, as agricultural practices evolve and new technology emerges, the models structures and algorithms need to be updated accordingly.

Moreover, utilizing crop models in practical agricultural decision-making would require thoughtful treatment for their uncertainties and limitations. The overreliance on model outputs absent contextual understanding can result in suboptimal management decisions and create perverse outcomes. Thus, although crop modeling will remain pivotal for examining agricultural contingencies and informing strategies of adaptation, continued research effort is required to ensure model reliability as well as precision in predicting diverse agronomic scenarios under wide climatic situations across the continuum.

5.8.1 Data Limitations and Model Uncertainties

Data limitations and model uncertainties are major challenges in cropping system modeling, affecting the reliability of predictions that are critical for effective agricultural decision-making (De Wit et al. 2019). A lack of sufficient or accurate inputs comprising soil properties, weather conditions, and management practices is known to impact model calibration/independent validation causing uncertainties in simulated crop responses (Seidel et al. 2018). The heterogeneity of data quality and availability across regions poses additional challenges to producing generalizable model outputs that could be extrapolated for a wide range of agroecological settings (Costa et al. 2019). Model uncertainties also stem from the simplifications and assumptions required in modeling extremely intricate biological and environmental interactions. Each of these factors (e.g., crop genotype variability, pest and disease dynamics, the stochastic nature of weather events) introduces additional sources of uncertainty which combined can make it difficult to predict accurately with crop models (Mehmood et al. 2020). Improving upon these challenges will require enhanced data collection, interdisciplinary collaboration, and model algorithms that provide a more accurate description of agricultural systems. The results have practical implications for improved utilization of crop models as decision support tools to realistically assess impacts, risk, and opportunities in agricultural practices and adaptation strategies.

5.8.2 Technical and Computational Challenges

Challenges of technical and computational nature are major stumbling blocks for advancing/deploying crop modeling, which impacts its accuracy, scalability, and utility in the practical sense (Rötter et al. 2018). The challenging part lies at the cutting edge, where these biological processes have to be integrated with soil dynamics and climatic variables in state-of-the-art model compounds requiring extensive computational resources (Vogel et al. 2018). Modeling at that fine scale, though, requires high-resolution simulations which tend to be computationally intensive and data-hungry practices that most researchers are unlikely able or willing to afford (Neumann et al. 2021). Complicating matters further is that these crop models typically have been developed and validated with complex algorithms, mathematical formulations, and require expertise in both agricultural science as well as computer modeling (Tian et al. 2020). This process can be costly and time consuming, as human intervention is necessary to fine tune the model against real data. In addition, rapid rates of technological improvements in sensor technologies for earth observations and big data analytics create possibilities to improve model capacities but demand continuous adjustment and integration into modeling frameworks as well (Sudmanns et al. 2020). Addressing these technical and computational challenges requires interdisciplinary collaboration, improvement of infrastructure, and continuous development of modeling methods to achieve the promise that crop modeling holds for sustainable agriculture as well as its potential impact on future food security.

5.8.3 Addressing the Challenges

Comprehensively addressing the issues and constraints downstream of crop modeling, including improvements in data collection and model development, represents a joint challenge that needs to be tackled through multidisciplinary collaboration (Hou et al. 2020). Foundational advances in improving data availability and quality before the actual field validation require expansion of monitoring networks—directly measuring soil properties, weather patterns, and agricultural practices pertinent for regional crop yields; this is true today far from any converging results. Newer remote sensing technologies, coupled with sensor networks can also be used in integration to address the information void and provide real-time data of crop health indicators land surface temperature, soil moisture conditions, or climate variables which can result in an improved input quality along with better spatial resolution (Gao and Zhang 2021). In addition, there remains work and innovation to be done refining the algorithms of these models for incorporation of complex biological processes as well as genetic variability and environmental interactions (Sakamoto 2018). However, the continued need for calibration and validation of empirically verified data will reinforce that a systematic inclusion of field trials or multi-group collaborative research initiatives would be critical for optimizing model robustness and predictive capacity. The implementation requires an interdisciplinary effort between agronomists, climatologists, computer scientists, and policymakers that may overcome technical hurdles to provide versatile modeling tools for policy and decision-makers in agriculture (Lambert et al. 2019). The agricultural community can overcome current constraints in part by employing these different approaches and thereby develop the full potential of crop modeling to address global food security challenges under changing climatic conditions.

5.9 Future Directions in Crop Modeling under Climate Change

The future of crop modeling under climate change represents a core frontier in agricultural research and innovation, offering potential for use in the fast-growing global landscape. On one hand, it presents challenges that require immediate attention upon rebuilding continuity and emerging strengths. With increasing climate variability and its more severe implications for agriculture, the demands to improve the predictive capabilities of crop models are intensifying. This includes the use of computational advancements in methods and data science, as well as interdisciplinary research allowing for using more realistic and robust modeling frameworks (Silva and Giller 2020). We present that future-oriented crop modeling should shift beyond predicting crop responses to climate change by also considering the need for adaptive strategies in optimizing resistance, resilience, and productivity across various agroecosystems. Furthermore, the incorporation of big data analytics and artificial intelligence on top of machine learning can further enhance model accuracy as well as scalability through real-time decision support systems for farmers and policymakers. In this time of tight budgets, we encourage partnerships and collaboration among scientists, stakeholders, and policymakers to catalyze cutting-edge discoveries through interdisciplinary research that continues to refine modeling approaches as they provide actionable results for climate-smart agriculture fostering food security in a rapidly changing world.

5.9.1 Advances in Modeling Techniques and Integration of Remote Sensing and GIS

Other transformative approaches in the advancement of crop modeling are leading to improving prediction and use across climate change which involves meeting new challenges. Geographic Information Systems (GIS) has no power as a model but function as a set of models. Advances in modeling techniques such as process-based models, machine-learning algorithms, and ensemble approaches are set to improve our understanding of intricate interactions among crops, soils, and climatic variables (Droutsas et al. 2022). These developments will result in more accurate simulations of crop growth, responses, and impacts under a wider spectrum of agroecological conditions. At the same time, satellite and UAV technology can be integrated into a GIS platform to achieve high spatial-temporal resolution capturing vital agricultural parameters, i.e., soil moisture, vegetation indices, or land-use patterns among others (Awais et al. 2022). A closer coupling of the data collection and modeling phases enables real-time monitoring of crop conditions, and early identification of stress factors, as well as adjusting to updated model inputs in response time windows that allow farmers and policymakers to act on decision support tools (Berger et al. 2022). Interdisciplinary efforts are needed to maximize the potential of these technologies, promoting cross-discipline collaboration that fuels innovative and sustainable climate-adjusted agricultural systems.

5.9.2 Multi-Model Ensembles

Multi-model ensembles are becoming an attractive frontier to further enhance the capabilities of crop models in dealing with their undeniable structural and parametric uncertainties at a larger scale, particularly when tackling climate change effects on agriculture (Iizumi et al. 2018). The approach involves coupling of multiple modeling frameworks, each with its algorithms and assumptions, to generate ensemble predictions over a range of possible outcomes (Troin et al. 2021). By combining different model types (process models and statistical or machine learning approaches), various sources of bias and uncertainty from the single methods can be compensated, making predictions more robust, e.g., future conditions (Laux et al. 2021). In addition, multi-model ensembles help to explore uncertainties in climate projections and their effect on crop productivity worldwide under diverse regions/strengths of cropping systems (ur Rahman et al. 2018). The reason for this is that it improves the accuracy of predictions of crop response to climate variability and also makes available risk management strategies, along with adaptation measures which will help decision-makers. Interdisciplinary Science, Scientists, and Policymakers jointly work together to generate globally standardized model outputs with validation using empirical data-based ensemble predictions and produce actionable knowledge for Global-scale Climate Smart Agriculture and Sustainable food production.

5.9.3 Policy Implications and Decision Support

These will become significant linkages to merge between scientific insights and practical applications in agricultural policy, as well as bond science-based know-how with decision support systems that should guide the futurescape of crop modeling under climate change. Climate variability is likely to become even more severe (Bathiany et al. 2018), and as a consequence, there is growing pressure for decision support tools that link advanced crop models to socio-economic information, environmental regulations, and climate adaptation strategies (Li et al. 2018). The results from these tools can be used to provide policymakers with evidence-based scenarios and future projections that will help inform policy development, resource allocation priorities, and mitigation strategies for food security objectives targeting enhanced agricultural resilience in the long term (Dhyani and Sood 2023). In addition, developments in data analytics, artificial intelligence (AI), and machine learning make it possible to continuously track crop conditions as well as any changes due to climate impacts which enables adaptive management measures for farmers along with early warning systems (Javaid et al. 2023). The collaborative work among researchers, policymakers, and stakeholders in co-developing easy-to-use decision support frameworks tailored toward the multifarious requirements of agricultural communities across the world is imperative. By leveraging scientific collaboration while capitalizing on the areas of policy making that is sought after by future research in crop modeling, this approach promotes sustainable agricultural practices along with food systems that can adapt to climate challenges. Under global climate change, decision support systems are indispensable for supporting functional and effective adaptive management practices or early warning responses that require farmer-specific information provision to deal with plausible agricultural risk impacts through resource use optimally. These are geared to realize global sustainable development impacts—climate resilience, food security, and equitable agricultural systems—as part of the UN 2030 Agenda. To address the complex climate change impacts on agriculture, this integrated framework needs to be co-developed and applied by researchers, policymakers, and stakeholders.

5.10 Conclusion

Based on the thorough review of crop modeling under climate change, it can be seen that although these models provide insights and predictions they are in no way bullet proof. These results suggest that the integration of some advanced modeling techniques (e.g., multi-model ensembles and input data from remote sensing) may improve prediction accuracy and robustness for better agricultural decision support. Additionally, these projects align with policy frameworks such as the UN 2030 Agenda and have scope for upscaling to promote climate-smart agricultural practices that guarantee food security in a changing world. Nevertheless, overcoming data challenges, technical difficulties, and the right robust decision support system continue to be prerequisites for translating scientific advancements into tangible solutions. Such issues highlight the need for continued collaboration between scientists, policymakers, and stakeholders to further develop this global framework of climate impact models into a local one, by validation against empirical data enabling better management decisions in response to anthropogenic Global Warming. Adoption of these future directions by the community can inform further evolution in crop modeling, and support resilient and adaptive agricultural systems capable of sustainably meeting global food demands.

References

Abbas AM, Abbas II, Agada IG (2018) Natural hazards and their impacts on the environment and agriculture in Nigeria. Int J Adv Res Publ 2(2):85–91

Abdollahipour A, Ahmadi H, Aminnejad B (2022) A review of downscaling methods of satellite-based precipitation estimates. Earth Sci Inf 15(1):1–20

Abraham M, Pingali P (2020) Transforming smallholder agriculture to achieve the SDGs. In: The role of smallholder farms in food and nutrition security. Springer, Cham, pp 173–209

Ahmed M, Ahmad S (2020) Systems modeling. Springer, Singapore

Ahmed M, Ahmad S, Raza MA, Kumar U, Ansar M, Shah GA et al (2020a) Models calibration and evaluation. In: Systems modeling. Springer, Singapore, pp 151–178

Ahmed M, Raza MA, Hussain T (2020b) Dynamic modeling. In: Systems modeling. Springer, Singapore, pp 111–150

Akhavizadegan F, Ansarifar J, Wang L, Huber I, Archontoulis SV (2021) A time-dependent parameter estimation framework for crop modeling. Sci Rep 11(1):11437

Ambethgar AS, Rameshkumar A (2024) Biological control of pests in major tropical vegetable crops: a review. Agri Rev. https://doi.org/10.18805/ag.R-2635

Anar MJ, Lin Z, Hoogenboom G, Shelia V, Batchelor WD, Teboh JM et al (2019) Modeling growth, development and yield of Sugarbeet using DSSAT. Agric Syst 169:58–70

Arora NK (2019) Impact of climate change on agriculture production and its sustainable solutions. Environ Sustain 2(2):95–96

Arora NK, Mishra I (2022) Current scenario and future directions for sustainable development goal 2: a roadmap to zero hunger. Environ Sustain 5(2):129–133

Aryal JP, Rahut DB, Maharjan S, Erenstein O (eds) (2018) Factors affecting the adoption of multiple climate-smart agricultural practices in the Indo-Gangetic Plains of India. Natural Resources Forum. Wiley Online Library

Awais M, Li W, Hussain S, Cheema MJM, Li W, Song R et al (2022) Comparative evaluation of land surface temperature images from unmanned aerial vehicle and satellite observation for agricultural areas using in situ data. Agriculture 12(2):184

Ayanlade A, Radeny M, Morton JF, Muchaba T (2018) Rainfall variability and drought characteristics in two agro-climatic zones: an assessment of climate change challenges in Africa. Sci Total Environ 630:728–737

Ayyam V, Palanivel S, Chandrakasan S, Ayyam V, Palanivel S, Chandrakasan S (2019) Approaches in land degradation management for productivity enhancement. In: Coastal ecosystems of the tropics-adaptive management. Springer, Singapore, pp 463–491

Bagheri A, Bondori A, Damalas CA (2019) Modeling cereal farmers' intended and actual adoption of integrated crop management (ICM) practices. J Rural Stud 70:58–65

Baker RE, Mahmud AS, Miller IF, Rajeev M, Rasambainarivo F, Rice BL et al (2022) Infectious disease in an era of global change. Nat Rev Microbiol 20(4):193–205

Basso B, Liu L (2019) Seasonal crop yield forecast: methods, applications, and accuracies. Adv Agron 154:201–255

Bathiany S, Dakos V, Scheffer M, Lenton TM (2018) Climate models predict increasing temperature variability in poor countries. Sci Adv 4(5):eaar5809
Behailu G, Ayal DY, Zeleke TT, Ture K, Bantider A (2021) Comparative analysis of meteorological records of climate variability and farmers' perceptions in sekota woreda, Ethiopia. Clim Serv 23:100239
Bellmann C (2019) Subsidies and sustainable agriculture: mapping the policy landscape. London, Chatham House
Berger K, Machwitz M, Kycko M, Kefauver SC, Van Wittenberghe S, Gerhards M et al (2022) Multi-sensor spectral synergies for crop stress detection and monitoring in the optical domain: a review. Remote Sens Environ 280:113198
Bhatla SC, Lal MA (2023) Plant physiology, development and metabolism. Springer Nature
Bhattacharya A (2022) Effect of low-temperature stress on germination, growth, and phenology of plants: a review. In: Physiological processes in plants under low temperature stress. Springer, Singapore, pp 1–106
Bisbis MB, Gruda N, Blanke M (2018) Potential impacts of climate change on vegetable production and product quality–a review. J Clean Prod 170:1602–1620
Boote KJ, Jones JW, Hoogenboom G (2018) Simulation of crop growth: CROPGRO model. In: Agricultural systems modeling and simulation. CRC Press, pp 651–692
Brouziyne Y, Abouabdillah A, Hirich A, Bouabid R, Zaaboul R, Benaabidate L (2018) Modeling sustainable adaptation strategies toward a climate-smart agriculture in a Mediterranean watershed under projected climate change scenarios. Agric Syst 162:154–163
Brown JN, Hochman Z, Holzworth D, Horan H (2018) Seasonal climate forecasts provide more definitive and accurate crop yield predictions. Agric For Meteorol 260:247–254
Burdett H, Wellen C (2022) Statistical and machine learning methods for crop yield prediction in the context of precision agriculture. Precis Agric 23(5):1553–1574
Caminade C, McIntyre KM, Jones AE (2019) Impact of recent and future climate change on vector-borne diseases. Ann N Y Acad Sci 1436(1):157–173
Carpentier CL, Braun H (2020) Agenda 2030 for sustainable development: a powerful global framework. J Int Counc Small Bus 1(1):14–23
Chahal I, Hooker D, Deen B, Janovicek K, Van Eerd L (2021) Long-term effects of crop rotation, tillage, and fertilizer nitrogen on soil health indicators and crop productivity in a temperate climate. Soil Tillage Res 213:105121
Challinor AJ, Müller C, Asseng S, Deva C, Nicklin KJ, Wallach D et al (2018) Improving the use of crop models for risk assessment and climate change adaptation. Agric Syst 159:296–306
Chaudhry S, Sidhu GPS (2022) Climate change regulated abiotic stress mechanisms in plants: a comprehensive review. Plant Cell Rep 41(1):1–31
Chokkavarapu N, Mandla VR (2019) Comparative study of GCMs, RCMs, downscaling and hydrological models: a review toward future climate change impact estimation. SN Appl Sci 1(12):1698
Cichota R, Vogeler I, Sharp J, Verburg K, Huth N, Holzworth D et al (2021) A protocol to build soil descriptions for APSIM simulations. MethodsX 8:101566
Clarke B, Otto F, Stuart-Smith R, Harrington L (2022) Extreme weather impacts of climate change: an attribution perspective. Environ Res Clim 1(1):012001
Colston JM, Ahmed T, Mahopo C, Kang G, Kosek M, de Sousa JF et al (2018) Evaluating meteorological data from weather stations, and from satellites and global models for a multi-site epidemiological study. Environ Res 165:91–109
Cooper M, Tang T, Gho C, Hart T, Hammer G, Messina C (2020) Integrating genetic gain and gap analysis to predict improvements in crop productivity. Crop Sci 60(2):582–604
Corbeels M, Berre D, Rusinamhodzi L, Lopez-Ridaura S (2018) Can we use crop modelling for identifying climate change adaptation options? Agric For Meteorol 256:46–52
Costa JM, Marques da Silva J, Pinheiro C, Barón M, Mylona P, Centritto M et al (2019) Opportunities and limitations of crop phenotyping in southern European countries. Front Plant Sci 10:1125

Courault D, Hossard L, Demarez V, Dechatre H, Irfan K, Baghdadi N et al (2021) STICS crop model and Sentinel-2 images for monitoring rice growth and yield in the Camargue region. Agron Sustain Dev 41:1–17
De Wit A, Boogaard H, Fumagalli D, Janssen S, Knapen R, van Kraalingen D et al (2019) 25 years of the WOFOST cropping systems model. Agric Syst 168:154–167
Delmar F (2019) Measuring growth: methodological considerations and empirical results. Routledge, Entrepreneurship and SME research, pp 199–215
Dhyani S, Sood R (2023) Modelling tools and plausible scenarios in science-policy to improve evidence-based decision-making for human well-being. In: Ecosystem and species habitat modeling for conservation and restoration. Springer, Singapore, pp 1–17
Dietz KJ, Zörb C, Geilfus CM (2021) Drought and crop yield. Plant Biol 23(6):881–893
Driesen E, Van den Ende W, De Proft M, Saeys W (2020) Influence of environmental factors light, CO_2, temperature, and relative humidity on stomatal opening and development: a review. Agronomy 10(12):1975
Droutsas I, Challinor AJ, Deva CR, Wang E (2022) Integration of machine learning into process-based modelling to improve simulation of complex crop responses. in silico Plant 4(2):diac017
Dusenge ME, Duarte AG, Way DA (2019) Plant carbon metabolism and climate change: elevated CO_2 and temperature impacts on photosynthesis, photorespiration and respiration. New Phytol 221(1):32–49
El Mokhtar MA, Anli M, Laouane RB, Boutasknit A, Boutaj H, Draoui A et al (2022) Food security and climate change. In: Research anthology on environmental and societal impacts of climate change. IGI Global, pp 44–63
Elavarasan D, Vincent DR, Sharma V, Zomaya AY, Srinivasan K (2018) Forecasting yield by integrating agrarian factors and machine learning models: a survey. Comput Electron Agric 155:257–282
El-Husseini M (2018) Impact of climate change induced by global weather engineering technology of "chemtrails" on plant protection. Arab J Plant Protect 36:80–85
Evers JB, Letort V, Renton M, Kang M (2018) Computational botany: advancing plant science through functional–structural plant modelling. Oxford University Press, Oxford, pp 767–772
Feleke HG, Savage M, Tesfaye K (2021) Calibration and validation of APSIM–Maize, DSSAT CERES–Maize and AquaCrop models for Ethiopian tropical environments. S Afr J Plant Soil 38(1):36–51
Filippi P, Jones EJ, Wimalathunge NS, Somarathna PD, Pozza LE, Ugbaje SU et al (2019) An approach to forecast grain crop yield using multi-layered, multi-farm data sets and machine learning. Precis Agric 20:1015–1029
Fowler MD, Kooperman GJ, Randerson JT, Pritchard MS (2019) The effect of plant physiological responses to rising CO_2 on global streamflow. Nat Clim Chang 9(11):873–879
GaneshamurthyA A, Kalaivanan D, Rupa T, Raghupathi H (2020) Groundwater decline and prolonged drought could reduce vigour, enhance vulnerability to diseases and pests and kill perennial horticultural crops: needs urgent policy intervention. J Horticult Sci 15(1):9–16
Gao F, Zhang X (2021) Mapping crop phenology in near real-time using satellite remote sensing: challenges and opportunities. J Remote Sens
Gao Y, Wallach D, Liu B, Dingkuhn M, Boote KJ, Singh U et al (2020) Comparison of three calibration methods for modeling rice phenology. Agric For Meteorol 280:107785
Gao F, Luan X, Yin Y, Sun S, Li Y, Mo F et al (2022) Exploring long-term impacts of different crop rotation systems on sustainable use of groundwater resources using DSSAT model. J Clean Prod 336:130377
Goyal G, Singh A, Kumar A, Kumar S, Gupta NK (2018) Impact of climate change in Indian agricultural context: a review article. IJCS 6(6):280–283
Gupta D, Gujre N, Singha S, Mitra S (2022) Role of existing and emerging technologies in advancing climate-smart agriculture through modeling: a review. Eco Inform 71:101805
Hanley ME, Bouma TJ, Mossman HL (2020) The gathering storm: optimizing management of coastal ecosystems in the face of a climate-driven threat. Ann Bot 125(2):197–212
Hanson H (2021) Wheat in the third world. Routledge

Hassan MU, Chattha MU, Khan I, Chattha MB, Barbanti L, Aamer M et al (2021) Heat stress in cultivated plants: nature, impact, mechanisms, and mitigation strategies—a review. Plant Biosyst Int J Deal All Aspect Plant Biol 155(2):211–234

Holzworth D, Huth NI, Fainges J, Brown H, Zurcher E, Cichota R et al (2018) APSIM next generation: overcoming challenges in modernising a farming systems model. Environ Model Softw 103:43–51

Hoogenboom G, Porter CH, Boote KJ, Shelia V, Wilkens PW, Singh U et al (2019) The DSSAT crop modeling ecosystem. In: Advances in crop modelling for a sustainable agriculture. Burleigh Dodds Science Publishing, pp 173–216

Hou D, Bolan NS, Tsang DC, Kirkham MB, O'connor D (2020) Sustainable soil use and management: an interdisciplinary and systematic approach. Sci Total Environ 729:138961

Huang J, Gómez-Dans JL, Huang H, Ma H, Wu Q, Lewis PE et al (2019) Assimilation of remote sensing into crop growth models: current status and perspectives. Agric For Meteorol 276:107609

Huffman GJ, Bolvin DT, Braithwaite D, Hsu K, Joyce R, Kidd C et al. (2018) NASA global precipitation measurement (GPM) integrated multi-satellite retrievals for GPM (IMERG). Algorithm Theoretical Basis Document (ATBD) Version, p 4

Iizumi T, Shin Y, Kim W, Kim M, Choi J (2018) Global crop yield forecasting using seasonal climate information from a multi-model ensemble. Clim Serv 11:13–23

Jagadish SK (2020) Heat stress during flowering in cereals–effects and adaptation strategies. New Phytol 226(6):1567–1572

Jägermeyr J, Müller C, Ruane AC, Elliott J, Balkovic J, Castillo O et al (2021) Climate impacts on global agriculture emerge earlier in new generation of climate and crop models. Nat Food 2(11):873–885

Jarquin D, De Leon N, Romay C, Bohn M, Buckler ES, Ciampitti I et al (2021) Utility of climatic information via combining ability models to improve genomic prediction for yield within the genomes to fields maize project. Front Genet 11:592769

Javaid M, Haleem A, Khan IH, Suman R (2023) Understanding the potential applications of artificial intelligence in agriculture sector. Adv Agrochem 2(1):15–30

Jayawardena DM, Heckathorn SA, Bista DR, Boldt JK (2019) Elevated carbon dioxide plus chronic warming causes dramatic increases in leaf angle in tomato, which correlates with reduced plant growth. Plant Cell Environ 42(4):1247–1256

Jin X, Kumar L, Li Z, Feng H, Xu X, Yang G et al (2018) A review of data assimilation of remote sensing and crop models. Eur J Agron 92:141–152

Jones MW, Peters GP, Gasser T, Andrew RM, Schwingshackl C, Gütschow J et al (2023) National contributions to climate change due to historical emissions of carbon dioxide, methane, and nitrous oxide since 1850. Sci Data 10(1):155

Kang Y, Özdoğan M (2019) Field-level crop yield mapping with Landsat using a hierarchical data assimilation approach. Remote Sens Environ 228:144–163

Keating BA (2024) APSIM's origins and the forces shaping its first 30 years of evolution: a review and reflections. Agron Sustain Dev 44(3):24

Keating BA, Thorburn PJ (2018) Modelling crops and cropping systems—evolving purpose, practice and prospects. Eur J Agron 100:163–176

Khan SM, Ali S, Nawaz A, Bukhari SAH, Ejaz S, Ahmad S (2019) Integrated pest and disease management for better agronomic crop production. In: Agronomic crops: Volume 2: management practices. Springer, Singapore, pp 385–428

Kivi MS, Blakely B, Masters M, Bernacchi CJ, Miguez FE, Dokoohaki H (2022) Development of a data-assimilation system to forecast agricultural systems: a case study of constraining soil water and soil nitrogen dynamics in the APSIM model. Sci Total Environ 820:153192

Kogan F (2019) Remote sensing for food security. Springer

Koutsoyiannis D (2021) Rethinking climate, climate change, and their relationship with water. Water 13(6):849

Kukal MS, Irmak S (2018) Climate-driven crop yield and yield variability and climate change impacts on the US Great Plains agricultural production. Sci Rep 8(1):1–18

Kumar V, Ranjan D, Verma K (2021) Global climate change: the loop between cause and impact. In: Global climate change. Elsevier, pp 187–211

Lambers H, Oliveira RS, Lambers H, Oliveira RS (2019) Photosynthesis, respiration, and long-distance transport: photosynthesis. In: Plant physiological ecology. Springer, Cham, pp 11–114

Lambert J, Sekhar NU, Chatrchyan A, DeGaetano A (2019) Agricultural decision support tools: a comparative perspective on these climate services. In: Sustainable solutions for food security: combating climate change by adaptation. Springer, Cham, pp 459–478

Langridge P, Reynolds M (2021) Breeding for drought and heat tolerance in wheat. Theor Appl Genet 134:1753–1769

Laurent C, Oger B, Taylor JA, Scholasch T, Metay A, Tisseyre B (2021) A review of the issues, methods and perspectives for yield estimation, prediction and forecasting in viticulture. Eur J Agron 130:126339

Laux P, Rötter RP, Webber H, Dieng D, Rahimi J, Wei J et al (2021) To bias correct or not to bias correct? An agricultural impact modelers' perspective on regional climate model data. Agric For Meteorol 304:108406

Legris M, Ince YÇ, Fankhauser C (2019) Molecular mechanisms underlying phytochrome-controlled morphogenesis in plants. Nat Commun 10(1):5219

Lewis E, Pritchard D, Villalobos-Herrera R, Blenkinsop S, McClean F, Guerreiro S et al (2021) Quality control of a global hourly rainfall dataset. Environ Model Softw 144:105169

Li S, Juhász-Horváth L, Pintér L, Rounsevell MD, Harrison PA (2018) Modelling regional cropping patterns under scenarios of climate and socio-economic change in Hungary. Sci Total Environ 622:1611–1620

Ling X, Huang Y, Guo W, Wang Y, Chen C, Qiu B et al (2021) Comprehensive evaluation of satellite-based and reanalysis soil moisture products using in situ observations over China. Hydrol Earth Syst Sci 25(7):4209–4229

Liu Y, Song W (2020) Modelling crop yield, water consumption, and water use efficiency for sustainable agroecosystem management. J Clean Prod 253:119940

Lohani N, Singh MB, Bhalla PL (2022) Short-term heat stress during flowering results in a decline in canola seed productivity. J Agron Crop Sci 208(4):486–496

Lu S, Bai X, Li W, Wang N (2019) Impacts of climate change on water resources and grain production. Technol Forecast Soc Chang 143:76–84

Maestrini B, Mimić G, van Oort PA, Jindo K, Brdar S, Athanasiadis IN et al (2022) Mixing process-based and data-driven approaches in yield prediction. Eur J Agron 139:126569

Manivasagam V, Rozenstein O (2020) Practices for upscaling crop simulation models from field scale to large regions. Comput Electron Agric 175:105554

Mehmood MZ, Afzal O, Aslam MA, Riaz H, Raza MA, Ahmed S et al (2020) Disease modeling as a tool to assess the impacts of climate variability on plant diseases and health. In: Systems modeling. Springer, Singapore, pp 327–351

Mendez M, Maathuis B, Hein-Griggs D, Alvarado-Gamboa L-F (2020) Performance evaluation of bias correction methods for climate change monthly precipitation projections over Costa Rica. Water 12(2):482

Misra V, Ansari MI (2022) Water logging tolerance and crop productivity. In: Augmenting crop productivity in stress environment. Springer, pp 161–175

Mitra R, McGough SF, Chakraborti T, Holmes C, Copping R, Hagenbuch N et al (2023) Learning from data with structured missingness. Nat Mach Intel 5(1):13–23

Monaco S (2024) SDG 2. End hunger, achieve food security and improved nutrition and promote sustainable agriculture. In: Identity, territories, and sustainability: challenges and opportunities for achieving the UN sustainable development goals. Emerald Publishing Limited, pp 25–32

Mulla S, Singh SK, Singh K, Praveen B (2020) Climate change and agriculture: a review of crop models. In: Global climate change and environmental policy: agriculture perspectives. Springer, Singapore, pp 423–435

Mwamahonje A, Eleblu JSY, Ofori K, Feyissa T, Deshpande S, Garcia-Oliveira AL et al (2021) Introgression of QTLs for drought tolerance into farmers' preferred sorghum varieties. Agriculture 11(9):883

Naz S, Fatima Z, Iqbal P, Khan A, Zakir I, Ullah H et al (2022) An introduction to climate change phenomenon. In: Building climate resilience in agriculture: theory, practice and future perspective. Springer, Cham, pp 3–16

Neumann A, Dong F, Shimoda Y, Arnillas CA, Javed A, Yang C et al (2021) A review of the current state of process-based and data-driven modelling: guidelines for Lake Erie managers and watershed modellers. Environ Rev 29(4):443–490

Nieder R, Benbi DK, Reichl FX, Nieder R, Benbi DK, Reichl FX (2018) Soil as a transmitter of human pathogens. In: Soil components and human health. Springer, Dordrecht, pp 723–827

Pasqualotto N, D'Urso G, Bolognesi SF, Belfiore OR, Van Wittenberghe S, Delegido J et al (2019) Retrieval of evapotranspiration from sentinel-2: comparison of vegetation indices, semi-empirical models and SNAP biophysical processor approach. Agronomy 9(10):663

Pasquel D, Roux S, Richetti J, Cammarano D, Tisseyre B, Taylor JA (2022) A review of methods to evaluate crop model performance at multiple and changing spatial scales. Precis Agric 23(4):1489–1513

Pastore MA, Lee TD, Hobbie SE, Reich PB (2020) Interactive effects of elevated CO_2, warming, reduced rainfall, and nitrogen on leaf gas exchange in five perennial grassland species. Plant Cell Environ 43(8):1862–1878

Paudel D, Dhakal S, Parajuli S, Adhikari L, Peng Z, Qian Y et al (2020) Use of quantitative trait loci to develop stress tolerance in plants. Elsevier, Plant life under changing environment, pp 917–965

Pei F, Wu C, Liu X, Li X, Yang K, Zhou Y et al (2018) Monitoring the vegetation activity in China using vegetation health indices. Agric For Meteorol 248:215–227

Pendrill F, Persson UM, Godar J, Kastner T, Moran D, Schmidt S et al (2019) Agricultural and forestry trade drives large share of tropical deforestation emissions. Glob Environ Chang 56:1–10

Peng B, Guan K, Tang J, Ainsworth EA, Asseng S, Bernacchi CJ et al (2020) Towards a multiscale crop modelling framework for climate change adaptation assessment. Nat Plant 6(4):338–348

Peterson CA, Bell LW, Carvalho PCF, Gaudin AC (2020) Resilience of an integrated crop–livestock system to climate change: a simulation analysis of cover crop grazing in southern Brazil. Front Sustain Food Syst 4:604099

Piao S, Liu Q, Chen A, Janssens IA, Fu Y, Dai J et al (2019) Plant phenology and global climate change: current progresses and challenges. Glob Chang Biol 25(6):1922–1940

Potgieter AB, Zhao Y, Zarco-Tejada PJ, Chenu K, Zhang Y, Porker K et al (2021) Evolution and application of digital technologies to predict crop type and crop phenology in agriculture. in silico Plant 3(1):diab017

Pouyat RV, Day SD, Brown S, Schwarz K, Shaw RE, Szlavecz K et al (2020) Urban soils. In: Forest and rangeland soils of the United States under changing conditions: a comprehensive science synthesis. Springer, Cham, pp 127–144

Prince AI, Nzechie O, Obiorah J, Ehi OE, Idakwoji AA (2023) Analyzing the critical impact of climate change on agriculture and food security in Nigeria. Int J Agric Earth Sci 9:42023

Rahn E, Vaast P, Läderach P, Van Asten P, Jassogne L, Ghazoul J (2018) Exploring adaptation strategies of coffee production to climate change using a process-based model. Ecol Model 371:76–89

Rawat J, Saxena J, Sanwal P (2019) Biochar: a sustainable approach for improving plant growth and soil properties. In: Biochar-an imperative amendment for soil and the environment. IntechOpen, pp 1–17

Rawe T, Antonelli M, Chatrchyan AM, Clayton T, Fanzo J, Gonsalves JF et al. (2019) Transforming food systems under climate change: local to global policy as a catalyst for change

Raymundo R, Sexton-Bowser S, Ciampitti IA, Morris GP (2021) Crop modeling defines opportunities and challenges for drought escape, water capture, and yield increase using chilling-tolerant sorghum. Plant Direct 5(9):e349

Raza A, Razzaq A, Mehmood SS, Zou X, Zhang X, Lv Y et al (2019) Impact of climate change on crops adaptation and strategies to tackle its outcome: a review. Plan Theory 8(2):34

Razavi S, Jakeman A, Saltelli A, Prieur C, Iooss B, Borgonovo E et al (2021) The future of sensitivity analysis: an essential discipline for systems modeling and policy support. Environ Model Softw 137:104954
Rienth M, Vigneron N, Darriet P, Sweetman C, Burbidge C, Bonghi C et al (2021) Grape berry secondary metabolites and their modulation by abiotic factors in a climate change scenario–a review. Front Plant Sci 12:643258
Rodriguez D, De Voil P, Hudson D, Brown J, Hayman P, Marrou H et al (2018) Predicting optimum crop designs using crop models and seasonal climate forecasts. Sci Rep 8(1):2231
Romeh AA (2019) Integrated pest management for sustainable agriculture. In: Sustainability of agricultural environment in Egypt: Part II: Soil-water-plant nexus. Springer, Cham, pp 215–234
Rötter RP, Hoffmann M, Koch M, Müller C (2018) Progress in modelling agricultural impacts of and adaptations to climate change. Curr Opin Plant Biol 45:255–261
Sakamoto T (2018) Refined shape model fitting methods for detecting various types of phenological information on major US crops. ISPRS J Photogramm Remote Sens 138:176–192
Savary S, Nelson AD, Djurle A, Esker PD, Sparks A, Amorim L et al (2018) Concepts, approaches, and avenues for modelling crop health and crop losses. Eur J Agron 100:4–18
Schauberger B, Jägermeyr J, Gornott C (2020) A systematic review of local to regional yield forecasting approaches and frequently used data resources. Eur J Agron 120:126153
Sehgal A, Sita K, Siddique KH, Kumar R, Bhogireddy S, Varshney RK et al (2018) Drought or/and heat-stress effects on seed filling in food crops: impacts on functional biochemistry, seed yields, and nutritional quality. Front Plant Sci 9:1705
Seidel SJ, Palosuo T, Thorburn P, Wallach D (2018) Towards improved calibration of crop models–where are we now and where should we go? Eur J Agron 94:25–35
Seneviratne SI, Zhang X, Adnan M, Badi W, Dereczynski C, Luca AD et al (2021) Weather and climate extreme events in a changing climate. In: Climate change 2021 – the physical science basis. Cambridge University Press
Shah F, Wu W (2019) Soil and crop management strategies to ensure higher crop productivity within sustainable environments. Sustain For 11(5):1485
Sharifzadeh MS, Abdollahzadeh G, Damalas CA, Rezaei R (2018) Farmers' criteria for pesticide selection and use in the pest control process. Agriculture 8(2):24
Sharma K, Salik F, Rather SA (2022) Responses and adaptation of photosynthesis and respiration under challenging environments. In: Plant abiotic stress physiology. Apple Academic Press, pp 119–160
Shelia V, Hansen J, Sharda V, Porter C, Aggarwal P, Wilkerson CJ et al (2019) A multi-scale and multi-model gridded framework for forecasting crop production, risk analysis, and climate change impact studies. Environ Model Softw 115:144–154
Shi W, Wang M, Liu Y (2021) Crop yield and production responses to climate disasters in China. Sci Total Environ 750:141147
Shirley R, Pope E, Bartlett M, Oliver S, Quadrianto N, Hurley P et al (2020) An empirical, Bayesian approach to modelling crop yield: maize in USA. Environ Res Commun 2(2):025002
Silva JV, Giller KE (2020) Grand challenges for the 21st century: what crop models can and can't (yet) do. J Agric Sci 158(10):794–805
Stöckle CO, Kemanian AR (2020) Can crop models identify critical gaps in genetics, environment, and management interactions? Front Plant Sci 11:737
Sudmanns M, Tiede D, Lang S, Bergstedt H, Trost G, Augustin H et al (2020) Big Earth data: disruptive changes in Earth observation data management and analysis? Int J Digit Earth 13(7):832–850
Sun Y, Yin Q, Crucifix M, Clemens SC, Araya-Melo P, Liu W et al (2019) Diverse manifestations of the mid-Pleistocene climate transition. Nat Commun 10(1):352
Swarup S, Cargill EJ, Crosby K, Flagel L, Kniskern J, Glenn KC (2021) Genetic diversity is indispensable for plant breeding to improve crops. Crop Sci 61(2):839–852
Talbot M-H, Monfet D (2024) Development of a crop growth model for the energy analysis of controlled agriculture environment spaces. Biosyst Eng 238:38–50

Tapiador FJ, Navarro A, Moreno R, Sánchez JL, García-Ortega E (2020) Regional climate models: 30 years of dynamical downscaling. Atmos Res 235:104785

Tian H, Wang T, Liu Y, Qiao X, Li Y (2020) Computer vision technology in agricultural automation—a review. Info Process Agri 7(1):1–19

Tian Z, Wang JW, Li J, Han B (2021) Designing future crops: challenges and strategies for sustainable agriculture. Plant J 105(5):1165–1178

Triana JSA, Chu ML, Guzman JA, Moriasi DN, Steiner JL (2019) Beyond model metrics: the perils of calibrating hydrologic models. J Hydrol 578:124032

Tribouillois H, Constantin J, Justes E (2018) Analysis and modeling of cover crop emergence: accuracy of a static model and the dynamic STICS soil-crop model. Eur J Agron 93:73–81

Trivedi A, Verma NS, Mishra S, Nandeha N, Gautam VK (2013) Crop modelling: a tool for agricultural research

Troin M, Arsenault R, Wood AW, Brissette F, Martel JL (2021) Generating ensemble streamflow forecasts: a review of methods and approaches over the past 40 years. Wiley Online Library

ur Rahman MH, Ahmad A, Wang X, Wajid A, Nasim W, Hussain M et al (2018) Multi-model projections of future climate and climate change impacts uncertainty assessment for cotton production in Pakistan. Agric For Meteorol 253:94–113

Velásquez AC, Castroverde CDM, He SY (2018) Plant–pathogen warfare under changing climate conditions. Curr Biol 28(10):R619–RR34

Velten K, Schmidt DM, Kahlen K (2024) Mathematical modeling and simulation: introduction for scientists and engineers. Wiley

Vogel H-J, Bartke S, Daedlow K, Helming K, Kögel-Knabner I, Lang B et al (2018) A systemic approach for modeling soil functions. Soil 4(1):83–92

Wallach D, Palosuo T, Thorburn P, Hochman Z, Gourdain E, Andrianasolo F et al (2021) The chaos in calibrating crop models: lessons learned from a multi-model calibration exercise. Environ Model Softw 145:105206

Walter J (2018) Effects of changes in soil moisture and precipitation patterns on plant-mediated biotic interactions in terrestrial ecosystems. Plant Ecol 219(12):1449–1462

Wang Y, Tao H, Tian B, Sheng D, Xu C, Zhou H et al (2019) Flowering dynamics, pollen, and pistil contribution to grain yield in response to high temperature during maize flowering. Environ Exp Bot 158:80–88

Webber H, Hoffmann M, Rezaei EE (2020) Crop models as tools for agroclimatology. In: Agroclimatology: linking agriculture to climate, vol 60. American Society of Agronomy, pp 519–546

Wiréhn L (2018) Nordic agriculture under climate change: a systematic review of challenges, opportunities and adaptation strategies for crop production. Land Use Policy 77:63–74

Yang X, Beyer K (2018) Propagation of climate model biases to biophysical modelling can complicate assessments of climate change impact in agricultural systems. Int J Climatol 39(1):424–444

Yang X, Zheng L, Yang Q, Wang Z, Cui S, Shen Y (2018) Modelling the effects of conservation tillage on crop water productivity, soil water dynamics and evapotranspiration of a maize-winter wheat-soybean rotation system on the Loess Plateau of China using APSIM. Agric Syst 166:111–123

Yao Y, Fu B, Liu Y, Wang Y, Song S (2021) The contribution of ecosystem restoration to sustainable development goals in Asian drylands: a literature review. Land Degrad Dev 32(16):4472–4483

Yin X, Struik PC, Goudriaan J (2021) On the needs for combining physiological principles and mathematics to improve crop models. Field Crop Res 271:108254

Yue Q, Guo P, Wu H, Wang Y, Zhang C (2022) Towards sustainable circular agriculture: an integrated optimization framework for crop-livestock-biogas-crop recycling system management under uncertainty. Agric Syst 196:103347

Zhang J, Huang Y, Pu R, Gonzalez-Moreno P, Yuan L, Wu K et al (2019) Monitoring plant diseases and pests through remote sensing technology: a review. Comput Electron Agric 165:104943

Zhang W, Sheng J, Li Z, Weindorf DC, Hu G, Xuan J et al (2021) Integrating rainwater harvesting and drip irrigation for water use efficiency improvements in apple orchards of northwest China. Sci Hortic 275:109728

Zhao C, Liu B, Xiao L, Hoogenboom G, Boote KJ, Kassie BT et al (2019) A SIMPLE crop model. Eur J Agron 104:97–106

Zhu X, Liu W, Chen J, Bruijnzeel LA, Mao Z, Yang X et al (2020) Reductions in water, soil and nutrient losses and pesticide pollution in agroforestry practices: a review of evidence and processes. Plant Soil 453:45–86

Zhu T, Fonseca De Lima CF, De Smet I (2021) The heat is on: how crop growth, development, and yield respond to high temperature. J Exp Bot 72(21):7359–7373

Climate Change and Crops: Adaptation and Strategies to Tackle Its Outcome

6

Faiza Munir, Tayyaba Bint Tariq, Saba Azeem, Sarah Fatima, Javed Iqbal, Banzeer Ahsan Abbasi, Shoaib Khan, Syeda Anber Zahra, and Majid Hussain

Abstract

Climate change is a major cause of variations in the average earth's temperature, changes in precipitation patterns, and surges in the incidence of weather conditions that, in turn, interrupt agricultural productivity and sustainability. The undesirable influence of climate change affects plant growth and yield, leading to a serious challenge to food security all over the world. Crops are usually prone to multifaceted effects of climate change, such as weather extremes (posing heat or cold stress), and variations in the pattern of precipitation lead to water scarcity

Faiza Munir and Javed Iqbal contributed equally with all other contributors.

F. Munir (✉) · T. B. Tariq · S. Azeem · S. Fatima
Department of Agricultural Sciences and Technology, Atta-ur-Rahman School of Applied Biosciences, National University of Sciences and Technology, Islamabad, Pakistan
e-mail: faiza.munir@asab.nust.edu.pk

J. Iqbal (✉)
Department of Botany, Bacha Khan University, Charsadda, Khyber Pakhtunkhwa, Pakistan

B. A. Abbasi
Department of Botany, Rawalpindi Women University, Rawalpindi, Pakistan

S. Khan
Department of Chemistry, Abbottabad University of Science and Technology (AUST), Abbottabad, Pakistan

S. A. Zahra
School of Food and Pharmacy, Shanghai Zhongqiao Vocational and Technical University, Shanghai, China

M. Hussain
Department of Forestry and Wildlife Management, University of Haripur, Haripur City, KP, Pakistan
e-mail: majid@uoh.edu.pk

© The Author(s), under exclusive license to Springer Nature Singapore Pte Ltd. 2025

S. Faiz et al. (eds.), *Climate Smart Agriculture for Future Food Security*,
https://doi.org/10.1007/978-981-96-4499-5_6

or flooding stress that eventually harms crop growth and decreases yield. These interruptions together undermine the agriculture sector to safeguard food security. Plants have evolved inbuilt mechanisms to adapt to harsh conditions, including physiological, morphological, biochemical, and molecular modifications to some extent. Understanding of natural mechanisms opted for by plants in response to environmental stress can be employed to devise strategies to alleviate the consequences of climate change. To deal with the intimidations caused by climate change, several agricultural adaptation strategies are used, such as crop rotation, intercropping, organic soil amendments, growing tolerant varieties, and genetic manipulation. Furthermore, advanced genetic engineering (CRISPR/Cas9) and high-throughput techniques must be integrated in combination with traditional methods to decrease the undesirable effects of climate change and ensure sustainable agriculture.

Keywords

Climate change · Agricultural productivity · Crop yield · Genetic manipulation · Sustainable agriculture

6.1 Introduction

Climate change pertains to long-term variations in the average weather patterns, particularly due to undesirable human activities, including fossil energy utilization, deforestation, and industrial practices. Such activities enhance the level of greenhouse gases (GHGs) such as carbon dioxide (CO_2), methane (CH_4), and nitrous oxide (N_2O) in the atmosphere, hence global warming and other climate interferences. Hence, the importance of climate change in the theory is given by the fact that it affects almost all facets of natural ecosystems, human health, and socioeconomic systems. It acts as a threat to biological diversity and enhances the number of cases of natural calamities. The current state of climate change is alarming. According to the Intergovernmental Panel on Climate Change (IPCC), the global average temperature has already risen by approximately 1.1 °C above pre-industrial levels as of 2020. This warming has led to observable changes in weather patterns, a rise in sea level, and increased frequency and intensity of extreme weather events, such as hurricanes, heatwaves, droughts, and floods (Fig. 6.1).

6.1.1 Current State and Projections

Projections indicate that if GHG emissions continue at the current rate, the world is likely to experience a temperature increase of 1.5°–2 °C by the mid-twenty-first century. The IPCC's Special Report on Global Warming of 1.5 °C emphasizes that the increase in global temperature should not exceed 1.5 °C necessitating drastic, systemic, and systemic solutions in all spheres of life immediately. Otherwise, it will cause more harm and dangerous effects such as frequent hurricanes, significant

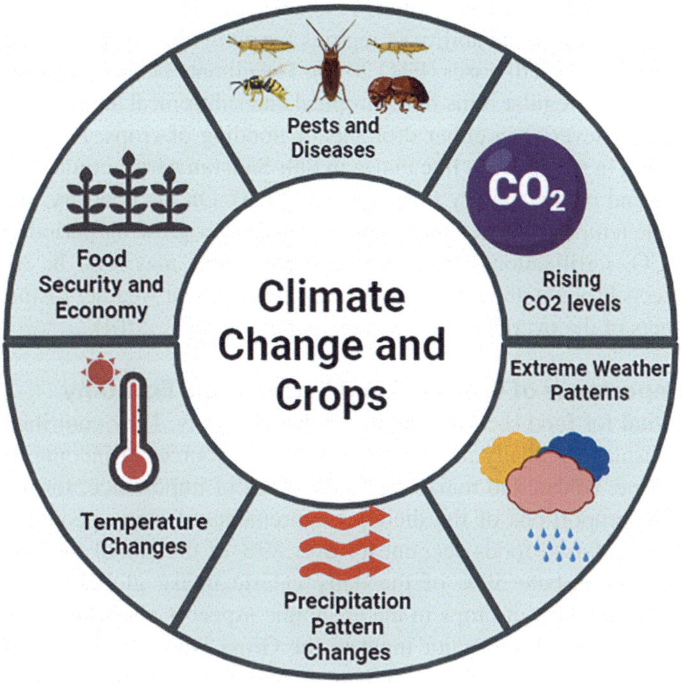

Fig. 6.1 Effect of climate change on crops

loss of species, decrease in food production, and low water provisions. Climate change measures can therefore be categorized into two broad categories of climate change measures: reduction measures and adaptation measures. The measures aimed at the reduction of GHG emissions are based on using renewable power sources, effective energy use, afforestation, and proper land utilization. Mitigation measures stand for the ability to enhance the effects of climate change on communities and ecosystems. These are breeding crops that can tolerate the effects of climate change, irrigation and water use efficiency, and disaster risk reduction and responding mechanisms. Thus, the definition, importance, contemporary state, and outlook for climate change should be known to create and modify measures that would help combat climate change and prepare for the consequences that cannot be prevented. The subsequent parts will elucidate in detail the effects of climate change on crops and how to make agriculture sustainable despite the prevailing unpredictable climate.

6.1.2 Impact on Agriculture

Climate change plays a vital role in altering agricultural situations worldwide as well as in the different continents. Some of the key climatic challenges that affect agriculture consist of an increase in temperature, changes in precipitation, and frequency of extreme climatic events. As stated by the FAO, such changes can decrease yields, change the ability of regions for particular crops, and enhance the occurrence of pests and diseases (FAO 2016). The climate factors, such as increased temperatures and irregular rains in the tropical and subtropical areas, affect production by causing severity in either droughts or flooding of crops. For example, the yield reduction in staple crops like maize in Sub-Saharan Africa could drop by up to 40% by the end of the century (World Bank 2013). On the contrary, some of the regions in the temperate zones may benefit from longer growing periods as well as increased CO_2 fertilizations, although these conditions may well be overlaid by adverse effects brought about by complicated and frequent weather disturbances or by bad effects of the overall increase in temperature (IPCC 2014).

6.1.2.1 Importance of Crops in Food Security and Economy

Crops are vital for food security and the global economy. They contribute most of the food consumed globally today and are essential sources of income for billions of people. Rice, wheat, and maize are staple crops of importance, for these grains contain large proportions of the dietary requirements of many people. FAO estimates that plant-based foods account for over 80% of the global food supply, and cereals alone contribute 50% of the daily calorie intake globally (FAO 2018). Therefore, the function of crops in the economic aspect is not limited to food production only. It provides a major input to the Gross Domestic Product (GDP) of many developing nations, offers employment, and is input to food processing, distribution, and selling. For instance, in India, it is a source of employment for around 42% of the population and contributes roughly 16% to the GDP (World Bank 2020). In the Sub-Saharan region, agriculture plays a significant role in economic growth and poverty reduction as over 60% of the workforce is into farming, and agriculture contributes about 23% to the regional GDP (FAO 2019). However, climate change affects the stability of the food and economy sectors and is considered a major threat to these sectors. Lower yields and higher variability in crop output can raise food costs and create higher levels of food vulnerability for the country's population, in particular for the population of the concerned region. According to the World Bank in 2015, it is suggested that around 130 million people will be victims of poverty by 2030 due to the effects of climate change on the food security and economic output of the world (World Bank 2015).

6.1.3 Effects of Climate Change on Crops

6.1.3.1 Temperature Changes

The increasing temperatures because of climate change are influencing the duration of the growing season around the world. The IPCC (2014) notes that average temperatures have increased by approximately 0.85 °C from 1880 to 2012, with further increases projected. This warming trend is influencing the phenology of crops, hence changing plants and periods of harvesting. Due to earlier springs in regions such as Europe and North America, the growing seasons have been extended, although this is accompanied by high risks of frost damage, which in turn affects crop production (Parry et al. 2005). However, in the tropical and subtropical regions of the world, this enhancement in the climate change impact has led to the shortening of growing seasons and elevated frequent drought occurrences. This reduction in growing season length can affect crop yield, especially for crops that are sensitive to heat, such as rice and wheat (Lobell et al. 2011). Furthermore, the factor of climate change affects the geographic distribution of crops differently; it can increase the area of certain crops' natural habitats and decrease the area of suitable climate for other crop species (Schlenker and Roberts 2009).

Rising temperatures are having a massive impact on cropping outcomes and overall productivity. High temperatures can cause heat stress that can impact the development stages, lead to reduced photosynthesis, and hinder floral/fruiting processes. Research has indicated that wheat and maize raise heat stress concerns as their yields are compromised even with small elevated temperatures. For example, Schlenker and Roberts (2009) showed that heat stress poses a huge risk to US wheat production as projected yield reductions range from a high of 70% by 2050 under the high emission scenarios, particularly during the flowering period. Further, it was noted that high temperatures are also bad for quality because they reduce nutrient content as well as increase pests and diseases. For instance, higher temperatures can also increase the growth rates of some fungi that cause diseases like wheat rust and rice blast; such a factor also decreases crop yield and quality. In addition, heat stress during the grain-filling stage decreases grain size and quality of cereal crops, thus decreasing market value and nutritional quality (Si et al. 2021; Ahmad et al. 2022).

6.1.3.2 Changes in Precipitation Patterns

Global warming is causing drastic changes in precipitation with more frequent and severe droughts in many parts of the world. Severities such as drought and water scarcity influence agriculture significantly, as water is an important production factor in crop farming. For instance, IPCC (2021a, b) states that climate models in ambient temperatures estimate reduced annual precipitation across arid and semi-arid regions, further enhancing water stress and drought adversity. Drought greatly affects crops in various ways such as reduced available water for plant uptake in the soil, lower levels of groundwater, and reduced water in the rivers that all work collectively to limit crop growth and productivity. For example, a study by Lesk et al. (2022) on the impacts of drought on the yield of different crops established that changing climatic conditions could lead to a decline in yields for major food crops, including maize and wheat. Also, water shortage can hamper the prospect of enhancing agricultural productivity, notably in areas where there is use of surface water and groundwater for irrigation needs, such as in the Indian sub-continent and China (Ahmad et al. 2024a). Drought stress negatively impacts plant metabolism and regular physiological functions, including photosynthesis, nutrient assimilation and uptake, and cellular metabolism. For instance, drought stress decreases nutrient contents and uptake in the soil, leading to crop stress and poor production (Farooq et al. 2009). Furthermore, very prolonged and severe conditions of drought negatively influence crop production, affecting food security and farmers' income, especially in developing countries where the rain-fed agriculture system is predominant (Schmidhuber and Tubiello 2021; Ahmad et al. 2024b).

On the other hand, climate change is also believed to result in frequent and intense cases of rainfall, resulting in flooding and waterlogging. These conditions are devastating to crops, the soil, and other forms of agricultural infrastructure and equipment. According to the IPCC (2021a, b), there has been an increase in the intensity and frequency of extreme rainfall, leading to increased risks to agriculture. One of the effects of flooding is that crops can be submerged in water; therefore, oxygen supply to the root system is limited, and root growth is severely affected. Stagnant water also results in the build-up of toxic nutrients as well as the promotion of anaerobic conditions, which are both detrimental to crop productivity (Pareek et al. 2020). For example, rice crops that are normally waterlogged are likely to be greatly affected by yield loss if they are submerged for a long time during various developmental stages (Colmer et al. 2014). Nevertheless, flooding causes the detrition of crops because of water accumulation, loss of fertile layers of the soil, and irregular planting seasons. For instance, the South Asian monsoon that results in heavy rains and flooding affects agricultural productivity by delaying planting activities and reducing crop yields (Shukla et al. 2019). The costs of floods and waterlogging can be significant, affecting not only farmers but also the whole agricultural food chain.

6.1.3.3 Increased Frequency of Extreme Weather Patterns

Storms, hurricanes, and heatwaves, among other severe conditions, are more frequent and severe due to climate change. These deviant events represent a major threat to the efficiency and survival of agriculture and crops in particular. For example, hurricanes directly affect crops and their physical environment through destruction and alteration of the physical properties of the land and infrastructure, hence exerting instantaneous input on negotiable farming productivity. Droughts make the air temperature rise, thus affecting the growing status of crops (Benitez-Alfonso et al. 2023a, b). Any extremes of climate affect crop preservation and crop productivity in several ways. Crops are affected by storms and hurricanes, which result in soil erosion, loss of topsoil, or overall deterioration of the quality of soil that is very important in producing crops (Corwin 2021). Also, direct contact with seawater from storm surges can saturate and eventually lead to making the fields saline, which would not support crop advancement and yields (Shinde Ganesh et al. 2023). While heat stress harms crops by causing high rates of transpiration, heat waves lower water use efficiency by the plants and reduce the rates of photosynthesis, thus decreasing crop yields (Yu et al. 2024). As a result, the challenges described above need to be approached with a focus on creating and implementing measures that strengthen the agricultural systems. This entails planting heat-tolerant crops, enhancing the regimes used to manage the soil or growing media, and implementing suitable water use systems to reduce the effects of climate change (Karri and Nalluri 2024).

6.1.3.4 Rising CO_2 Levels

The global increase of atmospheric CO_2 also affects photosynthesis, which is the conversion of light energy into chemical energy. Increased availability of CO_2 normally acts to increase photosynthesis in C3 plants, including rice, wheat, and soybeans, by increasing the rate of carbon fixation (Leung et al. 2022). This effect is called the CO_2 fertilization effect, and when all the other growing conditions are optimum, it leads to higher biomass and therefore potential crop yields (Ainsworth and Rogers 2007). However, the degree of this advantage depends on other conditions of the environment, including the amount of nutrients and water (KhokharVoytas et al. 2023). Although high levels of CO_2 are known to promote plant growth and productive efficiency through inducing photosynthesis, they affect crop nutrient composition and productivity. Research has revealed that the levels of nutrients in foods are lowered when they are grown under a high CO_2 environment with reduced content of nitrogen, zinc, and iron which are vital for health (Myers et al. 2014). This nutrient dilution effect can thus pose a threat to the quality of staple foods, nutritional security, and human health (Gojon et al. 2023). Moreover, the increase in growth arising from higher CO_2 levels may not be a tensorial increase in yield provided other factors like water and nutrients are not well regulated (Baslam et al. 2020).

6.1.3.5 Pests and Diseases

Global warming regarding its effect on temperature and precipitation can affect the location and density of pests. Higher temperatures can therefore influence the geographic distribution of many pests because it will bring into their favorable range areas that were previously uninhabitable (Deutsch et al. 2018). This shift can cause pest problems to crop up, as new areas get infested with the pests, or the previous areas get more problems with the pests. For instance, the movement of the fall armyworm pest that infests maize has shifted in terms of geographic range due to the temperatures that greened up because of the climate changes (Yan et al. 2022). Apart from changing pests' distribution and abundance, climate change has the potential to both increase the incidence and the impact of plant diseases. These conditions can lead to the proliferation of fungal pathogens and their effects, especially rusts and mildews that can lead to the destruction of crops (Bebber et al. 2014). Also, through altering the climate, the sequence and development of plant viral diseases and bacterial diseases may increase the susceptibility of crops to diseases (Krishna 2023). Consequently, the farmers might experience higher risks related to plant health management and crop yield performance under the prevailing and altered climate patterns (Chakraborty and Newton 2011).

6.2 Crop Adaptation Strategies

Agricultural sustainability and climate change are naturally interrelated in various ways. Climate change is the key factor causing biotic and abiotic stresses in agricultural regions. These strains include a change in the pattern of annual rainfall, an increase or decrease in average temperature, and the emergence of weeds, insects, and microbial communities (Raza et al. 2020). To deal with environmental stresses, plants have an intrinsic, highly variable, and complex mechanism including physiological, biochemical, molecular, and transcriptional responses. However, there is a limitation that plants naturally can only cope with stress up to a certain extent (Lavinia et al. 2022). The detrimental impacts of climate change on agriculture drive the attention of researchers to formulate adaptation strategies to mitigate the detrimental effects. Multifaceted strategies are required for adapting agriculture to climatic shifts that require the integration of agricultural modification, traditional approaches, and modern techniques (Janni et al. 2024).

6.2.1 Plant Adaptation Strategies to Climate Change

Crop adaptation entails the capacity of the plant to sustain typical physiological processes despite the environmental conditions. Plants have an inbuilt system to cope with environmental stresses, including abiotic and biotic stresses. The adaptations include stress signaling to activate molecular response, antioxidant defense system, osmotic adjustment, ion homeostasis, and morphological adaptation (Fig. 6.2).

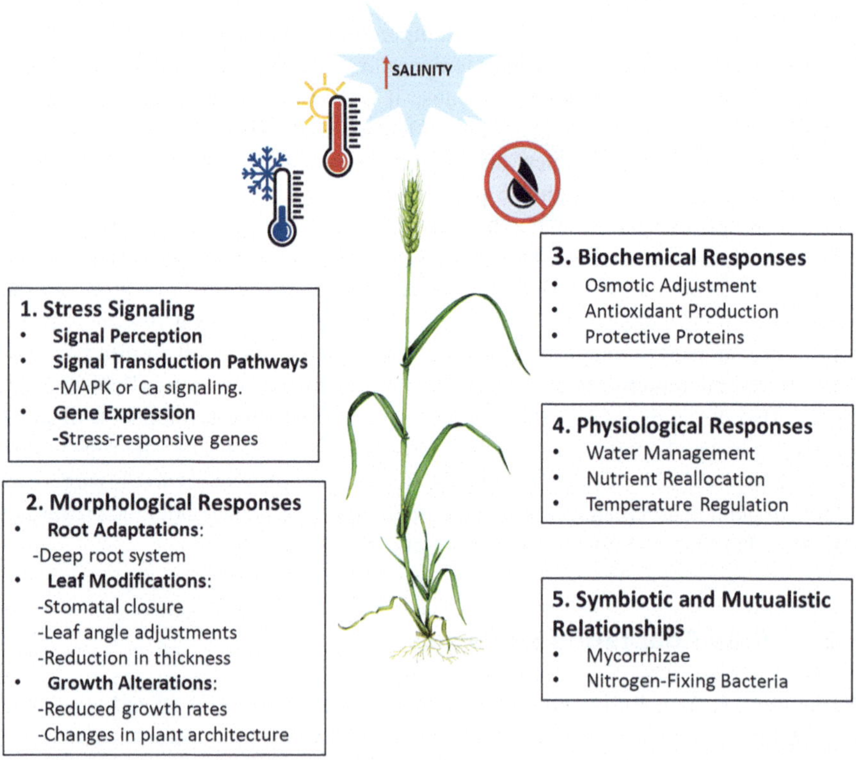

Fig. 6.2 Inherent adaptation approaches of plants to mitigate the effects of stress that occurred due to climatic change

Stress-inducing signals activate genome-wide reprogramming, initiating defensive mechanisms such as osmoregulation, ROS scavenging, and restoration of injury caused by stress. Multiple transcription factors, including DREBs, WRKY, MYB, and NAC, are actively involved in responding to environmental stresses. Upon activation, these transcription factors stimulate the downstream genes to produce a specific set of proteins to cope with stress. These stress-associated proteins are typically related to the MAPK cascade, calcium signaling, ROS production, and osmoprotectant proteins (Zhang et al. 2023a, b; Ullah et al. 2023a). Abiotic stresses usually trigger the ABA biosynthesis pathway, involving a phosphorylation-dependent signaling cascade. Among the different transcription factors, some are stress-specific factors. For instance, ABA-responsive cis-elements (ABRE) are in the promoter region of ABA-responsive genes, dehydration responsive elements (DRE) are in drought and cold-responsive genes, and C-repeat binding factor (CBF) is specifically related to low-temperature stress (Yamaguchi-Shinozaki and Shinozaki 2006).

Key Mechanisms in Plant Adaptation to Climate Change

Translational regulation is a critical mechanism for cellular response to various stress conditions. Translational regulation involves a 5′ ribosomal pause that leads to translational failure. Consequently, mRNA is degraded. Translational regulation is more pronounced under heat stress, and it is facilitated by extracellular cytoplasmic ribonuclease XRN4 and aided by RNA-binding protein LARP1. LARP1 is a high-temperature-specified cofactor of XRN4 to target polysomes. Under heat stress, the initiation process of translation is hindered, and the mRNA that encodes for ribosomal proteins (PR) is primarily deposited and sequestered in stress granules. In contrast, throughout the recovery phase, these mRNAs are released and translated swiftly to reinitiate the translation (Merret et al. 2017; Ullah et al. 2023b). The alterations in transcription and translation under stress conditions activate other defense mechanisms.

The molecular modifications facilitate the activation of several stress-associated genes and proteins that protect plants from harsh environmental conditions. In combination with these changes, a plant's antioxidant defense system is crucial in alleviating oxidative stress. The environmental stresses, including water-deficient conditions, salinity, and fluctuating temperatures (low/high), in plants badly impact plant growth, development, and overall production. Under stress conditions, the homeostasis between ROS production and their detoxification mechanism is disrupted. The degree of disruption depends upon the intensity and length of stress. To maintain the precise balance, an antioxidant defense system is activated, which involves the production of enzymatic and non-enzymatic antioxidants, where enzymatic antioxidants such as the production of superoxide dismutase (SOD), peroxidase (POD), catalase (CAT), ascorbate peroxidase (APX), and so on.

The non-enzymatic antioxidants like ascorbic acid, proline, anthocyanin, glutathione, etc. These antioxidants tend to scavenge the ROS to maintain cellular integrity (Mishra et al. 2023). SOD is known to catalyze the dismutation of O^{2-} alleviating its chance to form OH ion and converting it into hydrogen peroxide; thus, it acts as the first line of protection against ROS. The other enzymatic antioxidants, i.e., APX,

POD, and CAT, further neutralize the produced hydrogen peroxide. The POD and CAT regulate the levels of hydrogen peroxide in the cell, so they are important in fine-tuning the concentration of ROS. Catalases regulate the hydrogen peroxide by splitting it into oxygen and water. Ascorbic acid is a non-enzymatic antioxidant that is commonly found in subcellular compartments like the apoplast. It is known to directly seize ionic forms of reactive oxygen, including O^{2-} and OH^- ions, and split hydrogen peroxide into water (Dumanović et al. 2021). Proline is an amino acid that functions as an osmo protectant; it regulates the osmotic adjustment and detoxification of free radicals. The accumulation of proline is commonly associated with abiotic constraints. In general, all the enzymatic and non-enzymatic antioxidants safeguard cellular integrity by neutralizing free radicals that are generated in plants upon the onset of environmental stress (Dutta et al. 2018; Haider et al. 2021a, 2021b).

6.2.1.1 Morphological Adaptations via Transpiration, Stomatal Conductance, and Osmotic Adjustment

Morphological and physiological adaptations are another layer of a plant's inbuilt adaptation strategies to deal with challenges posed by climatic change. The morpho-physiological changes, together with molecular and biochemical adaptations, are vital to make plants resilient to harsh climatic conditions. Morphological changes in response to different stresses lead to reduced internode growth, small leaf size, reduced leaf area, change in branching pattern, and impaired roots and shoots. Sometimes, environmental stress, especially drought conditions, leads to root elongation. When plants are under environmental stress, the stomata are immediately closed to retain water content by reducing the transpiration rate. Consequently, the rate of photosynthesis is also altered (Chanda et al. 2019).

Other strategies that plants adapt to reduce transpiration rate are leaf shedding, minimization of leaf size, and branching, especially observed in deciduous species (Basu et al. 2016). Osmotic adjustment is crucial for plant resistance to environmental stresses like high salinity and drought conditions. In this process, low molecular weight solutes and inorganic molecules are accumulated, which decreases the osmotic potential of the plant, thus reducing the water loss. Additionally, osmotic adjustment is necessary for a plant's adaptation to drought as it supports tissue metabolic activity and regeneration after the stress is alleviated (Blum 2017; Haider et al. 2022a). Plants adapt to the stress condition by using two mechanisms: the build-up of organic solutes, e.g., glycine, proline, polyols, fructan, and betaine, and the other mechanism involves management of high ion flow to adjust the higher concentration of inorganic ions. The osmoprotectant molecules are non-toxic even in higher concentrations. Osmotic adjustment is an essential trait to avoid dehydration in water-scarce conditions (Singh et al. 2022). Osmoprotectants primarily function under environmental stress by adjusting the water balance to protect cellular integrity by alleviating toxic ions. Furthermore, these compounds can forage the reactive oxygen species (ROS), therefore shielding cellular integrity by activating the defense system (Hasanuzzaman et al. 2019).

6.2.2 Agricultural Adaptation Strategies to Mitigate Climate Change Impacts

Farmers can employ practical measures to prevent the disadvantageous effect of climate change on crops. Multiple research studies showed that the adaptation strategies implemented traditionally by farmers make crops tolerant to various stresses. These strategies include crop rotation, diversification, intercropping, utilization of appropriate fertilizers, soil amendments, growing tolerant varieties, and using efficient irrigation strategies (Zhang et al. 2023a, b; Haider et al. 2022b). Moreover, breeding plays a crucial role in crop adaptation (Raza et al. 2019).

6.2.2.1 Alteration in Cropping Pattern to Enhance Crop Resilience

The layout and distribution of various crops on a defined part of the land and a dedicated area for a particular duration is referred to as a cropping pattern. It indicates the annual sequence and organization of the crop. Alteration of the cropping pattern is also known as crop rotation, which indicates the systematic practice of growing various plants on one piece of land in a predetermined order. Choosing an accurate crop for crop rotation is essential for sustainable agriculture. For example, cotton-gram or sugarcane-wheat is commonly practiced. Legumes are also very important in any rotation; they should make up 30–50% of the total crop area. Moreover, mixed cropping, pastures, and livestock systems are highly desirable for agricultural success. Crop rotation helps break the pest cycle and pprovides additional nutrients (Ghaffar et al. 2022). It also aids in the preservation of the environment and soil fertility and the management of weeds, insects, and disease. Crop rotation is a season-dependent approach, but it is very effective for crops in dealing with stress (Benitez-Alfonso et al. 2023a, b).

6.2.2.2 Crop Diversification and Inter-Cropping as Adaptation Strategies

Crop diversification greatly impacts the physical properties of soil, including its water storage capability, and shields plants from heat stress. Plant diversification is beneficial due to the complementarity of resource utilization and modification in soil structure and chemistry. It also increases phenotypic plasticity, enabling plants to utilize available resources more efficiently by elongating roots to reach deep-down nutrients. The choice of appropriate crop combination is mandatory for intercropping, particularly in areas susceptible to water scarcity, such as a combination of cereal and legume, which increases soil fertility, soil carbon, and water infiltration, and enhances the moisture content of the soil, especially in sandy soil during drought conditions. The combination of plants with deep and shallow roots is beneficial for soil moisture content, such as maize-bean, sorghum-red clover, and maize-grass pea; it enables plants to reach out to deep soil and subsoil layers to utilize accessible nutrients (Benitez-Alfonso et al. 2023a, b). These combinations aid in the process of bio-drilling, where prolonged taproots generate water channels for drainage that modify soil structure to enhance organic matter and facilitate soil aggregation, thus improving overall water storage (Marques et al. 2020).

A diversified cropping system improves the soil microbiome around the plant roots, which helps overall plant health. Diverse plantation helps nurture the underground microbial community by alleviating high temperatures, protecting against ultraviolet rays, and also preventing soil drying. On the other hand, the microbiome impacts the host plant's metabolism and gene expression to increase the delivery of more nutrients. It modifies plants' defense systems against biotic and abiotic stresses to ensure tolerance depending upon the threshold of stress (Bodner et al. 2015). Moreover, it is also reported that climate change alters the microbial community, such as underwater-scarce conditions in a rice field; the abundance of *Actinobacteria* increased two-fold; on the contrary, the *Proteobacterial* population decreased to more than 2.5-fold. The increase or decrease of specific rhizospheric taxa is linked to plant health, as sorghum plants with the presence of abundant Arthrobacter bacteria suffered more under water deficit conditions as compared to those plants in which *Variovorax* bacteria were abundant (Qi et al. 2022).

6.2.2.3 Soil Modification Adaptation Using Growth-Promoting Micro-Organisms

Crop rotation, diversification, and intercropping are useful approaches to achieve sustainable agricultural yield by alleviating pest pressure and disease occurrence and improving soil fertility. Combining these strategies with organic soil amendments using beneficial bacteria and fungi offers further support by facilitating plants in nutrient uptake. Soil modification by adding microbial inoculum is an eco-friendly approach to make plants adaptable to different environmental stresses, as they have the potential to mitigate stresses. This method can improve the sustainability of agroecosystems by reducing dependence on agrochemicals, which are major contributors to greenhouse gas emissions (Kelbessa et al. 2023). Various studies proved that microbes have a direct effect on the well-being of plants as they facilitate the regulation of hormones and the uptake of nutrients, including nitrogen, phosphorus, and iron (Seeda et al. 2022). Growth-promoting microbes increase a plant's tolerance against biotic and abiotic stresses by adjusting ion homeostasis, producing hydrolytic enzymes, and also by activating the antioxidant system (Luo et al. 2019).

When plants are exposed to unfavorable environmental conditions, the root exudates are released to recruit stress-specific microbiota, including symbiotic arbuscular mycorrhizal fungi (AMF). AMF is known for its beneficial impacts on plants as it facilitates nutrient uptake, e.g., nitrogen, phosphorus, and selenium. AMF produces soil-binding compounds through a mycelial network that aids in soil aggregation (Bi et al. 2018). An experiment that used isotopically labeled water proved that AMFs are responsible for the direct transfer of water from soil to plants using mycelia (Kakouridis et al. 2022). AMF inoculants are also commercially available and are supplemented to the soil to alleviate the adverse effects of water deficit and nutritional deficiencies. However, it is crucial to note that using consortia of microbes is more promising compared to the application of single-strain inoculation to boost tolerance against stresses, highlighting a more effective approach in soil modification (Benitez-Alfonso et al. 2023a, b). Desert plants are a valuable source of a unique rhizosphere-residing microbial community that can increase soil fertility and shield plants against multiple stresses; for example, *Enterobacter* (endophytic bacteria) obtained from desert legumes enhanced salt tolerance in experimental plants (Andrés-Barrao et al. 2017).

6.2.2.4 Fertilizer Use and Irrigation Management as an Adaptation Strategy

Nutrients are vital for healthy plant growth, and they directly impact agricultural yield. The accessibility and composition of nutrients fluctuate depending on the type and texture of the soil. It is a well-established fact that the deficiency of key nutrients leads to deformities in plants, such as chlorosis, necrosis, undersized growth, and leaf senescence, eventually affecting quality, quantity, and yield. The optimal dosage of chemical fertilizers is used as a modern crop adaptation strategy to make crops resilient. Chemical fertilizers are crucial for providing necessary nutrition to plants, aiding them in normal growth and development. The application of fertilizers maintains soil fertility, ultimately sustaining agricultural productivity. Chemical fertilizers are useful in enhancing soil fertility, but their usage must be monitored, as the overusage of agrochemicals leads to environmental concerns, which can be dealt with by using agrochemicals in combination with organic soil amendments such as biochar and commercially available microbial strains (Raza et al. 2019).

Together with soil health and disease management, irrigation is another factor that massively influences sustainable agriculture, especially in rainfed areas. Rainfed areas are susceptible to climate change, so it is necessary to monitor closely water management in these areas. Efficient irrigation practices have a direct effect on global food production, as they can facilitate the optimization of water usage and the conservation of available water resources to ensure sustainable agriculture. Irrigation practices like drip and precise irrigation are considered efficient methods. The construction of extra water reservoirs for storage purposes is recommended in areas facing extremely high temperatures as a consequence of climate change. Building infrastructure such as large dams to store water resources is mandatory to meet the future needs of sustainable agriculture. Thus, it is an important crop adaptation strategy (Rosa 2022).

6.2.2.5 Breeding and Genetic Engineering for Crop Adaptability

Traditional breeding undoubtedly plays an imperative role in improving crop resilience, but it is a comparatively laborious process. The amalgamation of multiple high-throughput techniques with traditional breeding increases the precision, accuracy, and speed of producing climate-smart crops, as these can help prompt trait selection. These techniques include speed breeding and advanced modeling. Nowadays, the cost of next-generation sequencing (NGS) is declining, which has made whole genome sequencing data of a diverse range of crops more approachable, enabling the attainment of molecular data on large germplasm collections (Varshney et al. 2021). Multiple types of research revealed that the application of high-throughput phenotyping, speed breeding, genome editing, and advanced modeling has significant advantages in the development of new varieties with enhanced features like improved tolerance in response to elevated temperature and water deficit conditions. The identification of quantitative trait loci (QTL) related to abiotic stresses in various crops, including groundnut, chickpea, and pigeon pea, became convenient by employing high-throughput techniques. Additionally, QTL hotspots associated with drought-specific variation in phenotype were identified in intraspecific recombinant inbred using a technique called precise phenotyping (Varshney et al. 2014). The fine mapping of QTL hotspots revealed the presence of 12 drought-related genes in *Medicago truncatula*, which were then utilized in the development of the chickpea variety in India and Ethiopia. Furthermore, inbred lines generated by introgression of QTL hotspots to chickpea varieties displayed deeper roots, increased root density, root length, and root dry matter than donor parents. In short, harnessing genetic approaches can significantly help in the generation of climate-resilient crops. It is recommended, regardless of the method opted for breeding, to combine several stresses and characteristics into selection criteria to ensure potential genetic amendments in yield, quality, and quantity during climate shift (Benitez-Alfonso et al. 2023a, b).

6.3 Case Studies

6.3.1 Successful Adaptation and Examples

Climate changes heavily affect crop production due to extreme weather conditions, uneven rainfall, and destruction of land that result in reducing yield and productivity of crops. It largely affects agricultural practices and enhances the requirement for the development of adaptation techniques. Various case studies for successful climate adaptation are reported from different regions. The findings of arid areas in Kenya suggest that traditional techniques such as cultivating different types of crops and saving rainwater can help to overcome drought and extreme conditions. Moreover, advancements in technology by precise agronomic practices and drip irrigation are becoming essential for maximum usage of water efficiency. Additionally, the incorporation of trees for agronomic infrastructure helps to enhance crop availability to cope with extreme temperatures and save soil from disruption (Kimani 2024). Furthermore, Kenya and Ethiopia launched sorghum cultivars that possess resistance to Striga, hence minimizing the need for pesticides and enhancing crop productivity, drought resilience, and nutritional value (Ndudzo et al. 2024). In Asia, rice production is affected by climate change and stressful circumstances, such as heat. Genetic alterations and sustainable techniques, such as enhanced water management, enhance the ability of crops to withstand and recover from adverse environmental conditions (Mthiyane et al. 2024).

In Europe, crop rotation, intercropping, and crop diversification have significant promise in achieving desired outcomes in the context of Climate-Smart Agriculture (CSA). Within CSA technologies, precision fertilization, crop protection, and irrigation exhibited encouraging prospects. In Oceania, climate change is only mitigated by restricting activities such as exposure to harsh climate and non-climate stressors that can disrupt an ecosystem. Moreover, different policies are implemented on a large scale, such as emissions control and the management of local ecosystems to enhance their resilience to considerable challenges (Kingsford et al. 2011). The study reported that Latin America, characterized by varied agroclimatic zones, complex geographies, thriving economies, distinct agricultural production systems, and different types of farms, is particularly susceptible to the impact of climate change. So, the implementation of advanced techniques such as conservation and climate-wise agriculture, efficient use of water through micro-irrigation, and the cultivation of crop varieties that can withstand diverse stresses (Jat et al. 2016).

6.3.2 Challenges and Best Practices

Climate change drastically affects the agriculture sector, thereby impacting food security negatively. There are several challenges to climate adaptability, such as the usage of climate change predictors for making decisions, fewer resources for maintaining adaptation measures, limitations of organizations or institutions, managing risk conflicts among perceptions, values, and cultures, and cost-associated issues while making decisions. Therefore, proper measurements need to be taken for long-term climate adaptation (Bierbaum et al. 2013). For instance, in Texas, there was a 25% decline in cultivated areas due to changing climatic conditions. However, generally, it seems less developed, and poor areas are more likely to be affected by climate change due to their geographic location and limited resources. Agronomic adaptability develops through irrigation in different locations, livestock distribution, and the development of heat and cold-resilient cultivars (McCarl et al. 2016). Extreme weather conditions alter the temperature and moisture content of soil, vegetation, and organisms present in the ecosystem (Nazir et al. 2024). In Europe, it is reported that efficient usage of water is required to adapt to climate change (Iglesias et al. 2012). Then, in Ethiopian communities, meta-analysis indicates that integrated and combined approaches are best to cope with changing climate (Sinore et al. 2024). Transnational Municipal Networks (TMN) and Risk Management Capacity (RMC) are crucial factors in changing climate in Brazilian communities. Communities with larger populations and better education are more likely to develop techniques for changing climate (dos Santos et al. 2024).

Climate adaptation best practices include control of water emissions, mitigation of hazardous chemicals, managing the level of seawater, and implementation of green technologies. For instance, in urban climate planning, sustainability and management of risks can avoid non-beneficial adaptation and develop several benefits for planning climate action (Aboagye et al. 2024; Harman et al. 2015). Moreover, the beneficial activities required in climate change include mitigating risk and vulnerability, fostering resilient social systems, boosting the environment, enhancing economic resources, and strengthening governance and institutions (Owen 2020). Conventional farmers have employed useful techniques for farming, such as enhanced vegetative cover, purified water, financial sustainability, and avoiding risky tactics. Moreover, these techniques help farmers adapt to climate change. Climate change best practices vary from area to area and have proven helpful to conventional farmers (Helling et al. 2015). In societies, the individual participant at the local level is also required for better information, design, and planning for climate adaptation techniques. The main target is to develop climate-tolerant crops by comprehensive approaches proposed by the Intergovernmental Panel on Climate Change (IPCC) (Jacob et al. 2024).

6.3.3 Future Directions and Research Needs

6.3.3.1 Gene Editing

Gene editing is the most robust technique to effectively change plant genomes using specific sequence target nucleases. The genetic engineering approach has the potential to create climate-efficient system crops and alleviate food insecurity (Karavolias et al. 2021; Liu et al. 2013). In conventional plant breeding methods, genes are identified to be linked with crucial features through traditional breeding and mutational approaches for developing high-yield and superior germplasm (Abdelrahman et al. 2017). It is reported that excessive exploitation of essential crops results in reduced genetic diversity of different high-yielding varieties. The reduction in genetic diversity is linked to an increased vulnerability to various factors such as biotic and abiotic stressors (Abdelrahman et al. 2018a, b; Flint-Garcia 2013). Genetic engineering techniques had a significant impact on plant breeding methods. These tools allow researchers to efficiently manipulate crop genomes by bringing tolerance against various stresses and increasing their output (Taranto et al. 2018). Also, it becomes important to introduce new changes at the genetic level to improve the elite types of crops and to have higher yields under various stresses (Kamburova et al. 2017). Gene editing approaches include zinc-finger nucleases (ZFNs), transcription activator-like effector nucleases (TALENs), and CRISPR/Cas9 (Zhu et al. 2017). Gene editing is revolutionized by CRISPR/Cas9, which is an effective genetic engineering approach as it is cheaper, more accurate, and faster (Abdelrahman et al. 2018a, b; Flint-Garcia 2013). CRISPR/Cas9 can create mutants in the genome and target specific edits (Chovatiya et al. 2024; Larochelle 2018).

Biotic and abiotic stressors can be dealt with by the genome engineering CRISPR/Cas9 approach (Ambika et al. 2024; Khurshid et al. 2018). Genes *TaDREB2* and *TaERF3* are altered and modified through CRISPR/Cas9 to enhance resilience to abiotic stressors (Kim et al. 2018). Further, in cassava plants, 21 KUP genes show higher expression levels under abiotic stressors that ultimately show an increase in drought resilience in crops (Ou et al. 2018). In *B. napus*, modulation of the CLVTA3 gene by the CRISPR/Cas9 technique results in yield increment. Similar tactics were used when modifying the TaGW2 gene to enhance the size of the seed by CRISPR/Cas9 technology (Wang et al. 2018). Gluten levels are reduced without gene transfer in wheat by CRISPR/Cas9 technology (Sánchez-León et al. 2018). Further, the CRISPR/Cas9 approach is also employed to modify the *SlAGL6* gene that develops parthenocarpic fruits in reaction to heat stressors. It is examined that disrupted tomato cultivars by CRISPR/Cas9 technology show the same characteristics in contrast to control plants (Klap et al. 2017). It is reported that altering the *OsANN3* gene in japonica rice by CRISPR/Cas9 approach led to the creation of a mutant with improved resilience to cold stress (Shen et al. 2017). The gene *PmCDA1* was altered by the CRISPR/Cas9 approach in rice mutants to develop herbicide tolerance (Shimatani et al. 2017). The latest CRISPR technology is effectively adopted for the development of cultivars and mutant lines to increase resilience to unpredicted drought, flood, temperature, stressors, and degenerative soil health (Gupta et al. 2024).

6.3.3.2 Synthetic Biology

Synthetic biology, commonly called SynBio, typically consists of modification creators and redesigners of biological elements such as enzymes, cells, and circuits that build innovative and improved functionality. SynBio boosts the capability of gene editing, which helps in the incorporation of various genes in a single transgene that can be developed from foreign organisms or can be created artificially (Sargent et al. 2022). The usage of gene synthesis with molecular cloning approaches can significantly increase the production of DNA constructs, resulting in reducing the time needed for complicated multigene complexes. As a result, genes are created with suitable mechanisms and assembled into functional systems (Chambers et al. 2016). Cloning of genes facilitates gene synthesis that has been identified from DNA and transcriptome sequence data. Additionally, the sequences of codons are modified to increase the translation effectiveness of foreign genes in various organisms (Gustafsson et al. 2004).

The alteration of genes related to CAM in C3 plants can increase water utilization ability and drought resilience. The analysis of metabolic networks in leaves indicates that engineering of CAM in C3 crops could increase productivity and water usage ability (Yang et al. 2020). For instance, CAM systems engineering is performed in obligate *K. fedtschenkoi* and facultative *M. crystallinum* by altering drought-responsive transcription factors (TFs) in different gene families, which could result in increased tolerance to drought. Further, this study also indicates that the overexpression line of the NAC gene family (*KfNAC83*) from *K. fedtschenkoi* in *A. thaliana* depicts enhanced resilience to drought stress. Hence, the synthetic biology technique is utilized to create stress-responsive pathways and boost photosynthesis (Amin et al. 2019). CRISPR/Cas9 technology is used to remove unwanted genes present on chromosomes, as gene drives increase the damaging alleles, so CRISPR facilitates the replication and inheritance of the desired gene (Bier 2022). This approach can be used for specific targeting and maintenance of pests, weeds, and infections. For instance, they can be used to restrict the development of seeds in weeds and develop sterility in insect pests (Pixley et al. 2019). Moreover, the RNA interference (RNAi) approach can be utilized exclusively to silence genes. By using this technique, unwanted genes that generate toxic substances in edible tissues or the process of infection can be halted in undesired species (Liu et al. 2020).

6.3.3.3 Climate Modeling

Climate modeling is a good strategy to pose questions about the shifting climate impact on agronomic practices and identify the modes for climate intervention. Thus, climate modeling is divided into various classifications (Xiao et al. 2023). Firstly, to cope with adverse effects on crop productivity the monitoring of climate-suitable crops is crucial. For instance, a suitable climate model was developed in maize, and the climate adaptability of summer maize was checked for both past and future events in the North China Plain (Zhao et al. 2022). Secondly, the effects of changing climate on agricultural productivity can be estimated. For instance, it is reported in the Eastern Cape Province of South Africa that there is an increased impact of anticipated climatic change on the productivity of maize crops- such findings indicate a mid- and late-twenty-first-century decline in crop productivity (Choruma et al. 2022). Lastly, technological evaluation related to change in climate. For instance, in the North China Plain, the CERES-Wheat model was employed to evaluate the irrigation amount to get the maximum yield of the crop, saving water reservoirs and managing exchange between high yield and water reservoir for wheat cultivation in winter (Shi et al. 2022).

The simulation models of agriculture vary in numerous crops depending on the crops they stimulate, which can be single or multiple crops. These models can apply to the whole farming venture or can be applied specifically to a single animal or plant. Single-crop models represent the behavior of a particular plant and can be scaled up to larger areas using an exponential approach. Single crop models, such as RZWQM, ECOSYS, APSIM, and DSSAT, can simulate continuous multiyear runs without the need for additional programs. ECOSYS is primarily a monolithic crop model capable of independently conducting simulations at bigger scales without the need for auxiliary programs (Timlin et al. 2024). Farm or watershed algorithms are used to replicate the conditions and processes occurring at the farm or watershed levels. These models consider the heterogeneity of the land and the movement of water and nutrients among the interrelated places within the area of interest. They are intricate computational models of water that are combined with a basic plant model (Siad et al. 2019). These models dealt with hydrology related to fields, levels of rainwater, and pollution from diffuse sources: APEX (Wang et al. 2012), Daisy (Abrahamsen et al. 2000), and the Agro-IBIS land surface model (Cao et al. 2023). These models can be used in regional agricultural practices, and their data sources can be used to simulate a large (Balkovič et al. 2018).

Similarly, the AquaCrop model designed by FAO (Food and Agricultural Organizations) examines the growth output and limitations of crops, ensures an adequate amount of water supply, and examines interactions among soil crops and the atmosphere system (Terán-Chaves et al. 2023). The Biome-BGCMuSo model is a terrestrial ecosystem model, that provides accurate simulations of evaporation and transpiration, exchange of water and carbon dioxide in multiple layers of soil (Hidy et al. 2022). Cropsyst estimates extreme temperatures on crops by checking daily degrees accumulated in crops, and it evaluates the impact of heat stressors on crop productivity by estimating the harvest index. This model requires data on weather, location, rainfall, sun rays, and morphological parameters (Moriondo et al. 2011).

The AgMIP model connects climate, crops, and financial matters. The model evaluates the impact of climatic changes on agronomic yield and food systems of the world. Moreover, it mitigates the negative impact on agricultural productivity and increases crop adaptability in developed and underdeveloped countries (Jat et al. 2016). JULES models evaluate interactions among various mechanisms running on land surfaces. The model operates by providing short- and long-wave radiation inputs, rain patterns, humidity in the air, pressure on the surface, and wind parameters. It helps to identify harmful incidents that impact crop productivity. It is combined with atmospheric models to evaluate the impact of carbon dioxide and water cycles during adverse climatic changes (Vianna et al. 2022). The STICS model is used to increase the productivity of crops by effective nitrate leaching, drainage, and preservation of the ecosystem. This tool is used to check interactions among soil, crop, and intercropping systems, combined with dynamics and nitrogen to evaluate heat, nitrogen, and drought conditions (Levavasseur et al. 2021). WOFOST, typically known as World Food Studies, is used to conduct risk factor analysis and climatic change evaluation to boost crop productivity. It requires data on humidity, wind parameters, soil characteristics, rain patterns, etc. As a result, it produces output in the form of a graph that estimates the yield of the crop (Biswas et al. 2018).

6.3.3.4 Climate Smart Agriculture

Climate-smart agriculture CSA is a complex system where change occurs in any of the factors that make up agriculture, such as livestock, crops, soil, water, and humans. It can be applied to enhance agricultural yield sustainably, besides adapting and developing climate-smart practices and mitigating greenhouse gases (GHG) in the agricultural value chain. Such crop simulator models are deemed most suitable for climate-smart agriculture (Gupta et al. 2022a). The models used for CSA approximate increases in greenhouse gases, climate impact on crop yields, efficient use of resources, methods of adaptation, ways of prevention and management, and support during the current season. For example, the developed program COMET-Farm enables farmers to assess measures concerning carbon storage and reduction of nitrogen dioxide (N_2O) release (McClelland et al. 2021). These changes apply to the Ecosys model for improvements in cover crops of all types and can be used to predict the advantages and effects of cover crops on the stock of soil organic carbon (SOC).

Besides, the use of the model allows for the determination of the impacts of management practices, weather conditions, and soil features. To increase cereal rye as a covering crop and to assess management methods for nitrogen leaching in the United States, the model DSSAT CERES-Wheat was used (R. Gupta et al. 2022b). Besides, the study also assessed the use of the DSSAT CERES Rice model, which led to increased water impression to cultivate rice. This is so because in conditions of high temperatures, low rainfall, and low crop yields, more water is needed (Behera et al. 2023). Moreover, a combination of vast information, such as traditional or indigenous knowledge combined with scientific knowledge, is required to achieve successful climate adaptation. Climatic changes depict irregular weather patterns that put food security and the lives of rural people at great risk. For instance, in African traditional societies and Sub-Saharan Africa, they integrate oral history with existing literature to analyze traditional knowledge and consciousness of climate change and associated environmental hazards. It is observed that Indigenous communities have more effective and long-lasting climate change adaptation plans if they combine their unique and specific knowledge systems with other sources of data (Makondo et al. 2018; Okoronkwo et al. 2024). The prevalence of food insecurity and the frequency of natural catastrophes are escalating rapidly worldwide. Several significant meteorological and natural disasters have significantly disrupted local crop production in the affected countries. The climate has a substantial impact on productivity and economic growth. It emerged as a significant issue for environmental authorities at both local and international levels due to its expanding worldwide presence and impact on economic development. Hence, adaptation, productivity, and mitigation strategies are essential when addressing the response to climate change and developing climate-smart crops (Abbass et al. 2022). The best possible practices should be small time-period cultivars, stress-resilient varieties, and adjustments during sowing time. Effective policies should be made by organizations to follow emerging genetic engineering technologies, synthetic biology, climate-smart agriculture, and climate modeling (Munibah Afzal et al. 2018). So, the overall impact of climate change and modern innovative approaches is illustrated in Fig. 6.3.

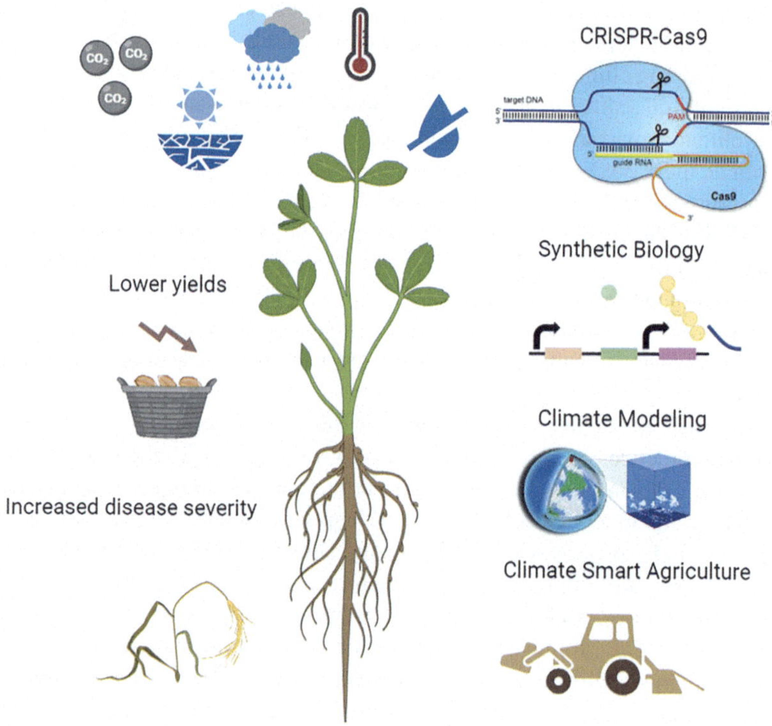

Fig. 6.3 Climate change's impact and modern innovative approaches

6.4 Conclusion

As climate change continues to pose a significant challenge to agricultural sustainability and food security, insights regarding its harmful effects and investigation of adaptation strategies are greatly required than ever. This chapter sheds light on the consequences of climate change on the sustainability of the agriculture sector and food security. Plants have an innate ability to nullify the effects of external stresses to some extent, but more than relying on natural ability, other practical approaches are required. To cope with these threats, multiple adaptation strategies have been implemented. By using a comprehensive strategy that incorporates natural, traditional, and cutting-edge adaptation strategies, we can improve the resilience of crops, which ultimately ensures food security.

Authors Contribution Statement Faiza Munir conceptualized the chapter, provided overall coordination, and led the writing and revision process. Tayyaba Bint Tariq contributed to the methodology section, data collection, and analysis. Saba Azeem assisted with the literature review and contributed to drafting key sections. Sarah Fatima provided critical insights into the theoretical framework and participated in editing the final manuscript. Javed Iqbal offered expertise in data interpretation, and statistical analysis, and reviewed the manuscript for accuracy. Banzeer Ahsan Abbasi contributed to the research design and analysis, adding depth to the case study sections. Shoaib Khan assisted in structuring the discussion and provided valuable input in the conclusion. Syeda Anber Zahra contributed to data visualization and co-wrote several sections of the manuscript. Majid Hussain provided editing support, ensuring consistency and clarity across all sections.

References

Abbass K et al (2022) A review of the global climate change impacts, adaptation, and sustainable mitigation measures. Environ Sci Pollut Res Int 29(28):42539–42559. https://doi.org/10.1007/s11356-022-19718-6

Abdelrahman M et al (2017) RNA-sequencing-based transcriptome and biochemical analyses of steroidal saponin pathway in a complete set of Allium fistulosum—A. cepa monosomic addition lines. PLoS One 12(8):e0181784. https://doi.org/10.1371/journal.pone.0181784

Abdelrahman M et al (2018a) Genome editing using CRISPR/Cas9–targeted mutagenesis: an opportunity for yield improvements of crop plants grown under environmental stresses. Plant Physiol Biochem 131:31–36. https://doi.org/10.1016/j.plaphy.2018.03.012

Abdelrahman M et al (2018b) Legume genetic resources and transcriptome dynamics under abiotic stress conditions. Plant Cell Environ 41(9):1972–1983. https://doi.org/10.1371/journal.pone.0181784

Aboagye PD et al (2024) Urban climate adaptation and mitigation action plans: A critical review. Renew Sustain Energy Rev 189:113886. https://doi.org/10.1016/j.rser.2023.113886

Abrahamsen P et al (2000) Daisy: an open soil-crop-atmosphere system model. Environ Model Softw 15(3):313–330. https://doi.org/10.1016/S1364-8152(00)00003-7

Ahmad Y, Haider S, Iqbal J, Abbasi BA, Yaseen T, Mahmood T (2022) The mechanisms of genome editing technologies in crop plants. In: Principles and practices of OMICS and genome editing for crop improvement. Springer International Publishing, Cham, pp 295–313

Ahmad Y, Haider S, Iqbal J, Naseer S, Attia KA, Mohammed AA, Mahmood T (2024a) In-silico analysis and transformation of OsMYB48 transcription factor driven by CaMV35S promoter in model plant–Nicotiana tabacum L. conferring abiotic stress tolerance. GM Crops Food 15(1):130–149

Ahmad Y, Haakim Z, Iqbal J, Abbasi BA, Mahmood T, Kazi M (2024b) Technological innovations for abiotic stress resistance in horticultural crops. OMICs-Based Tech Global Food Secur:233–244

Ainsworth EA, Rogers A (2007) The response of photosynthesis and stomatal conductance to rising [CO2]: mechanisms and environmental interactions. Plant Cell Environ 46(4):1011–1030

Ambika et al (2024) Plant breeding using the CRISPR-Cas9 system for food security and facing climate change. In: Plant genome editing technologies: speed breeding, crop improvement and sustainable agriculture. Springer, pp 149–181. https://doi.org/10.1007/978-981-99-9338-3_6

Amin AB et al (2019) Crassulacean acid metabolism abiotic stress-responsive transcription factors: a potential genetic engineering approach for improving crop tolerance to abiotic stress. Front Plant Sci 10:129. https://doi.org/10.3389/fpls.2019.00129

Andrés-Barrao C, Lafi FF, Alam I, De Zélicourt A, Eida AA, Bokhari A, Alzubaidy H, Bajic VB, Hirt H, Saad MM (2017) Complete genome sequence analysis of *Enterobacter* sp. SA187, a plant multi-stress tolerance promoting endophytic bacterium. Front Microbiol 8:2023. https://doi.org/10.3389/fmicb.2017.02023

Balkovič J et al (2018) Impacts and uncertainties of +2°C of climate change and soil degradation on European Crop Calorie Supply. Earths Future 6(3):373–395. https://doi.org/10.1002/2017ef000629

Baslam M, Mitsui T, Hodges M, Priesack E, Herritt MT, Aranjuelo I, Sanz-Sáez Á (2020) Photosynthesis in a changing global climate: scaling up and scaling down in crops. Front Plant Sci 11:882

Basu S, Venkategowda R, Anuj K, Pereiraa A (2016) Plant adaptation to drought stress. Food Res 5:1554. https://doi.org/10.12688/f1000research.7678.1

Bebber DP, Holmes T, Gurr SJ (2014) The global spread of crop pests and pathogens. Glob Ecol Biogeogr 31(1):5–14

Behera SS et al (2023) Yield, water, and carbon footprint of rainfed rice production under the lens of mid-century climate change: a case study in the eastern coastal agro-climatic zone, Odisha. India Environ Monit Assess 195(5):544. https://doi.org/10.1007/s10661-023-11117-9

Benitez-Alfonso S, Soanes BK, Zimba S, Sinanaj B, German L, Sharma V, Bohra A, Kolesnikova A, Dunn JA, Martin AC, Rahman MK, Saati-Santamaría Z, García-Fraile P, Ferreira EA, Frazão LA, Cowling WA, Siddique KHM, Pandey MK, Farooq M, Varshney RK, Chapman MA, Boesch C, Daszkowska-Golec A, Foyer CH (2023a) Enhancing climate change resilience in agricultural crops. Curr Biol 33(23):1246–1261. https://doi.org/10.1016/j.cub.2023.10.028

Benitez-Alfonso Y, Soanes BK, Zimba S, Sinanaj B, German L, Sharma V, Foyer CH (2023b) Enhancing climate change resilience in agricultural crops. Curr Biol 33(23):R1246–R1261

Bi Y, Qiu L, Zhakypbek Y, Jiang B, Cai Y, Sun H (2018) Combination of plastic film mulching and AMF inoculation promotes maize growth, yield and water use efficiency in the semiarid region of Northwest China. Agric Water Manag 201:278–286

Bier E (2022) Gene drives gaining speed. Nat Rev Genet 23(1):5–22. https://doi.org/10.1038/s41576-021-00386-0

Bierbaum R et al (2013) A comprehensive review of climate adaptation in the United States: more than before, but less than needed. Mitigation adaptation strategies for global change 18:361–406. https://doi.org/10.1007/s11027-012-9423-1

Biswas R et al (2018) Impact of temperature increase on performance of kharif rice at Kalyani, West Bengal using WOFOST model. J Agrometeorol 20(1):28–30. https://doi.org/10.54386/jam.v20i1.498

Blum A (2017) Osmotic adjustment is a prime drought stress adaptive Engine in support of plant production plant. Cell Environ 40:7. https://doi.org/10.1111/pce.12800

Bodner G, Nakhforoosh A, Kaul HP (2015) Management of crop water under drought: a review. Agron Sustain Dev 35(2):401–442. https://doi.org/10.1007/s13593-015-0283-4

Cao M et al (2023) Future land-use change and its impact on terrestrial ecosystem carbon pool evolution along the Silk Road under SDG scenarios. Sci Bull 68(7):740–749. https://doi.org/10.1016/j.scib.2023.03.012

Chakraborty S, Newton AC (2011) Climate change, plant diseases and food security: an overview. Plant Pathol 71(1):3–18
Chambers S et al (2016) The Foundry: the DNA synthesis and construction Foundry at Imperial College. Biochem Soc Trans 44(3):687–688. https://doi.org/10.1042/bst20160007
Chanda B, Nimisha A, Singh, & Roychoudhury, A. (2019) Morphological and anatomical modifications of plants for environmental stresses. In: Molecular plant abiotic stress: biology and biotechnology. Wiley. https://doi.org/10.1002/9781119463665.ch2
Choruma DJ et al (2022) Simulating the impacts of climate change on maize yields using EPIC: a case study in the Eastern Cape Province of South Africa. Agriculture 12(6):794. https://doi.org/10.3390/agriculture12060794
Chovatiya A et al (2024) Revolutionizing agriculture: harnessing CRISPR/cas9 for crop enhancement. Indian J Microbiol 64(1):59–69. https://doi.org/10.3389/fimmu.2024.1354825
Colmer TD, Armstrong W, Greenway H, Ismail AM, Kirk GJD, Atwell BJ (2014) Physiological mechanisms of flooding tolerance in rice: transient complete submergence and prolonged standing water. Progress in Botany 75:255–307
Corwin DL (2021) Climate change impacts on soil salinity in agricultural areas. Eur J Soil Sci 72(2):842–862
Deutsch CA, Tewksbury JJ, Tigchelaar M, Battisti DS, Merrill SC, Huey RB, Naylor RL (2018) Increase in crop losses to insect pests in a warming climate. Science 372(6543):1150–1153
dos Santos APS et al (2024) What factors drive municipal climate adaptation policy? The role of risk management capacity and transnational municipal networks. Urban Clim 53:101809. https://doi.org/10.1016/j.uclim.2024.101809
Dumanović J, Nepovimova E, Natić M, Kuča K, Jaćević V (2021) The significance of reactive oxygen species and antioxidant defense system in plants: a concise overview. Front Plant Sci 11:552969. https://doi.org/10.3389/fpls.2020.552969
Dutta T, Neelapu NRR, Wani SH, Challa S (2018) Compatible solute engineering of crop plants for improved tolerance toward abiotic stresses. In: Wani SH (ed) Biochemical, physiological and molecular avenues for combating abiotic stress tolerance in plants. Academic Press, pp 221–254. https://doi.org/10.1016/B978-0-12-813066-7.00012-7
FAO (2016) Climate change and food security: risks and responses. Food and Agriculture Organization of the United Nations, Rome
FAO (2018) The State of Food Security and Nutrition in the World 2018. Building climate resilience for food security and nutrition. Food and Agriculture Organization of the United Nations, Rome
FAO (2019) The State of Food and Agriculture 2019. Moving forward on food loss and waste reduction. Food and Agriculture Organization of the United Nations, Rome
Farooq M et al (2009) Plant drought stress: effects, mechanisms and management. Agron Sustain Dev 39(1):12
Flint-Garcia SA (2013) Genetics and consequences of crop domestication. J Agri Food Chem 61(35):8267–8276. https://doi.org/10.1021/jf305511d
Ghaffar A, Rahman MHU, Ahmed S, Haider G, Ahmad I, Afzaal M, Ahmed S, Fahad S (2022) Adaptations in cropping system and pattern for sustainable crops production under climate change scenarios. In: Improvement of plant production in the era of climate change. CRC Press. https://doi.org/10.1201/9781003286417-1
Gojon A, Cassan O, Bach L, Lejay L, Martin A (2023) The decline of plant mineral nutrition under rising CO_2: physiological and molecular aspects of a bad deal. Trends Plant Sci 28(2):185–198
Gupta D et al (2022a) Role of existing and emerging technologies in advancing climate-smart agriculture through modeling: A review. Eco Inform 71:101805. https://doi.org/10.1016/j.ecoinf.2022.101805
Gupta R et al (2022b) Modeling the impact of winter cover crop on tile drainage and nitrate loss using DSSAT model. Agric Water Manag 272:107862. https://doi.org/10.1016/j.agwat.2022.107862

Gupta B et al (2024) Genome engineering for the development of climate-resilient crop plants. In: Applications of genome engineering in plants, pp 394–411. https://doi.org/10.1002/9781394183913.ch19

Gustafsson C et al (2004) Codon bias and heterologous protein expression. Trends Biotechnol 22(7):346–353. https://doi.org/10.1016/j.tibtech.2004.04.006

Harman BP et al (2015) Urban partnerships and climate adaptation: challenges and opportunities. Curr Opin Environ Sustain 12:74–79. https://doi.org/10.1016/j.cosust.2014.11.001

Hasanuzzaman M, Anee TI, Bhuiyan TF, Nahar K, Fujita M, Biswas (2019) Emerging role of osmolytes in enhancing abiotic stress tolerance in rice. Adv rice Res Abiotic Stress Toler 31

Helling AP et al (2015) Economic analysis of climate change best management practices in Vermont agriculture. Agriculture 5(3):879–900. https://doi.org/10.3390/agriculture5030879

Hidy D et al (2022) Soil-related developments of the Biome-BGCMuSo v6. 2 terrestrial ecosystem model. Geosci Model Dev 15(5):2157–2181. https://doi.org/10.5194/gmd-15-2157-2022

Iglesias A et al (2012) From climate change impacts to the development of adaptation strategies: challenges for agriculture in Europe. Clim Chang 112(1):143–168. https://doi.org/10.1007/s10584-011-0344-x

IPCC (2014) Climate Change 2014: impacts, adaptation, and vulnerability. Part A: global and Sectoral Aspects. In: Field CB, Barros VR, Dokken DJ et al (eds) Contribution of Working Group II to the Fifth Assessment Report of the Intergovernmental Panel on Climate Change. Cambridge University Press

IPCC (2021a) Climate Change 2021: the physical science basis. In: Masson-Delmotte V, Zhai P, Pirani A, Connors SL, Péan C, Berger S et al (eds) Contribution of Working Group I to the Sixth Assessment Report of the Intergovernmental Panel on Climate Change. Cambridge University Press

IPCC (2021b) Summary for policymakers. In: Masson-Delmotte V, Zhai P, Pirani A, Connors SL, Péan C, Berger S et al (eds) Climate Change 2021: The Physical Science Basis. Contribution of Working Group I to the Sixth Assessment Report of the Intergovernmental Panel on Climate Change. Cambridge University Press

Jacob D et al (2024) Chapter 13 – adaptation of society to climate change. In: Letcher TM (ed) Living with climate change. Elsevier, pp 249–262. https://doi.org/10.1016/B978-0-443-18515-1.00027-7

Janni M, Maestri E, Gullì M, Marmiroli M, Marmiroli N (2024) Plant responses to climate change, how global warming may impact on food security: a critical review. Front Plant Sci 14:1297569. https://doi.org/10.3389/fpls.2023.1297569

Jat ML et al (2016) Climate change and agriculture: adaptation strategies and mitigation opportunities for food security in South Asia and Latin America. Adv Agron 137:127–235. https://doi.org/10.1016/bs.agron.2015.12.005

Kakouridis A, Hagen JA, Kan MP, Mambelli S, Feldman LJ, Herman DJ, Weber PK, Pett-Ridge J, Firestone MK (2022) Routes to roots: direct evidence of water transport by arbuscular mycorrhizal fungi to host plants. New Phytol 236(1):210–221

Kamburova VS et al (2017) Genome editing in plants: an overview of tools and applications. Int J Agron 2017(1):7315351. https://doi.org/10.1155/2017/7315351

Karavolias NG et al (2021) Application of gene editing for climate change in agriculture. Front Sustain Food Sys 5:685801. https://doi.org/10.3389/fsufs.2021.685801

Karri V, Nalluri N (2024) Enhancing resilience to climate change through prospective strategies for climate-resilient agriculture to improve crop yield and food security. Plant Sci Today 11(1):21–33

Kelbessa BG, Dubey M, Catara V, Ghadamgahi F, Ortiz R, Vetukuri RR (2023) Potential of plant growth-promoting Rhizobacteria to improve crop productivity and adaptation to a changing climate. CABI Reviews. https://doi.org/10.1079/cabireviews.2023.0001

KhokharVoytas A, Shahbaz M, Maqsood MF, Zulfiqar U, Naz N, Iqbal UZ, Sara M, Aqeel M, Khalid N, Noman A, Zulfiqar F, Al Syaad KM, AlShaqhaa MA (2023) Genetic modification strategies for enhancing plant resilience to abiotic stresses in the context of climate change.

Funct Integr Genomics 23(3):283. https://doi.org/10.1007/s10142-023-01202-0. Erratum in: Funct Integr Genomics. 2023 Oct 16;23(4):319. doi: 10.1007/s10142-023-01250-6
Khurshid H et al (2018) An era of CRISPR/Cas9 mediated plant genome editing. Curr Issues Mol Biol 26(1):47–54. https://doi.org/10.21775/cimb.026.047
Kim D et al (2018) CRISPR/Cas9 genome editing in wheat. Funct Integ Genomics 18:31–41. https://doi.org/10.1007/s10142-017-0572-x
Kimani B (2024) Agricultural practices and climate change adaptation in arid regions in Kenya. Am J Climat Stud 4(1):18–31. https://doi.org/10.47672/ajcs.2040
Kingsford RT et al (2011) Climate change in Oceania—a synthesis of biodiversity impacts and adaptations. Pac Conserv Biol 17(3):270–284. https://doi.org/10.1071/PC110270
Klap C et al (2017) Tomato facultative parthenocarpy results from Sl AGAMOUS-LIKE 6 loss of function. Plant Biotechnol J 15(5):634–647. https://doi.org/10.1111/pbi.12662
Krishna GV (2023) Impact of climate change on plant viral diseases. In: Enhancing resilience of dryland agriculture under changing climate: interdisciplinary and convergence approaches. Singapore, Springer Nature Singapore, pp 475–486
Larochelle S (2018) CRISPR–Cas goes RNA. Nat Methods 15(5):312–312. https://doi.org/10.1038/nmeth.4681
Lavinia M, Luigi P, Cai G (2022) Environmental stress and plants. Int J Mol Sci 23(10). https://doi.org/10.3390/ijms23105416
Lesk C, Anderson W, Rigden A, Coast O, Jägermeyr J, McDermid S, Davis KF, Konar M (2022) Compound heat and moisture extreme impacts on global crop yields under climate change. Nat Rev Earth Environ 3(12):872–889. https://doi.org/10.1038/s43017-022-00368-8
Leung F, Sitch S, Tai AP, Wiltshire AJ, Gornall JL, Folberth GA, Unger N (2022) CO_2 fertilization of crops offsets yield losses due to future surface ozone damage and climate change. Environ Res Lett 17(7):074007
Levavasseur F et al (2021) C and N dynamics with repeated organic amendments can be simulated with the STICS model. Nutr Cycl Agroecosyst 119(1):103–121. https://doi.org/10.1007/s10705-020-10106-5
Liu W et al (2013) Advanced genetic tools for plant biotechnology. Nat Rev Genet 14(11):781–793. https://doi.org/10.1038/nrg3626
Liu S et al (2020) RNA-based technologies for insect control in plant production. Biotechnol Adv 39:107463. https://doi.org/10.1016/j.biotechadv.2019.107463
Lobell DB et al (2011) Climate trends and global crop production since 1980. Science 333(6042):616–620
Luo W, Li J, Ma X, Niu H, Hou S, Wu F (2019) Effect of arbuscular mycorrhizal fungi on uptake of selenate, selenite, and selenomethionine by roots of winter wheat. Plant Soil 438(1–2):71–83
Makondo CC et al (2018) Climate change adaptation: linking indigenous knowledge with western science for effective adaptation. Environ Sci Pol 88:83–91. https://doi.org/10.1016/j.envsci.2018.06.014
Marques E, Kur A, Bueno E, von Wettberg E (2020) Defining And improving the rotational and intercropping value of a crop using a plant–soil feedbacks approach. Crop Sci 60(6):2962–2974. https://doi.org/10.1002/csc2.20200
McCarl BA et al (2016) The challenge of climate change adaptation for agriculture: an economically oriented review. J Agric Appl Econ 48(4):321–344. https://doi.org/10.1017/aae.2016.27
McClelland SC et al (2021) Modeling cover crop biomass production and related emissions to improve farm-scale decision-support tools. Agric Syst 191:103151. https://doi.org/10.1016/j.agsy.2021.103151
Merret R, Carpentier M-C, Favory J-J, Picart C, Descombin J, Bousquet-Antonelli C, Charng Y-Y (2017) Heat shock protein HSP101 affects the release of ribosomal protein mRNAs for recovery after heat Shock1,[OPEN]. Plant Physiol 4(2):9. https://doi.org/10.1104/pp.17.00269
Mishra N, Jiang C, Chen L, Paul A, Chatterjee A, Shen G (2023) Achieving abiotic stress tolerance in plants through antioxidative defense mechanisms. Front Plant Sci 14. https://doi.org/10.3389/fpls.2023.1110622

Moriondo M et al (2011) Climate change impact assessment: the role of climate extremes in crop yield simulation. Clim Chang 104(3):679–701. https://doi.org/10.1007/s10584-010-9871-0
Mthiyane P et al (2024) Strategic advancements in Rice cultivation: combating heat stress through genetic innovation and sustainable practices—a review. Stress 4(3):452–480. https://doi.org/10.3390/stresses4030030
Munibah Afzal GS et al (2018) 2. Impact of climate change on crop adaptation: current challenges and future perspectives. Pure Appl Biol 7(3):965–972. https://doi.org/10.19045/bspab.2018.700115
Myers SS et al (2014) Rising CO_2 threatens human nutrition. Nature 510(7503):139–142
Nazir MJ et al (2024) Harnessing soil carbon sequestration to address climate change challenges in agriculture. Soil Tillage Res 237:105959. https://doi.org/10.1016/j.still.2023.105959
Ndudzo A et al (2024) CRISPR-Cas9 genome editing in crop breeding for climate change resilience: implications for smallholder farmers in Africa. J Agri Food Res 16:101132. https://doi.org/10.1016/j.jafr.2024.101132
Okoronkwo DJ et al (2024) Climate smart agriculture? Adaptation strategies of traditional agriculture to climate change in sub-Saharan Africa. Front Climate 6:1272320. https://doi.org/10.3389/fclim.2024.1272320
Ou W et al (2018) Genome-wide identification and expression analysis of the KUP family under abiotic stress in cassava (Manihot esculenta Crantz). Front Physiol 9:17. https://doi.org/10.3389/fphys.2018.00017
Owen G (2020) What makes climate change adaptation effective? A systematic review of the literature. Glob Environ Chang 62:102071. https://doi.org/10.1016/j.gloenvcha.2020.102071
Pareek A, Dhankher OP, Foyer CH (2020) Mitigating the impact of climate change on plant productivity and ecosystem sustainability. J Exp Bot 71(2):451–456. https://doi.org/10.1093/jxb/erz518
Parry M, Rosenzweig C, Livermore M (2005) Climate change, global food supply and risk of hunger. Philos Trans R Soc Lond B Biol Sci 360(1463):2125–2138. https://doi.org/10.1098/rstb.2005.1751
Pixley KV et al (2019) Genome editing, gene drives, and synthetic biology: will they contribute to disease-resistant crops, and who will benefit? Annu Rev Phytopathol 57:165–188. https://doi.org/10.1146/annurev-phyto-080417-045954
Qi M, Berry JC, Veley KW, O'Connor L, Finkel OM, Salas-González I, Kuhs M, Jupe J, Holcomb E, Glavina del Rio T (2022) Identification of beneficial and detrimental bacteria impacting sorghum responses to drought using multi-scale and multi-system microbiome comparisons. ISME J 16(9):1957–1969. https://doi.org/10.1038/s41396-022-01151-7
Raza A, Razzaq A, Mehmood SS, Zou X, Zhang X, Lv Y, Xu J (2019) Impact of climate change on crops adaptation and strategies to tackle its outcome: a review. Plants (Basel, Switzerland) 8(2):34. https://doi.org/10.3390/plants8020034
Raza A, Farwa A, Zou X, Xuekun Z, Tosif H (2020) Plant adaptation and tolerance to environmental stresses: mechanisms and perspectives. In: Plant ecophysiology and adaptation under climate change: mechanisms and perspectives. Springer Nature. https://doi.org/10.1007/978-981-15-2156-0_5
Rosa L (2022) Adapting agriculture to climate change via sustainable irrigation: biophysical potentials and feedbacks. Environ Res Lett 17(6):063008. https://doi.org/10.1088/1748-9326/ac7408
Sánchez-León S et al (2018) Low-gluten, nontransgenic wheat engineered with CRISPR/Cas9. Plant Biotechnol J 16(4):902–910. https://doi.org/10.1111/pbi.12837
Sargent D et al (2022) Synthetic biology and opportunities within agricultural crops. J Sustain Agri Environ 1(2):89–107. https://doi.org/10.1002/sae2.12014
Schlenker W, Roberts MJ (2009) Nonlinear temperature effects indicate severe damages to U.S. crop yields under climate change. Proc Natl Acad Sci 106(37):15594–15598
Schmidhuber J, Tubiello FN (2021) Global food security under climate change. Proc Natl Acad Sci 118(8):e2100329118. https://doi.org/10.1073/pnas.2100329118
Seeda A, Abou El-Nour EZ, Abdallah M, El-Bassiouny H, El-Monem A (2022) Impacts of salinity stress on plants and their tolerance strategies: a review. Middle East J Appl Sci 12:282–400

Shen C et al (2017) Knock out of the annexin gene OsAnn3 via CRISPR/Cas9-mediated genome editing decreased cold tolerance in rice. J Plant Biol 60:539–547. https://doi.org/10.1007/s12374-016-0400-1
Shi X et al (2022) Optimal irrigation under the constraint of water resources for winter wheat in the North China plain. Agriculture 12(12):2057. https://doi.org/10.3390/agriculture12122057
Shimatani Z et al (2017) Targeted base editing in rice and tomato using a CRISPR-Cas9 cytidine deaminase fusion. Nat Biotechnol 35(5):441–443. https://doi.org/10.1038/nbt.3833
Shinde Ganesh S, Sapkale Shivam J, Sapkale Jagdish B (2023) The disastrous effects of soil salinity and pH on environmental systems. Disas Adv 16(6):53–62
Shukla PR et al (2019) Climate change and land: an IPCC special report on climate change, desertification, land degradation, sustainable land management, food security, and greenhouse gas fluxes in terrestrial ecosystems. Intergovernmental Panel on Climate Change
Si H, Ma C et al (2021) Effects of heat stress on growth, physiology of plants, yield, and grain quality of different spring wheat (Triticum aestivum L.) genotypes. Sustain For 13(5):2972. https://doi.org/10.3390/su13052972
Siad SM et al (2019) A review of coupled hydrologic and crop growth models. Agric Water Manag 224:105746. https://doi.org/10.1016/j.agwat.2019.105746
Singh P, Choudhary KK, Chaudhary N, Gupta S, Sahu M, Tejaswini B, Sarkar S (2022) Salt stress resilience in plants mediated through osmolyte accumulation and its crosstalk mechanism with phytohormones. Front Plant Sci 13. https://doi.org/10.3389/fpls.2022.1006617
Sinore T et al (2024) Impact of climate change on agriculture and adaptation strategies in Ethiopia: a meta-analysis. Heliyon 10:e26103. https://doi.org/10.1016/j.heliyon.2024.e26103
Taranto F et al (2018) Biotechnological and digital revolution for climate-smart plant breeding. Agronomy 8(12):277. https://doi.org/10.3390/agronomy8120277
Terán-Chaves CA et al (2023) Water productivity indices of onion (Allium cepa) under drip irrigation and mulching in a semi-arid tropical region of Colombia. Horticulturae 9(6):632. https://doi.org/10.3390/horticulturae9060632
Timlin D et al (2024) The role of crop simulation modeling in assessing potential climate change impacts. Agrosystems Geosci Environ 7(1):e20453. https://doi.org/10.1002/agg2.20453
Varshney RK, Thudi M, Nayak SN, Gaur PM, Kashiwagi J, Krishnamurthy L, Jaganathan D, Koppolu J, Bohra A, Tripathi S (2014) Genetic dissection of drought tolerance in chickpea (Cicer arietinum L.). Theor Appl Genet 127(3):445–462. https://doi.org/10.1007/s00122-013-2230-4
Varshney RK, Bohra A, Yu J, Graner A, Zhang Q, Sorrells ME (2021) Designing future crops: genomics-assisted breeding comes of age. Trends Plant Sci 26(7):631–649. https://doi.org/10.1016/j.tplants.2021.01.003
Vianna MS et al (2022) Improving the representation of sugarcane crop in the Joint UK Land Environment Simulator (JULES) model for climate impact assessment. GCB Bioenergy 14(10):1097–1116. https://doi.org/10.1111/gcbb.12989
Wang X et al (2012) EPIC and APEX: model use, calibration, and validation. Trans ASABE 55(4):1447–1462. https://doi.org/10.13031/2013.42253
Wang W et al (2018) Transgenerational CRISPR-Cas9 activity facilitates multiplex gene editing in allopolyploid wheat. The CRISPR journal 1(1):65–74. https://doi.org/10.1089/crispr.2017.0010
World Bank (2013) Turn down the heat: climate extremes, regional impacts, and the case for resilience. World Bank, Washington, DC
World Bank (2015) Shock waves: managing the impacts of climate change on poverty. World Bank, Washington, DC
World Bank (2020) Employment in agriculture (% of total employment) (Modeled ILO Estimate). Retrieved from: https://data.worldbank.org/indicator/SL.AGR.EMPL.ZS
Xiao D et al (2023) Modeling the adaptation of agricultural production to climate change. Agriculture 13(2):414. https://doi.org/10.3390/agriculture13020414
Yamaguchi-Shinozaki K, Shinozaki K (2006) Transcriptional regulatory networks in cellular responses and tolerance to dehydration and cold stresses. Annu Rev Plant Biol 57:21. https://doi.org/10.1146/annurev.arplant.57.032905.105444

Yan X-R, Wang Z-Y, Feng S-Q, Zhao Z-H, Li Z-H (2022) Impact of temperature change on the fall armyworm, Spodoptera frugiperda, under global climate change. Insects 13(11):981. https://doi.org/10.3390/insects13110981

Yang X et al (2020) Editorial: systems biology and synthetic biology in relation to drought tolerance or avoidance in plants. Front Plant Sci 11:394. https://doi.org/10.3389/fpls.2020.00394

Yu H, Yang T, Wang C, Huang S, Chen H, Ao X (2024) Impacts of global climate change on agricultural production: A comprehensive review. Agronomy 14(7):1360. https://doi.org/10.3390/agronomy14071360

Zhang J, Deng M, Han Y et al (2023a) Spatiotemporal variation of irrigation water requirements for grain crops under climate change in Northwest China. Environ Sci Pollut Res 30(45):45711–45724. https://doi.org/10.1007/s11356-023-25438-2

Zhang Y, Xu J, Li R, Ge Y, Li Y, Li R (2023b) Plants' response to abiotic stress: mechanisms and strategies. Int J Mol Sci 24(13). https://doi.org/10.3390/ijms241310915

Zhao Y et al (2022) Future projection for climate suitability of summer maize in the North China Plain. Agriculture 12(3):348. https://doi.org/10.3390/agriculture12030348

Zhu C et al (2017) Characteristics of genome editing mutations in cereal crops. Trends Plant Sci 22(1):38–52. https://doi.org/10.1016/j.tplants.2016.08.009

The Potential Effect of Climate Change in Agricultural Insect and Pests

7

Sadia Zafar, Javed Iqbal, Maria Bilal, Banzeer Ahsan Abbasi, Shoaib Khan, Shumaila Ijaz, Zakir Ullah, Ghulam Murtaza, Rashid Iqbal, Lubna Ansari, Rabia Kazmi, and Hafsa Ansar

Abstract

Climate change is manifesting through extreme weather events, altered rainfall patterns, and a rising global temperature. These changes are poised to have significant effects on flora, fauna, and insect pest populations. Abiotic factors, such as temperature, precipitation, and atmospheric CO_2 levels, directly influence insect population dynamics by affecting development rates, survival, fecundity, voltinism, and dispersal. Uncontrolled pest populations can lead to increased crop stress, higher crop losses, and reduced agricultural productivity. As climate

Sadia Zafar and Javed Iqbal contributed equally with all other contributors.

S. Zafar (✉) · M. Bilal
Department of Botany, Division of Science and Technology, University of Education, Lahore, Punjab, Pakistan
e-mail: sadia.zafar@ue.edu.pk

J. Iqbal (✉)
Department of Botany, Bacha Khan University, Charsadda, Khyber Pakhtunkhwa, Pakistan

B. A. Abbasi · R. Kazmi · H. Ansar
Department of Botany, Rawalpindi Women University, Rawalpindi, Pakistan

S. Khan
Department of Chemistry, Abbottabad University of Science and Technology (AUST), Abbottabad, Pakistan

S. Ijaz
Department of Burn and Plastic Surgery, Department of Wound Repair, Shenzhen Institute of Translational Medicine, The First Affiliated Hospital of Shenzhen University, Second People's Hospital, School of Biomedical Engineering, Shenzhen University Medical School, Shenzhen University, Shenzhen, China

© The Author(s), under exclusive license to Springer Nature Singapore Pte Ltd. 2025

S. Faiz et al. (eds.), *Climate Smart Agriculture for Future Food Security*,
https://doi.org/10.1007/978-981-96-4499-5_7

change progresses, the stress on crops will intensify, exacerbated by the direct effects of these climatic variables on plant health. Temperature, in particular, plays a critical role, with rising temperatures potentially altering the population dynamics of agricultural pests and insects. Global warming may also shift the geographical distribution of species, favoring multivoltine species that can complete multiple generations per year. Elevated $C0_2$levels further contribute to this dynamic by altering the nutritional quality of crops, which can lead to increased insect herbivory, greater crop damage, reduced yields, and higher costs for crop protection, all of which have significant economic implications. Therefore, climate change is expected to cause fluctuations in insect pest populations. This chapter delves into the potential impacts of climate change on insect pests.

Keywords

Climate change · Agricultural insects · Agricultural pests · Abiotic factors · Global warming

7.1 Introduction

In recent decades, significant scientific and technological advancements have accelerated economic growth and increased agricultural productivity globally (Pingali and Abraham 2022). These developments have driven what can be termed "agricultural revolutions," marked by human, cultural, and technological progress. However, the rapid population growth over the last century, coupled with environmental changes, has introduced challenges that threaten food security. To meet the demands of a growing global population, agricultural production will likely need to double by 2050 (Tilman et al. 2011). Numerous studies suggest that increasing agricultural output, rather than expanding arable land, is the most sustainable approach to ensuring food security (Godfray et al. 2010).

Z. Ullah
Department of Plant Sciences, Faculty of Biological Sciences, Quaid-i-Azam University Islamabad, Islamabad, Pakistan
e-mail: zakirullah@bs.qau.edu.pk

G. Murtaza
School of Agriculture, Yunnan University, Kunming, Yunnan, China

R. Iqbal
Department of Agronomy, Faculty of Agriculture and Environment, The Islamia University of Bahawalpur, Bahawalpur, Pakistan

Department of Life Sciences, Western Caspian University, Baku, Azerbaijan
e-mail: rashid.iqbal@iub.edu.pk

L. Ansari
Department: Forestry and Range Management, Pir Mehr Ali Shah- Arid Agriculture University Rawalpindi, Rawalpindi, Pakistan
e-mail: lubna.ansari@uaar.edu.pk

Climate change has the potential to impact agricultural production both directly, through its effects on pests and diseases, and indirectly, by altering plant productivity (Rubab et al. 2023). There is substantial evidence that phytophagous insects, particularly Lepidoptera, are sensitive to climate variations (Wilson and Maclean 2011). Climate change is also affecting insect phenology, including the timing of Lepidoptera arrivals (Parmesan 2007; Ahmad et al. 2022) and the emergence of various insect species (Whittaker and Tribe 1996). Many of these impacted species are herbivorous pests, which can lead to the disappearance of some pests, the emergence of new ones, and shifts in the pressure exerted by existing pest species (Ananthakrishnan and Sivaramakrishnan 2017).

The increasing complexity of harmful invasive insect infestations is becoming a more significant challenge, with climate change being a primary driver of this issue in agricultural systems worldwide (Ward and Masters 2007). Climate change, driven by increased concentrations of atmospheric greenhouse gases (GHGs), is leading to more frequent and severe weather patterns, contributing to global warming (Florentine et al. 2020). These combined effects of climate change are placing immense pressure on agricultural systems and food production sectors (Ahmad et al. 2022).

As noted by Kurukulasuriya and Rosenthal (2003), the most significant climatic pressure on agricultural output stems from the increasing frequency of extreme weather events. These events include temperature fluctuations that lead to heatwaves and frosts, as well as altered rainfall patterns that cause prolonged droughts or floods (Haider et al. 2021a, b). Climate change pressures can be categorized into direct effects on host plants and indirect effects on insect pests and other harmful species that impact agricultural productivity (Florentine et al. 2020; Ahmad et al. 2024). Climate change significantly influences the relationships between pests, their natural enemies, and their environment, affecting reproduction, survival, dispersal, and population dynamics (Prakash et al. 2014; Raja et al. 2022).

A considerable proportion of agricultural pests are also invasive species (Ward and Masters 2007). Invasive species, defined as organisms that exist outside their natural range and have the potential to cause harm to the environment, economy, or human and animal health, are expected to expand geographically under various climate change scenarios (Kistner 2017). Climate change will influence the distribution of these species due to their responses to variations in temperature, humidity, and biotic interactions (Finch et al. 2021). The likelihood of invasive species establishing themselves in new areas is influenced by climatic conditions, with a greater chance of establishment in regions with similar climates to their native range (Walther et al. 2009). However, unfavorable climatic conditions can limit the successful establishment of invasive species. Nonetheless, climate change has created new ecological niches, increasing the likelihood of invasive pests establishing themselves outside their typical geographic range (FAO 2021a, b).

It is important to note that biological invasions are driven by factors beyond climate change alone. Most invasive species spread naturally or through human activities such as migration and international trade (FAO 2021a, b). After arriving in a new habitat, alien species may experience favorable or unfavorable effects from the local environment and ongoing climate change during the stages of introduction, establishment, and dispersal (FAO 2021a, b). Characterizing these effects is crucial for food security, as climate change could alter the success of insect pest invasions with significant implications for agricultural productivity (Ziska et al. 2011). This essay will explore anticipated changes in temperature and their potential impact on the ecology and biology of harmful insects, particularly invasive pest species that pose a significant risk to agricultural productivity. These changes include rising temperatures, increasing atmospheric CO_2 concentrations, and variable precipitation patterns. The current challenges in plant production will be addressed, focusing on modified integrated pest management (IPM) techniques as potential solutions, combining sustainable food production practices, predictive modeling tools, and monitoring procedures.

7.2 Climate under Change

Climate significantly influences the distribution and characteristics of both controlled and natural systems, including cryology, hydrology, freshwater and marine ecosystems, forestry, agriculture, and water resources (Rosenzweig et al. 2007). Climate change is characterized by long-term alterations in precipitation, humidity, temperature, and other climatic variables. The global food supply is under serious threat from rising temperatures, climate extremes, increased CO_2 and other greenhouse gases (GHGs), and changing precipitation patterns (Shrestha 2019).

Global warming, a major issue facing the world today, has reached unprecedented levels, as evidenced by rapid increases in sea level and air temperature (Field et al. 2014). According to the World Meteorological Organization (WMO), global temperatures have risen by approximately one degree since the onset of widespread industrialization. The Intergovernmental Panel on Climate Change (IPCC) reported that the 2000s were the warmest decade of the past three decades (Field et al. 2014). Various global climate models and development scenarios predict that Earth's temperature could increase by 1.4 to 5.8 degrees Celsius over the next century (Pachauari and Reisinger 2007).

The primary driver of global warming is the increase in atmospheric greenhouse gases. The most prevalent of these gases are nitrous oxide (N_2O), carbon dioxide (CO_2), and methane (CH_4), produced by anthropogenic activities such as fossil fuel combustion and land-use changes (Yoro and Daramola 2020). These gases disrupt the long-wave balance at Earth's surface, leading to more energy being available for latent and sensible heat flux, which in turn raises air temperatures (Streck 2005).

7.3 Impact of Climate Change on Crop Production

Agriculture, often referred to as an "open-air factory," is an economic activity heavily dependent on climate and specific weather patterns for the production of food and other essential goods. However, agriculture is highly vulnerable to the uncertainties associated with climate change (Dessai and Der Sluijs 2007). Global agricultural systems are expected to experience both positive and negative impacts from climate change, with the negative effects likely to outweigh the positives (Müller 2013). From species to ecosystem levels, temperature increases, altered precipitation patterns, and elevated CO_2 concentrations are significantly affecting ecosystems (Stevens et al. 2004). Understanding the impact of climate change on crop productivity is crucial since insect pests depend on and thrive on the plant species they consume.

7.4 Impact of Increasing Temperature on Crop Production

Temperature is a key factor determining the distribution and abundance of plants due to the physiological limitations of each species (Parmesan and Yohe 2003). It influences where different crops can be grown and affects their growth, development, and yield (Rehman et al. 2015). Basic temperature requirements are necessary for crops to complete their life cycle and specific phenophases. Extreme temperatures, particularly during critical phenophases such as anthesis, can have detrimental effects on crop growth and yield (Luo 2011; Li et al. 2021). Warmer temperatures in spring and summer are expected to benefit crop production in northern regions where the growing season is currently limited (Tubiello et al. 2002). However, these benefits may be offset by other environmental factors such as water availability, wind, and sunlight intensity, which are also influenced by rising temperatures (Rehman et al. 2015).

The indirect effects of temperature on these environmental factors can further impact crop yields. For instance, higher temperatures increase atmospheric water demand, potentially leading to greater water stress due to increased evaporative demand. This, in turn, reduces soil moisture, ultimately lowering crop yields (Zhao et al. 2017; Ali et al. 2022).

7.5 Impact of Increasing CO_2 Concentration

Photosynthesis, a process powered by sunlight, converts CO_2 and water into carbohydrates and starch, with CO_2 playing a crucial role. Rötter and Van De Geijn (1999) emphasized that without CO_2 entering through stomatal openings, photosynthesis cannot occur in the green pigments of leaves. Higher CO_2 concentrations enhance plant growth by accelerating carbon uptake, which is vital for plant structure (Woodward 1990). The key effects of elevated CO_2 levels on plants include increased photosynthetic rates, improved water and light use efficiency, and reduced transpiration and stomatal conductance. Consequently, higher atmospheric CO_2 levels can directly influence ecosystems by promoting plant growth and development (Reddy et al. 2010).

7.6 Impact of Climate Change on Insect Pests

Global environmental changes significantly affect agricultural productivity, including the dynamics of insect pests. Climate change impacts crops and the pests that attack them both directly and indirectly. It directly influences pest development, reproduction, survival, and distribution, while also indirectly affecting interactions between pests, their habitats, and other insect species, including natural enemies, competitors, vectors, and mutualists (Prakash et al. 2014). Insects, being poikilothermic, have body temperatures that fluctuate with external temperatures. Thus, temperature is a crucial environmental factor governing insect development, distribution, behavior, and reproduction (Kocmánková et al. 2010). Climate change, driven by increasing atmospheric CO_2, rising temperatures, and decreasing soil moisture, is likely to significantly alter insect pest populations and crop loss rates (Fand et al. 2012). Additionally, climate change has created new biological niches, providing insect pests with greater opportunities to spread and migrate across different geographic regions (FAO 2020). The complex physiological effects of rising temperatures and CO_2 can significantly influence interactions between crops and insect pests (Roth and Lindroth 1995). As the environment continues to change, farmers must prepare for emerging and serious pest issues, as the spread of crop pests across national and international borders poses a global threat to food security (FAO 2020).

Therefore, it is very likely that the main contributors to climate change increasing atmospheric CO_2, rising temperatures, and decreasing soil moisture will significantly affect the dynamics of insect pest populations and, in turn, the percentage of crop losses (Fand et al. 2012). Due to the Because of the additional biological niches that climate change has created, insect pests have greater opportunities to spread, disseminate, and move between different geographic locations (FAO 2020). The complicated physiological impacts of rising temperatures and CO_2 can have a substantial impact on interactions between agricultural crops and insect pests (Roth and Lindroth 1995). Farmers need to get ready for new and serious pest problems in the years to come as a result of the changing environment. The spread of crop pests over national and international borders is a global problem that impacts all countries and regions and poses a threat to food security (FAO 2020).

7.7 Impact of Temperature on Survival Rate of Insects

Temperature plays a critical role in influencing insect species in various complex ways. Since insects are poikilothermic, their growth, development, and reproduction are largely governed by temperature, as illustrated in Fig. 7.1. For instance, the breeding cycle of the spring beetle, *Dentroctonus rufipennis*, was reduced by half under warmer temperatures (McKee and Aukema 2015). At −16 °C, a temperature lethal to many, the pine beetle population significantly declines; however, during milder winters, their survival and dispersal rates increase. Aphids, for example, go through an additional 1–5 life cycles per season when temperatures

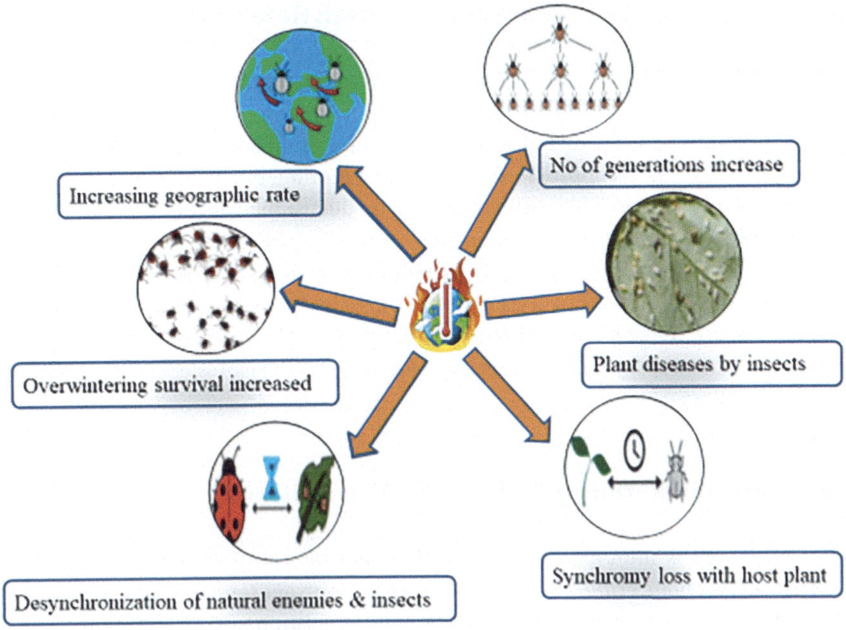

Fig. 7.1 Effects of temperature on agricultural insects and pests

rise by 2 °C (Yamamura and Kiritani 1998). In the case of the brown plant hopper, *Nilapavata lugens*, adult survival remained largely unchanged between 25 °C and 35 °C, but saw a significant drop at 40 °C. Although female oviposition was higher at 35 °C and 40 °C compared to 25 °C and 30 °C, egg survival markedly decreased at 35 °C, with pre-oviposition periods also shortening at these higher temperatures.

These studies indicate that temperatures above 35 °C are likely to limit the development of Brown Plant Hopper (BPH). In areas where temperatures remain below 30 °C, the prevalence of BPH is expected to increase due to global warming (Heong et al. 1995). Insects may also adapt to climate change by altering their life cycles. Insects from warmer climates may exhibit higher fecundity and shorter growth stages. For instance, Dhillon and Sharma (2007) found that at 25 °C, the egg stage of *Helicoverpa armigera* lasted 10.4 days, but shortened to 7.9 days at 27–28 °C. An increase in both temperature and egg age, from 0–27 °C and three days, respectively, resulted in reduced egg viability and fewer days needed for hatching. The degree days (DD) required were highest at 10 °C and lowest at 27 °C, with a hatchability rate exceeding 75% at 27 °C (Dhillon and Sharma 2007). In the case of the rice ear head bug, *Leptocorisa acuta*, a daily average increase of 0.5–2 °C had no effect on generation time, but a 3 °C increase extended it by 1–3 days. Meanwhile, the population of adults, wing-padded nymphs, and non-wing-padded nymphs decreased with a temperature rise of 0.5–1 °C (Karuppaiah and Sujayanad 2012).

7.8 Impact of Temperature on Growth Rate of Insects

Pests' rapid development rates allow them to adapt more swiftly to temperature changes. For example, the population of *Nephotettix cincticeps* is expected to rise by three to four times. There appears to be a similar trend between the abundance of *N. cincticeps* and *Chilo suppressalis* and global warming, but the increase in *N. cincticeps* is notably higher. This difference may be due to the varying number of generations each species produces annually. With its biannual generation cycle, *N. cincticeps* may be more sensitive to temperature changes (Yumamura et al. 2006). Predictions suggest that for every 1 °C increase in temperature, the winter mortality of adult *Nezara viridula* and *Halyomorpha halys* will decrease by 15%, while *C. suppressalis* is projected to produce two generations of survivors each year (Streck 2005).

7.9 Global Warming vs. Species Distributions

Climate change is anticipated to cause the distribution of the spruce budworm (*Choristoneura fumiferana*) to shift poleward, as warmer temperatures and changing weather patterns create new environmental conditions (Regniere et al. 2010). This shift in distribution may have mixed implications depending on the species involved and the regions they move into. Some areas might experience increased pest pressures, while others could see a reduction or complete absence of these pests.

In North American forests, both native and non-native insect pests are already responding to climate change, influencing their range and impact (Ijaz et al. 2023, 2024). For example, the eastern spruce budworm (*C. fumiferana*) has been observed to expand its range and increase in abundance in response to warmer temperatures (Gray 2008). Similarly, the western spruce budworm (*Choristoneura occidentalis*) has shown changes in distribution patterns and population dynamics due to shifting climatic conditions (Williams and Liebhold 1995).

Furthermore, studies have indicated that the range expansions of these pests could lead to increased forest damage and altered ecosystem dynamics. For instance, warmer temperatures may accelerate the life cycles of these insects, resulting in more frequent outbreaks and increased defoliation of forest trees. The altered distributions may also lead to higher interactions with other pest species, potentially exacerbating the overall impact on forest health (Logan et al. 2003; Volney and Fleming 2000a, b).

7.10 Precipitation and Drought

Numerous studies on climate change have predicted that modifications in rainfall in different areas may be impacted by patterns brought on by climate change (Cumhur and Cresser 2008). The population of soil-dwelling *Agriotes lineatus* (wireworms) increases quickly in the upper soil due to increased summer rainfall and dry conditions (Staley et al. 2007; Haider et al. 2022a, b). The desert locust (*Schistocerca gregaria*) is migrating according to patterns determined by changes in rainfall in Sub-Saharan Africa (Cheke and Tratalos 2007; Ullah et al. 2023a, b). Higher November rainfall favored higher infestation, according to a study on the variation in rainfall between the monsoon and November and its link to the degree of *Helicoverpa armigera* damage (Srivastava et al. 2010).

Some places are expected to get lower average rainfall, and there will probably be more summer droughts. Different species of root herbivores reacted differently to the manipulations of summer rainfall. Increased rainfall led to an increase in the number of larvae of the main root-chewing species, *Agriotes lineatus*; however, the manipulation of rainfall had no effect on the abundance of *Coccoidea lecanopsis formicarum*. Therefore, based on their life history and manner of feeding, root herbivores are likely to respond differently to an increased incidence of summer droughts (Staley et al. 2007). *Scopelosaurus lepidus* eggs were significantly damaged by drought conditions, and in extremely dry conditions, the eggs did not hatch at all (Johnson et al. 2010).

7.11 Indirect Effects of Climate Change on Insect Pests via Host Plants

The CO_2 in the atmosphere is probably going to increase plant resistance and defense against the invasion of phytophagous insects. Turgor pressure and plant tissue water content both drop during moderate water stress, resulting in harder foliage. Plants respond to higher tropospheric CO_2 concentrations by being more productive because of the direct fertilization effects (Lindroth et al. 1993).

Two major populations of soybean aphids (*Aphis glycines*) and adult western corn rootworms (*Diabrotica virgifera*) (foliage chewer) are increased upon exposure to increasing CO_2. In the soybean agroecosystem, herbivory may increase in response to higher CO_2 concentrations in the atmosphere (Dermody et al. 2008).

7.12 Effects of Rising CO_2 Levels on Insects

In general, it is believed that the effects of CO_2 on insects are indirect and that changes in the host crop have an impact on insect damage. Insect damage to soybeans cultivated in high CO_2 environments increased by 57%. The most common insects that caused damage were the Japanese beetle, potato leafhopper, western corn rootworm, and Mexican bean beetle. It's possible that measured rises in the amounts of simple sugars in the soybean leaves prompted the extra insect feeding. In order to get enough nitrogen for their metabolism, insects will occasionally feed more on leaves with reduced nitrogen concentration (Coviella and Trumble 1999; Hunter 2001). Elevated CO_2 levels in plant tissue can lead to higher carbon to

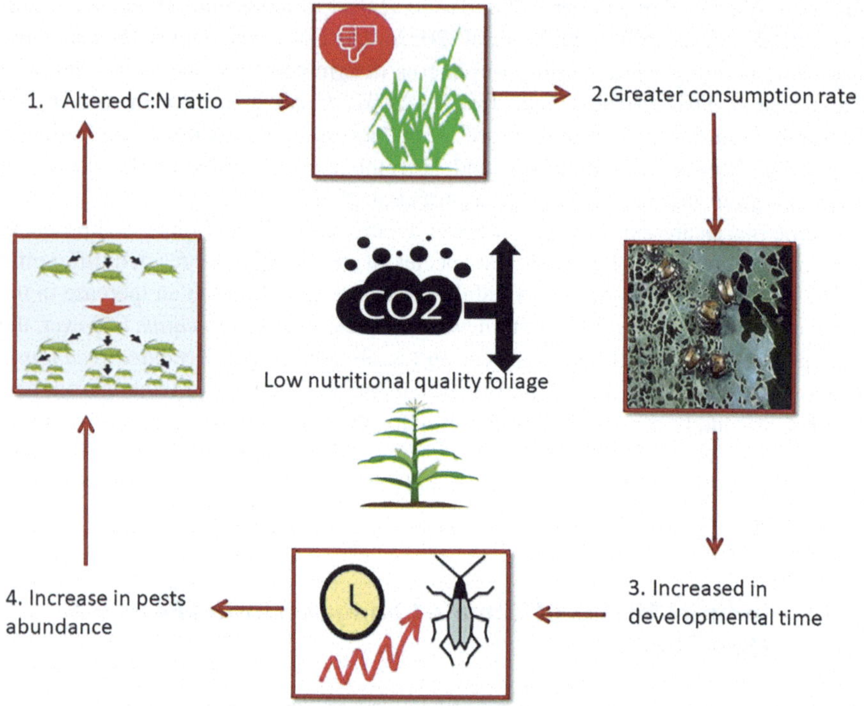

Fig. 7.2 Rising level of CO_2 on agricultural insects and pests

nitrogen ratios (C:N), which can impede insect development and lengthen the life stages that are susceptible to parasitoids, as shown in Fig. 7.2.

7.13 Potential Responses of Insects to Climate Change

The weather and climate have a significant impact on the growth and dispersal of insects. According to current climate change projections. The average yearly temperature worldwide will increase by 1 °C by 2025 and 3 °C by the end of the following century. Temperature-dependent insects are affected in many ways by such temperature increases, particularly in the Middle-European region. A changing climate may have an impact on a variety of factors, including geographic distribution, overwintering, population growth rates, generational increases, elongation of the development season, synchronicity between crop and pest phenology, interspecific interactions, and an increased risk of invasion by migrant pests (Parmesan 2007; Porter et al. 1991).

A northward shift the global circulation model of the Goddard Institute for Space Studies estimates a potential dispersion of the European corn borer of up to 1220 kilometers; nearly all areas where it is now known to exist have an extra generation present (Porter et al. 1991). Since insects cause significant wood loss, which negatively affects publication in the realm of forestry sciences regarding the equilibrium of carbon stored by forests and several discoveries regarding the effects of climate change on insects.

According to Volney and Fleming (2000a, b), pests are important but frequently disregarded elements of the forest ecosystem, with far-reaching effects on the composition and operations of forests in the future. Pest outbreak frequency and intensity will clearly alter due to changes in the environment, particularly in the areas where host ranges are expanding.

7.14 Response of Insect Pests to Increased Temperature

Insects usually double their metabolic rate for every 10 °C increase in temperature due to their extremely sensitive physiology (Dukes et al. 2009). Several studies have shown that rising temperatures in this specific environment tend to accelerate the rate at which insects feed, grow, and migrate. By affecting variables including fecundity, survival, generation time, population size, and geographic range, this may affect population dynamics (Bale et al. 2002). While certain species can multiply rapidly and thrive, most species find it difficult to maintain their populations when they are incapable of evolving and adapting to warmer climates.

According to Deutsch et al. (2018), their predictions indicate that insect populations in maize planted in both temperate and tropical regions will rise, while those in rice grown in tropical zones will see their populations decline.

Future developments in insect population dynamics will depend on how quickly global temperatures rise in the next years. Climate estimates indicate that the average world temperature will increase by 1.8–4 °C by the end of the current century (Collins et al. 2007). As general temperatures generally rise toward the ideal conditions for the development and growth of many insect pest species, the extent of pest infestations is predicted to increase under global warming scenarios, potentially reducing thermal limitations on the dynamics of populations (Deutsch et al. 2008). However, because of the particular ecological niche requirements, physiological tolerances of insects, and varying effects of temperature on their phenology and life cycle, global warming may not necessarily result in an increase in pest population and, consequently, in financial crop losses (Lehmann et al. 2020).

7.15 Response of Insect Pests to Increased CO_2 Concentration

Since insects require nitrogen for growth, rising CO_2 levels may cause some pest groups to consume plants more quickly (Bezemer et al. 1998). Because pests must consume more tissue in order to obtain the same amount of nutrients, plants may sustain greater harm as a result. Caterpillars, miners, and chewers are examples of foliage feeders that exhibit this phenomenon of increased consumption rates as a way of making up for the nitrogen reduction that CO_2 fertilization is expected to cause (Hamilton et al. 2005). Elevated CO_2 levels in plants can diminish the higher feeding rates do not always make up for a meal of inferior quality, nor do the efficiency of the arthropods that eat them (Zvereva and Kozlov 2005).

The type of pest and how it feeds determine the impact of CO_2 fertilization. For example, populations of whole-cell feeders, such as thrips, frequently grow (Bezemer et al. 1998). Pest insects that feed on phloem, such as aphids and whiteflies, may exhibit both a decrease in population density and an increase in population growth rates (Sutherst et al. 2011). There is mixed data regarding how higher CO_2 affects sucking insects; some research indicates that it increases their fecundity and abundance (Sun and Ge 2011). Thus, even after a great deal of research has been done on how aphids react to increased atmospheric CO_2 levels, it is still hard to forecast how they will behave in the future and to provide general guidelines for how different aphid populations will react to climate change (Sun and Ge 2011).

7.16 Response of Insect Pests to Changeable Precipitation Patterns

Precipitation variations, encompassing shifts in amount, intensity, and frequency, are essential markers of climate change. The majority of research has demonstrated that although the intensity of precipitation has increased, the frequency of precipitation occurrences has reduced. Droughts and floods are occurring more frequently as a result of this tendency. Insect eggs and larvae can potentially be moved by heavy rains and flooding (Shrestha 2019). During periods of intense rainfall, small-bodied pests including aphids, mites, jassids, and whiteflies may be carried away (Pathak et al. 2012).

Variations in rainfall can have a big effect on insect populations. For instance, Staley et al. (2007) investigated how the soil-dwelling wireworm (Agriotes lineatus L.) in grassland plots was affected by drought and increasing summer rainfall. Climate change is predicted to make wireworms an even more severe pest, particularly for crops often produced in grassland plots, such as potatoes, corn, and sugar beets (Johnson et al. 2008). According to Staley et al. (2007), wireworm populations in the upper soil layers rose quickly when summer rainfall exceeded ambient and dry conditions (Gregory et al. 2009).

7.17 Adaptation and Mitigation Strategies for Pest Management in a Changing Climate

The mitigating potential risks and using contemporary risk management strategies are essential components of the ongoing process of adapting to climate change (Howden et al. 2007). It is widely expected that as a result of climate change, insect infestations will become more erratic and disperse regionally. Furthermore, the direct effects of climate change on crop yields and the interactions between insects and plants in ecosystems remain uncertain (Gregory et al. 2009).The degree of adaptability of agricultural production systems will depend on a variety of biological, economic, and social factors. The ability of local communities to adapt their pest management strategies will depend on their financial, social, and physical resources (Sutherst et al. 2011).

The speed of international trade combined with climate change will increase the uncertainties around the introduction of both new and old pests. Consequently, increasing resilience to disturbances and climate change will become increasingly more important (Barzman et al. 2015).

7.18 Modified Integrated Pest Management Practices

Modified Integrated Pest Management (IPM), or integrated pest management, addresses illnesses, weeds, and dangerous creatures that feed on plants, mainly insects and mites. Plant protection in sustainable agriculture places a high emphasis on preventative or indirect strategies, which have to be fully employed before turning to direct control measures. Scientifically established thresholds and forecasting approaches are two examples of the most recent instruments that must be used when making decisions on whether control measures are necessary. Only when indirect measures fail to prevent losses that are economically undesirable should direct pest management methods be taken into consideration (Boller et al. 2004).

The FAO advocates for a dual approach that combines global and regional actions with substantial financial investment to improve current early detection and control systems. This approach requires developing innovative farming techniques, introducing new crop species, and implementing integrated pest management principles to prevent the spread of pests (Gomez-Zavaglia et al. 2020).

In a changing climate, agricultural pests can have detrimental effects on crops that must be mitigated by improved farming practices and adaptive management strategies. These tactics could involve, according to Andrew and Hill (2017), growing a variety of crop kinds, modifying planting schedules to avoid pest outbreaks at their height, and improving biodiversity at field edges to boost the number of natural enemies. Insect perception relies heavily on pheromones and allelochemicals, which are also essential for many IPM techniques, such as biological control, mating disruption, push-pull tactics, trapping, and monitoring (Heuskin et al. 2011).

Pheromones and allelochemicals may become less effective in their current form as the temperature heats and microclimates become more varied. According to Andrew and Hill (2017), in order to lessen their volatility under high temperatures, a synergist or other adjuvant may be required.

Moreover, several biopesticides based on bacteria, nematodes, fungi, or entomopathogenic viruses are highly sensitive to environmental changes. Rising temperatures and decreasing relative humidity could reduce the effectiveness of these control strategies, and synthetic insecticides might perform similarly under such conditions (Nihal 2020).

We provide a brief summary of the most popular ecological models that are targeted at insect populations. For around 30 years, the Forest Vegetation Simulator (FVS), an individual-tree forest growth model that is reliant on geographic region but not distance, has been extensively utilized in the US to assist in management decision-making. It has been regularly expanded, enhanced, and modified to handle more management responsibilities, such as the implications of climate change prediction. Individual tree growth and mortality are predicted by component models, while disturbance agents such as insects, diseases, and fire are represented by extensions to the base model. Regionally specific model versions depict the various geographic regions. The availability of data and the suitability of current models account for the discrepancies. According to Crookston and Dixon (2005), the model facilitates the defining of management rules in the input. The Phenology and Population SIM (INSIM) is an age-structured model that provides estimates of population growth and number of individuals based on biological data about the insect species. It also incorporates a sophisticated model of natural enemies vs. pests (Mols 1992).

A stochastic simulation model called the Boll Weevil DISPersal Model (BWDISP) forecasts how boll weevil populations will disperse across cotton. This insect is temperature-sensitive in both development and dissemination, so it's critical to comprehend how it might react to climate change. Furthermore, if this pest is not well managed, additional secondary pests may attack the crop (McKibben et al. 1991).

The changing patterns of the corn rootworm population in the northern United States are examined with a process-oriented simulation model known as the Northern Corn ROOTWORM Model (ROOTWORM). The rootworm's attack on the plant's roots and tassels reduces corn production. The model examines the effects of planting date on the population dynamics of insects. It gives information on the number of individuals in each growth stage as well as corn phenology. The model can investigate the impact of global change on the distribution and abundance of insect populations, as well as any potential economic consequences (Norango and Sawyer 1989). An example of a simulation model is the Agro-ECOsystem ManagemenT and OPtimization Model (ECOTOPE), which depicts the dynamics of soil nitrogen, crop growth, and insect population in an agricultural ecosystem. The best management techniques are derived from it (Seppelt 2001). The Boundary LAYER Model (BLAYER) is a weather simulation that has been modified to predict when and where insect pests would migrate into the maize belt of the United States. Researching potential alterations in pest populations, such as migration or dispersal patterns brought on by climate change, is highly beneficial (Paegle 1983).

Mechanistic species distribution models examine how the local environment influences physiological performance, as opposed to correlative models. Then, regions that restrict physiological performance to the point where it hinders survival, development, or reproduction are removed from the final distribution in order to anticipate future species distributions (Kearney et al. 2010). Moreover, it has been proposed that the bulk of management problems are best addressed by mechanistic models, which have the ability to identify elements that drive biogeography and extrapolate beyond known conditions (Cuddington et al. 2013).

7.19 Conclusion

Temperate insects are more affected by climate change because it allows their range to expand. Temperature is one of the major abiotic factors that drives population growth. It affects survival, growth and development, voltinism, and dispersal. Drought and precipitation are important factors that affect soil insect abundance. The CO_2 is causing indirect effect through host nutrient alteration, which has both positive and negative effects. Because insect species are highly heterogeneous, as are their host plants, and because global climate variability is present, diverse responses of insect species to global warming are anticipated in different regions of the world. Furthermore, a general increase in the frequency of pest outbreaks involving a wider variety of insect pests is anticipated. The ability to produce more generations and an enhanced overwintering survival rate will lead to an increase in the abundance of some pests. Plant illnesses spread by insects will probably increase, and invasive pest species will probably spread more easily to new locations. It is thought that in the future, invaders may benefit even more from climate change and global warming. Consequently, it makes sense to create strategies and actions for adapting to climate change that will stop the entry and spread of invasive insect pests. The prevention, eradication, and assessment of biological invasion through the use of predictive modeling techniques are the most crucial particular actions that can stop the harmful effects of climate change and invasive insect pests.

Author Contribution Statement Sadia Zafar led the conceptualization, writing, and revision of the chapter, ensuring its overall structure and coherence. Javed Iqbal provided key insights into data interpretation and statistical analysis, contributing to the manuscript's clarity and accuracy. Maria Bilal co-wrote several sections, focusing on the methodology and literature review. Banzeer Ahsan Abbasi contributed to the research design and analysis of case studies, enriching the chapter's depth. Shoaib Khan supported the writing process, particularly in the discussion and conclusion sections. Shumaila Ijaz provided assistance in the collection and analysis of data, contributing to the methodology section. Zakir Ullah contributed to the theoretical framework and assisted in structuring the manuscript. Ghulam Murtaza contributed to data visualization and collaborated in refining the final draft. Rashid Iqbal provided editorial support, ensuring consistency and clarity throughout. Lubna Ansari assisted with research analysis and helped in reviewing and proofreading the manuscript. Rabia Kazmi contributed to the writing of specific sections and provided valuable feedback. Hafsa Ansar contributed to the final revisions and editing, ensuring the chapter's quality and coherence.

References

Ahmad Y, Haider S, Iqbal J, Abbasi BA, Yaseen T, Mahmood T (2022) The mechanisms of genome editing technologies in crop plants. In: Principles and practices of OMICS and genome editing for crop improvement. Springer International Publishing, Cham, pp 295–313

Ahmad Y, Haakim Z, Iqbal J, Abbasi BA, Mahmood T, Kazi M (2024) Technological innovations for abiotic stress resistance in horticultural crops. In: OMICs-based techniques for global food security. Wiley, pp 233–244

Ali I, Salah KBH, Sher H, Ali H, Ullah Z, Ali A, Mahmood T (2022) Drought stress enhances the efficiency of floral dip method of Agrobacterium-mediated transformation in Arabidopsis thaliana. Braz J Biol 84:e259326

Ananthakrishnan TN, Sivaramakrishnan KG (2017) Ecological entomology: insect life in odd environments. Scientific Publishers

Andrew NR, Hill SJ (2017) Effect of climate change on insect pest management. In: Coll M, Wajnberg E (eds) Environmental pest management: challenges for agronomists, ecologists, economists and policymakers, 1st edn. John Wiley and Sons Ltd, Hoboken, NJ, pp 197–215, ISBN 978-1-119-25555-0

Bale JS, Masters GJ, Hodkinson ID, Awmack C, Bezemer TM, Brown VK, Butterfield J, Buse A, Coulson JC, Farrar J et al (2002) Herbivory in global climate change research: direct effects of rising temperature on insect herbivores. Glob Chang Biol 8:1–16

Barzman M, Bàrberi P, Birch ANE, Boonekamp P, Dachbrodt-Saaydeh S, Graf B, Hommel B, Jensen JE, Kiss J, Kudsk P et al (2015) Eight principles of integrated pest management. Agron Sustain Dev 35:1199–1215

Bezemer TM, Jones TH, Knight KJ (1998) Long-term effects of elevated CO_2 and temperature on populations of the peach potato aphid *Myzus persicae* and its parasitoid *Aphidius matricariae*. Oecologia 116:128–135

Boller EF, Avilla J, Joerg E, Malavolta C, Wijnands FG, Esbjerg P (2004) Integrated Production: Principles and Technical Guidelines. IOBC/WPRS, Dijon. Available online: https://www.iobc-wprs.org/ip_ipm/01_IOBC_Principles_and_Tech_Guidelines_2004.pdf (accessed on 13 Mar 2015)

Cheke RA, Tratalos JA (2007) Migration, patchiness and population processes illustrated 57: 145-154. By two migrant pests. Biosci 57:145–154

Collins W, Colman R, Haywood J, Manning MR, Mote P (2007) The physical science behind climate change. Sci Am 297:64–73

Coviella CE, Trumble JT (1999) Effects of elevated atmospheric carbon dioxide on insect-plant interactions. Conserv Biol 13:700–712

Crookston NL, Dixon GE (2005) The forest vegetation simulator: a review of its structure, content, and applications. Comput Electron Agric 49(1):60–80

Cuddington K, Fortin MJ, Gerber LR, Hastings A, Liebhold A, O'Connor M, Ray C (2013) Process-based models are required to manage ecological systems in a changing world. Ecosphere 4:1–12

Cumhur A, Cresser MS (2008) The effects of global climate change on agriculture. Am Eurasian J Agric Environ Sci 3(5):672–676

Dermody O, O'Neill BF, Zangrel AR, Berenbaum MR, Delucia EH (2008) Effects of elevated CO and O on leaf damage and 2 3 insect abundance in soybean ecosystem. Arthropod Plant Interact 2:125–135

Dessai S, Van Der Sluijs J (2007) Uncertainty and climate change adaptation: a scoping study, vol 2007. Copernicus Institute for Sustainable Development and Innovation, Department of Science Technology and Society, Utrecht

Deutsch CA, Tewksbury JJ, Huey RB, Sheldon KS, Ghalambor CK, Haak DC, Martin PR (2008) Impacts of climate warming on terrestrial ectotherms across latitude. Proc Natl Acad Sci USA 105:6668–6672

Deutsch CA, Tewksbury JJ, Tigchelaar M, Battisti DS, Merrill SC, Huey RB, Naylor RL (2018) Increase in crop losses to insect pests in a warming climate. Science 361:916–919

Dhillon MK, Sharma HC (2007) Effect of storage temperature and duration on viability of eggs of *Helicoverpa armigera* (Lepidoptera: Noctuidae). Bull Entomol Res 97:55–59

Dukes JSDS, Pontius J, Orwig D, Garnas JRGR, Rodgers VL, Brazee N, Cooke B, Theoharides KATA, Stange EESE, Harrington R et al (2009) Responses of insect pests, pathogens, and invasive plant species to climate change in the forests of northeastern North America: what can we predict? This article is one of a selection of papers from NE Forests 2100: A Synthesis of Climate Change Impacts on Forests of the Northeastern US and Eastern Canada. Can J For Res 39:231–248

Fand BB, Kamble AL, Kumar M (2012) Will climate change pose serious threat to crop pest management: a critical review. Int. J. Sci. Res. 2:1–14

FAO. Climate related transboundary pests and diseases. Available online: http://www.fao.org/3/a--ai785e.pdf (accessed on 19 December 2020)

FAO. Climate related transboundary pests and diseases. Available online: http://www.fao.org/3/a--ai785e.pdf (accessed on 19 April 2021a)

FAO. Food and Agriculture Organization Plant Pests and Diseases in the Context of Climate Change and Climate Variability, Food Security and Biodiversity Risks. Available online: http://www.fao.org/3/nb088/nb088.pdf (accessed on 25 April 2021b)

Field CB, Barros VR, Dokken DJ, Mach KJ, Mastrandrea MD, Bilir TE, Chatterjee M, Ebi KL, Estrada YO, Genova RC et al (2014) IPCC summary for policymakers. In: Climate change 2014: impacts, adaptation, and vulnerability, part a: global and sectoral aspects; contribution of working group II to the fifth assessment report of the intergovernmental panel on climate change. Cambridge University Press, Cambridge, UK; New York, pp 1–32. ISBN 978-92-9169-143-2

Finch DM, Butler JL, Runyon JB, Fettig CJ, Kilkenny FF, Jose S, Frankel SJ, Cushman A, Cobb RC, Dukes JS et al (2021) Effects of climate change on invasive species. In: Poland TM, Patel-Weynand T, Finch D, Miniat CF, Lopez V (eds) Invasive species in forests and rangelands of the United States: a comprehensive science synthesis for the United States forest Sector. Springer, Berlin, pp 57–85

Florentine S, Chauhan BS, Jabran K (2020) Agriculture and crop protection; its global importance and relationship with climate change. In: Jabran K, Chauhan BS, Florentine S (eds) Crop protection under changing climate. Springer, Cham, pp 1–16

Godfray HCJ, Beddington JR, Crute IR, Haddad L, Lawrence D, Muir JF, Pretty J, Robinson S, Thomas SM, Toulmin C (2010) Food security: the challenge of feeding 9 billion people. Science 327:812–818. [CrossRef]

Gomez-Zavaglia A, Mejuto JC, Simal-Gandara J (2020) Mitigation of emerging implications of climate change on food production systems. Food Res Int 134:109256

Gray DR (2008) The relationship between climate and outbreak characteristics of the spruce budworm in eastern Canada. Climate Change 87:361–383

Gregory PJ, Johnson SN, Newton AC, Ingram JSI (2009) Integrating pests and pathogens into the climate change/food security debate. J Exp Bot 60:2827–2838

Haider S, Iqbal J, Naseer S, Yaseen T, Shaukat M, Bibi H, Mahmood T (2021a) Molecular mechanisms of plant tolerance to heat stress: current landscape and future perspectives. Plant Cell Rep 40:1–25

Haider S, Iqbal J, Shaukat M, Naseer S, Mahmood T (2021b) The epigenetic chromatin-based regulation of somatic heat stress memory in plants. Plant Gene 27:100318

Haider S, Raza A, Iqbal J, Shaukat M, Mahmood T (2022a) Analyzing the regulatory role of heat shock transcription factors in plant heat stress tolerance: a brief appraisal. Mol Biol Rep 49(6):5771–5785

Haider S, Iqbal J, Naseer S, Shaukat M, Abbasi BA, Yaseen T, Mahmood T (2022b) Unfolding molecular switches in plant heat stress resistance: a comprehensive review. Plant Cell Rep 41(3):775–798

Hamilton JG, Dermody O, Aldea M, Zangerl AR, Rogers A, Berenbaum MR, DeLucia EH (2005) Anthropogenic changes in tropospheric composition increase susceptibility of soybean to insect herbivory. Environ Entomol 34:479–485

Heong KL, Song YH, Pimsamarn S, Zhang R, Bae SD (1995) Global warming and rice arthropod communities. In: Peng S, Ingram KT, Neue HU, Ziska LH (eds) Climate change and rice. Springer Publications, Berlin, pp 327–335

Heuskin S, Verheggen FJ, Haubruge E, Wathelet JP, Lognay G (2011) The use of semiochemical slow-release devices in integrated pest management strategies. Biotechnol Agron Soc Environ 15:459–470

Howden SM, Soussana JF, Tubiello FN, Chhetri N, Dunlop M, Meinke H (2007) Adapting agriculture to climate change. Proc Natl Acad Sci USA 104:19691–19696

Hunter MD (2001) Effects of elevated atmospheric carbon dioxide on insect-plant interactions. Agric Forest Entomol 3:153–159

Ijaz S, Iqbal J, Abbasi BA, Tufail A, Ullah Z, Sharifi-Rad J, Uddin S (2023) Biofortification: lessons from the past and strategies for future food security. In: Legumes biofortification. Springer International Publishing, Cham, pp 521–545

Ijaz S, Iqbal J, Abbasi BA, Yaseen T, Rehman S, Kazi M, Mahmood T (2024) Role of OMICS-based technologies in plant sciences. In: OMICs-based techniques for global food security, pp 45–66

Johnson SN, Anderson EA, Dawson G, Griffiths DW (2008) Varietal susceptibility of potatoes to wireworm herbivory. Agric For Entomol 10:167–174

Johnson SN, Gregory PJ, McNicol JW, Oodally Y, Zhang X, Murray PJ (2010) Effects of soil conditions and drought on egg hatching and larval survival of the clover root weevil (*Sitona lepidus*). Appld Soil Ecol 44:75–79

Karuppaiah V, Sujayanad GK (2012) Impact of climate change on population dynamics of insect pests. World J Agri Sci 8(3):240–246

Kearney MR, Wintle BA, Porter WP (2010) Correlative and mechanistic models of species distribution provide congruent forecasts under climate change. Conserv Lett 3:203–213

Kistner EJ (2017) Climate change impacts on the potential distribution and abundance of the Brown Marmorated Stink Bug (Hemiptera: Pentatomidae) with special reference to North America and Europe. Environ Entomol 46:1212–1224

Kocmánková E, Trnka M, Juroch J, Dubrovský M, Semerádová D, Možný M, Žalud Z, Pokorný R, Lebeda A (2010) Impact of climate change on the occurrence and activity of harmful organisms. Plant Prot Sci 45:S48–S52

Kurukulasuriya P, Rosenthal S (2003) Climate change and agriculture: a review of impacts and adaptations. The World Bank, Washington, DC

Lehmann P, Ammunét T, Barton M, Battisti A, Eigenbrode SD, Jepsen JU, Kalinkat G, Neuvonen S, Niemelä P, Terblanche JS et al (2020) Complex responses of global insect pests to climate warming. Front Ecol Environ 18:141–150

Li W, Huai X, Li P, Raza A, Mubarik MS, Habib M, Faiz S, Zhang B, Pan J, Khan RSA (2021) Glutathione peroxidase (GPX) gene family in rapeseed (Brassica napus L.): genomics analysis and expression profiling against different stresses. Antioxidant 10(9):1481. https://doi.org/10.3390/antiox10091481

Lindroth RL, Kinney KK, Platz CL (1993) Responses of deciduous trees to elevated atmospheric CO productivity, phytochemistry and 2 insect performance. Ecology 74:763–777

Logan JA et al (2003) Assessing the effects of climate change on forest pest dynamics. Ecol Appl 13(2):457–471

Luo Q (2011) Temperature thresholds and crop production: a review. Clim Chang 109:583–598

McKee FR, Aukema BH (2015) Influence of temperature on the reproductive success, brood development and brood fitness of the eastern larch beetle Dendroctonus simplex LeConte. Agric For Entomol 17(1):102–112

McKibben GH, Willers JL, Smith JW, Wagner TL (1991) Stochastic model for studying boll weevil dispersal. Environ Ent 20(5):1327–1332

Mols PJM (1992) Forecasting an indispensable port of IPM in apple orchards. Acta Phytop Entom H 27(1–4):449–460

Müller C (2013) African lessons on climate change risks for agriculture. Annu Rev Nutr 33:395–411

Nihal R (2020) Global climate change and its impact on integrated pest management. Agro Econ Int J 7:133–137

Norango SE, Sawyer AJ (1989) A simulation model of northern corn rootworm. Diabrotica barberi, population dynamics, oviposition: significance of host-plant phenology. Can Ent 121:169–191

Pachauari RK, Reisinger A (2007) Climate change 2007: synthesis report. In: Contribution of working groups I, II and III to the fourth assessment report on intergovernmental panel on climate change. Intergovernmental Panel on Climate Change (IPCC), Geneva. ISBN 92-9169-122-4

Paegle ML (1983) Correlation of nocturnal thunderstorms and boundary-layer convergence. Mon Weather Rev 101:877–933

Parmesan C (2007) Influences of species, latitudes and methodologies on estimates of phenological response to global warming. Glob Chang Biol 13:1860–1872

Parmesan C, Yohe G (2003) A globally coherent fingerprint of climate change impacts across natural systems. Nat Cell Biol 421:37–42

Pathak H, Aggarwal PK, Singh SD (2012) Climate change impact, adaptation and mitigation in agriculture: methodology for assessment and applications. Indian Agricultural Research Institute, New Delhi. ISBN 978-81-88708-82-6

Pingali P, Abraham M (2022) Food systems transformation in Asia–a brief economic history. Agric Econ 53(6):895–910. https://doi.org/10.1111/agec.12734

Porter JH, Parry ML, Carter TR (1991) The potential effects of climatic change on agricultural insect pests. Agric For Meteorol 57(1–3):221–240

Prakash A, Rao J, Mukherjee AK, Berliner J, Pokhare SS, Adak T, Munda S, Shashank PR (2014) Climate change: impact on crop pests. Applied Zoologists Research Association (AZRA), Central Rice Research Institute, Odisha. ISBN 81-900947-2-7

Raja S, Mahmood A, Alamery S, Cheema HMN, Javaid A, Alvi AK, Siddique F, Aslam MM, Khan MT, Attia KA, Kimiko I, Fiaz S (2022) Physicochemical, molecular and cultural identification of microbial pathogens in wastewater irrigated crops. Pol J Environ Stud 31(4):1–9. https://doi.org/10.15244/pjoes/145605

Reddy AR, Rasineni GK, Raghavendra AS (2010) The impact of global elevated CO_2 concentration on photosynthesis and plant productivity. Curr Sci 99:46–57

Regniere J, Amant R, Duval P (2010) Predicting insect distributions under climate change from physiological responses: spruce budworm as an example. Biol Invasions 14:1571

Rehman MU, Rather GH, Gull Y, Mir MR, Mir MM, Waida UI, Hakeem KR (2015) Effect of climate change on horticultural crops. In: Hakeem KR (ed) Crop production and global environmental issues. Springer International Publishing, Cham, pp 211–239

Rosenzweig C, Major DC, Demong K, Stanton C, Horton R, Stults M (2007) Managing climate change risks in New York City's water system: assessment and adaptation planning. Mitig Adapt Strateg Glob Chang 12:1391–1409

Roth SK, Lindroth RL (1995) Elevated atmospheric CO_2: effects on phytochemistry, insect performance and insect-parasitoid interactions. Glob Chang Biol 1:173–182

Rötter R, Van De Geijn S (1999) Climate change effects on plant growth, crop yield and livestock. Clim Chang 43:651–681

Rubab M, Jannat S, Freeg H et al (2023) Evaluation of functional kompetitive allele-specific PCR (KASP) markers for selection of drought-tolerant wheat (Triticum aestivum) genotypes. Funct Plant Biol 51:FP23032. https://doi.org/10.1071/FP23032

Seppelt R (2001) Agro-ecosystem management. In: Pinter J (ed) Global optimization & selected case studies. Kluver, Dordrecht, New York

Shrestha S (2019) Effects of climate change in agricultural insect pest. Acta Sci Agric 3:74–80

Srivastava CP, Joshi N, Trivedi TP (2010) Forecasting of Helicoverap armigera population and impact of climate change. Ind J Agrl Sci 80(1):3–10

Staley JT, Hodgson CJ, Mortimer SR, Morecroft MD, Masters GJ, Brown VK, Taylor ME (2007) Effects of summer rainfall manipulations on the abundance and vertical distribution of herbivorous soil macro-invertebrates. Eur J Soil Biol 43:189–198

Stevens CJ, Dise NB, Mountford JO, Gowing DJ (2004) Impact of nitrogen deposition on the species richness of grasslands. Science 303:1876–1879

Streck NA (2005) Climate change and agroecosystems: the effect of elevated atmospheric CO_2 and temperature on crop growth, development, and yield. Ciênc Rural 35:730–740

Sun Y, Ge F (2011) How do aphids respond to elevated CO_2? J Asia Pac Entomol 14:217–220

Sutherst RW, Constable F, Finlay KJ, Harrington R, Luck J, Zalucki MP (2011) Adapting to crop pest and pathogen risks under a changing climate. Wiley Interdiscip Rev Clim Chang 2:220–237

Tilman D, Balzer C, Hill J, Befort BL (2011) Global food demand and the sustainable intensification of agriculture. Proc Natl Acad Sci USA 108:20260–20264

Tubiello F, Rosenzweig C, Goldberg R, Jagtap S, Jones J (2002) Effects of climate change on US crop production: simulation results using two different GCM scenarios. Part I: wheat, potato, maize, and citrus. Clim Res 20:259–270

Ullah Z, Iqbal J, Abbasi BA, Ijaz S, Sharifi-Rad J, Yaseen T, Mahmood T (2023a) How genome editing can be helpful in the biofortification of legumes. In: Legumes biofortification. Springer International Publishing, Cham, pp 207–232

Ullah Z, Iqbal J, Abbasi BA, Akhtar W, Kanwal S, Ali I, Mahmood T (2023b) Assessment of Gus expression induced by anti-sense Os PPO gene promoter and antioxidant enzymatic assays in response to drought and heavy metal stress in transgenic Arabidopsis thaliana. Sustain For 15(17):12783

Volney WJA, Fleming RA (2000a) Climate change and forest pests. For Chron 76(2):273–279

Volney WJA, Fleming RA (2000b) Climate change and impacts of boreal forest insects. Agric Ecosyst Environ 82(1–3):283–294

Walther GR, Roques A, Hulme PE, Sykes MT, Pyšek P, Kühn I, Zobel M, Bacher S, Botta-Dukát Z, Bugmann H et al (2009) Alien species in a warmer world: risks and opportunities. Trends Ecol Evol 24:686–693

Ward NL, Masters GJ (2007) Linking climate change and species invasion: an illustration using insect herbivores. Glob Chang Biol 13:1605–1615

Whittaker JB, Tribe NP (1996) An altitudinal transect as an indicator of responses of a spittlebug (Auchenorrhyncha: Cercopidae) to climate change. European Journal of Entomology 93:319–324

Williams DW, Liebhold AM (1995) Forest defoliators and climatic change: potential changes in spatial distribution of outbreaks of western spruce budworm (Lepidoptera: Tortricidae) and gypsy moth (Lepidoptera: Lymantriidae). Environ Entomol 24(1):1–9

Wilson RJ, Maclean IM (2011) Recent evidence for the climate change threat to Lepidoptera and other insects. J Insect Conserv 15:259–268

Woodward FI (1990) Global change: translating plant ecophysiological responses to ecosystems. Trends Ecol Evol 5:308–311

Yamamura K, Kiritani K (1998) A simple method to estimate the potential increase in the number of generations under global warming in temperate zones. Appld Entomol Zool 33:289–298

Yoro KO, Daramola MO (2020) Chapter 1—CO_2 emission sources, greenhouse gases, and the global warming effect. In: Rahimpour MR, Farsi M, Makarem MA (eds) Advances in carbon capture, 1st edn. Woodhead Publishing, Sawston, UK, pp 3–28. ISBN 9780128196571

Yumamura K, Yokazawa M, Nishimori M, Ueda Y, Yokosuka T (2006) How to analyse long- term insect population dynamics under climate change: 50 year data of three insect pests in paddy fields. Popln ulation Ecol 48:38–48

Zhao C, Liu B, Piao S, Wang X, Lobell DB, Huang Y, Huang M, Yao Y, Bassu S, Ciais P et al (2017) Temperature increase reduces global yields of major crops in four independent estimates. Proc Natl Acad Sci USA 114:9326–9331

Ziska LH, Blumenthal DM, Runion GB, Hunt ER, Diaz-Soltero H (2011) Invasive species and climate change: an agronomic perspective. Clim Chang 105:13–42

Zvereva EL, Kozlov MV (2005) Consequences of simultaneous elevation of carbon dioxide and temperature for plant-herbivore interactions: a metaanalysis. Glob Chang Biol 12:27–41

Legumes Production Under Changing Climate

8

Iahtisham-Ul-Haq and Noor Shahid

Abstract

Legumes, essential for global food security and nutrition, face significant challenges in production due to climate change. This chapter explores the multifaceted impacts of changing climate conditions on legume cultivation, focusing on temperature fluctuations, altered precipitation patterns, and increased incidences of extreme weather events. The discussion highlights how these climatic factors affect legume growth, yield, and nutritional quality, emphasizing the need for adaptive agricultural practices. Advances in crop breeding, including the development of drought-resistant and heat-tolerant legume varieties, are examined as crucial strategies to mitigate climate-related risks. Additionally, the chapter delves into the role of sustainable farming practices, such as crop rotation and intercropping, in enhancing the resilience of legume production systems. Case studies from diverse geographic regions illustrate practical applications and successes in adapting legume farming to changing climates. Furthermore, the chapter addresses the socio-economic implications of climate change on legume farmers, particularly in developing countries, and underscores the importance of policy support and international cooperation in fostering sustainable legume production. By integrating scientific research with practical solutions, this chapter aims to provide a comprehensive understanding of the challenges and opportunities in legume production under changing climate conditions, offering valuable insights for researchers, policymakers, and farmers.

Iahtisham-Ul-Haq (✉) · N. Shahid
Kauser Abdulla Malik School of Life Sciences, Forman Christian College (A Chartered University), Lahore, Punjab, Pakistan
e-mail: iahtisham@fccollege.edu.pk

© The Author(s), under exclusive license to Springer Nature Singapore Pte Ltd. 2025
S. Faiz et al. (eds.), *Climate Smart Agriculture for Future Food Security*,
https://doi.org/10.1007/978-981-96-4499-5_8

Keywords

Climate irregularities · Agricultural strategies · Sustainable mechanization · Homeostasis · Photorespiration

8.1 Introduction to Climate Resilience in Agriculture

8.1.1 A Brief Overview: Climate Change and Its Impact on Agriculture

Climate change is the center of most modern-day research across the world. This is because of the vast impact it has over different aspects of the Earth's systems which go on to affect biodiversity, ecosystems, agriculture, food security, and consequently, human societies. Most climate change is caused by anthropogenic activities which includes the burning of fossil fuel for energy production, transportation, agricultural practices, and industrial processes to sustain the largely growing human population (Abbass et al. 2022). Since human activities are a significant driver of climate change, even more so than natural drivers, it can have far-reaching consequences for the Earth's systems. Understanding and addressing the effects of human-induced environmental changes is therefore crucial to achieve sustainable development, agricultural prosperity and food safety. In the last 100 years or so, the pace of climate change has exacerbated many folds, increasing the average temperature of the earth by 0.9 °C, mainly due to greenhouse gas (GHC) emissions. With the current pace, this is estimated to reach 1.5 °C or more by 2050, owing to the rates at which deforestation, pollution, and climate irregularities are taking place (Arora 2019). According to the United Nations, the world population is likely to increase to 9.7 billion by 2050 which would further coerce the agricultural sector to meet the increasing food demands which are already taking a drastic hit due to climate change. Another publication in 2016 by the Food and Agriculture Organization (FAO), stated that if the ongoing situation of greenhouse gas emissions and climate change continue, then by the year 2100 there would be a major decrease in the production of cereal crops; 20–45% decrease in maize yields, 20–30% decrease in rice, and 5–50% decrease in yield of wheat. This would result in a catastrophic food shortage along with a steep increase in crop prices. A World Food Program (WFP) report carried out in 2018 disclosed that the increase in crop yield/hectare is remarkably lower than the rate at which the world population is increasing.

8.1.2 Importance of Crop Resilience and Climate Resilient Cropping Systems in Ensuring Global Food Security

The ever-increasing demand of food as a result of a growing population has resulted in the development and application of extensive agricultural practices, including the prodigious use of agrichemicals, abuse of water resources, and livestock production. Expectedly, this further worsened the release of greenhouse gases, causing more pollution by increasing the carbon footprint and inducing uneven climate patterns resulting in adverse effects for agricultural production. This is where crop resilience proves as a successful technique to remedy global food security crisis. Crop resilience is known as the capacity of a crop variety to withstand and adapt to adverse environmental conditions, stresses, and shocks while maintaining quality and productivity (Azeem et al. 2024). Some key aspects of crop resilience include:

- Tolerance to abiotic stress such as heat, cold, drought, flooding, soil salinity, and poor soil conditions.
- Resistance to biotic stress involving resistance to diseases, pests, and weeds either through natural defenses or genetic traits that reduce the adverse effects of the stress on crop productivity.
- Genetic diversity, which allows for the selection of traits that enhance resilience allowing better varieties to flourish and provide a genetic bank for breeding better and more resilient varieties.
- Adaptation and phenotypic plasticity, which allows for a varied response to changing environmental cues.
- Regenerative capacity, which gives crops the ability to regenerate after exposure to prolonged stress, such as re-growing foliage after drought or flooding.
- Resource efficiency, which allows resilient crops to effectively and efficiently use naturally available resources such as water, nutrients, and light to reduce agricultural input therefore promoting sustainability.
- Yield stability, which allows crops to produce stable yields despite fluctuations in environmental conditions to promote food security and socio-economic stability (Gupta et al. 2019) (Dhanker and Foyer 2018).

The above-mentioned crop resilience traits are achieved through breeding programs to select desirable traits with the help of genomic tools and genetically modified crop varieties to enable greater global food security.

The greatest sufferers of climate change are small-scale farmers who have limited resources. Their agricultural strategies are barely profitable and are often unsustainable. The United Nations Food and Agriculture Organization has worked on enabling a new approach which uses climate resilient cropping systems, making adaptation and mitigation to climate change possible through climate-smart crop production practices (Shahzadi et al. 2024). This requires small scale farmers to use high quality seeds which are adapted to flourish in local climate and pest varieties and by diversifying their cropping systems by using sustainable mechanization. Moreover, this strategy enables them to conserve soil and water resources by improving resource management through the application of technical knowledge, which improves crop resilience to certain climatic stresses, thereby reducing yield gaps (FAO 2024).

8.2 Significance of Legumes in Human Diet

Human consumption of legumes dates back to centuries and has since then played a significant role in fulfilling our dietary needs. Their nutritional versatility and demand lead to the development of various agricultural practices over different cultures and time periods. Legumes are distributed and grown throughout the world except for the Arctic and Antarctic regions. They are the third highest crop in production after cereals and oilseeds. Legumes which come from the Fabaceae family, contain a variety of pulses with different origins of cultivation. The wide geographic distribution of legumes formed distinct areas of predominant use leading to a variety of products that are prepared in zone-specific cuisines (Amin and Borchgrevink 2022). In addition to supporting sustainable productivity, legumes offer a wholesome nutritional profile and have a long shelf life, making them a durable food choice (Miklas et al. 2022; Uebersax et al. 2022). 2016 was named as the International Year of Pulses by the Food and Agriculture Organization of the United Nations, in which they laid special emphasis on the potential of legumes to eradicate hunger and malnutrition. The objectives of this initiative were to (i) advance the value and use of legumes throughout the food system, (ii) promote awareness of their health as well as socio-economic benefits, (iii) aid advanced research, (iv) cultivate better utilization of legumes in crop rotations, and (v) address trade challenges. The FAO project was also associated with an increasing awareness of the role that pulses play in achieving important targets under Sustainable Development Goal 2, specifically in relation to small-scale farmer incomes, malnutrition and food access, in addition to resilient and sustainable agriculture.

8.2.1 Nutritional Composition of Legumes

Globally, pulses and legumes are valued as a nutritionally rich and economically feasible part of the human diet, secondary to cereal crops. Their dietary importance is established through the diversification of its components which includes proteins (20–45%), complex carbohydrates (60%), dietary fiber (5–37%), and a low-fat content (~5%), with the exception of chickpeas containing a fat content of 15%, as well as soybean and peanuts with a higher fat content of 45–50%. Legumes also contain many minerals, vitamins, and bioactive phytochemicals. They possess the benefit of being cholesterol free and have a high antioxidant capacity. Legumes have high nutritional values, on average they contain 60.76 g of carbohydrates, 23.10 g of proteins, 15.55 g of dietary fiber, and 1.86 g of fat per 100 g of the grain. Moreover, they also contain notable quantities of essential minerals, including potassium (1244 mg), phosphorus (367 mg), magnesium (177 mg), calcium (113 mg), iron (6.23 mg), sodium (17 mg), zinc (3.15 mg), copper (0.763 mg), manganese (2.392 mg), and selenium (7.9 μg) per 100 g (Affrifah et al. 2023). The low-fat content of legumes, not only adds a marketing advantage when it comes to choosing healthy dietary options, but also makes them a fitting choice for product development or for use in a variety of food applications. Additionally, legumes are known to have a higher quantity of essential amino acids such as lysine, as compared to animal proteins (Kumar et al. 2022). Apart from their nutritional diversity and richness, legumes also contain some antinutritional factors (ANFs) such as α-galactosides, polyphenols, phytates, lectins, trypsin and chymotrypsin inhibitors, which have some antinutrient capacities which restrict the absorption of some nutrients in the body; however, the negative effects of these substances can be reversed through various cooking and processing methods such as microwaving, pressure cooking, soaking, germination, fermentation, and extrusion which enhances the digestibility of the nutrients in legumes (Alonso et al. 1998; Fernández et al. 1997; El-Hady and Habiba 2003; Arif et al. 2012; Wiesinger et al. 2022).

8.2.2 Health Benefits of Consuming Legumes

Legumes are important contributors to a fiber-rich and low-glycemic index diet (Collins 2020). The presence of dietary fiber and complex carbohydrates like oligosaccharides in legumes, along with resistant (RS) and slow digestible starch (SDS) make it a suitable choice for celiac and diabetic patients due to their minimal elicitation of blood glucose response (Maphosa and Jideani 2017). In addition to being cholesterol-free and having a high antioxidant capacity, legumes are also rich in bioactive compounds which is proven to play a significant role in the prevention of coronary heart disease and other life-threatening metabolic or degenerative illnesses such as diabetes mellitus and colon cancer (Raju and Mehta 2009). Bioactive compounds are non-nutrient chemicals that are available in small quantities but possess many health benefits which can reduce the incidence of chronic disease development (Singh et al. 2018). These bioactive compounds consist mainly of phenolic compounds including phenolic acids, flavonoids, condensed tannins, phytosterols, bioactive carbohydrates including dietary fiber, resistant starch, oligosaccharides, saponins, carotenoids and antioxidants (Venkidasamy et al. 2019). In addition to these bioactive substances, legumes also contain phytic acid, lectins, and some amino acids which were previously thought of as antinutritional factors; however, current literature also recognizes these substances as nutritionally active factors with beneficial health properties (Pedrosa et al. 2021; Martín-Cabrejas 2019). A comprehensive study by Conti et al. (2021) linked the bioactive compounds found in legumes to their respective properties. In this, many peptides were found to have anti-inflammatory, antioxidant, and antimicrobial activities. Phenolic compounds were recognized for their roles in modulating cell proliferation, metabolism, and homeostasis. Flavonoids from the polyphenol group like anthocyanins and anthoxanthins were found to have the highest antioxidant activity along with free-radical scavenging capacity, anticancer, and anti-inflammatory properties. Lectins, which are found in legumes are known to have both negative effects, as they inhibit nutrient absorption, and positive effects as multiple studies have found them to be cytotoxic against cancer cell lines in skin, breast, colon, liver, and bone (Gautam et al. 2018; Shi et al. 2014). These bioactive compounds in legumes altogether play a cardinal role in a variety of metabolic cellular activities and can even slow down the natural aging process of the body by working as cofactors for antioxidative enzymes (Höhn et al. 2017).

8.2.3 Role of Legumes in Addressing Malnutrition

Despite being one of the most important crops, secondary to cereals, legumes have been an underdog of the agricultural produce. This lack of attention stems from the insufficient knowledge of the nutritional and functional benefits, along with a lag in research and development, of legumes. Legumes harness many desirable properties and can be utilized in the formation of inexpensive products that are available and accessible for people of low-income groups (Qayyum et al. 2012). Developing countries like India, Pakistan, Bangladesh, Nigeria, Ethiopia, and Kenya with high rates of malnutrition have a dire need for affordable, nutritionally rich and readily available food varieties which could cater to their increasing populations. Legumes could offer the perfect solution to this demand as their diverse nutriments as well as abundant bioactive benefits could be used to develop and produce affordable and therapeutic functional foods. Educating people of the nutritional richness of legumes along with the techniques to properly detoxify the grain from any antinutrients, would make it more attractive to consumers. Moreover, encouraging communities to cultivate legumes on a small-scale level would not only promote financial stability and food security but would also help to meet increased crop demands (Maphosa and Jideani 2017).

8.2.4 Cultural and Culinary Significance of Legumes

Legumes have a predominant place in the culinary and cultural experience throughout the world. People of different cultural ethnicities have legume-based cuisines, for example a majority of Indian people rely on legumes for their protein requirements owing to religious restrictions. Middle eastern and Mexican cuisines also rely heavily on legumes, incorporating the food group into salads, curries, soups, and stews (Jannat et al. 2022). Previously, there was a barrier between people choosing legume-based food options since beans were often challenging to prepare requiring longer soaking, cooking, and processing times. However, in light of recent research, the health-benefits of legumes are being reassessed and the knowledge gaps in terms of legume nutrients are being bridged in the food industry. This is leading to legumes being recognized as a healthy food choice, one that is free from artificially produced or genetically modified ingredients. Over the past decade, consumer preferences have also largely shifted to vegetarian and plant-based food options, shifting focus from a largely meat-based protein diet. With an advancement in cooking technology and product development, legumes are now available in forms that are ready to consume such as canned beans, legume products such as hummus and legume-based pasta varieties which are considered much healthier than traditional flour pasta since it has a higher content of protein and fiber. Legume based food choices also cater to celiac patients who require gluten free alternatives, as well as lactose intolerant people who can switch to plant-based milk such as soymilk and even consumers that are interested in low-carbohydrate foods to uplift their general health (Affrifah et al. 2023).

8.3 Effect of Climate Change on Legume Production

Globally, legumes are grown on an area of approximately 81 million ha with a produce of more than 92 million tonnes. Reportedly, India is the largest legume producer and consumer in the world, accounting for 1/4th production worldwide. Other major producers include China, Canada, Brazil, Australia, USA, Argentina, and Russia. Legumes are adapted to growing in a wide range of environmental conditions so they can grow in both cool and warm seasons. Dry bean is the highest cultivated legume in the world with an area of 34.5 million ha and a production of 30.4 million tonnes, chickpea comes second with a cultivated area of 17.7 million ha (FAOSTAT 2018). Legumes play a pivotal role in the ecosystem as their biological nitrogen fixation capacity makes them an environmentally beneficial crop. For this reason, as well as its deep root systems, low input requirements, ability to survive in problem soils and withstand abiotic stresses make legumes a popular cultivating choice (Kumar et al. 2018). Long-term cereal cropping systems not only have high inputs and intensive practices but are also responsible for greenhouse gas emission and ground water pollution by chemical fertilizers (Hazra et al. 2018). This was remedied by the introduction of cropping system heterogeneity which enriched the soil and was fundamental for the development of crop resilience against climate change and biotic or abiotic stress (Newton et al. 2011).

Legumes play an important role in carbon sequestration owing to their deep root system, symbiotic nitrogen fixation, leaf fall, and carbon-rich root exudates (Ghosh et al. 2012). Legumes are also known to perform well under moisture and high temperature stress conditions. Important mechanisms of withstanding stress include the generation of anti-oxidants by legumes while other factors such as agronomic management practices, nutrient supplementation, water availability and frequency of biotic or abiotic stress occurrence, play a decisive role in adapting climate change. Research using open top chambers and FACE studies has shown the impact of elevated CO_2 levels due to climate change, on the growth of legumes (Bindumadhava et al. 2018; Jin et al. 2012; Srinivasarao et al. 2016). Studies showed that elevated CO_2 has a negligible impact on legumes under poor management practices and nutrient deficient soils. Elevated CO_2 levels demonstrated positive impacts with an increase in biomass, height, and productivity in legumes, especially when supplemented with nitrogen and phosphorus; however, negative impacts were also observed such that elevated CO_2 reduced photosynthetic efficiency known as photosynthetic acclimation and this cynical impact of climate change caused decreased protein and micronutrient content in legumes (Sharma et al. 2016). The effect of elevated CO_2 levels due to climate change showed a variable impact on water use efficiency (WUE) depending on the soil moisture content. In moderate drought conditions, it increased the WUE of the crop by lowering the transpiration rate and increasing the leaf area; however, it was also suggested that the productivity of legumes would significantly reduce under arid and semi-arid conditions where water availability is very scarce, therefore moisture stress in combination with elevated temperature and CO_2 levels would have a catastrophic effect on legume growth and crop productivity (Jagadish et al. 2016; Palit et al. 2020). Climate change is also altering the geographical distribution of pests subsequently leading to increasing pest outbreaks which negatively affect legume production. Legumes can adapt both positively or negatively under climate change depending on the physiological adjustment and crop management practices. Identifying the phenological triggers and adaptation strategies that empower legumes to flourish under changing climatic conditions would require extensive research addressing heterogeneous aspects and variable stressors (CO_2, temperature, biotic, and abiotic stress) through multi-disciplinary research using agronomical, physiological, and microbiological studies (Abdullah et al. 2022).

8.3.1 Overview of the Legume Lifecycle

To comprehend the impacts of climate change on legumes, it is important to develop an understanding of the stages and life cycle of legumes. These are as follows:

1. Germination—The process in which a seed sprouts and develops into a seedling.
2. Vegetative growth—The phase in which the plant develops roots, shoots, and leaves.
3. Flowering—The plant produces flowers which are often pollinated by insects.
4. Pod and seed development—Fertilized flowers develop into pods containing seeds.
5. Seed dispersal—Mature pods release seeds which then germinate to start the cycle again.

There are different types of legumes which complete their lifecycle in varied time periods. Annual legumes complete their lifecycle in one growing season. Examples include soyabean, peas and lentils. Biennial legumes live for 2 years, in the first year they focus on vegetative growth and then in the second year they proceed to the flowering stage in which they produce seeds and then start a new cycle. An example of this is Sweetclover. Perennial legumes live for multiple years producing seeds annually. Well-known examples of this are Alfalfa and clover.

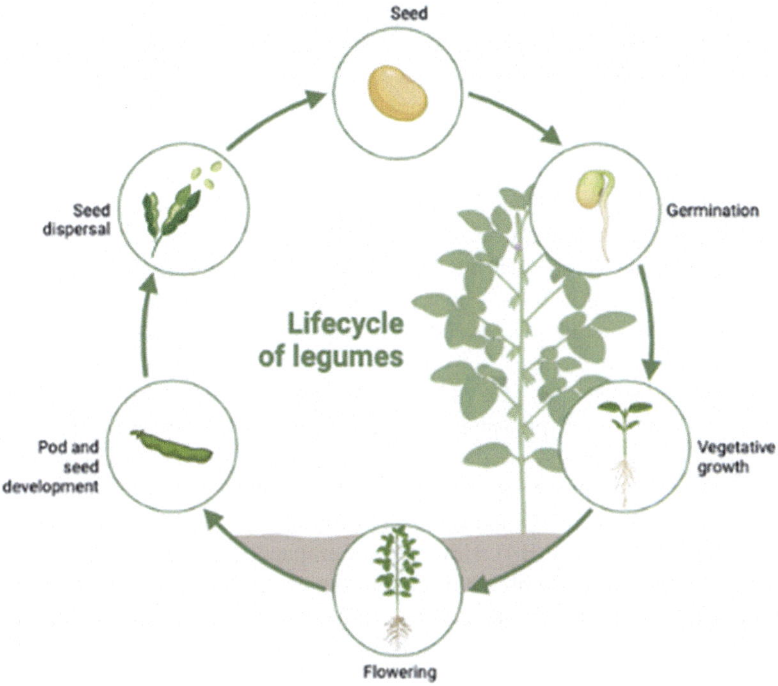

8.3.2 Impact of Climate Change on Rooting, Shooting, and Nodulation

Studies show that rising levels of temperature and CO_2 concentrations have varied direct and indirect effects on the physiological and bio-chemical properties of legumes. Considerable research was carried out to understand the effects of climate change on legume growth which showed the basic anatomical difference of C_3 crops to sequester atmospheric carbon which makes legumes far more adaptable than C_4 crops, with the extent of exposure as a key variable factor (Palit et al. 2020). This difference in growth response of C_3 and C_4 crops resulted because of differential CO_2 saturation point which favors C_3 plants to maintain their positive growth rate by obtaining higher photosynthetic response (Jin et al. 2012; Lee 2011). A study carried out by Srinivasarao et al. (2016) revealed an increase in shoot and overall biomass of some legume types in the presence of elevated CO_2 conditions, and also highlighted the failure of some C_4 crops to perform better under these elevated CO_2 conditions. Higher substrate or carbon dioxide availability in legumes rapidly increased the rate of photosynthesis via carboxylase activity, concurrently also reducing photorespiration. Legumes undoubtedly make a suitable choice for climate-resilience; however, to achieve this an optimum supply of essential nutrients such as nitrogen, phosphorus, potassium, and soil moisture must be present. A systematic review of multiple FACE experiments showed the relationship between crop performance, elevated CO_2, and applied inputs. Results from Kimball (2016) showed that C_3 legumes showed up to a 25% increase in shoot biomass in the presence of nitrogen and water, whereas C_3 cereals showed a 19% increase, this difference was negligible in case of C_4 crops. In the absence of nitrogen, C_3 cereals showed a 10% increase in shoot biomass while legumes showed a positive growth rate without any yield penalty due to their capacity to carry out biological nitrogen fixation. The importance of innate traits like biological nitrogen fixation and rhizobial or mycorrhizal infection in sustaining crop performance by reducing photosynthetic down regulation under elevated CO_2 levels was also highlighted in a previous study by Thomas et al. (2006). The results showed that legumes performed better than cereals under increased CO_2 and are also less prone to water stress especially during the reproductive stage (Newton et al. 1996). The role of potassium in plant growth under elevated CO_2 also proved to be important due to its central role in synthesizing adenosine triphosphate (ATP) and other biochemical cellular components. FACE experiments showed a 6–7 times increase in shoot biomass for legumes in elevated CO_2 levels that were also rich in supply of phosphorus. This was because of the higher photosynthetic and biomass accumulation rates (Jin et al. 2012; Jin et al. 2013; Zhang et al. 2014). However, the impact of elevated CO_2 on legumes could vary depending on the duration of exposure to increased CO_2 levels. Photosynthetic downregulation could invalidate the positive impact of climate change due to poor stomatal conductance and scaled down activity of RuBisCo (Rogers et al. 2006, 2009).

Studies show how elevated CO_2 levels elicit a higher photosynthetic response which significantly impacts the yield, growth rate, biomass, pod number, and even seed yield. Intrinsic physiological responses such as photosynthetic rate, canopy temperature, and carbon partitioning, along with better synchrony between source and sink, lower harvest index, larger reproductive organs, and indeterminate growth habits facilitate legumes to operate smartly and as a result build stronger climate-resilience (Ainsworth et al. 2006). Dissimilar from elevated CO_2 levels, elevated temperature has an unfavorable impact on the productivity of legumes, the extent of which depends on the growth stage of the plant and magnitude of temperature. Since higher temperatures deflate the sink size and relocate carbohydrates from the grains, the seeds do not get sufficient time to develop resulting in lower yields (Wang et al. 2006). This heat stress results in stomatal closure, flower and pod termination as well as hindrance in physiological functions like respiration or electron chain transport (Saha et al. 2016). Fortunately, legumes can recover from about 10 days of heat stress without any significant yield loss. Their indeterminate growth pattern and high plasticity in crop phenology makes legumes an ideal candidate for climate-resilient crops to remedy food insecurity (Liu et al. 2003).

Elevated CO_2 increases carbon assimilation which has a substantial effect on root morphology, redox potential and pH modification, as well as symbiotic relationship with mycorrhizal fungi (Haling et al. 2013). Research showed that below-ground parameters such as root growth, nodulation and plant-microbe interaction, respond variably with elevations in CO_2 and temperature and although it has a promising impact on nodule size, number, and biomass in different legume varieties, fixed benefits can only be anticipated under optimal conditions (Bertrand et al. 2011). Legumes outperform other crops against climate change because they have adaptation strategies like allocating more photo assimilates below the ground and can also proliferate their roots to greater depths to utilize nutrients beyond the surface which are solubilized via root mediated secretions (Nie et al. 2013). Phosphorus is known to play a significant role in promoting root growth and strengthening the symbiotic relationship between microbes and roots. Experimental evidence showed a 2–2.5 times increase in root length in elevated CO_2 supplemented with an optimal dose of phosphorus (Campbell and Sage 2002; Wasaki et al. 2005). Veneklaas et al. (2012) reported a higher photosynthetic response with phosphorus use due to an increased rate of transformation from inorganic phosphorus to organic phosphorus which sustained the production of RuBisCo, allowing for a greater portion of phosphorus to be utilized in photosynthetic assimilation and metabolism. Additionally, it was found that apart from CO_2, temperature, and micro-nutrient availability, the genotype and management conditions were also strong determinants of legume growth under climate change (Butterly et al. 2015). It was found that auxin-responsive promoters such as transport inhibitor response 1 (TIR1) and auxin transport genes act as a precursor for the initiation of root hair elongation, especially under elevated CO_2 levels with phosphorus deficiency (Jin et al. 2015). Imaging technologies, statistical tools, and various learning models can be used to assess the genotype, environment, and managing interactions to select for suitable varieties of climate resilient legumes to ensure nutritional security.

Elevated CO_2 has a positive impact on root growth. It also has an impact on the root-shoot ratio of legumes which needs to be optimally balanced in order to have an equal distribution of photosynthates to different parts of the plant to ensure uniform growth under changing climate. Jin et al. (2013) found that neither elevated CO_2 alone, nor in supplementation with phosphorus had any noteworthy impact on the root-shoot ratio. This variation could have resulted from a difference in the C-sink capacity as legumes reserve more carbon to support root growth and other symbiotic soil interactions (Niu et al. 2012). Further studies suggested that the higher root-shoot ratio was observed in the growth stages of the plant as compared to later stages in the legume cycle (Pandey et al. 2016). This resulted from a modification of carbon allocation to the roots, resulting in a significant root-shoot ratio reduction due to altered metabolic activity.

Elevated CO_2 has also shown to have a positive impact on nodule growth of legumes; however, this impact is closely linked with legume species, symbiotic bacteria, nutritional status of plant as well as the soil or environmental conditions (Bertrand et al. 2011; Rogers et al. 2006). The exposure time of elevated CO_2 was a significant determinant for nodule growth. Short-term exposure promotes nodule functions like increased nitrogenase activity and symbiotic nitrogen fixation, whereas prolonged exposure impacts the nodule development by determining biomass and quantity per plant. Nodule biomass is also positively influenced by phosphorus supplementation. Elevated CO_2 levels have a favorable impact on root exudates secretion which substantially assisted the rhizobium to infect, colonize, and form nodules. Contradictory to the favorable impact of elevated CO_2 levels, elevated temperature has a seriously detrimental effect on nodule metabolism and biochemical properties. Higher temperature levels disturb the nodule metabolism by adjusting the carbon allocation, increasing the risk of oxidative stress, exposing Bacteroides to oxygen and by disrupting normal nitrogen metabolism, which collectively hinders nitrogenase from functioning properly. Additionally, elevated temperature also reduces the photosynthetic rate of legumes by forcefully closing guard cells, decreasing the soluble protein content in leaves, slowing down carboxylase activity and eventually inducing water stress (Aranjuelo et al. 2014). Increased temperature also alters gene regulation controlling nodulation and rhizobium infection. Since rhizobia is temperature sensitive and works optimally between 20 and 25 °C, even minor fluctuations of +2 °C can damage the symbiotic relationship by slowing down root growth and infection rate. Nodulation and biological nitrogen fixation are highly interconnected processes so even slight single-step adjustments, radically impact the plant activity. On account of this, even though elevated CO_2 has a beneficial effect on nodulation, this is likely to be interrupted by soil properties such as temperature, micronutrient, and water availability.

8.3.3 Impact of Climate Change on Crop Water Use

Climate change results in two main phenomena: The increase in CO_2 in the environment and rise of temperature. As mentioned before, both have contrasting effects on legume production. As the global temperature rises, it is bound to change many weather parameters like wind speed and direction, solar radiation, precipitation, relative humidity, pan evaporation and vapor pressure deficit. Though all these factors are linked, it is crucial to understand and examine the effects of evapotranspiration rate and vapor pressure deficit to assess the water-use efficiency of legumes. The vapor pressure deficit (VPD) is the difference between the amount of moisture in the air and the amount of moisture that the air can hold at saturation. The normal range for VPD is between 0.8 and 0.95 KPa (Vadez et al. 2012). Evapotranspiration (ET) rate denotes the amount of water lost from a cropped surface in units of water depth. It is expressed in millimeters per unit time. The water-use efficiency (WUE) is the ratio of biomass produced to water used by a plant. High vapor pressure density means drier air, which creates a steeper gradient between the legume plant and the atmosphere. This gradient drives transpiration eventually increasing evapotranspiration rates of a plant. A certain level of evapotranspiration is necessary for plants to cool themselves and transport nutrients, however an excess of this leads to water stress, reduced photosynthesis and lower biomass production, thus decreasing the water-use efficiency. High water-use efficiency can be achieved through balancing evapotranspiration with plant growth. Generally, optimal conditions involve a moderate vapor pressure deficit that allows for sufficient transpiration for cooling and nutrient transport without excessive water loss. WUE changes throughout the crops' life cycle. In case of high VPD, moisture from a plant gets depleted through transpiration resulting in soil moisture exhaustion. Under changing climate, a large amount of water has to be transpired to sustain the biomass accumulation rate of legumes;

however, this is not environmentally or economically sustainable. To remedy this, legume genotypes that have sufficient biomass and a lower transpiration rate, especially when VPD is high should be screened and utilized to increase the water-use efficiency of a plant under conditions of stress (Sinclair et al. 2008). Mainly, the tactic is to preserve water for the consequent grain filing stage in the legume lifecycle. An increase in evapotranspiration due to elevated temperature results in a loss of WUE in legumes (Roderick et al. 2007), whereas reduced evapotranspiration caused a significant reduction in water loss, saving about 5–10% of soil moisture. This also undesirably increased the chance of soil erosion (Rogers et al. 2009). Another promising strategy to increase WUE is by screening legume genotypes, depending on soil conditions using advanced molecular and physiological tools for traits such as drought tolerance such as chickpea (Shah et al. 2020), mungbean (Kumar et al. 2020), and lentils (Biju et al. 2018). The use of modern technology such as crop simulation modelling, artificial intelligence, and geographic positioning systems are essential for predicting climatic conditions and detecting novel traits in legumes that are able to endure abiotic stress and offer high WUE to ensure the capitalization of potential genotypic variability.

Stomatal activity and root conductance also determine the water requirement of a plant under changing climate. In elevated CO_2, depending upon the plant size, leaf cover, and soil condition, stomata activity and transpiration are reduced to lower the water intake (Rasul and Sharma 2016). Therefore, ideal legume varieties should have better adaptation, marginal leaf conductance, and the capacity to save water particularly in the early growing phase. In elevated CO_2 levels, the plant adjusts its guard cell activity by regulating the rate of photosynthesis and leaf water balance. Research shows that aquaporins and abscisic acid play a crucial role in maintaining optimal hydraulic conductivity in roots and leaves. Physiological and molecular interventions should be considered to obtain high WUE in legumes besides their readiness and capability to adjust naturally to elevated CO_2 conditions.

8.3.4 Impact of Climate Change on Insect, Pest, and Diseases Incidences

Insects being cold-blooded have roughly the same temperature as the environment. This makes temperature the most important factor that determines insect behavior, distribution, reproduction, and survival. Climate change can impact ecosystems enough to cause the extinction or revival of insect species (Vitousek et al. 2017). It does so by adapting the physiology and behavior of insects due to changes in host plant resistance, natural threats, and other insect interactions (Delcour et al. 2015). Elevated CO_2 conditions tend to intensify infestations in legumes by lepidopteron pod-borers, coleopterous defoliators, as well as sucking pests like alphids and jassids (Ryalls et al. 2017; Sharma 2010). Climate change which caused elevated CO_2 encourages canopy growth which forms ideal conditions for pathogens to multiply and experience evolution, eventually affecting virulence. Another important determinant of insect, pest, and disease incidence is the soil moisture dynamic. Variations in temperature and humidity increase the risk of damage by pests, alongside strengthening their survival mechanisms. Overcoming challenges brought about by insect and disease occurrences, requires an in-depth understanding of the dynamics that allow insects and pests to flourish under different ecosystems. To avoid disease infestation, it is crucial to be aware of the pathogen virulence, infection rate, or lifecycle and also the susceptibility of a host under changing climate. Moreover, it is important to understand that the application of modern therapies such as new age integrated pest management (IPM) technologies, including fungi produced spinosads and Bt toxins, must be applied along with other conventional chemical methods which are detrimental to the environment. Since legumes are ideal adaptors to climate change, they can overcome biotic stresses also with the right strategies. Research by Palit et al. (2020) suggests that legumes have the ability to develop different plant defense chemicals like peroxidase, polyphenol oxidase, tyrosine ammonia lyase, and phenylalanine ammonia lyase, upon exposure to elevated CO_2 levels which prevents the severity of attack by pests and also greatly increases host resistance. The selection of legume genotypes that are resistant to biotic stress is necessary to conform with contemporary management practices.

8.4 Adaptive Agriculture and Sustainable Farming

Future increases in ambient temperatures are anticipated due to various factors related to global climate change. These rising temperatures lead to heat and water stress, which pose a significant threat to crop production worldwide. Legumes are recognized for their role in agricultural sustainability and their nutritional and health benefits. However, abiotic stress creates challenges for legume crops, negatively affecting their morphology, physiology, and reproductive growth. As legume cultivation expands into warmer regions and temperature variability rises due to climate change, high-temperature stress during the reproductive stage is becoming a major obstacle. This critical period in a plant's life cycle is highly vulnerable to heat stress, which disrupts various metabolic processes and decreases crop yields. Food legumes subjected to high temperatures during reproduction may experience flower abortion, infertility of pollen and ovules, impaired fertilization, and reduced seed filling, resulting in smaller seeds and lower yields. Enhancing climate resilience in major legumes through modern agricultural practices and breeding techniques can improve field performance.

Abiotic stresses, including temperature fluctuations, drought, and salinity, impact the growth of legumes at various stages of development (Suzuki et al. 2014). These stresses are the main contributors to global crop losses, often reducing yields by more than 50% (Rodríguez et al. 2006). Abiotic stress leads to a range of changes in plant morphology, physiology, biochemistry, and molecular processes, adversely affecting plant growth, productivity, and overall yield (Wang et al. 2001; Bita and Gerats 2013). Plants subjected to these stresses experience disruptions in essential physiological functions such as photosynthesis, respiration, nitrogen fixation, reproduction, and oxidative metabolism (Farooq et al. 2008).

After looking at the impacts of climate change on legume production, we now look into adaptive agriculture and sustainable farming as a strategy to address the pressing challenges of climate change, food crop degradation, and food security. Sustainable farming focuses on producing food using methods that protect the environment and support the economy while improving the quality of life of producers and consumers. It involves practices like soil management, water management, reduced chemical inputs, and biodiversity conservation. Adaptive agriculture on the other hand involves the adjustment of farming practices to cope with changing environmental conditions. It involves strategies like crop diversification, development of climate resilient varieties and the application of precision agriculture which uses technology to optimize resource use to increase yields.

8.4.1 Crop Rotation Strategies for Legume Production

Crop rotation is an agricultural practice which increases the efficiency of farming systems throughout the world by improving the soil condition and augmenting system productivity. Enriched soils with a higher water uptake and storage capacity, and increased soil biodiversity to promote symbiotic relationships, not only improve yield but also increase tolerance to abiotic stress in a variety of crop rotations. Farmers highly benefit from crop rotations as they reduce production risk and work toward the long-term development of soil health by reducing insect, weed and disease incidence. The role of crop rotation includes:

1. Soil health management—Researchers and farmers lead the way in advancing soil health practices. Their efforts are crucial in preventing soil depletion and ensuring long-term sustainability. Agricultural practices like diverse crop cultivation, crop rotation, and intercropping significantly impact soil quality from various spatial and temporal angles. Implementing crop rotation not only optimizes crop production but also boosts soil health by improving fertility, nutrient efficiency, and reducing the spread of soil-borne diseases.
2. Resistance to Diseases—Crop rotation is highly effective in breaking the disease cycle within the soil, which is crucial for managing plant diseases. Monoculture, or growing the same crop repeatedly on the same land, can lead to the accumulation of pathogens. Without rotation, these pathogens can build up and spread, leading to more severe plant disease outbreaks. Monoculture farming can act as a host for diseases, allowing them to thrive. In contrast, rotating crops with those from different plant families disrupts the pathogen cycle, as the pathogens cannot infect plants from unrelated families, causing their population in the soil to decline. For instance, a 7-year rotation involving 3 years of alfalfa, 2 years of chile pepper, and 2 years of cotton before returning to alfalfa can effectively break the disease cycle. Alfalfa, chile pepper, and cotton belong to different plant families, which helps in reducing pathogen populations and enhancing productivity (Idowu et al. 2020; Omer et al. 2018). Lupwayi et al. (2006) expanded on how crop rotation contributes to pest management by breaking pest cycles,

reducing weeds and diseases, improving soil quality, and protecting the ecosystem. It also helps in shortening the life cycles of soil-borne pathogens related to specific plant species, fostering beneficial soil microorganisms (Yang et al. 2020). Moreover, rotating different crop genotypes, such as various chickpea cultivars or alternating between pea and chickpea, can enhance soil microbial functionality and improve the performance of subsequent crops like wheat. Different plant species produce unique residues and root exudates that boost the diversity and efficiency of soil microorganisms and support soil carbon and nitrogen cycling (Mayer et al. 2019). The aim of implementing crop rotation is to increase microbial diversity and soil heterogeneity, as various microbes interact with different plant roots, enriching the soil microbiome and promoting overall soil health (Vukicevich et al. 2016).

3. Enhanced soil quality—Diversified crop rotation, including the use of cover crops between cash crops, enhances soil physical and chemical properties by integrating root exudates and biomass. These processes help maintain the soil's physical and chemical balance and mitigate erosion. The roots and residues from these plants contribute to soil quality through various mechanisms, such as creating resilient biopores and influencing rhizosphere interactions, which can stabilize or destabilize soil components.
4. Soil Quality Maintenance—Maintaining soil quality is crucial for an effective farming system, so optimal physical, chemical, and biological characteristics enhance the overall soil health. Crop rotations with wheat and pulses are known to improve soil conditions and boost productivity (Campbell 2005; Gan et al. 2003; Yang et al. 2014). Research indicates that a 4-year rotation of wheat-fallow-field pea (Pisum sativum) can sufficiently protect soil from deterioration by preserving its physical structure (O'Dea et al. 2015; Blanco-Canqui et al. 2013). A study by Gan et al. (2015) suggests that incorporating diverse cropping systems and rotations can enhance soil water conservation and increase nitrogen availability. For example, cover crops like rye and phacelia (Phacelia tanacetifolia L.) have been effective in absorbing nitrogen and water, reducing nitrate leaching in irrigated broccoli (Brassica oleracea var. italica) by 70%, and lowering soil fertilizer needs. Globally, diversified cropping systems have shown numerous benefits. Combining cereals with broadleaf crops, for instance, improves water use efficiency and grain production. Various studies have linked crop rotation diversification to higher soil organic carbon (SOC) levels compared to monoculture practices (Gan et al. 2015; Li et al. 2019). Crop rotations generally lead to increased soil enzyme activity due to the diversity and distribution of crop residues. High-quality soil components positively impact crop productivity (Lal 2009). Crop rotation helps manage weeds, insects, and diseases, reduces soil depletion, and promotes nitrogen fixation. For example, in long-term rotational trials comparing consistent corn, consistent soybean, corn-soybean, and corn-soybean-wheat rotations, the corn-soybean-wheat rotation increased soil organic carbon by 7% compared to consistent soybean rotations. Additionally, rotational and tilling interactions were 7% higher than conventional farming methods (Zuber et al. 2018). Utilizing a variety of crops can enhance carbon

sequestration, improve soil composition and texture, and increase production while mitigating soil and environmental damage.

Various studies show that some of the major factors affecting the adoption of crop rotation practices includes land fragmentation and inadequate resource allocation. With high production costs, suboptimal input resource planning, reluctance to use machinery, and lack of incentives to increase the productivity of small-scale farmers, crop productivity is seriously hindered (Alemu et al. 2017; Beuchelt 2016; Bowman and Zilberman 2013; Lawless and Anderson 1996; Manjunatha et al. 2013). According to Asmah et al. (2016), farmers with numerous small plots tend to use fewer modern technologies, practice monocropping, and reduce the distance between their plots. In contrast, as farm size increases, farmers are more likely to diversify their crops and employ crop rotation, which helps lower the production cost per unit. The study also emphasizes the factors influencing the adoption of these practices, including diffusion of information (such as cultivation knowledge and adoption techniques) and the effects of interventions (which improve production and benefit farmers). However, there remain significant gaps in knowledge, particularly concerning training and outreach programs.

8.4.2 Intercropping Techniques and Benefits

Intercropping involves cultivating two or more crops simultaneously on the same land within a single growing season (Lithourgidis et al. 2011). This practice has significant potential to enhance productivity compared to sole cropping systems, particularly in resource-limited conditions like restricted light, water, and nutrients (Yin et al. 2020). Legume intercropping offers a promising approach to crop diversification and intensification. It enriches soil through biologically fixed nitrogen via legume-rhizobia symbiosis, which improves soil quality and increases soil organic carbon (Raji and Dörsch 2020). Moreover, intercropping different crop species, such as cereals and legumes with contrasting root systems and rhizosphere activities, often results in better ground cover and more efficient resource uptake (Layek et al. 2018). Common intercropping combinations in cereal-legume systems include maize (Zea mays) with soybean (Glycine max) (Fu et al. 2019), maize with pigeon pea (Cajanus cajan), maize with groundnuts (Arachis hypogaea), maize with cowpea (Vigna unguiculata), and maize with beans (Phaseolus vulgaris) (Mucheru-Muna et al. 2010). Other examples are millet with groundnuts, sorghum with cowpea (Nyoki and Ndakidemi 2016), rice with pulses, and barley with faba bean (Vicia faba) (Dwivedi et al. 2015; Mouradi et al. 2018). The effectiveness of intercropping systems depends on factors such as the choice of cultivars, crop competition, and seeding ratios (Layek et al. 2018).

In arid regions, particularly in Africa, agriculture under drought and rainfed conditions faces significant challenges, including limited freshwater resources (Tanwar et al. 2014). Legume-based intercropping has been found to enhance productivity compared to sole cropping systems in such environments. This method can boost resource utilization, improve growth performance of intercrops, and increase stress tolerance and resource use efficiency under water-scarce conditions (Bargaz et al. 2016; Tsubo et al. 2005). Intercropping systems are effective in increasing water use efficiency because they optimize water uptake through complementary root interactions and reduce evaporation losses through improved ground cover (Layek et al. 2018).

Moreover, legume intercropping enriches soil organic matter, which enhances water retention in the root zone and improves soil fertility (Tirado and Cotter 2010). For example, intercropping cereals such as millet and sorghum with legumes like cowpea and groundnut can lead to higher yields due to increased availability of nutrients like nitrogen and phosphorus (Alvey et al. 2001). Organic fertilizers have been shown to significantly boost maize yields in resource-poor areas like Malawi (Makumba et al. 2006).

Intercropping offers benefits such as reduced risk of crop failure, increased soil organic carbon, diversified farm income, and decreased need for herbicides (Asgharipour and Rafiei 2010). Legume intercropping positively influences soil nutrient availability and biogeochemical cycling, which enhances plant growth and stress tolerance (Dai et al. 2019). This practice is recommended for small-scale farmers in semi-arid zones (Tsubo et al. 2005). For instance, pigeon pea intercropped with maize improves soil structure and nutrient storage, increasing phosphorus and organic phosphorus levels significantly compared to sole cropping (Garland et al. 2017). Similarly, cowpea-maize intercropping enhances soil phosphorus availability and maize yields in alkaline soils (Latati et al. 2014).

In soybean-wheat intercropping systems, root microbial diversity, and phosphorus acquisition are improved, contributing to better nutrient acquisition (Bargaz et al. 2017). Intercropping durum wheat with chickpeas or barley with faba beans has been shown to increase phosphorus availability and enzyme activity in the rhizosphere (Betencourt et al. 2012; Mouradi et al. 2018. In contrast, some studies have found that intercropping does not always improve water uptake compared to sole cropping, and may affect nutrient status differently for different crops (Ariel et al. 2013; Hauggaard-Nielsen et al. 2001).

Overall, intercropping can enhance agro-ecosystem services by increasing yields, soil carbon sequestration, and soil quality. It promotes below-ground interactions that improve nutrient bioavailability for companion crops (Xue et al. 2016).

8.4.3 Minimum Tillage and Mulching as Soil Conservation Practices

Minimum tillage and mulching offer numerous benefits for soil health and agricultural productivity. Minimum tillage reduces soil disturbance, which helps maintain soil structure, promotes the formation of beneficial soil aggregates, and enhances stability while minimizing erosion. Similarly, mulching provides a protective layer that guards against wind and water erosion, preventing surface crusting and topsoil loss. Both practices improve soil moisture retention through similar methods;

minimum tillage by reducing surface evaporation and enhancing water infiltration, and mulching by significantly cutting evaporation rates and helping to maintain soil moisture. They also boost soil fertility. Minimum tillage encourages organic matter buildup and better decomposition of plant residues, while mulching adds valuable organic material and supports a healthy microbial community. In terms of weed control, minimum tillage disrupts weed growth less frequently, and mulching acts as a physical barrier to suppress weeds and reduce herbicide use. These methods also regulate soil temperature, with minimum tillage stabilizing temperature extremes and mulching providing insulation to maintain consistent soil temperatures. Furthermore, minimum tillage improves soil aeration by maintaining structure and reducing compaction, while mulching can enhance soil structure and indirectly aid aeration through decomposition. Both practices also support soil biological activity by fostering beneficial microorganisms and earthworms. Collectively contributing to healthier soil, improved crop yields, and more sustainable farming practices.

8.4.4 Breeding of Heat-Resistant Legumes

The negative impact of abiotic stresses on agricultural productivity can be mitigated through a combination of cultural practices and genetic improvements. Genetic advancements can produce cultivars that perform better under high temperatures, resulting in increased economic yields (Tilman et al. 2002; Varshney et al. 2011). However, screening for heat stress tolerance in the field is challenging due to the influence of other environmental factors, though several traits are available for effective selection (Hall 2011). Heat-tolerant genotypes can be identified under controlled conditions, which, despite being costly, avoid interference from other variables affecting high-temperature tolerance mechanisms in field conditions (Souza et al. 2012). Developing effective thermotolerance markers is crucial for breeders to enhance tolerance (Bita and Gerats 2013). To create superior varieties with increased heat tolerance, it is essential to understand the stress response mechanisms in legumes, including variations in gene expression and the resulting changes in the transcriptome, metabolome, and proteome (Ramalingam et al. 2015). There is still limited understanding of the genetic basis for high-temperature tolerance in grain legumes due to the few genetic inheritance studies conducted (Jha et al. 2017). Various genetic analyses, including Mendelian and quantitative genetics, have been used to explore the genetic basis of heat stress tolerance in legumes. Initially, the genetic inheritance of key agronomic traits related to yield performance under high-temperature stress, which are primarily controlled by major or single genes, was studied (Patel and Hall 1988; Hall 1993). For example, in cowpea, heat tolerance was attributed to a single gene based on analysis of traits such as the number of pods per peduncle and the proportion of tolerant plants under high temperatures in different populations derived from heat-sensitive and heat-tolerant genotypes (Marfo and Hall 1992). In common bean, multiple mechanisms for thermotolerance were revealed through analysis of traits such as pods per plant, seeds per plant, and seed weight in heat-tolerant genotypes (Rainey and Griffiths 2005). Understanding the genetic basis of thermotolerance can thus improve plant performance under stress conditions, leading to better overall performance.

Traditional breeding programs aim to develop crop cultivars with high yield under non-stress conditions, which has successfully increased both per-unit crop production and total agricultural output. However, improving heat stress tolerance through genetic engineering offers a more cost-effective approach for enhancing crop production under adverse conditions. The challenge lies in the varying sensitivity to heat across different developmental stages, making the creation of heat-tolerant crops complex. Despite significant progress in breeding heat-tolerant lines for various crops, the genetic foundations and extent of heat tolerance remain largely unknown, particularly in legumes. Developing new varieties is both time-consuming and expensive; therefore, understanding heat tolerance mechanisms is crucial for devising effective strategies to screen legume germplasm for heat-tolerance traits. Incorporating wild varieties and landraces into breeding programs can increase genetic diversity (Driedonks et al. 2016).

Conventional breeding for high-temperature tolerance is one method to mitigate the adverse effects of heat stress on crop yields (Krishnamurthy et al. 2011). Breeding programs are usually conducted in climatic conditions similar to those where the crop will be cultivated. For hot regions, breeding lines are selected under high-temperature conditions (Mickelbart et al. 2015). This approach has proven somewhat successful, as crops grown in warmer areas tend to be more heat-tolerant than those from cooler regions (Kugblenu et al. 2013; Gaur et al. 2014). For example, the chickpea genotype ICCV 92944, which has demonstrated heat tolerance in trials, has been released as JG14 in India, Yezin6 in Myanmar, and Chinadesi2 in Kenya. Additionally, two heat-tolerant faba bean varieties, Shendi and Manami, have been released in Sudan (Gaur et al. 2014). A new cowpea variety has been developed to yield more grain under high temperatures during the reproductive stage (Ehlers and Hall 1998). Heat-tolerant genotypes for various legumes have been created using conventional breeding techniques. For instance, rapid generation advancement and empirical methods have led to the development of heat-tolerant chickpea genotypes (Krishnamurthy et al. 2011). Techniques such as the stress tolerance index (STI), geometric mean (GM), and recurrent selection have been used to develop heat-tolerant common beans (Porch 2006). Heat-tolerant lentil genotypes have been developed using rapid initial growth and early maturity (Sultana et al. 2012). Additionally, mungbean, pea, and snap bean have been made heat-tolerant using temperature-induction response and pedigree methods (Porch and Hall 2013; Bindumadhava et al. 2017). Breeding methods and technologies hold promise for enhancing global legume production which is eminent through the success of breeding in developing heat-tolerant lines.

8.5 Socio-Economic Effects of Climate Resilient Legumes

Adopting climate-resilient legumes and promoting their growth and agriculture has a range of socio-economic benefits which not only impacts individuals or farmers but also communities and eventually economies. By reducing the risk of a crop's failure to grow due to climate variability, climate-resilient legumes provide a much more consistent and reliable income stream for farmers. Additionally, it also increases the yield and reduces input costs which leads to better profitability, especially if farmers are benefitting from financial incentives for sustainable agricultural practices. This stability helps in improving financial security and reduces economic vulnerability. Another major economic benefit of climate-resilient legumes is the employment it provides to rural workers. The cultivation, harvesting, processing, and supplying of these resilient legumes create job opportunities in the market for roles in farming, agribusiness, local processing, and logistics. This increased agricultural activity and associated value chain development acts as a rural economic stimulus which results in the contribution to the overall community and infrastructure development of rural areas. A considerable social effect of climate-resilient legumes is that it promotes gender equality by empowering women. This is because of the central role women play in agriculture as most of the female population of rural areas is involved in agricultural practices as daily-wagers to earn for their families, therefore improving their economic conditions and giving them more decision-making power within their households and communities. In this way it promotes inclusivity and reduces social inequality by empowering the women and marginalized groups of an area. Apart from positively impacting the environment and ecosystem, climate-resilient legumes improve the soil health by enhancing soil fertility via nitrogen fixation which reduces the need for synthetic fertilizers, thus reducing input costs and offering higher profitability. Since climate-resilient legumes require fewer resources, including the use of less water in comparison to non-resilient crops, their agriculture promotes sustainable resource use. On a national level, climate-resilient legumes foster food security by contributing to stable food production which ensures a consistent supply of nutritious food, especially under challenging climatic conditions or catastrophic events. The rich nutrient profile of legumes along with increased production and consumption enhances nutritional outcomes and dietary diversity nationally. On a domestic and international level, the production of climate-resilient legumes increases market access through the development of value chains which creates new economic opportunities. These legumes also offer more export potential since climate-resilient legumes that meet quality and sustainability standards can tap into niche markets and trade opportunities, allowing the inflow of more foreign exchange. The diversified income sources along with collaborative efforts in adopting climate-resilient agriculture helps to build social cohesion and community resilience by developing stronger community networks and cooperation between farmers, private organizations, and support services. It also increases social stability by improving livelihoods and addressing economic vulnerabilities which reduces the risk of migration due to financial and environmental pressures or challenges. Overall, climate-resilient legumes offer substantial and multi-dimensional benefits including new income opportunities, increased economic stability, rural development, empowerment of marginalized communities, improved food security, and environment sustainability, significantly facilitating individual and community well-being.

8.6 Policies, Advises, and Future Directions for Climate Smart Legume Production

8.6.1 Current Policy Landscape and Challenges

Climate resilient legume production is an important strategy to address food insecurity and climate change. The policies and landscape in this regard are evolving as governments, organizations, and stake-holders adapt to the increasing urgency of climate change and sustainable agriculture. Current international agreements and frameworks that are in place today include the Paris Agreement, which encourages countries to adopt climate-smart agricultural practices to minimize greenhouse gas emissions and to adapt to the impacts of climate change globally. Additionally, the United Nations Sustainable Development Goals (SDG), particularly SDG 2—Zero hunger and SDG 13—Climate Action are currently in place to promote sustainable agriculture and climate resilience. On a national level, many countries have incorporated climate-smart agriculture and the promotion of legume production for the benefit of soil health, which also leads to reduced reliance on synthetic fertilizers, in the National Climate Action Plans. Moreover, some nations offer financial incentives for adopting sustainable practices. Governments and international bodies also often fund research and offer innovation support for climate resilience agriculture, to support farmer education and to ensure the dissemination of knowledge, including ideal practices for climate smart crop cultivation.

Currently, the key challenges faced include trade policies which affect market access for farmers and other stakeholders. This includes economic barriers which incur the high initial cost to transition from traditional to climate-smart practices which are often a hurdle for smallholder farmers. Another obstacle is the mismatch between the market demand and production of climate-resilient legumes which eventually affects the profitability of the crop. Despite the efforts, there is a huge gap in the technical knowledge and technological capacity required for climate smart agriculture to flourish. Additionally, the fragmentation of policies across sectors including agriculture, environment, and trade sectors leads to inconsistent and inefficient results, requiring more work on policy integration and implementation, as execution of plans is often disrupted by bureaucratic hurdles, insufficient funding, or lack of coordination.

8.6.2 Policy Recommendations for Promoting Climate-Smart Agriculture

Addressing these challenges in the future requires a multifaceted approach. Effective policy recommendations would focus on incentivizing innovation, supporting research and ensuring equitable access to resources. Some comprehensive policy recommendations include:

- **Incentivizing Climate-smart Practices**
  This includes financial incentives such as tax subsidies for farmers adopting climate-smart practices such as reduced tollage, cover cropping and improved nutrient management. Payment schemes could also be implemented for ecosystem services which would reward farmers for any practices that enhance carbon sequestration, improve soil and water quality, and increase biodiversity.
- **Supporting Research and Innovation**
  An increase in funding for research and development on climate-resilient legumes, soil health and sustainable pest management. This could also be accelerated by encouraging public-private partnerships to drive innovation. Organizations should also promote knowledge sharing to facilitate the dissemination of the most suitable practices and technological advancements through farmer networks, digital platforms, curriculum-based trainings and certifications or other extension services.
- **Farmer Training Programs**
  The development of targeted training programs that address specific regional challenges and promote practical on-ground solutions. This could also help with capacity building by providing farmers the necessary knowledge and skills for implementing climate-smart agriculture. The introduction of agriculture in formal education curricula at various levels, from primary schools to universities and specialized vocational training programs would help with technical understanding and expertise.
- **Development and Implementation of Integrated Policies**
  Policy coordination would ensure that agricultural, environmental, and climate policies are in harmony to support the implementation of climate-smart agriculture.

  Cross-sector collaboration would also foster an alliance between government and non-government agencies, as well as research institutions and the private sector.
- **Market Access and Value Chains**
  The creation of market incentives for climate-smart products, like certifications schemes and eco-labelling would help consumers identify and support sustainable agriculture. Additionally, investing in infrastructure that supports the processing, storage and transport of climate-smart products would lead to a minimization of losses.

- **Facilitated Access to Finance**
  Increasing access to sustainable climate finance for farmers and agricultural businesses through grants, low-interest loans, and insurance schemes would foster growth. The development of risk management tools, like finance and insurance schemes would help stakeholders avoid risks, uncertainties, and losses.
- **Ensuring Equity and Inclusivity**
  Tailoring policies to address the needs of small-scale and marginalized farmers who are most vulnerable to climate change incurred impacts. An inclusive or participatory approach should be used to engage local communities in the design and implementation of climate-smart agricultural policies to ensure equity.
- **Monitoring and Evaluating Impact**
  Data collection systems should be used to evaluate the impact of policies through metrics for environmental, economic, and social outcomes. Adaptive management techniques should also be used to monitor data to adapt and refine policies, ensuring improvement and responsiveness to transpiring challenges.
- **Awareness and Public Support**
  Awareness campaigns should be conducted to educate stakeholders about the importance and outcome of their practices. Stakeholders like farmers and consumers should also be engaged to build public support and understanding.

References

Abbass K, Qasim MZ, Song H et al (2022) A review of the global climate change impacts, adaptation, and sustainable mitigation measures. Environ Sci Pollut Res 29:42539–42559. https://doi.org/10.1007/s11356-022-19718-6

Abdullah et al (2022) Physiological, biochemical and yield responses of linseed (Linum usitatissimum L.) in alpha tocopherol mediated alleviation of salinity stress. Front Plant Sci 13:867172

Affrifah NS, Uebersax MA, Amin S (2023) Nutritional significance, value-added applications, and consumer perceptions of food legumes: a review. Legume Sci 5(4):e192. https://doi.org/10.1002/leg3.192

Ainsworth EA, Rogers A, Vodkin LO, Walter A, Schurr U (2006) The effects of elevated CO_2 concentration on soybean gene expression. An analysis of growing and mature leaves. Plant Physiol 142:135–147. https://doi.org/10.1104/pp.106.086256

Alemu GT, Berhanie Ayele Z, Abelieneh Berhanu A (2017, 2017) Effects of land fragmentation on productivity in northwestern Ethiopia. Adv Agri:4509605, 9 pages

Alonso R, Orue E, Marzo F (1998) Effects of extrusion and conventional processing methods on protein and antinutritional factor contents in pea seeds. Food Chem 63:505–512

Alvey S, Bagayoko M, Neumann G, Buerkert A (2001) Cereal/legume rotations affect chemical properties and biological activities in two West African soils. Plant Soil 231:45–54. https://doi.org/10.1023/A:1010386800937

Amin S, Borchgrevink CP (2022) A Culinology® perspective of dry beans and other pulses. In: Siddiq M, Uebersax MA (eds) Dry beans and pulses: production, processing, and nutrition, 2nd edn. John Wiley, pp 453–480. https://doi.org/10.1002/9781119776802.ch18

Aranjuelo I, Cabrerizo PM, Aparicio-Tejo PM, Arrese-Igor C (2014) Unravelling the mechanisms that improve photosynthetic performance of N_2-fixing pea plants exposed to elevated [CO_2]. Environ Exp Bot 1:167–174. https://doi.org/10.1016/j.envexpbot.2013.10.020

Ariel CE, Eduardo OA, Benito GE, Lidia G (2013) Effects of two plant arrangements in corn (Zea mays L.) and soybean (Glycine max L. Merrill) intercropping on soil nitrogen and phosphorus status and growth of component crops at an Argentinean Argiudoll. Am J Agric 1:22–31

Arif S, Ahmad A, Masud T, Khalid N, Hayat I, Siddique F, Muhammad A (2012) Effect of flour processing on the quality characteristics of a soy-based beverage. Int J Food Sci Nut 63(8):940–946

Arora NK (2019) Impact of climate change on agriculture production and its sustainable solutions. Environ Sustain 2:95–96. https://doi.org/10.1007/s42398-019-00078-w

Asgharipour M, Rafiei M (2010) Intercropping of Isabgol (*Plantago ovata* L.) and Lentil as influenced by drought stress. Am Eurasian J Sustain Agric 4(3):341–348

Asmah S, Syafiq M, Yahya MS et al (2016) Effects of monoculture and polyculture farming in oil palm smallholdings on terrestrial arthropod diversity. J Asia Pac Entomol 19(2):415–421

Azeem A et al (2024) In enhancing crop resilience in the face of a changing climate: strategies for sustainable agricultural production. IGI Global. https://doi.org/10.4018/978-1-6684-9838-5.ch006

Bargaz ME, Isaac ES, Jensen GC (2016) Nodulation and root growth increase in lower soil layers of water-limited faba bean intercropped with wheat. J. Plant. Nutr. Soil Sci 179:537–546. https://doi.org/10.1002/jpln.201500533

Bargaz A, Noyce GL, Fulthorpe R, Carlsson G, Furze JR, Jensen ES, Dhiba D, Isaac ME (2017) Species interactions enhance root allocation, microbial diversity and P acquisition in intercropped wheat and soybean under P deficiency. Appl Soil Ecol 120:179–188. https://doi.org/10.1016/j.apsoil.2017.08.011

Bertrand A, Prévost D, Juge C, Chalifour FP (2011) Impact of elevated CO_2 on carbohydrate and ureide concentrations in soybean inoculated with different strains of Bradyrhizobiumjaponicum. Botany 89:481–490. https://doi.org/10.1139/b11-034

Betencourt E, Duputel M, Colomb B, Desclaux D, Hinsinger P (2012) Intercropping promotes the ability of durum wheat and chickpea to increase rhizosphere phosphorus availability in a low P soil. Soil Biol Biochem 46:181–190. https://doi.org/10.1016/j.soilbio.2011.11.015

Beuchelt TD (2016) Gender, social equity and innovations in smallholder farming systems: pitfalls and pathways. In: Gatzweiler FW, von Braun J (eds) Technological and institutional innovations for marginalized smallholders in agricultural development. Springer International Publishing, Cham, pp 181–198

Biju S, Fuentes S, Gupta D (2018) The use of infrared thermal imaging as a non-destructive screening tool for identifying drought-tolerant lentil genotypes. Plant Physiol Biochem 127:11–24. https://doi.org/10.1016/j.plaphy.2018.03.005

Bindumadhava H, Nair RM, Nayyar H, Riley JJ, Easdown W (2017) Mungbean production under a changing climate-insights from growth physiology. Mysore J Agric Sci 51:21–26

Bindumadhava H, Sharma L, Nair RM, Nayyar H, Riley JJ, Easdown W (2018) High-temperature-tolerant mungbean (Vignaradiata L.) lines produce better yields when exposed to higher CO_2 levels. J Crop Imp 4:418–430. https://doi.org/10.1080/15427528.2018.1439132

Bita CE, Gerats T (2013) Plant tolerance to high temperature in a changing environment: scientific fundamentals and production of heat tolerance crops. Front Plant Sci 4:273. https://doi.org/10.3389/fpls.2013.00273

Blanco-Canqui H, Holman JD, Schlegel AJ, Tatarko J, Shaver TM (2013) Replacing fallow with cover crops in a semiarid soil: effects on soil properties. Soil Sci Soc Am J 77(3):1026–1034

Bowman MS, Zilberman D (2013) Economic factors affecting diversified farming systems. Ecol Soc 18(1):180133

Building climate resilient cropping systems (2024) Food and Agriculture Organization of the United Nations. Available at: https://www.fao.org/in-action/save-grow-climate-smart/our-approach/en/

Butterly CR, Armstrong R, Chen D, Tang C (2015) Carbon and nitrogen partitioning of wheat and field pea grown with two nitrogen levels under elevated CO_2. Plant Soil 391:367–382. https://doi.org/10.1007/s11104-015-2441-5

Campbell C (2005) Carbon storage in soils of the North American great plains: effect of cropping frequency. Agron J 97:363

Campbell CD, Sage RF (2002) Interactions between atmospheric CO_2 concentration and phosphorus nutrition on the formation of proteoid roots in white lupin (Lupinusalbus L.). Plant Cell Environ 25:1051–1059. https://doi.org/10.1046/j.1365-3040.2002.00883.x

Collins K (2020) Carbohydrate quality part 2: pulses and healthy eating. Nutrition research in perspective. Retrieved from https://karencollinsnutrition.com/pulses-healthy-eating/ (September 18, 2022)

Conti MV, Guzzetti L, Panzeri D, De Giuseppe R, Coccetti P, Labra M, Cena H (2021) Bioactive compounds in legumes: implications for sustainable nutrition and health in the elderly population. Trends Food Sci Technol 117:139–147. https://doi.org/10.1016/j.tifs.2021.02.072

Delcour P, Spanoghe P, Uyttendaele M (2015) Literature review: impact of climate change on pesticide use. Food Res Int 68:7–15. https://doi.org/10.1016/j.foodres.2014.09.030

Dhanker, O.P. and Foyer, C.H. (2018) Climate resilient crops for improving global food security and safety – dhankher – 2018 – plant, cell & environment – Wiley Online Library, Climate resilient crops for improving global food security and safety. Available at: https://onlinelibrary.wiley.com/doi/full/10.1111/pce.13207

Driedonks N, Rieu I, Vriezen WH (2016) Breeding for plant heat tolerance at vegetative and reproductive stages. Plant Reprod 29:67–79. https://doi.org/10.1007/s00497-016-0275-9

Dwivedi I, Dev V, Kumar RS, Yadav M, Yadav D, Gupta A, Singh SST (2015) Potential role of maize-legume intercropping systems to improve soil fertility status under smallholder farming systems for sustainable agriculture in India. Int J Life Sci 4:145

Ehlers JD, Hall AE (1998) Heat tolerance of contrasting cowpea lines in short and long days. Field Crop Res 55:11–21. https://doi.org/10.1016/S0378-4290(97)00055-5

El-Hady EA, Habiba R (2003) Effect of soaking and extrusion conditions on antinutrients and protein digestibility of legume seeds. LWT-Food Sci Technol 36:285–293

FAOSTAT (Food and Agriculture Organization of the United Nations) (2018) FAOSTAT (Food and Agriculture Organization of the United Nations) Statistics database (2018) http://www.fao.org/faostat/en/#data

Farooq M, Basra SMA, Wahid A, Cheema ZA, Cheema MA, Khaliq A (2008) Physiological role of exogenously applied glycine betaine to improve drought tolerance in fine grain aromatic rice (Oryza sativa L.). J Agron Crop Sci 194:325–333. https://doi.org/10.1111/j.1439-037X.2008.00323.x

Fernández M, Aranda P, López-Jurado M, GarcíaFuentes MA, Urbano G (1997) Bioavailability of phytic acid phosphorus in processed Vicia faba L. var. Major. J Agri Food Chem 45:4367–4371

Fu Z, Zhou L, Chen P, Du Q, Pang T, Song C, Wang X, Liu W, Yang W, Yong T (2019) Effects of maize-soybean relay intercropping on crop nutrient uptake and soil bacterial community. J Integr Agric 18:2006–2018. https://doi.org/10.1016/S2095-3119(18)62114-8

Gan YT, Miller PR, Mcconkey BG, Zentner RP, Stevenson FC, McDonald CL (2003) Influence of diverse cropping sequences on durum wheat yield and protein in the semiarid northern great plains. Agron J 95(2):245

Gan Y, Hamel C, O'Donovan JT et al (2015) Diversifying crop rotations with pulses enhances system productivity. Sci Rep 5(1):14625

Gaur PM, Samineni S, Krishnamurthy L, Varshney RK, Kumar S, Ghanem ME et al (2014) High temperature tolerance in grain legumes. Legume Perspect 7:23–24. https://doi.org/10.3389/fphys.2012.00179

Gautam AK, Shrivastava N, Sharma B, Bhagyawant S (2018) Current scenario of legume lectins and their practical applications. J Crop Sci Biotechnol 21:217–227. https://doi.org/10.1007/s12892-018-0002-0

Ghosh PK, Venkatesh MS, Hazra KK, Kumar N (2012) Long-term effect of pulses and nutrient management on soil organic carbon dynamics and sustainability on an inceptisol of indo-gangetic plains of India. Exp Agric 48:473–487. https://doi.org/10.1017/S0014479712000130

Gupta D et al. (2019) Hub: toward climate-resilient lentils: challenges and opportunities. Genomic designing of climate-smart pulse crops, 165–234: https://doi.org/10.1007/

978-3-319-96932-9_4, Genomic Designing of Climate-Smart Pulse Crops. Available at: https://sci-hub.se/https://link.springer.com/chapter/10.1007/978-3-319-96932-9_4

Haling RE, Brown LK, Bengough AG, Young IM, Hallett PD, White PJ, George TS (2013) Root hairs improve root penetration, root–soil contact, and phosphorus acquisition in soils of different strength. J Exp Bot 39:2208–2221. https://doi.org/10.1093/jxb/ert200

Hall AE (1993) Physiology and breeding for heat tolerance in cowpea, and comparisons with other crops. In: Kuo CG (ed) Adaptation of food crops to temperature and water stress. Asian Vegetable Research and Development Center, Tainan, pp 271–284

Hall AE (2011) Breeding cowpea for future climates. In: Yadav SS, Redden R, Hatfield JL, Lotze-Campen H, Hall AJW (eds) Crop adaptation to climate change. John Wiley and Sons, Hoboken, NJ

Hauggaard-Nielsen H, Ambus P, Jensen ES (2001) Interspecific competition, N use and interference with weeds in pea–barley intercropping. Field Crop Res 70(2):101–109. https://doi.org/10.1016/S0378-4290(01)00126-5

Hazra KK, Ghosh PK, Venkatesh MS, Nath CP, Kumar N, Singh M, Singh J, Nadarajan N (2018) Improving soil organic carbon pools through inclusion of summer mung bean in cereal–cereal cropping systems in Indo-Gangetic plain. Arch Agron Soil Sci 64:1690–1704. https://doi.org/10.1080/03650340.2018.1451638

Höhn A, Weber D, Jung T, Ott C, Hugo M, Kochlik B, Kehm R, König J, Grune T, Castro JP (2017) Happily (n)ever after: aging in the context of oxidative stress, proteostasis loss and cellular senescence. Redox Biol 11:482–501. https://doi.org/10.1016/j.redox.2016.12.001

Idowu O, Ghimire R, Flynn RP, Ganguli A (2020) SOIL HEALTH-importance, assessment, and management. College of Agricultural, Consumer and Environmental Sciences, New Mexico State University, Las Cruces, NM

Jagadish SV, Bahuguna RN, Djanaguiraman M, Gamuyao R, Prasad PV, Craufurd PQ (2016) Implications of high temperature and elevated CO_2 on flowering time in plants. Front Plant Sci 7:913. https://doi.org/10.3389/fpls.2016.00913

Jannat S, Shah AH, Hassan M, Ahmad Sher A, Fiaz S et al (2022) Genetic diversity of common bean (Phaseolus vulgaris L.) ecotypes from Pakistan using simple sequence repeats. Saudi J Biol Sci 29:103300

Jha UC, Bohra A, Parida SK, Jha R (2017) Integrated "omics" approaches to sustain global productivity of major grain legumes under heat stress. Plant Breed 136:437–459. https://doi.org/10.1111/pbr.12489

Jin J, Tang C, Armstrong R, Sale P (2012) Phosphorus supply enhances the response of legumes to elevated CO_2 (FACE) in a phosphorus-deficient vertisol. Plant Soil 1:91–104. https://doi.org/10.1007/s11104-012-1270-z

Jin J, Tang C, Armstrong R, Butterly C, Sale P (2013) Elevated CO_2 temporally enhances phosphorus immobilization in the rhizosphere of wheat and chickpea. Plant Soil 368:315–328. https://doi.org/10.1007/s11104-012-1516-9

Jin J, Tang C, Sale P (2015) The impact of elevated carbon dioxide on the phosphorus nutrition of plants: a review. Ann Bot 116:987–999. https://doi.org/10.1093/aob/mcv088

Kimball BA (2016) Crop responses to elevated CO_2 and interactions with H_2O, N, and temperature. Curr Opin Plant Biol 31:36–43. https://doi.org/10.1016/j.pbi.2016.03.006

Krishnamurthy L, Gaur PM, Basu PS, Chaturvedi SK, Tripathi S, Vadez V et al (2011) Large genetic variation for heat tolerance in the reference collection of chickpea (Cicer arietinum L.) germplasm. Plant Genet Resour 9:59–69. https://doi.org/10.1017/S1479262110000407

Kugblenu YO, Oppong Danso E, Ofori K, Andersen MN, AbenneyMickson S, Sabi EB et al (2013) Screening tomato genotypes for adaptation to high temperature in West Africa. Acta Agric. Scand Sect B Soil Plant Sci 63:516–522. https://doi.org/10.1080/09064710.2013.813062

Kumar N, Hazra KK, Nath CP, Praharaj CS, U. (2018) Singh Grain legumes for resource conservation and agricultural sustainability in South Asia. Legumes for soil health and sustainable management. Springer, Singapore, pp 77–107

Kumar S, Ayachit G, Sahoo L (2020) Screening of mungbean for drought tolerance and transcriptome profiling between drought-tolerant and susceptible genotype in response to drought stress. Plant Physiol Biochem 157:229–238. https://doi.org/10.1016/j.plaphy.2020.10.021

Kumar Y, Basu S, Goswami D, Devi M, Shivhare US, Vishwakarma RK (2022) Anti-nutritional compounds in pulses: implications and alleviation methods. Legume Sci 4:e111

Lal R (2009) Soil degradation as a reason for inadequate human nutrition. Food Secur 1(1):45–57

Latati M, Blavet D, Alkama N, Laoufi H, Drevon JJ, Gérard F, Pansu M, Ounane SM (2014) The intercropping cowpea-maize improves soil phosphorus availability and maize yields in an alkaline soil. Plant Soil 385:181–191. https://doi.org/10.1007/s11104-014-2214-6

Lawless MW, Anderson PC (1996) Generational technological change: effects of innovation and local rivalry on performance. Acad Manag J 39(5):1185–1217

Layek J, Das A, Mitran T, Nath C, Meena RS, Yadav GS, Shivakumar BG, Kumar S, Lal R (2018) Cereal+Legume intercropping: an option for improving productivity and sustaining soil health. In: Meena RS, Das A, Yadav GS, Lal R (eds) Legumes for soil health and sustainable management. Springer, Singapore, pp 347–386. https://doi.org/10.1007/978-981-13-0253-4_11

Lee JS (2011) Combined effect of elevated CO_2 and temperature on the growth and phenology of two annual C_3 and C_4 weedy species. Agric Ecosyst Environ 140:484–489. https://doi.org/10.1016/j.agee.2011.01.013

Li J, Huang L, Zhang J, Coulter JA, Li L, Gan Y (2019) Diversifying crop rotation improves system robustness. Agron Sustain Dev 39(4):38

Lithourgidis AS, Dordas CA, Damalas CA, Vlachostergios D (2011) Annual intercrops: an alternative pathway for sustainable agriculture. Aust J Crop Sci 5:396

Liu PH, Gan Y, Warkentin T, McDonald C (2003) Morphological plasticity of chickpea in a semiarid environment. Crop Sci 43:426–429. https://doi.org/10.2135/cropsci2003.4260

Lupwayi NZ, Clayton GW, Rice WA (2006) Rhizobial inoculants for legume crops. J Crop Improv 15(2):289–321

Makumba W, Janssen B, Oenema O, Akinnifesi F, Mweta D, Kwesiga F (2006) The long-term effects of a gliricidia–maize intercropping system in Southern Malawi, on gliricidia and maize yields, and soil properties. Agric Ecosyst Environ 116:85–92. https://doi.org/10.1016/j.agee.2006.03.012

Manjunatha AV, Anik AR, Speelman S, Nuppenau EA (2013) Impact of land fragmentation, farm size, land ownership and crop diversity on profit and efficiency of irrigated farms in India. Land Use Policy 31:397–405

Maphosa Y, Jideani VA (2017) The role of legumes in human nutrition. InTech. https://doi.org/10.5772/intechopen.69127

Marfo KO, Hall AE (1992) Inheritance of heat tolerance during pod set in cowpea. Crop Sci 32:912–918. https://doi.org/10.2135/cropsci1992.0011183X003200040015x

Martín-Cabrejas MÁ (2019) Legumes: nutritional quality, processing and potential health benefits. Royal Society of Chemistry, pp. 152–176. https://doi.org/10.1039/9781788015721

Mayer Z, Sasvari Z, Szentpéteri V, Rethati BP, Vajna B, Posta K (2019) Effect of long-term cropping systems on the diversity of the soil bacterial communities. Agronomy 9(12):878

Mickelbart MV, Hasegawa PM, Bailey-Serres J (2015) Genetic mechanisms of abiotic stress tolerance that translate to crop yield stability. Nat Rev Genet 16:237–251. https://doi.org/10.1038/nrg3901

Miklas PN, Kelly JD, Cichy KA (2022) Dry bean breeding and production technologies. In: Siddiq M, Uebersax MA (eds) Dry beans and pulses: production, processing and nutrition, 2nd edn. John Wiley, pp 29–56. https://doi.org/10.1002/9781119776802.ch2

Mouradi M, Farissi M, Makoudi B, Bouizgaren A, Ghoulam C (2018) Effect of faba bean (Vicia faba L.)–rhizobia symbiosis on barley's growth, phosphorus uptake and acid phosphatase activity in the intercropping system. Ann. Agrar. Sci. 16:297–303. https://doi.org/10.1016/j.aasci.2018.05.003

Mucheru-Muna M, Pypers P, Mugendi D, J. Kung'u, J. Mugwe, R. Merckx, B. Vanlauwe. (2010) A staggered maize–legume intercrop arrangement robustly increases crop yields and economic returns in the highlands of Central Kenya. Field Crop Res 115:132–139. https://doi.org/10.1016/j.fcr.2009.10.013

Newton PCD, Clark H, Bell CC, Glasgow EM (1996) Interaction of soil moisture and elevated CO_2 on the above-ground growth rate, root length density and gas exchange of turves from temperate pasture. J Exp Bot 47:771–779. https://doi.org/10.1093/jxb/47.6.771
Newton AC, Johnson SN, Gregory PJ (2011) Implications of climate change for diseases, crop yields and food security. Euphytica 179:3–18. https://doi.org/10.1007/s10681-011-0359-4
Nie M, Lu M, Bell J, Raut S, Pendall E (2013) Altered root traits due to elevated CO_2: a meta-analysis. Glob Ecol Biogeogr 22:1095–1105. https://doi.org/10.1111/geb.12062
Niu YF, Chai RS, Jin GL, Wang H, Tang CX, Zhang YS (2012) Responses of root architecture development to low phosphorus availability: a review. Ann Bot 112:391–408. https://doi.org/10.1093/aob/mcs285
Nyoki D, Ndakidemi PA (2016) Intercropping system, rhizobia inoculation, phosphorus and potassium fertilization: a strategy of soil replenishment for improved crop yield. Int J Curr Microbiol Appl Sci 5(10):504–522. https://doi.org/10.20546/ijcmas.2016.510.056
O'Dea JK, Jones CA, Zabinski CA, Miller PR, Keren IN (2015) Legume, cropping intensity, and N-fertilization effects on soil attributes and processes from an eight-year-old semiarid wheat system. Nutr Cycl Agroecosyst 102(2):179–194
Omer M, Idowu O, Ulery A, VanLeeuwen D, Guldan S (2018) Seasonal changes of soil quality indicators in selected arid cropping systems. Agriculture 8(8):124
Palit P, Kudapa H, Zougmore R, Kholova J, Whitbread A, Sharma M, Varshney RK (2020) An integrated research framework combining genomics, systems biology, physiology, modelling and breeding for legume improvement in response to elevated CO_2 under climate change scenario. Curr. Plant Biol 22:100149. https://doi.org/10.1016/j.cpb.2020.100149
Pandey R, Meena S, Krishnapriya V, Khetarpal S, Pal M (2016) Interactive effects of phosphorus nutrition and atmospheric carbon dioxide levels on growth, nitrogen fixation and yield of green gram. Ind J Fertil 12:56–64
Patel PN, Hall AE (1988) Inheritance of heat-induced brown discoloration in seed coats of cowpea. Crop Sci 28:929–932. https://doi.org/10.2135/cropsci1988.0011183X002800060011x
Pedrosa MM, Guillamón E, Arribas C (2021) Autoclaved and extruded legumes as a source of bioactive phytochemicals: a review. Food 10:379. https://doi.org/10.3390/foods10020379
Porch TG (2006) Application of stress indices for heat tolerance screening of common bean. J Agron Crop Sci 192:390–394. https://doi.org/10.1111/j.1439-037x.2006.00229.x
Porch TG, Hall AE (2013) Heat tolerance. In: Kole C (ed) Genomics and breeding for climate-resilient crops. Springer, Berlin, pp 167–202
Qayyum MMN, Butt MS, Anjum FM, Nawaz H (2012) Composition analysis of some selected legumes for protein isolates recovery. J Animal Plant Sci 22(4):1156–1162. ISSN: 1018-7081
Rainey KM, Griffiths PD (2005) Differential response of common bean genotypes to high temperature. J Am Soc Hortic Sci 130:18–23
Raji SG, Dörsch P (2020) Effect of legume intercropping on N_2O emissions and CH_4 uptake during maize production in the Great Rift Valley, Ethiopia. Biogeosciences 17:345–359. https://doi.org/10.5194/bg-17-345-2020
Raju J, Mehta R (2009) Cancer chemopreventive and therapeutic effects of diosgenin, a food saponin. Nutr & Cancer 61:27–35
Ramalingam A, Kudapa H, Pazhamala LT, Weckwerth W, Varshney RK (2015) Proteomics and metabolomics: two emerging areas for legume improvement. Front Plant Sci 6:1116. https://doi.org/10.3389/fpls.2015.01116
Rasul G, Sharma B (2016) The nexus approach to water–energy–food security: an option for adaptation to climate change. Clim Pol 16:682–702. https://doi.org/10.1080/14693062.2015.1029865
Roderick ML, Rotstayn LD, Farquhar GD, Hobbins MT (2007) On the attribution of changing pan evaporation. Geophys Res Lett 34. https://doi.org/10.1029/2007GL031166
Rodríguez M, Canales E, Borroto CJ, Carmona E, Lopez J, Pujol M et al (2006) Identification of genes induced upon water-deficit stress in a drought-tolerant rice cultivar. J Plant Physiol 163:577–584. https://doi.org/10.1016/j.jplph.2005.07.005
Rogers A, Gibon Y, Stitt M, Morgan PB, Bernacchi CJ, Ort DR, Long SP (2006) Increased C availability at elevated carbon dioxide concentration improves N assimilation in a legume. Plant Cell Environ 29:1651–1658. https://doi.org/10.1111/j.1365-3040.2006.01549

Rogers A, Ainsworth EA, Leakey AD (2009) Will elevated carbon dioxide concentration amplify the benefits of nitrogen fixation in legumes? Plant Physiol 151:1009–1016. https://doi.org/10.1104/pp.109.144113

Ryalls JM, Moore BD, Riegler M, Bromfield LM, Hall AA, Johnson SN (2017) Climate and atmospheric change impacts on sap-feeding herbivores: a mechanistic explanation based on functional groups of primary metabolites. Funct Ecol 31:161–171. https://doi.org/10.1111/1365-2435.12715

Saha S, Chakraborty D, Sehgal VK, Nain L, Pal M (2016) Long-term atmospheric CO_2 enrichment impact on soil biophysical properties and root nodule biophysics in chickpea (Cicer arietinum L.). Eur J Agron 75:1–11. https://doi.org/10.1016/j.eja.2015.12.005

Shah TM, Imran M, Atta BM, Ashraf MY, Hameed A, Waqar I, Shafiq M, Hussain K, Naveed M, Aslam M, Maqbool MA (2020) Selection and screening of drought tolerant high yielding chickpea genotypes based on physio-biochemical indices and multi-environmental yield trials. BMC Plant Biol 20:1–16. https://doi.org/10.1186/s12870-020-02381-9

Shahzadi A, Noreen Z, Alamery S et al (2024) Effects of biochar on growth and yield of Wheat (Triticum aestivum L.) under salt stress. Sci Rep 14:20024. https://doi.org/10.1038/s41598-024-70917-2

Sharma, HC (2010) Effect of climate change on IPM in grain legumes. Proceedings of the 5th international food legumes research conference (IFLRC V), and the 7th European conference on grain legumes (AEP VII) (pp. 26-30), Anatalaya, Turkey

Sharma HC, War AR, Pathania M, Sharma SP, Akbar SM, Munghate RS (2016) Elevated CO_2 influences host plant defense response in chickpea against Helicoverpa armigera. Arthropod Plant Interact 10:171–181. https://doi.org/10.1007/s11829-016-9422-3

Shi Z, Chen J, Li CY, An N, Wang ZJ, Yang SL, Huang KF, Bao JK (2014) Antitumor effects of concanavalin A and Sophora flavescens lectin in vitro and in vivo. Acta Pharmacol Sin 35:248–256. https://doi.org/10.1038/aps.2013.151

Sinclair TR, Zwieniecki MA, Holbrook NM (2008) Low leaf hydraulic conductance associated with drought tolerance in soybean. Physiol Plantarum 132:446–451. https://doi.org/10.1111/j.1399-3054.2007.01028.x

Singh B, Singh JP, Kaur A, Singh N (2018) Insights into the phenolic compounds present in jambolan (Syzygium cumini) along with their health-promoting effects. Int J Food Sci Technol 53:2431–2447

Souza MA, Pimentel AJB, Ribeiro G (2012) Breeding for heat-stress tolerance. In: Fritsche-Neto R, Borém A (eds) Plant breeding for abiotic stress tolerance. Springer, Berlin, pp 137–156

Srinivasarao C, Kundu S, Shanker AK, Naik RP, Vanaja M, Venkanna K, Rao VUM (2016) Continuous cropping under elevated CO_2: differential effects on C4 and C3 crops, soil properties and carbon dynamics in semi-arid alfisols. Agric Ecosyst Environ 15:73–86. https://doi.org/10.1016/j.agee.2015.11.016

Sultana R, Vales MI, Saxena KB, Rathore A, Rao S, Rao SK et al (2012) Water-logging tolerances in pigeonpea (Cajanus cajan L. Millsp.): genotypic variability and identification of tolerant genotypes. J Agric Sci 151:659–671. https://doi.org/10.1017/S0021859612000755

Suzuki N, Rivero RM, Shulaev V, Blumwald E, Mittler R (2014) Abiotic and biotic stress combinations. New Phytol 203:32–43. https://doi.org/10.1111/nph.12797

Tanwar SPS, Rao SS, Regar PL, Datt S, Praveen Kumar BS, Jodha P, Santra R, Kumar RR (2014) Improving water and land use efficiency of fallow-wheat system in shallow Lithic Calciorthid soils of arid region: introduction of bed planting and rainy season sorghum–legume intercropping. Soil Tillage Res 138:44–55. https://doi.org/10.1016/j.still.2013.12.005

Thomas RB, Van Bloem SJ, Schlesinger WH (2006) Climate change and symbiotic nitrogen fixation in agroecosystems in agroecosystems in a changing climate. CRC Press, Boca Raton, FL, pp 85–116

Tilman D, Cassman KG, Matson PA, Naylor R, Polasky S (2002) Agricultural sustainability and intensive production practices. Nature 418:671–677. https://doi.org/10.1038/nature01014

Tirado R, Cotter J (2010) Ecological farming: drought-resistant agriculture 16. Greenpeace Research Laboratories, Exeter

Tsubo M, Walker S, Ogindo HO (2005) A simulation model of cereal–legume intercropping systems for semi-arid regions. Field Crop Res 93:10–22. https://doi.org/10.1016/j.fcr.2004.09.002
Uebersax MA, Cichy KA, Gomez FE, Porch TG, Heitholt J, Osorno JM, Kamfwa K, Snapp SS, Bales S (2022) Dry beans (Phaseolus vulgaris L.) as a vital component of sustainable agriculture and food security – a review. Legume Sci 2022:e155
Vadez V, Berger JD, Warkentin T, Asseng S, Ratnakumar P, Rao KPC, Sharma HC (2012) Adaptation of grain legumes to climate change: a review. Agron Sustain Dev 132:31–44. https://doi.org/10.1007/s13593-011-0020-6
Varshney RK, Bansal KC, Aggarwal PK, Datta SK, Craufurd PQ (2011) Agricultural biotechnology for crop improvement in a variable climate: hope or hype? Trends Plant Sci 16:363–371. https://doi.org/10.1016/j.tplants.2011.03.004
Veneklaas EJ, Lambers H, Bragg J, Finnegan PM, Lovelock CE, Plaxton WC, Raven JA (2012) Opportunities for improving phosphorus-use efficiency in crop plants. New Phytol 195:306–320. https://doi.org/10.1111/j.1469-8137.2012.04190.x
Venkidasamy B, Selvaraj D, Nile AS, Ramalingam S, Kai G et al (2019) Indian pulses: a review on nutritional, functional and biochemical properties with future perspectives. Trends Food Sci Technol 88:228–242
Vitousek PM, Loope LL, Westbrooks R (2017) Biological invasions as global environmental change. Am Sci 84:468
Vukicevich E, Lowery T, Bowen P, Urbez-Torres JR, Hart M (2016) Cover crops to increase soil microbial diversity and mitigate decline in perennial agriculture. a review. Agron Sustain Dev 36(3):48
Wang WX, Vinocur B, Shoseyov O, Altman A (2001) Biotechnology of plant osmotic stress tolerance: physiological and molecular considerations. Acta Hortic 560:285–292. https://doi.org/10.17660/ActaHortic.2001.560.54
Wang J, Gan YT, Clarke F, McDonald CL (2006) Response of chickpea yield to high temperature stress during reproductive development. Crop Sci 46:2171–2178. https://doi.org/10.2135/cropsci2006.02.0092
Wasaki J, Rothe A, Kania A, Neumann G, Römheld V, Shinano T, Kandeler E (2005) Root exudation, phosphorus acquisition, and microbial diversity in the rhizosphere of white lupine as affected by phosphorus supply and atmospheric carbon dioxide concentration. J Environ Qual 34:2157–2166. https://doi.org/10.2134/jeq2004.0423
Wiesinger JA, Marsolais F, Glahn RP (2022) Health implications and nutrient bioavailability of bioactive compounds in dry beans and other pulses. In: Siddiq M, Uebersax MA (eds) Dry beans and pulses: production, processing, and nutrition, 2nd edn. John Wiley, pp 505–529. https://doi.org/10.1002/9781119776802.ch20
Xue Y, Xia H, Christie P, Zhang Z, Li L, Tang C (2016) Crop acquisition of phosphorus, iron and zinc from soil in cereal/legume intercropping systems: a critical review. Ann Bot 117:363–377. https://doi.org/10.1093/aob/mcv182
Yang X, Gao W, Zhang M, Chen Y, Sui P (2014) Reducing agricultural carbon footprint through diversified crop rotation systems in the North China Plain. J Clean Prod 76:131–139
Yang T, Siddique KHM, Liu K (2020) Cropping systems in agriculture and their impact on soil health-A review. Global Ecol Conserv 23:e01118
Yin W, Chai Q, Zhao C, Yu A, Fan Z, Hu F, Fan H, Guo Y, Coulter JA (2020) Water utilization in intercropping: a review. Agric Water Manag 241:106335. https://doi.org/10.1016/j.agwat.2020.106335
Zhang Y, Chen X, Zhang C, Pan G, Zhang X (2014) Availability of soil nitrogen and phosphorus under elevated [CO_2] and temperature in the Taihu Lake region. China J Plant Nutr Soil Sci 177:343–348. https://doi.org/10.1002/jpln.201200526
Zuber SM, Behnke GD, Nafziger ED, Villamil MB (2018) Carbon and nitrogen content of soil organic matter and microbial biomass under long-term crop rotation and tillage in Illinois, USA. Agriculture 8(3):37

9 Promoting the Adaptation of Climate-Smart Agriculture Practices Among Farming Communities

Imtiaz Hussain, Muhammad Zubair Khan, Nagina Rafiq, and Saiqa Bashir

Abstract

Climate Smart Agriculture (CSA) is a comprehensive approach aimed at transforming agricultural systems to enhance resilience to climate change while promoting environmental sustainability. This strategy integrates context-specific techniques and technologies that enhance resource use efficiency, thereby addressing critical challenges such as food security, biofuel demand, and climate adaptation. Despite its potential, the widespread adoption of CSA practices, particularly in developing nations, faces significant barriers. Limited exposure to innovative CSA technologies, financial constraints, and inadequate infrastructure hinder farmers from fully implementing these practices. Moreover, the complex nature of climate impacts necessitates a high level of expertise to tailor solutions to local conditions. Effective execution of CSA requires substantial capacity building, including training for farmers and the development of skilled personnel and infrastructure. Additionally, creating incentive systems to reward sustainable behaviors is crucial for encouraging the adoption of CSA practices. While there is general consensus on the importance of CSA for enhancing agricultural resilience, the intricacies of its implementation, including the costs of oversight and the need for localized solutions, present ongoing challenges. By addressing these barriers and leveraging the potential of CSA, the agricultural sector can contribute significantly to climate change mitigation and adaptation, ensuring food security in an increasingly uncertain climate.

I. Hussain (✉) · N. Rafiq · S. Bashir
Department of Food Science & Technology, University of Poonch Rawalakot, Rawalakot, Pakistan

M. Z. Khan
Department of Plant Breeding and Molecular Genetics, University of Poonch Rawalakot, Rawalakot, Pakistan

© The Author(s), under exclusive license to Springer Nature Singapore Pte Ltd. 2025

S. Faiz et al. (eds.), *Climate Smart Agriculture for Future Food Security*,
https://doi.org/10.1007/978-981-96-4499-5_9

Keywords

Climate Smart Agriculture (CSA) · Resilience · Food security · Sustainable practices · Capacity building

9.1 Introduction

9.1.1 Overview of Climate-Smart Agriculture (CSA) and Its Significance

The combined effect of various factors such as population growth, agricultural resources, food security, biofuel demand, climate adaptation and mitigation, oil prices, and food costs has resulted in the emergence and progress of "Climate Smart Agriculture" (CSA) (Balo and Mahata 2022). CSA encompasses an approach aimed at changing the food and agriculture sectors to become more resilient to climate change side by side promoting environmental sustainability (Lipper et al. 2014).

Climate-smart agriculture is a broad approach which integrates context-specific techniques and technologies, such as sustainable enhancement of resources. Current agriculture related policies have preferred these approaches to discuss both environmental and economic challenges (Lipper et al. 2014; Jayne et al. 2018). The strategy is to bring a solution for minimizing and adapting global warming while ensuring food security (FAO 2013). Developing nations have adopted climate-smart agricultural methods and technologies. However, still there are several challenges which hinder farmers from adopting CSA practices fully (FAO 2013; Nkonya et al. 2015; Hansen et al. 2019).

CSA is a strategy which neglects the specific background and community-appropriate adaptable results in favor of improving farming practices and safeguarding food security in any uncertain climate (Lipper et al. 2014; McCarthy and Brubaker 2014; McCarthy et al. 2011). It focuses to determine suitable behaviors in this situation and creates a system to reward experts who stick to these behaviors. A principle element of CSA can be finding acceptable behaviors in this setting and generating a system for rewarding workers who adopt these routines. Although this approach holds significant potential, execution is challenged by intricate obstacles and the expense of oversight. (Hamilton and Zilberman 2006). According to FAO (2013), farmers had limited exposure to CSA technologies and practices due to their innovative nature.

Similarly, CSA interventions require a substantial improvement of a person's or an organization's capability regarding technology, skilled personnel, infrastructure, and financial resources. A high level of expertise is also needed to fully understand site-specific conditions. Studies show the difficulties even if there is general consensus regarding the significance of CSA for agriculture's climate change adaptation (Zougmoré et al. 2016; Campbell et al. 2014).

9.1.2 The Impact of Climate Change on Agriculture and Farming Communities

Climate change has a profound impact on food systems and agricultural productivity, heightening vulnerability and uncertainty for both farmers and policymakers (Stafford 2000). The threat to global food security is expected to intensify in the coming years, exacerbated by factors such as population growth, economic development, urbanization, and the increasing frequency of natural disasters like floods, droughts, and extreme temperatures (Javaid et al. 2022; Jamil et al. 2023). By 2050, the global population is expected to surpass 9 billion, with poverty and hunger becoming more prevalent, further complicating efforts to ensure food security (Saiz-Rubio and Rovira-Más 2020; Friha et al. 2021; Subeesh and Mehta 2021).

In response, international organizations are partnering with the World Bank and the Food and Agriculture Organization (FAO) to enhance food production across various dimensions. This collaboration involves a broad array of actors, including public and private sectors, civil society, researchers, investors, and policymakers, all supporting the global transition toward climate-smart agriculture (CSA) (Foughali et al. 2018; Lakshmi et al. 2023; Khilji et al. 2022). Climate change disrupts agricultural productivity and increases the unpredictability of crop yields, which contributes to food and nutrition insecurity and destabilizes the global food supply.

Climate change poses serious threats to food production through water shortages, pest infestations, and soil degradation, resulting in substantial crop losses and severe risks to global food security (Mirón et al. 2023; Läderach et al. 2017; Zizinga et al. 2022; Arora 2019; Kang et al. 2009). The United Nations estimates that by 2050, the global population will reach 9.7 billion, requiring a 70% increase in food-calorie production to meet the rising demand (Secretary-General UN 2021). Consequently, the adoption of effective mitigation strategies is essential for minimizing the adverse effects of climate change while enhancing the resilience and adaptability of smallholder farming systems.

9.1.3 The Need for Adaptation to Ensure Food Security and Sustainable Livelihoods

In response to the increasing demand for agricultural production and the growing challenges posed by climate change, farmers are increasingly adopting climate-smart agriculture (CSA) techniques. Numerous studies have explored the impacts of community-supported agricultural strategies within the CSA framework (Neuenschwander et al. 2023; Adonadaga et al. 2022; Zizinga et al. 2022). For instance, research by Fentie and Beyene (2019) demonstrated that row planting significantly boosts maize yields and enhances commercialization in Ghana. Similarly, Aryal et al. (2020) highlighted that soil, water, and nutrient management practices in South Asia help reduce greenhouse gas (GHG) emissions, contributing to climate change mitigation.

As global climate challenges and the demand for agricultural production intensify, farmers are increasingly compelled to adopt CSA practices. Several studies have examined diverse CSA methods and their effects (Zizinga et al. 2022; Neuenschwander et al. 2023; Adonadaga et al. 2022). Fentie and Beyene (2019) found that row planting increases maize productivity and strengthens commercialization in Ghana, while Aryal et al. (2020) noted that managing soil, water, and nutrients in South Asia helps lower GHG emissions. Additionally, Zizinga et al. (2022) conducted field trials on varying mulch thicknesses (0, 2, 4, and 6 cm), finding that CSA techniques enhanced water-use efficiency and maize productivity, with optimal results at 6 cm mulch depth. However, these studies were limited to specific CSA techniques, and their findings may not be applicable to other contexts.

9.2 Understanding Climate-Smart Agriculture

During the 2010 Hague Conference on Agriculture, the United Nations' Food and Agriculture Organization (FAO) officially advocated for addressing climate change, ensuring food security, and promoting climate-smart Agriculture (CSA). A sustainable approach to agricultural development and efficiency, CSA can boost adaptation, lower greenhouse gas emissions, and guarantee national food security (Taylor 2018). In order to address the dangers that climate change poses to food security and sustainable agricultural growth, it integrates the three elements of economics, society, and environment (Lipper et al. 2014).

9.2.1 Definition and Principles of Climate-Smart Agriculture

Both the Food and Agriculture Organization (FAO) and the World Bank view climate-smart agriculture (CSA) as essential for enhancing food production, encompassing a variety of agricultural technologies and practices. The World Bank describes climate-smart approaches as a "governance framework" that incorporates the necessary tools, technologies—such as energy efficiency, carbon capture, and next-generation renewable energy—and institutions to address climate-related challenges (WB 2009). This innovative funding model aligns with the FAO's emphasis on the importance of CSA as a prerequisite for its effective implementation. The FAO defines climate-smart agriculture as a strategy aimed at achieving three primary objectives: (a) sustainably increasing agricultural productivity and rural incomes, (b) promoting adaptation and resilience to climate change across different scales, from local to global, and (c) mitigating or reducing greenhouse gas emissions from agricultural activities (Erlangga et al. 2023).

9.2.2 Key Components of CSA: Productivity, Resilience, and Mitigation

The three main areas of focus for CSA are: increasing the ability of agricultural systems to sequester carbon; minimizing or completely eliminating greenhouse gas emissions from agricultural systems; and sustainably increasing the productivity efficiencies of agricultural systems to support farmers' income, guarantee food security, and improve agricultural systems' resilience to climate change. The tripartite win-win goals of the agricultural systems—increasing productivity, enhancing stress tolerance, and lowering greenhouse gas emissions—are ultimately achieved (Zhao et al. 2023; Lipper et al. 2014). While accomplishing all three goals is the purpose of CSA, not every practice used in every situation may result in three victories. To arrive at locally acceptable solutions, CSA needs to consider these three goals from local to global sizes from both short- and long-term perspectives. Each goal will have a varying relative relevance depending on the situation and environment. Achieving the three CSA objectives should be the primary focus, and a balance between them is required (Matteoli et al. 2020). According to these studies, there is a statistically a strong and favorable relationship between crop yields or production and CSA implementation. A notable exception was made by Gorst et al. (2018), who showed that while CSA adoptions such as crop switching, adjusted crop scheduling, and conservation technologies statistically significantly increase rice productivity in Pakistan, their effect on wheat productivity is negligible. Note that due to differences in CSA classifications, crop kinds, and contextual contexts, there is an insufficient evidence to support direct comparisons of productivity-enhancing benefits or yields across known research. However, the overall data lends credence to the claim that the implementation of CSAs positively impacts a critical aspect of the initial CSA goal, which is the long-term enhancement of agricultural output. A further aspect of the initial CSA goal is the long-term growth of revenue. Examining how CSA adoption affects income-related metrics including income, net returns, and profit is essential. Although increased productivity and crop yields mainly support the food supply and help rural dwellers escape hunger, an increase in income has the ability to alleviate poverty. Additionally, it gives rural dwellers access to a wider range of food varieties than only those they grow (Zheng et al. 2024).

9.2.3 Examples of CSA Practices in Different Farming Systems

Increasing the ability of agrifood systems and communities to adjust to and recover from the consequences of climate change is the second goal of climate-smart agriculture, or CSA. Resilience can be understood via a variety of perspectives, including psychology, society, and biology, therefore there is disagreement about the best way to describe it practically (Herrman et al. 2011). Building physical resilience is the main emphasis of a large portion of CSA research, according to our analysis of the literature. To improve physical health, this entails increasing food consumption, dietary diversity, and food security (Bazzana et al. 2022; Teklewold et al. 2019; Hasan et al. 2018). Consequently, we investigate research that looks at how CSA adoption affects outcomes related to food. Resilience also describes an agri-food system's capacity to control possible hazards and minimize vulnerabilities. Potential examples of CSA practices in different farming systems, reported by Zhao et al. (2023) are shown in Table 9.1.

Table 9.1 Examples of CSA practices in different farming systems (Zhao et al. 2023)

Country	Major issues	CSA practices adopted
Maharashtra (India), Asia	Major climate risks, insufficient irrigation water, and the emission of greenhouse gases pose significant challenges to agricultural sustainability and productivity.	Water management technologies include methods such as well digging, pipe wells, rainwater harvesting, drip irrigation, and various groundwater extraction techniques. Nutrient management practices encompass the application of farmyard manure, vermicompost, and the incorporation of straw residues. When combined, these water and nutrient management strategies enhance agricultural productivity and sustainability (Zhao et al. 2023).
Mekong Delta (Vietnam), Asia	Notable climate risks Low crop yields Greenhouse gas emissions.	Utilizing high-quality rice varieties, adjusting sowing and harvest schedules, minimizing the use of chemical fertilizers, and altering irrigation strategies are key practices (Zhao et al. 2023).
Nepal, Asia	Notable climate-risks; low-yield, GHG emissions.	Implementing management measures, e.g., no tillage, crop rotation, and straw returning to the field; Refining soil biological activity, water-use-efficiency, and soil physical properties (Zhao et al. 2023).
Pakistan, Asia	Notable climate risks; low yield; GHG emissions.	Using bed seeders and lasers to level land; actions for indirect use of water and drainage management, lowering tillage and fallow, use of improved varieties resistant to waterlogging and drought (Zhao et al. 2023).
Zambia, Africa	Significant levels of climate risks; poor yield; GHG emissions.	Organic-mulching in farmlands, rotation of cereals and legumes, and better crop varieties (Zhao et al. 2023).
Malawi, Africa	Significant levels of climate risks; low yield; GHG emissions.	Adoption of continuous agroforestry inter-cropping of two major fertilizer species; Agroforestry complex system (Zhao et al. 2023).
Namibia, Africa	Significant levels of climate risks; ow yield; GHG emissions.	Collection of nutrient-rich earthworm compost leachate; hydroponic cultivation; mushroom cultivation along the coastal sides of the Namib Desert; Collecting fog-water (Zhao et al. 2023).
California, United States	Flexible agricultural system; GHG emissions; low production efficiency.	Improving underground water pumps; installation of drip irrigation and/or micro sprinkler irrigation; policies formulation and implementation (Zhao et al. 2023).
France, Europe	Influences of climate change on agricultural growth; GHG emissions.	Focusing on the service functions of agricultural ecosystems; developing precision agriculture (Zhao et al. 2023).

(continued)

Table 9.1 (continued)

Country	Major issues	CSA practices adopted
Switzerland, Europe	Effects of climate change on agricultural progress; GHG emissions.	Recycling of the waste generated by farm to the biogas plant; renewable energy production (Zhao et al. 2023).
Netherlands, Europe	Impacts of climate-change on agricultural development; GHG emissions.	LED horticultural technology adoption; enhancing the viability of horticulture (Zhao et al. 2023).
Cyprus, Asia	Impacts of climate change on agricultural development; GHG emissions.	Using agricultural robots for spraying pesticides on crops; supporting crop production and protection; reduced use of pesticides; Enhancing the sustainability of the environment of the available bioresources (Zhao et al. 2023).

Table data show original practices previously reported by Zhao et al. (2023) for Asia, Europe, and Africa

In the context of climate-smart agriculture (CSA) adoption for mitigation, agri-food systems are significant contributors to greenhouse gas emissions, including both carbon dioxide (CO_2) and non-CO_2 gases, e.g., nitrous oxide (N_2O) and methane (CH_4). Numerous studies have explored the effects of CSA practices on these emissions, addressing not just CO_2 but also other gases. For example, research by Chitakira and Ngcobo (2021) highlighted that row planting in South Africa could cut CO_2 emissions by 887 million tons, while composting could further reduce emissions by 1026 MT. Jamil et al. (2023) presented that zero-tillage methods could prevent the release of 94,000 tons of CO_2, and planting genetically modified (GM) potatoes, resistant to cold and late blight, could lower emissions by 740 million pounds.

Studies also emphasize strategies for sequestering carbon within CSA in reducing carbon emissions (Sikka et al. 2018; Kumar and Nath 2019). For instance, according to Sikka et al. (2018), carbon sequestration in India was improved by using participatory integrated watershed management. Furthermore, Iqbal et al. (2023) discovered that applying fertilizers, medium-textured soils, and biochar might cut N_2O emissions by 18% and CH_4 emissions by 25%.

Numerous studies have investigated how implementing Climate-Smart Agriculture (CSA) influences global greenhouse gas (GHG) emissions (Jamil et al. 2023; Acosta-Alba et al. 2019; Ariani et al. 2018). These diverse findings underscore the global nature of GHG emissions, which require coordinated international efforts to address. In particular, Kakraliya et al. (2021) focused on the role of CSA adoption in mitigating GHG emissions within India. Adoption of CSAs is routinely found to lower GHG emissions in India.

9.3 Challenges for Adoption

Adopting CSA is a crucial step in the shift to a sustainable food system by farmers (Data 2022). To prepare for the switch to CSA, all stakeholders in the food value chain must collaborate to make the required adjustments (Aare et al. 2021; Hamilton and Zilberman 2006). To make this transition, adjustments may be required at numerous technological, social, organizational, institutional, political, and economic levels (Eliasson et al. 2022; Reidsma et al. 2023; Resnick and Swinnen 2023; Zaharia et al. 2021). Understanding this process will make it easier to create and implement policy initiatives meant to remove barriers to CSA implementation (EEA 2023; Wreford et al. 2017).

9.3.1 Barriers to the Adoption of Climate-Smart Agriculture Practices

The literature on CSA technologies has highlighted a number of barriers, but most of them are specific to the poor world and pay little attention to Europe (29 Fusco et al. 2020; 30 Gemtou et al. 2024). A number of barriers such as personal, economic, institutional, social, and informational that require more attention are highlighted based on the themes of barriers.

9.3.2 Socio-Economic and Cultural Factors Influencing Adoption

Individual conduct is one type of barrier that falls under the cultural and socioeconomic categories and can keep farmers from adopting sustainable agriculture techniques. Bias against the adoption of CSA includes risk aversion attitudes, the belief that there are no benefits from CSA adoption, poor interest in implementation, and the preference for immediate gratification. The main psychological challenges for CSA adoption include lack of motivation, aversion to change, and prior experience. The adoption of CSA may also be hampered by farmers' risk-averse mindset brought on by concern over lower yield or higher implementation costs of new technologies or techniques. The rate at which sustainable agricultural technology and practices are adopted by farmers is determined by their cognitive ability (21 Dessart et al. 2019; Massfeller et al. 2022; Westermann et al. 2018). Stakeholders who were included in the discussion revealed that, while not all farmers share this perspective, young farmer families tend to be quite entrepreneurial and have high intentions for adopting new technologies. Another obstacle keeping producers from using CSA is a lack of confidence.

The absence of substantial financial incentives to cover adoption costs, the relatively low premium prices for sustainably produced food, the extended supply chains for agricultural products, the expenses associated with implementing smart farming technologies, and farmers' inclination toward short-term gains over long-term benefits are several barriers to the adoption of Community Supported Agriculture (CSA) that have been noted (Pedersen et al. 2024).

In interviews with stakeholders who were not interested in CSA, various social barriers emerged, including expectations and peer attitudes regarding CSA adoption, culturally ingrained local customs and traditions related to farming practices, and apprehension about social disapproval when adopting new CSA initiatives. While social norms can positively influence adoption behavior, farmers may fall prey to confirmation bias by interpreting information in ways that align with their existing beliefs (Le Coent et al. 2021). If a farmer's social network does not support CSA practices, this may diminish their motivation to explore innovative CSA techniques. Additionally, farmers may feel pressured by neighbors to adopt CSA methods, which could lead to resistance. Furthermore, if CSA technologies are perceived as divergent from societal norms, farmers may be reluctant to embrace them due to fears of social rejection.

9.3.3 Knowledge and Information-Gaps Within Farming Communities

Knowledge and information gaps within farming communities often hinder adoption of agriculture that is climate-smart (CSA). Key barriers include limited awareness, the lack of accessible scientific knowledge, the need for further training related to CSA technologies, inconsistent information on CSA benefits across various sources, and information asymmetry along the CSA supply chain. While young farmer families generally show entrepreneurial enthusiasm toward acquiring the knowledge needed for CSA implementation, this is not universally true (Pedersen et al. 2024). To address these gaps, farmers need encouragement to participate more actively in cooperatives and educational programs, gaining a broader understanding beyond just vendor and salesperson information. Additionally, promoting educational initiatives focused on sustainable farming is crucial, as the adoption of CSA practices is often hampered by insufficient specific instruction on climate-smart solutions by a lack of specific teaching on climate smart solutions.

9.4 Strategies for Promoting CSA

Equipping farmers with the knowledge to adopt climate-smart agriculture (CSA) is vital for effectively addressing and mitigating the impacts of climate change. Training and educational initiatives play a central role in enhancing farmers' understanding of climate change, its effects on agriculture, and the benefits of CSA methods. These programs provide farmers with the necessary tools and information to make informed decisions, utilize resources efficiently, and build resilience against climate-related challenges.

Effective education and training programs use a farmer-centered and participatory approach, tailoring their content to the specific needs and contexts of farmers (Archana et al. 2017). Such programs include practical, hands-on training in areas such as drought-resistant crops, water conservation techniques, integrated pest management, and climate-adaptive livestock care (FAO, Climate-Smart Agriculture 2023). By incorporating interactive elements—such as field demonstrations, peer exchanges, and on-farm trials—these initiatives enhance farmers' understanding and encourage the adoption of CSA practices.

Furthermore, educational programs can disseminate climate information and early warning systems, enabling farmers to better anticipate and respond to climate fluctuations and extreme weather events (Vincent and Balasubramani 2021). Access to timely and accurate climate data allows farmers to make informed decisions about planting times, crop selection, and irrigation scheduling, thereby reducing potential losses and improving productivity. By bridging knowledge gaps and providing climate-smart strategies, education and training programs are essential for developing resilient and sustainable agricultural systems.

Investing in farmer education and training also yields long-term benefits by nurturing a new generation of climate-smart farmers who are prepared to confront the challenges of a changing climate (Addaney et al. 2021). Integrating CSA principles into agricultural education, vocational training, and extension services enables nations to foster widespread adoption and sustainability of these practices. By imparting crucial knowledge, skills, and a forward-thinking approach, education and training programs are pivotal in transforming agricultural systems and securing a climate-resilient future.

9.4.1 Education and Training Programs for Farmers

In addition to hands-on training, it is essential to cultivate a deep understanding of climate change and its diverse effects on agriculture. This includes integrating principles of climate science into educational programs for farmers and elucidating the connections between global warming, greenhouse gas emissions, and shifting weather patterns (Chhetri 2012). A strong grasp of the scientific elements of climate change helps farmers comprehend the rationale behind climate-smart agriculture (CSA) practices, enabling them to make informed choices that support both adaptation and mitigation efforts.

Effective educational initiatives also highlight the economic advantages of adopting CSA (Boudalia et al. 2024). They illustrate how sustainable farming methods can lead to increased yields, lower input costs, and better market opportunities (Deshmukh and Dhuware 2021). By emphasizing these economic benefits, training programs can motivate farmers to implement CSA practices, thereby bolstering their long-term financial resilience.

Collaborative efforts among research institutions, extension services, and farmer organizations are crucial for developing and sharing CSA knowledge that is tailored to specific local contexts (Shahid and Ahmed 2014). This collaboration may involve participatory research, the establishment of demonstration farms, and fostering farmer-to-farmer knowledge exchanges (Adamsone-Fiskovica and Grivins 2022). By bridging scientific research with practical applications, these partnerships ensure that farmers receive relevant and effective CSA practices suited to their particular agro-ecological environments.

9.4.2 Developing Extension Services and Support Systems

Developing robust extension services and support systems is crucial in developing countries in Asia and Africa, where agriculture is a primary source of livelihood and communities are highly susceptible to climate change impacts (Thottadi and Singh 2024). These regions encounter challenges such as limited access to information, financial constraints, and inadequate infrastructure, which hinder the widespread adoption of climate-smart agriculture.

A key strategy involves enhancing the capabilities of national agricultural research and extension systems (Brewer et al. 2005). This requires investing in the training of extension agents, updating their knowledge and skills on climate-smart practices, and providing the necessary resources to effectively communicate this information to farmers. International organizations like the Food and Agriculture Organization (FAO) significantly contribute to these capacity-building efforts through technical support, training sessions, and knowledge-sharing platforms (FAO 2020).

Given that traditional top-down extension methods often fail to reach marginalized farmers, there is a shift toward participatory and farmer-centered approaches (Corpuz et al. 2022). Methods like farmer-to-farmer exchanges, demonstration plots, and farmer field schools have shown to be successful in encouraging hands-on education and promoting the use of context-specific community support strategies (Maiangwa et al. 2010). These methods empower farmers to actively participate in the learning process, share experiences, and adapt technologies to local conditions.

Digital technologies offer significant potential for transforming extension services, especially in remote and underserved areas (Witherspoon and Harris 2011). Mobile apps, online platforms, and interactive voice response systems can provide farmers with real-time weather updates, market information, and expert advice on climate-smart agriculture (Duquinal et al. 2020). Initiatives like the FAO's Digital Village Initiative aim to bridge information gaps and deliver timely and relevant data to farmers.

Financial inclusion is also essential for enabling farmers, particularly smallholders, to adopt climate-smart practices (Swanson and Davis 2014). Access to credit, insurance, and other financial services can help mitigate climate change risks and provide resources for investing in resilient technologies and practices (Jha and Singh 2021). Governments and financial institutions play a crucial role in developing financial products and services tailored to the needs of smallholder farmers.

In summary, enhancing extension services and support systems is vital for promoting the widespread adoption of climate-smart agriculture in developing countries across Asia and Africa. By focusing on capacity building, adopting participatory approaches, leveraging digital technologies, and ensuring financial access, we can support farmers in adapting to climate change, improving their livelihoods, and contributing to a more sustainable and food-secure future.

9.4.3 Encouraging Community Involvement and Collaboration

Fostering community engagement and cooperation is essential for the successful implementation of climate-smart agriculture (CSA), particularly in the diverse and often resource-constrained communities of Asia and Africa. Leveraging local knowledge, promoting collective efforts, and developing strong social networks are crucial for driving sustainable progress.

A key component of this strategy is recognizing the value of traditional knowledge and practices that communities have developed for their specific ecosystems (Walker and Sanz 2024). Extension services and development agencies should focus on integrating this indigenous knowledge with scientific approaches to create CSA solutions that are both contextually relevant and culturally sensitive (World Bank 2012). Such collaborative efforts build trust, foster a sense of ownership, and contribute to the long-term success of interventions.

Community-based organizations and farmer cooperatives play a critical role in facilitating collective action and peer-to-peer learning (World Bank 2016). These platforms enable farmers to share experiences, pool resources, and scale up CSA practices. Supporting and strengthening these institutions is vital to ensure they remain inclusive, democratic, and responsive to the needs of their members (Mizik 2021).

Collaboration should extend beyond the farming community to include local leaders, policymakers, and traditional authorities (Zhang et al. 2011). Their support is crucial for implementing policy changes, allocating resources, and resolving conflicts related to land and water (Binswanger-Mkhize et al. 2010). Building strong partnerships between communities and local governments ensures that CSA initiatives align with broader development goals and support sustainable rural livelihoods.

South-South cooperation offers valuable opportunities for knowledge sharing and technology transfer between developing countries facing similar climate challenges (Stafford-Smith et al. 2017). Programs that facilitate farmer-to-farmer exchanges, study tours, and joint research enable communities in Asia and Africa to learn from successful CSA practices implemented in other developing regions (Walker and Sanz 2024). This cross-cultural exchange fosters innovation and accelerates the adoption of locally appropriate solutions.

In conclusion, enhancing community involvement and collaboration is crucial for the effective implementation of climate-smart agriculture in the developing regions of Asia and Africa. By valuing local knowledge, strengthening community organizations, engaging local leaders, and promoting South-South cooperation, we can empower communities to adopt sustainable agricultural practices, improve resilience, enhance livelihoods, and contribute to a food-secure future.

9.5 Policy and Institutional Support

The effective promotion of climate-smart agriculture (CSA) depends on a supportive policy environment and strong institutional frameworks. These components are crucial for creating incentives, removing barriers, and fostering conditions that encourage the widespread adoption of CSA practices.

National governments play a key role in integrating CSA into agricultural policies, plans, and investment strategies (Dinesh et al. 2017). This includes embedding CSA principles into national climate adaptation and mitigation strategies, aligning agricultural subsidies and incentives with CSA goals, and advancing the development of climate-resilient technologies (Zougmoré et al. 2016). Clear policy guidelines and supportive regulations offer long-term direction and motivate stakeholders across the agricultural sector to embrace CSA.

Strong institutional frameworks are essential for the effective coordination, implementation, and monitoring of CSA initiatives (Vermeulen et al. 2012). This involves defining explicit roles and responsibilities for government agencies engaged in agriculture, climate change, and rural development, fostering inter-ministerial collaboration, and establishing platforms for stakeholder engagement (Aryal et al. 2019). Well-defined institutional arrangements ensure efficient resource use, reduce duplication of efforts, and promote a cohesive approach to CSA promotion.

Access to finance is another critical factor in enabling farmers to adopt climate-smart practices. Governments can encourage private sector investment in CSA by creating favorable investment conditions, offering tax incentives or subsidies for CSA technologies, and establishing dedicated funds for CSA projects (World Bank Group 2024). Cutting-edge financing techniques that manage climate-related risks and mobilize additional resources include green bonds and CSA insurance programs. Monitoring and evaluation frameworks are vital for assessing the effectiveness of CSA policies and programs, identifying areas for improvement, and ensuring accountability (Aryal et al. 2019). Governments should invest in robust data collection systems, establish clear indicators for measuring CSA outcomes, and promote transparency in reporting progress. Regular evaluations are essential for recognizing successful strategies, addressing challenges, and ensuring that policies and institutions adapt to the evolving needs of farmers and the agricultural sector.

The transition to climate-smart agriculture requires a collaborative effort from multiple stakeholders, with government policies and institutional support forming the backbone of successful implementation. This supportive framework is characterized by well-defined policy directives, active organizational involvement.

9.5.1 Government Policies Promoting CSA

Governments are crucial in fostering a supportive environment for climate-smart agriculture (CSA) by integrating it into national policies and development strategies (Bhattacharyya et al. 2020). This includes incorporating CSA into national climate change action plans, agricultural policies, and investment strategies. For example, issuing clear policy guidelines that encourage sustainable land management practices—such as conservation agriculture and agroforestry—can motivate farmers to adopt these climate-smart approaches (Zougmoré et al. 2016). Additionally, governments can implement regulations that improve sustainable water management, such as promoting efficient irrigation technologies or establishing limits on water extraction for agricultural purposes, thereby ensuring the long-term sustainability of agricultural production.

9.5.2 Role of Local and International Organizations

Local and international organizations are essential in promoting the adoption of climate-smart agriculture (CSA) by connecting policy with practical implementation. International entities like the Food and Agriculture Organization (FAO) and the World Bank provide technical support, facilitate knowledge exchange, and enhance capacity-building among governments and farming organizations (FAO.org 2020). They also play a vital role in fostering South-South cooperation, allowing countries to share experiences and strategies related to CSA implementation (Mrabet and Moussadek 2022). At the local level, non-governmental organizations and community-based groups engage directly with farmers, offering training on CSA practices, helping with access to resources and markets, and reinforcing local institutions (Vincent and Balasubramani 2021).

9.5.3 Financial Incentives and Subsidies for Adoption

Financial incentives and subsidies are effective tools for encouraging farmers to overcome initial challenges in adopting climate-smart practices. Governments can offer direct subsidies for climate-smart technologies, such as drought-resistant seeds, water-efficient irrigation systems, or renewable energy solutions for farms (Mwongera et al. 2019). In addition, tax incentives or lower interest rates on loans for CSA-related investments can further stimulate private sector involvement. Furthermore, governments can partner with financial institutions to develop innovative financing solutions, such as index-based agricultural insurance, which protects farmers against climate risks and encourages greater investment in CSA (Sahu et al. 2020).

9.6 Role of Technology and Innovation

9.6.1 Advancements in Agricultural Technologies Supporting CSA

Agriculture is the primary economic driver of many emerging nations. In poor nations, food security and agricultural productivity are threatened by climate change in a number of ways. Thus, when these developing countries set their CSA development goals, they should prioritize ensuring food security, promoting economic growth, and improving agricultural production efficiency. Increased funding should be directed toward agricultural systems to progressively reduce greenhouse gas (GHG) emissions (Lipper et al. 2014). The threats posed by climate change and variability to the agriculture sector are significant, especially given its crucial role in supporting livelihoods across the continent and underpinning African economies (WMO 2020). It is essential to develop an agricultural system capable of enhancing food production despite the challenges posed by climate change. To this end, climate-smart agriculture (CSA) was introduced as a sustainable approach to help agri-food systems transition to environmentally friendly and climate-resilient practices (FAO 2022).

Examples of climate-smart technologies include nutrient management, landscaping, mulching, reduced tillage, intercropping, agroecology, small-scale irrigation, livestock diversity, and soil and water conservation (Senyolo et al. 2018; Chandra et al. 2018). Recent research indicates that several African nations have formulated strategic plans to integrate climate-smart agricultural technologies to address agricultural productivity challenges. These efforts are expected to mitigate the impacts of greenhouse gas emissions, enhance adaptability to climate change, and increase food yields (Barasa et al. 2021).

In contrast, developed countries have advanced agricultural practices, high-efficiency agricultural products, relatively abundant land resources per capita, and sophisticated mechanized production systems. As a result, the primary objectives for the growth of CSA in these industrialized nations include enhancing agricultural capacity and reducing greenhouse gas emissions to adapt to climate change. This requires the incorporation of innovative technologies, careful policy formulation and implementation, increased resilience of agricultural systems, and improved production efficiency while minimizing emissions (Long et al. 2016).

California, recognized as one of the most productive and resource-rich agricultural regions globally, emphasizes reducing greenhouse gas emissions and sustainably managing water resources (Lewis and Rudnick 2019). Through a range of laws, regulations, and relevant agricultural technologies supported by the public research system, the California government has successfully reduced GHG emissions.

9.6.1.1 Utilizing Remote Sensing Methods to Protect Agricultural Information

Remote-sensing technology is extensively employed in numerous fields due to its rapid, large-scale, dynamic, real-time, and cost-effective observation capabilities. With the use of remote sensing, which collects information from satellites or unmanned aerial vehicles, the physical features of the earth's surface can be identified and seen (Meier et al. 2020).

9.6.1.2 Use of the Internet of Things (IoT)

The Internet of Things (IoT) is mostly utilized in agriculture to gather data from many sensor types about crop and environmental characteristics including temperature, humidity, pH level, leaf color, etc. A wide range of topics related to the deployment of IoT in agriculture have been researched, such as assessing IoT applications (Jusoh et al. 2021), creating IoT architectures for food control (Dias et al. 2020), and confirming how IoT and agriculture unmanned aerial vehicles (UAVs) are integrated in smart agriculture (Boursianis et al. 2022; Islam et al. 2021).

9.6.1.3 Artificial Intelligence Applications

Artificial intelligence (AI) applications are poised to play a significant role in the future of climate-smart agriculture (CSA). AI refers to the use of digital computers and controlled technologies to simulate and enhance human intelligence, enabling the detection of environments, and the collection of relevant data. Its benefits have been recognized across various fields (Subeesh and Mehta 2021). For instance, diverse cropping patterns such as rice-wheat rotation and rice-potato-sesame systems, along with strategies like crop diversification and no-till agriculture can enhance agricultural productivity while reducing greenhouse gas emissions (Dandois and Ellis 2013; Datta et al. 2011). Additionally, introducing suitable dryland crops can help minimize the submergence period within the annual planting cycle (Adhya et al. 2000). Agriculture is increasingly shifting toward modernization and mechanization, which will likely create more stringent demands for meteorological services. It is essential to integrate farmers' needs with these services efficiently through internet technologies. By leveraging a technology platform, stakeholders—including growers, transport companies, service providers, and consumers—can collaborate to establish a comprehensive meteorological service system tailored to agriculture. The adoption of advanced internet technologies to protect agricultural data, enhance cropping patterns and management practices, provide an "internet + weather" service, improve the quality of agricultural services, and offer insurance based on agricultural weather indices is expected to be a key development direction for CSA (Zhao et al. 2023).

9.6.2 The Use of Data Analytics and Decision-Making Tools

According to the World Bank, CSA is a practical strategy for accomplishing the Sustainable Development Goals (SDGs) (World Bank 2015a). The majority of developing countries are looking for different ways to create dependable, low-cost weather monitoring and forecasting systems that can be combined with cutting-edge smart technologies (like biosensors, agricultural drones, remote sensing, and Internet of Things sensors) in order to increase food security (Tenzin et al. 2017; Adoghe et al. 2017).

Frequently called "smart farming" or "smart agriculture," the digital revolution in agriculture incorporates cutting-edge digital tools and technologies into conventional farming methods. In the United States and Australia, precision agricultural technologies were first adopted in the late 1980s (Stafford 2000). This innovative approach enhances crop yield and sustainability through the use of automation, artificial intelligence (AI), and the Internet of Things (IoT). Automated equipment streamlines planting, irrigation, and harvesting processes, while AI systems examine information from multiple sources to facilitate well-informed decision-making. Through the collection of real-time data from sensors, drones, and other linked devices, IoT devices increase operational efficiency (Lakshmi et al. 2023; Friha et al. 2021; Saiz-Rubio and Rovira-Más 2020; Subeesh and Mehta 2021; Foughali et al. 2018).

Central to smart farming is sensor technology, which enables real-time monitoring of agricultural data. The role of soil sensors in measuring soil temperature, pH, moisture content, and nutrient levels has been discussed by Adamchuk et al. (2004). Utilizing data from these sensors helps farmers make informed decisions about fertilization, irrigation, and other management practices, thus improving crop yields and sustainability. Singh et al. (2023) highlighted the importance of remote sensing for efficient irrigation, while unmanned aerial vehicles (UAVs) provide aerial imagery that allows for precise intervention and early problem detection, optimizing resource use and enhancing crop management (Radoglou-Grammatikis et al. 2020).

Advanced data analytics combined with sensor-based monitoring systems has the potential to revolutionize agricultural operations and develop sustainable food production systems. Thorough data gathering is essential for making well-informed decisions. Farmers can make data-driven decisions that optimize resource efficiency, boost agricultural yields, and promote environmental sustainability by obtaining precise and up-to-date data. Collaborations with academic institutions and continuous multidisciplinary research enable these technologies' scalability and impact (Weraikat et al. 2024).

An automated irrigation system was proposed by Pawar et al. (2021), incorporating sensors to measure humidity, pH, and soil moisture. This technology allows farmers to access field data remotely, saving time and physical effort while enhancing output and resource efficiency. The system has a Global System for Mobile Communication (GSM) module that uses Short Message Service (SMS) to provide real-time field condition updates. Furthermore, an IoT-enabled soil nutrient system was introduced by Murali et al. (2023) to help farmers make educated decisions at every stage of farming. According to Sharma et al. (2020), digital platforms and technology can enhance supply chain participants' coordination and communication, resulting in higher output, effectiveness, and efficiency. It is imperative to augment the data science proficiencies of agricultural producers, necessitating instructional and training initiatives to empower stakeholders to utilize data efficiently and cooperate to enhance farming methodologies (Devare et al. 2023).

It is essential to ensure data accuracy and understand the factors affecting it before making decisions. Thus, data collection alone is not enough; the data must be cleansed, organized, and prepared for analysis (Saiz-Rubio and Rovira-Más 2020).

9.6.3 Integration of Digital Platforms and Mobile Apps

One of the biggest technological advances of the last 10 years has been recognized as digitization, owing to its profound influence on many facets of our daily existence, including farming.

(Mondejar et al. 2021; MacPherson et al. 2022). Innovation presents potential answers to agricultural problems in the digital age, particularly the Rural farmers' deficiency of knowledge before and after farming works (Trendov et al. 2019). For instance, farmers can use technology to improve their decision-making and planning skills with reference to farm operations. According to theory, information asymmetry results from parties having different approaches and access to information, which usually causes welfare inequality. Information provision has traditionally been the primary tactic used in both developed and developing nations to minimize information asymmetries, enabling consumers to mitigate the detrimental impacts of asymmetric information that typically surface during transactions (Anderson and Feder 2007). Thus, innovations in information, communication, and technology (ICT) can reduce the incidence of asymmetric information and the disparity in welfare across stakeholders.

Africa's move from analog to digital mobile phone usage could be a big help in addressing some of the issues the agriculture industry is currently facing. Africa may lessen its sensitivity to climate variability and lessen the effects of climate variability by using mobile phones to acquire climatic information. According to research by Henriksson et al. (2021), radio is still the most popular medium for climate forecasting, even if short message services (SMS), the internet, and WhatsApp are popular, especially among women. Additionally, the study shows that better educated farmers are more likely to get their weather forecasts via SMS, WhatsApp, the internet, or newspapers. According to Oyinbo et al. (2020), digital technologies are being used to help agricultural expansion in the Nigerian states of Kaduna, Katsina, and Kano.

Africa has benefited greatly from digitization, particularly in light of the continent's agriculture's struggles with issues including degraded land, fluctuating demand brought on by climate change, and insufficient market systems. According to Abdul-Rahaman and Abdulai (2022), mobile money users in Ghana fared better than non-users. Mobile money is a digital payment system that allows users to move money between mobile devices. Mobile phones can enhance farmers' livelihoods both directly and indirectly by reducing the costs associated with agricultural damage from wildlife, improving communication among farmer groups, and facilitating access to vital resources and information, hence reducing conflict between humans and elephants in Laikipia, Uganda (Graham et al. 2012),

ICT for agriculture has been the subject of several studies that have looked at how technology may help farmers get the information they need, like producer-buyer interactions and pricing information (Baumüller 2012). According to Oke et al. (2022), for example, in Ogun State, Nigeria, the use of radio and television channels enhanced technical efficiency and considerably decreased inefficiencies in fish production, even when controlling for model variables like age and farming experience. This implies that expanding access to media outlets could help to improve fish farming. Due to its lower levels of digital literacy than other sectors, the agriculture industry is nonetheless vulnerable. A survey by the International Finance Corporation states that in a few African countries, the percentage of the agriculture sector with digital skills was no higher than 20% in 2019 and is expected to drop to less than 40% by 2030.

Perhaps a greater focus on actionable recommendations and insights for farmers, development specialists, stakeholders engaged in boosting agricultural sustainability, including decision-makers and ICT developers and productivity in Africa should be placed on future research projects (Erlangga et al. 2023). The use of technology in climate-smart agriculture initiatives is justified by the significant reliance of small-scale farmers and agricultural sector players on ICT (World Bank 2015a, b). Climate-smart ICTs disseminate information on opportunities for agricultural development. Women are more involved in community leadership, income use control, and agricultural production decision-making as a result (Huyer 2012).

When climate-smart technologies are able to reach marginalized groups and are customized to their socio-cultural traits, their influence is much larger. Women's participation in CSA activities and commodity value chains has been shown to benefit from CSA initiatives based on information technology, such as mobile phones, radio, and social media (World Bank 2015a, b). For example, in Senegal, women smallholder farmers find that the most helpful communication channels include radio, public information broadcasts, and SMS messaging written in their native tongues. Furthermore, a study on the potential of information techny psychology to assist women-owned and operated agricultural businesses in Zambia and Kenya finds that women and men use ICT tools differently (World Bank 2015b).

9.7 Community-Based Approaches

9.7.1 Case Studies of Successful Community-Based CSA Projects

9.7.1.1 Case Study 01

Mali—Enhancing Agricultural Policies Through the Climate-Smart Agriculture Investment Plan

The Ministry of Agriculture of Mali facilitated an extensive consultation process that led to the development of the 2019 Climate-Smart Agriculture Investment Plan (CSAIP). This process involved a wide range of stakeholders, including government agencies, civil society organizations, and scientific institutions. The World Bank supported this multi-country initiative through the Adaptation of African Agriculture program, in alignment with its IDA18 commitments (International Development Association 2017). To ensure that the CSAIP is based on robust scientific evidence, various organizations provided technical assistance, including the International Center for Tropical Agriculture (CIAT), World Agroforestry, the Alliance of Bioversity International, and the Research Program on Climate Change, Agriculture, and Food Security (CCAFS) of the Consortium of International Agricultural Research Centers (CGIAR). The CSAIP focuses on practical strategies aimed at reducing greenhouse gas emissions, enhancing resilience, mitigating risks, and adapting to the impacts of climate change. It also emphasizes the potential positive effects of these investments on both the environment and the economy.

Rationale

The Climate-Smart Agriculture Investment Plan (CSAIP) provides a comprehensive overview of Mali's agricultural sector and its challenges related to climate change. Agriculture occupies over 30% of Mali's land area, with most farmers working small plots primarily for subsistence purposes. Additionally, more than 85% of agricultural households raise livestock. There are significant disparities in production across most crops, resulting in overall low agricultural output. Rapid population growth and increasing pressure on land resources have led to land degradation. More than 72% of Mali's population is affected by droughts, floods, and other adverse weather events, placing the country in the medium-to-high risk category for climate-related hazards.

The poverty rate in rural areas is alarmingly high, around 90%, with limited opportunities for productivity and resilience improvements due to insufficient access to agricultural inputs, financial services, and extension support. CSA represents a multisectoral approach aimed at enhancing the economy, ensuring food security, improving livelihoods, and promoting environmentally sustainable agricultural practices while building resilience. Any CSA investment plan developed and implemented is designed to influence government policies, attract funding for priority CSA initiatives, and strengthen institutional and human resources to facilitate widespread CSA adoption.

Description

The connections among the five action points for implementing climate-smart agriculture (CSA) are evident in the formulation of Mali's Climate-Smart Agriculture Investment Plan (CSAIP), though not necessarily in a specific sequence. The plan outlines project concepts for 12 key investment areas and emphasizes the need for developing a monitoring and evaluation (M&E) framework (action point 1). It builds on and enhances Mali's agricultural policies and objectives (action point 2). As a signatory to the Paris Agreement under the United Nations Framework Convention on Climate Change (UNFCCC), Mali has also submitted its first Nationally Determined Contribution (NDC). The CSAIP directs investments based on four components of CSA planning and implementation: situation analysis (i), interventions and prioritization (ii), program/project design (iii), and monitoring, evaluation, and learning (iv). Each of these components contributes directly to the evidence base. They establish a framework for selecting a theory of change and determining how to modify and execute plans to achieve desired outcomes. The M&E process is essential for creating and implementing CSA strategies. It enables the Malian government, development partners, and implementing agencies to track progress and identify issues that may require adaptive responses by monitoring activities, outputs, results, and impacts against set targets. M&E initiatives foster accountability, provide opportunities for learning, and generate data that can underpin success stories.

Outcomes

In Mali, the impacts of climate change are widespread and significant, particularly as shifting weather patterns adversely affect food security and livelihoods in rural areas (World Bank 2019). Through stakeholder engagement and assessments related to agricultural policies, economic conditions, and climate change scenarios, Mali's Climate-Smart Agriculture Investment Plan (CSAIP) has identified 12 key investment areas. The CSAIP outlines four national investments, two restoration projects, and six commodity-specific initiatives, with a total estimated investment ranging from USD 300 to 500 million.

These investments aim to enhance agricultural resilience to climate change, increase crop yields, and assist approximately 1.8 million beneficiaries and their families in adapting to these challenges. The four national programs focus on developing agroclimatic information systems, monitoring soil fertility, integrating CSA into national agricultural extension services, and enhancing remote-sensing monitoring capabilities. By boosting agricultural productivity and reducing climate change-related risks, these projects are expected to contribute to the achievement of Sustainable Development Goals (SDG) 1 and SDG 2, which aim to alleviate poverty and hunger. Furthermore, beneficiaries will gain better access to information and resources, thereby strengthening value chains and promoting equitable employment and economic growth (SDG 8).

9.7.2 Leveraging Local Knowledge and Traditions

Mali can use CSA investment to support national policies through the CSAIP, including at least 13 Malian programs and policies related to climate change adaptation. The plan strongly aligns with Mali's NDC in terms of targeted adaptation activities (e.g., water management) and overarching goals (e.g., national planning) (World Bank 2019).

9.7.2.1 Case Study 02

Mapping Climate-Smart Farming Methods for Rural Women in Somalia

Introduction

According to a 2019 study meant to support the argument for CSA, three of the more than 80 CSA practices utilized by rural women in Somalia are raising mixed herds of cattle, planting trees to protect soil and water resources, and growing food varieties resistant to drought. The IFAD funded the research and used it as a model for its agricultural initiatives, which help farmers who raise livestock and crops adjust to changing climate conditions.

Rationale

Climate change is impacting Somalia significantly, primarily through the increasing frequency and severity of droughts and floods, which is particularly concerning for a nation already susceptible to conflict. The Notre Dame Global Adaptation Initiative identifies Somalia as one of the countrie's most vulnerable to climate change and among the least equipped to cope with its effects (Chen et al. 2015). Research on the role of women in Somalia's agricultural sector is limited, and even less is known about the climate-smart agriculture (CSA) practices they employ. To address this gap, the International Fund for Agricultural Development (IFAD) funded a study aimed at exploring the CSA practices of rural Somali women. The findings indicate that these CSA methods can significantly improve the living standards of low-income smallholder households in Somalia, providing a strong empirical basis for future IFAD-supported initiatives.

Description

In 2019, the Somali Disaster Resilience Institute, a national research organization located in Mogadishu, conducted the study referenced in this chapter. The institute performed a comprehensive literature review, facilitated focus groups with over 120 rural women, and consulted 20 experts. Data on climate-smart agriculture (CSA) practices were gathered from five Somali states: Hirshabelle, Jubaland, Mudug, Puntland, and South West. This research is part of a larger initiative by the International Fund for Agricultural Development (IFAD) in Somalia.

In the 1980s, IFAD invested USD 140 million in agricultural development projects in the country, directly benefiting 1,780,000 individuals. The organization is committed to increasing its funding to support smallholder farmers in adapting to the impacts of climate change. Additionally, IFAD aligns its programs with the climate adaptation and mitigation goals established by countries in their Nationally Determined Contributions (NDCs) under the Paris Agreement.

Outcomes

The study addresses a knowledge gap by identifying over 80 climate-smart agriculture (CSA) practices. A forthcoming compendium accompanying the study report will provide brief summaries of more than 60 techniques, categorized into seven key areas with various subcategories: crops, livestock, energy, soils, water, forestry, and aquaculture.

9.7.3 Building Resilient and Adaptive Farming Communities

The study identifies a variety of practices that correspond with the three main pillars of climate-smart agriculture (CSA): reducing greenhouse gas emissions, enhancing resilience and adaptation, and increasing productivity and income. The findings illustrate how agricultural methods can be modified to lessen farmers' vulnerability to climate change. Most CSA practices adopted by rural women center around crop production. These practices include cultivating native, drought- and disease-resistant crop varieties, implementing crop rotation to maintain soil fertility, and utilizing traditional seed storage techniques, all of which have significant potential to boost productivity and resilience.

Some farmers indicated that they align their field preparations with weather forecasts obtained via radio, which the study identified as an effective means of disseminating information about CSA practices. While nomadic herding is typically viewed as a male-dominated activity, no women in the study participated in this practice. Instead, women manage livestock near their homes, grow fodder crops such as Sudan grass and alfalfa, fence grazing areas, and gather wild fodder plants. These adaptive strategies are crucial for helping farmers cope with diminishing grasslands caused by drought.

The research uncovered varying perspectives on gender roles within farming. Some women felt that men should be responsible for irrigation management, while others indicated that both genders share these responsibilities. Irrigation plays a critical role in enabling farmers to cultivate crops during dry seasons and off-peak periods. Focus groups in regions like Galkayo emphasized practices such as fodder production, water storage, and herd and grazing management as essential for community well-being.

Despite the promise of many resilience-building techniques, their adoption is infrequent. Soil conservation practices, including conservation agriculture and erosion control, were acknowledged by some groups, but their implementation remains limited. The study also highlighted several barriers to the widespread adoption of CSA methods, such as limited access to training and essential services like veterinary care and agricultural extension. Expert interviews noted that rural women often facing financial constraints are excluded from decision-making processes, and lack ownership rights to fixed assets.

9.8 Monitoring and Evaluation

Monitoring and evaluation (M&E) play a key role in the successful execution of climate-smart agriculture (CSA), serving as essential mechanisms to ensure its effectiveness. A well-designed M&E framework offers a structured method for tracking progress, assessing impact, identifying successful practices, and responding to emerging challenges. This systematic approach is critical for guaranteeing that CSA initiatives achieve meaningful and enduring outcomes.

9.8.1 Methods for Tracking the Progress and Impact of CSA Adoption

To effectively monitor the progress and impact of climate-smart agriculture (CSA), a comprehensive approach that considers both environmental and socioeconomic factors is essential (Price-Kelly et al. 2016; Vallejo 2017). Quantitative techniques, including remote sensing and crop modeling, are useful for tracking changes in land use, vegetation, and water productivity, offering valuable data on the ecological effects of CSA practices (Rosenstock et al. 2016). In addition, household surveys and economic assessments can evaluate shifts in farm income, food security, and livelihoods, helping to highlight the socioeconomic advantages of CSA implementation (Mburu et al. 2015). Qualitative methods, such as focus group discussions and participatory rural appraisals, further deepen understanding by capturing farmers' perspectives, challenges, and responses to CSA practices.

9.8.2 Identifying Best Practices and Lessons Learned

Monitoring and evaluation (M&E) are crucial for identifying effective practices and drawing valuable lessons from climate-smart agriculture (CSA) initiatives, which in turn supports the sharing and expansion of successful strategies (FAO.org 2020). Case studies of successful CSA projects can highlight effective implementation methods, innovative partnerships, and supportive policy frameworks (Westermann et al. 2018). Evaluating performance data from various CSA practices across different agro-ecological zones and farming systems helps identify context-specific best practices (Debray et al. 2019). Furthermore, documenting challenges, failures, and unintended consequences provides important.

9.8.3 Continuous Improvement and Adaptation Strategies

Monitoring and evaluation (M&E) should be viewed as an ongoing process rather than a one-time task. It plays a crucial role in the continuous enhancement and adaptation of climate-smart agriculture (CSA) policies, programs, and practices (Njuguna et al. 2024). By regularly tracking key indicators, deviations from desired outcomes can be promptly identified, leading to corrective actions and adaptive management (Kalele et al. 2021). For instance, if monitoring data show low adoption rates of a specific CSA practice, further analysis may uncover underlying issues such as difficulties with access to finance, technical knowledge, or market connections. This iterative learning approach helps ensure that CSA initiatives effectively respond to evolving challenges, including climate variability, market shifts, and technological advancements.

9.9 Conclusion

9.9.1 Summary of Key Points and Strategies for Promoting CSA

Climate-smart agriculture (CSA) provides a crucial strategy for adapting farming to the interconnected issues of climate change, food security, and sustainable development. It targets three primary objectives: enhancing agricultural productivity to address increasing food demands, boosting the resilience of farming systems to withstand climate-related shocks, and minimizing greenhouse gas emissions through sustainable practices. CSA stresses the importance of customizing strategies to fit local conditions, taking into account factors such as climate, soil type, and socio-economic contexts. An integrated, holistic approach is essential, incorporating climate considerations into every facet of agricultural planning, from policy formulation to technological advancements.

Several strategies support the promotion of CSA. Expanding research and fostering innovation in climate-smart technologies are critical, as is sharing knowledge among farmers. Effective policy frameworks play a key role by incentivizing climate-smart practices and removing barriers to their adoption. Strengthening institutions to implement CSA programs and providing technical assistance to farmers is equally important. Mobilizing financial resources from both public and private sectors is needed to support research, technology transfer, and risk management.

On the ground, CSA practices vary depending on the specific needs of each region. Conservation agriculture, for instance, focuses on reducing tillage, using cover crops, and diversifying cropping systems to enhance soil health and conserve water. Efficient water management, such as improved irrigation techniques and drought-tolerant crops, helps optimize water use. Livestock management practices that reduce methane emissions and improve animal health are also integral, alongside integrating forestry into agricultural systems for carbon sequestration and ecosystem benefits.

Achieving CSA's goals requires the collaboration of governments, farmers, researchers, and the private sector. By embracing CSA principles and applying context-specific strategies, agriculture can become more resilient, productive, and sustainable, helping to feed a growing population while mitigating the effects of climate change.

9.9.2 Importance of Ongoing Support and Collaboration for Successful Adaptation

Adapting to complex challenges like climate change requires more than just initial efforts—it calls for continuous support, collaboration, and improvement. Success in adaptation relies on ongoing efforts that help strategies evolve with changing conditions. This includes providing technical assistance, securing financial resources, and investing in skill development to enable long-term adaptation. Collaboration plays a critical role in fostering shared learning, which helps avoid repeating mistakes, sparks innovation, and builds resilience through strengthened social networks. Many challenges are rooted in systemic issues, so addressing them requires partnerships across sectors, policy advocacy, and the co-production of knowledge between communities and experts.

In short, sustained support and collaboration are essential to building resilient communities that can adapt to the ever-changing demands of the future.

9.9.3 Future Directions for Enhancing Climate-Smart Agriculture Practices

While Climate-Smart Agriculture (CSA) has made progress, there is still untapped potential for further improvement. Key areas for advancing CSA include closing data gaps by investing in better climate impact monitoring and using digital tools like remote sensing and AI to improve agricultural decision-making. Ensuring equitable access to CSA practices is crucial, with particular focus on supporting smallholder farmers, women, and marginalized groups. This can be achieved through participatory approaches that involve farmers in the development and implementation of CSA methods.

To strengthen CSA, market incentives such as carbon credits and ecosystem service payments should be expanded, alongside improving access to climate-smart financial products. Collaboration across sectors is also essential, integrating agriculture, climate, and development initiatives while engaging private sector innovation. Policy efforts must prioritize climate adaptation and mitigation in agriculture, embedding climate risk considerations into decision-making processes and creating adaptable governance systems.

By focusing on these areas, CSA can evolve from a promising concept into a globally adopted practice, helping agriculture thrive in the face of climate change while supporting food security and sustainability goals.

References

Aare AK, Lund S, Hauggaard-Nielsen H (2021) Exploring transitions towards sustainable farming practices through participatory research–the case of Danish farmers' use of species mixtures. Agric Syst 189:103053. https://doi.org/10.1016/j.agsy.2021.103053

Abdul-Rahaman A, Abdulai A (2022) Mobile money adoption, input use, and farm output among smallholder rice farmers in Ghana. Agribusiness 38(1):236–255. https://doi.org/10.1002/agr.21721

Acosta-Alba I, Chia E, Andrieu N (2019) The LCA4CSA framework: using life cycle assessment to strengthen environmental sustainability analysis of climate smart agriculture options at farm and crop system levels. Agric Syst 171:155–170. https://doi.org/10.1016/j.agsy.2019.02.001

Adamchuk VI, Hummel JW, Morgan MT, Upadhyaya SK (2004) On-the-go soil sensors for precision agriculture. Comput Electron Agric 44(1):71–91

Adamsone-Fiskovica A, Grivins M (2022) Knowledge production and communication in on-farm demonstrations: putting farmer participatory research and extension into practice. J Agric Educ Ext 28(4):479–502

Addaney M, Sarpong GE, Akudugu JA (2021) Climate change adaptation in Akropong, Ghana: experiences of female smallholder farmers. J Land Rural Stud 9(2):344–367

Adhya TK, Mishra SR, Rath AK, Bharati K, Mohanty SR, Ramakrishnan B et al (2000) Methane efflux from rice-based cropping systems under humid tropical conditions of eastern India. Agric Ecosyst Environ 79(1):85–90

Adoghe AU, Popoola SI, Chukwuedo OM, Airoboman AE, Atayero AA (2017) Smart weather station for rural agriculture using meteorological sensors and solar energy. In: Proceedings of the world congress on engineering, London, UK, 5–7 July 2017, vol I, pp 1–4. http://eprints.covenantuniversity.edu.ng/8584/#.Xt0880VKiM8

Adonadaga MG, Ampadu B, Ampofo S, Adiali F (2022) Climate change adaptation strategies towards reducing vulnerability to drought in Northern Ghana. Eur J Environ Earth Sci 3:1–6. https://doi.org/10.24018/ejgeo.2022.3.4.294

Anderson JR, Feder G (2007) Agricultural extension. Handb Agric Econ 3:2343–2378. https://doi.org/10.1016/S1574-0072(06)03044-1

Archana T, Rani VS, Nagasree K, Devi KS (2017) Strategies for effective transfer of Climate Resilient Agriculture (CRA) technologies. Int J Curr Microbiol App Sci 6(9):3660–3664

Ariani M, Hervani A, Setyanto P (2018) IOP Conf Ser Earth Environ Sci 200:Climate smart agriculture to increase productivity and reduce greenhouse gas emission-a preliminary study. https://doi.org/10.1088/1755-1315/200/1/012024

Arora NK (2019) Impact of climate change on agriculture production and its sustainable solutions. Environ Sustain 2(2):95–96. https://doi.org/10.1007/s42398-019-00078-w

Aryal JP, Sapkota TB, Rahut DB, Jat ML (2019) Agricultural sustainability under emerging climatic variability: role of climate smart agriculture and relevant policies in India. In: International Journal of Innovation and Sustainable Development. (in press). https://www.inderscience.com/info/ingeneral/forthcoming.php?jcode=IJISD

Aryal JP, Rahut DB, Sapkota TB, Khurana R, Khatri-Chhetri A (2020) Climate change mitigation options among farmers in South Asia. Environ Dev Sustain 22(4):3267–3289

Balo S, Mahata D (2022) A review of climate-smart agriculture is a new approach to farming system. Int J Environ Clim Change 12(11):2682–2692. (Past name: British Journal of Environment & Climate Change, Past ISSN: 2231–4784)

Barasa PM, Botai CM, Botai JO, Mabhaudhi T (2021) A review of climate-smart agriculture research and applications in Africa. Agronomy 11(6):1255. https://doi.org/10.3390/agronomy11061255

Baumüller H (2012) Facilitating agricultural technology adoption among the poor: the role of service delivery through mobile phones. In: ZEF working paper series 93. https://doi.org/10.2139/ssrn.2237987

Bazzana D, Foltz J, Zhang Y (2022) Impact of climate smart agriculture on food security: an agentbased analysis. Food Policy 111:102304. https://doi.org/10.1016/j.foodpol.2022.102304

Bhattacharyya P, Pathak H, Pal S, Bhattacharyya P, Pathak H, Pal S (2020) Mainstreaming of climate-smart agriculture. In: Climate smart agriculture: concepts, challenges, and opportunities, pp 169–188

Binswanger-Mkhize HP, de Regt JP, Spector S (eds) (2010) Local and community driven development: moving to scale in theory and practice. World Bank Publications

Boudalia S, Dawit Teweldebirhan M, Ariom TO, Diouf NS, Nambeye E, Mwatitha Gondwe T et al (2024) Gendered gaps in the adoption of climate-smart agriculture in Africa and how to overcome them. Sustain For 16(13):5539

Boursianis AD, Papadopoulou MS, Diamantoulakis P, Liopa-Tsakalidi A, Barouchas P, Salahas G et al (2022) Internet of things (IoT) and agricultural unmanned aerial vehicles (UAVs) in smart farming: a comprehensive review. Internet Things 18:100187

Brewer E, Demmer M, Du B, Ho M, Kam M, Nedevschi S et al (2005) The case for technology in developing regions. Computer 38(6):25–38

Campbell BM, Thornton P, Zougmoré R, Van Asten P, Lipper L (2014) Sustainable intensification: what is its role in climate smart agriculture? Curr Opin Environ Sustain 8:39–43

Chandra A, McNamara KE, Dargusch P (2018) Climate-smart agriculture: perspectives and framings. Clim Pol 18(4):526–541

Chen C, Noble I, Hellmann J, Coffee J, Murillo M, Chawla N (2015) University of Notre Dame global adaptation index. University of Notre Dame

Chhetri NB (2012) Adapting agriculture to climate variability and change: capacity building through technological innovation. In: Climate variability—Some aspects, challenges and prospects, pp 182–192

Chitakira M, Ngcobo NZP (2021) Uptake of climate smart agriculture in Peri-Urban areas of South Africa's economic hub requires up-scaling. Front Sustain Food Syst 5:1–13. https://doi.org/10.3389/fsufs.2021.706738

Corpuz DA, Time MJC, Afalla BT (2022) Empowering the community through the extension services of a teacher education institution in The Philippines. Cogent Education 9(1):2149225

Dandois JP, Ellis EC (2013) High spatial resolution three-dimensional mapping of vegetation spectral dynamics using computer vision. Remote Sens Environ 136:259–276

Data NTT (2022) Transforming food systems with farmers: a pathway for the EU. World Economic Forum, 2022. https://www.weforum.org/reports/transforming-food-systems-with-farmers-a-pathway-for-the-eu/

Datta A, Rao KS, Santra SC, Mandal TK, Adhya TK, Datta A, Rao KS, Santra SC, Mandal TK, Adhya TK (2011) Greenhouse gas emissions from rice based cropping systems: economic and technologic challenges and opportunities. Mitig Adapt Strateg Glob Chang 16:597–615

Debray V, Wezel A, Lambert-Derkimba A, Roesch K, Lieblein G, Francis CA (2019) Agroecological practices for climate change adaptation in semiarid and subhumid Africa. Agroecol Sustain Food Syst 43(4):429–456

Deshmukh KK, Dhuware SR (2021) Impact of front line demonstration on yield enhancement of transplanted rice in Seoni District of Madhya Pradesh. Int J Curr Microbiol App Sci 10(02):3369–3372. https://doi.org/10.20546/ijcmas.2021.1002.370

Dessart FJ, Barreiro-Hurlé J, Van Bavel R (2019) Behavioural factors affecting the adoption of sustainable farming practices: a policy-oriented review. Eur Rev Agric Econ 46(3):417–471. https://doi.org/10.1093/erae/jbz019

Devare M, Arnaud E, Antezana E, King B (2023) Governing agricultural data: challenges and recommendations. In: Towards responsible plant data linkage: Data challenges for agricultural research and development. Springer, Berlin/Heidelberg, p 201

Dias RM, Marques G, Bhoi AK (2020, March) Internet of things for enhanced food safety and quality assurance: a literature review. In: International conference on emerging trends and advances in electrical engineering and renewable energy. Springer Nature Singapore, Singapore, pp 653–663

Dinesh D, Aggarwal P, Khatri-Chhetri A, Rodríguez AML, Mungai C, Sebastian L, Zougmore RB (2017) The rise in climate-smart agriculture strategies, policies, partnerships and investments across the globe. Agri Dev 30:4–9

Duquinal MI, Riboroso RA, Domondon CS, Reotutar MAC, Llagas RM Jr (2020) TRAINING NEEDS OF THE COMMUNITY RESIDENTS OF BANTAY, ILOCOS SUR: AN ASSESSMENT. International Journal of Advanced Research 8:2320–5407. https://doi.org/10.21474/IJAR01/11526

EEA (2023) Transforming Europe's food System—Assessing the EU Policy Mix. European Environment Agency, 2023. https://www.eea.europa.eu/publications/transforming-europes-food-system

Eliasson K, Wiréhn L, Neset TS, Linnér BO (2022) Transformations towards sustainable food systems: contrasting Swedish practitioner perspectives with the European Commission's Farm to Fork Strategy. Sustain Sci 17(6):2411–2425. https://doi.org/10.1007/s11625-022-01174-3

Erlangga E, Machuku O, Dahino CJ (2023) A review article on the impact and challenges of mobile phone usage on agricultural production in Africa. Cogent Food Agri 9(2):2273634. To link to this article: https://doi.org/10.1080/23311932.2023.2273634

FAO (2013) Climate-smart agriculture sourcebook. Rome, 570 pp

FAO (Food and Agriculture Organization) (2022) Climate-smart agriculture. https://www.fao.org/climate-smart-agriculture/en/
FAO.org (2020) Food and Agriculture Organization of the United Nations. FAO.org
Fentie A, Beyene AD (2019) Climate-smart agricultural practices and welfare of rural smallholders in Ethiopia: does planting method matter? Land Use Policy 85:387–396. https://doi.org/10.1016/j.landusepol.2019.04.020
Foughali K, Fathallah K, Frihida A (2018) Using Cloud IOT for disease prevention in precision agriculture. Procedia Comp Sci 130:575–582
Friha O, Ferrag MA, Shu L, Maglaras L, Wang X (2021) Internet of things for the future of smart agriculture: a comprehensive survey of emerging technologies. IEEE/CAA J Automatica Sinica 8(4):718–752
Fusco G, Melgiovanni M, Porrini D, Ricciardo TM (2020) How to improve the diffusion of climate-smart agriculture: what the literature tells us. Sustain For 12(12):5168. https://doi.org/10.3390/su12125168
Gemtou M, Kakkavou K, Anastasiou E, Fountas S, Pedersen SM, Isakhanyan G et al (2024) Farmers' transition to climate-smart agriculture: a systematic review of the decision-making factors affecting adoption. Sustain For 16(7):2828. https://doi.org/10.3390/su16072828
Gorst A, Dehlavi A, Groom B (2018) Crop productivity and adaptation to climate change in Pakistan. Environ Dev Econ 23:679–701. https://doi.org/10.1017/S1355770X18000232
Graham MD, Adams WM, Kahiro GN (2012) Mobile phone communication in effective human elephant–conflict management in Laikipia County. Kenya Oryx 46(1):137–144. https://doi.org/10.1017/S0030605311001104
Hamilton SF, Zilberman D (2006) Green markets, eco-certification, and equilibrium fraud. J Environ Econ Manag 52(3):627–644
Hansen J, Hellin J, Rosenstock T, Fisher E, Cairns J, Stirling C et al (2019) Climate risk management and rural poverty reduction. Agric Syst 172:28–46
Hasan MK, Desiere S, D'Haese M, Kumar L (2018) Impact of climate-smart agriculture adoption on the food security of coastal farmers in Bangladesh. Food Secur 10:1073–1088. https://doi.org/10.1007/s12571-018-0824-1
Henriksson R, Vincent K, Archer E, Jewitt G (2021) Understanding gender differences in availability, accessibility and use of climate information among smallholder farmers in Malawi. Climate and Development 13(6):503–514. https://doi.org/10.1080/17565529.2020.1806777
Herrman H, Stewart DE, Diaz-Granados N et al (2011) What is resilience? Can J Psychiatr 56:258–265. https://doi.org/10.1177/070674371105600504
Huyer S (2012) Gender mainstreaming in ICT for agriculture. USAID Briefing Paper. Fostering Agriculture Competitiveness Employing Information Communication Technologies (FACET) Project.
International Development Association (IDA) (2017) Report from the Executive Directors of the International Development Association to the Board of Governors. Additions to IDA resources: Eighteenth Replenishment. Towards 2030: investing in growth, resilience and opportunity. http://documents1.worldbank.org/curated/en/348661486654455091/pdf/112728-correct-file-PUBLIC-Rpt-from-EDs-Additions-to-IDA-Resources-2-9-17-For-Disclosure.pdf
Iqbal S, Xu J, Khan S et al (2023) Regenerative fertilization strategies for climate-smart agriculture: consequences for greenhouse gas emissions from global drylands. J Clean Prod 398:136650. https://doi.org/10.1016/j.jclepro.2023.136650
Islam N, Rashid MM, Pasandideh F, Ray B, Moore S, Kadel R (2021) A review of applications and communication technologies for internet of things (Iot) and unmanned aerial vehicle (UAV) based sustainable smart farming. Sustain For 13(4):1821
Jamil S, Kanwal S, Kanwal R et al (2023) Climate-smart agriculture: a way to ensure food security. Pakistan J Bot 55:1157–1167. https://doi.org/10.30848/PJB2023-3(26)
Javaid MM, AlGwaiz HIM, Waheed H, Ashraf M, Mahmood A, Li F-M, Attia KA, Nadeem MA, AlKahtani MDF, Fiaz S, Nadeem M, Ahmad HB (2022) Ridge-furrow mulching enhances capture and utilization of rainfall for improved maize production under rain-fed conditions. Agronomy 12:1187. https://doi.org/10.3390/agronomy12051187

Jayne TS, Mason NM, Burke WJ, Ariga J (2018) Taking stock of Africa's second-generation agricultural input subsidy programs. Food Policy 75:1–14

Jha S, Singh S (2021) Role of agriculture extension for climate smart agriculture. Implications for climate smart agriculture. Biotech Books

Jusoh MF, Muttalib MFA, Krishnan KT, Katimon A (2021, May) An overview of the internet of things (IoT) and irrigation approach through bibliometric analysis. In: IOP conference series: Earth and environmental science, vol 756, No. 1. IOP Publishing, p 012041

Kakraliya SK, Jat HS, Sapkota TB et al (2021) Effect of climate-smart agriculture practices on climate change adaptation, greenhouse gas mitigation and economic efficiency of rice-wheat system in India. Agric 11. https://doi.org/10.3390/agriculture11121269

Kalele DN, Ogara WO, Oludhe C, Onono JO (2021) Climate change impacts and relevance of smallholder farmers' response in arid and semi-arid lands in Kenya. Scientific African 12:e00814

Kang Y, Khan S, Ma X (2009) Climate change impacts on crop yield, crop water productivity and food security–a review. Prog Nat Sci 19(12):1665–1674. https://doi.org/10.1016/j.pnsc.2009.08.001

Khilji SA, Sajid ZA, Fayyaz S, Shah AA, Shah AN, Rauf M, Arif M, Yang SH, Fiaz S (2022) Fulvic acid alleviates paper sludge toxicity in Canola (Brassica napus L.) by Reducing Cr, Cd, and Pb Uptake. Front Plant Sci 13:874723. https://doi.org/10.3389/fpls.2022.874723

Kumar N, Nath CP (2019) Impact of zero-till residue management and crop diversification with legumes on soil aggregation and carbon sequestration. Soil Tillage Res 189:158–167. https://doi.org/10.1016/j.still.2019.02.001

Läderach P, Ramirez-Villegas J, Navarro-Racines C, Zelaya C, Martinez-Valle A, Jarvis A (2017) Climate change adaptation of coffee production in space and time. Clim Chang 141(1):47–62. https://doi.org/10.1007/s10584-016-1788-9

Lakshmi GP, Asha PN, Sandhya G, Sharma SV, Shilpashree S, Subramanya SG (2023) An intelligent IOT sensor coupled precision irrigation model for agriculture. Measurement: Sensors 25:100608

Le Coent P, Préget R, Thoyer S (2021) Farmers follow the herd: a theoretical model on social norms and payments for environmental services. Environ Resour Econ 78(2):287–306. https://doi.org/10.1007/s10640-020-00532-y

Lewis J, Rudnick J (2019) The policy enabling environment for climate smart agriculture: a case study of California. Front Sustain Food Sys 3:31

Lipper L, Thornton P, Campbell BM, Baedeker T, Braimoh A, Bwalya M et al (2014) Climate-smart agriculture for food security. Nat Clim Chang 4(12):1068–1072

Long TB, Blok V, Coninx I (2016) Barriers to the adoption and diffusion of technological innovations for climate-smart agriculture in Europe: evidence from The Netherlands, France, Switzerland and Italy. J Clean Prod 112:9–21

MacPherson J, Voglhuber-Slavinsky A, Olbrisch M, Schöbel P, Dönitz E, Mouratiadou I, Helming K (2022) Future agricultural systems and the role of digitalization for achieving sustainability goals. A review. Agron Sustain Dev 42(4):70. https://doi.org/10.1007/s13593-022-00792-6

Maiangwa MG, Adeniji B, Hassan AA, Hassan MB (2010) Challenges of decentralized, farmer-led and fee-for-service extension. J Agri Exten 14(1)

Massfeller A, Meraner M, Hüttel S, Uehleke R (2022) Farmers' acceptance of results-based agri-environmental schemes: a German perspective. Land Use Policy 120:106281. https://doi.org/10.1016/j.landusepol.2022.106281

Matteoli F, Schnetzer J, Jacobs H (2020) Climate-smart agriculture (CSA): an integrated approach for climate change management in the agriculture sector. In: Handbook of climate change management: research, leadership, transformation, pp 1–29

Mburu BK, Kung'u JB, Muriuki JN (2015) Climate change adaptation strategies by small-scale farmers in Yatta District, Kenya. Afr J Environ Sci Technol 9(9):712–722

McCarthy N, Brubaker J (2014) Climate-smart agriculture and resource tenure in Sub-Saharan Africa: a conceptual framework. Sidalc.net

McCarthy N, Lipper L, Branca G (2011) Climate-smart agriculture: smallholder adoption and implications for climate change adaption and mitigation, vol 3. FAO, Rome, pp 1–37
Meier J, Mauser W, Hank T, Bach H (2020) Assessments on the impact of high-resolution-sensor pixel sizes for common agricultural policy and smart farming services in European regions. Comput Electron Agric 169:105205
Mirón IJ, Linares C, Díaz J (2023) The influence of climate change on food production and food safety. Environ Res 216:114674. https://doi.org/10.1016/j.envres.2022.114674
Mizik T (2021) Climate-smart agriculture on small-scale farms: a systematic literature review. Agronomy 11(6):1096
Mondejar ME, Avtar R, Diaz HLB, Dubey RK, Esteban J, Gómez-Morales A, Hallam B, Mbungu NT, Okolo CC, Prasad KA, She Q, Garcia-Segura S (2021) Digitalization to achieve sustainable development goals: steps towards a smart Green planet. Sci Total Environ 794:148539. https://doi.org/10.1016/j.scitotenv.2021.148539
Mrabet R, Moussadek R (2022) Development of climate smart agriculture in Africa. In: Conservation agriculture in Africa: climate smart agricultural development. CABI, GB, pp 17–65
Murali KS, Abhishek R, Neelamadhab P (2023) IoT-enabled soil nutrient analysis and crop recommendation model for precision agriculture. Computers 12:61
Mwongera C, Nowak A, Notenbaert AM, Grey S, Osiemo J, Kinyua I et al (2019) Climate-smart agricultural value chains: risks and perspectives. In: The climate-smart agriculture papers: investigating the business of a productive, resilient and low emission future, pp 235–245
Neuenschwander P, Borgemeister C, De Groote H et al (2023) Perspective article: food security in tropical Africa through climate-smart plant health management; a new framework for sustainable plant health management. Heliyon 9:1–8. https://doi.org/10.1016/j.heliyon.2023.e15116
Njuguna L, Uri I, Beauchamp E (2024) Front matter. In: National Monitoring, evaluation, and learning Systems for Climate Change Adaptation: a comparative analysis of nine countries. International Institute for Sustainable Development (IISD), p ii
Nkonya E, Place F, Kato E, Mwanjololo M (2015) Climate risk management through sustainable land management in Sub-Saharan Africa. In: Sustainable intensification to advance food security and enhance climate resilience in Africa, pp 75–111
Oke FO, Arowolo AO, Olorunsogo GO, Akerele D (2022) Information communication technologies utilization and profitability of catfish farming in Ijebu-Ode Zone of the agricultural development programme. Ogun State J Agri Exten 26(2):53–60. https://doi.org/10.4314/jae.v26i2.6
Oyinbo O, Chamberlin J, Maertens M (2020) Design of digital agricultural extension tools: perspectives from extension agents in Nigeria. J Agric Econ 71(3):798–815. https://doi.org/10.1111/1477-9552.12371
Pawar RY, Deshmukh PB, Metre VA, Ghogardare SV (2021) Improving plant yield by smart irrigation system using pH and NPK sensors. In: Techno-Societal 2020: Proceedings of the 3rd international conference on advanced technologies for societal applications—Volume 2. Springer International Publishing, Cham, pp 293–303
Pedersen SM, Erekalo KT, Christensen T, Denver S, Gemtou M, Fountas S et al (2024) Drivers and barriers to climate-smart agricultural practices and technologies adoption: insights from stakeholders of five European food supply chains. Smart Agri Technol 8:100478
Price-Kelly H et al. (2016) Developing national adaptation monitoring and evaluation systems: a guidebook, GIZ, Canada. Retrieved from https://coilink.org/20.500.12592/8sgvdt on 15 Sep 2024
Radoglou-Grammatikis P, Sarigiannidis P, Lagkas T, Moscholios I (2020) A compilation of UAV applications for precision agriculture. Comput Netw 172:107148
Reidsma P, Accatino F, Appel F, Gavrilescu C, Krupin V, Tasevska GM et al (2023) Alternative systems and strategies to improve future sustainability and resilience of farming systems across Europe: from adaptation to transformation. Land Use Policy 134:106881
Resnick D, Swinnen J (2023) Introduction: Political economy of food system transformation. In: Resnick D, Swinnen J (eds) The political economy of food system transformation, 1st edn. Oxford University Press, Oxford, pp 1–31. https://doi.org/10.1093/oso/9780198882121.003.0001

Rosenstock TS, Lamanna C, DeRenzi B, Chesterman S, Kadiyala S, Ng'endo M et al (2016) Surveillance of climate-smart agriculture for nutrition (SCAN): innovations for monitoring climate, agriculture and nutrition at scale. CCAFS Info Note

Sahu G, Rout PP, Mohapatra S, Das SP, Pradhan PP (2020) Climate smart agriculture: a new approach for sustainable intensification. Curr J App Sci Technol 39(23):138–147

Saiz-Rubio V, Rovira-Más F (2020) From smart farming towards agriculture 5.0: a review on crop data management. Agronomy 10(2):207

Secretary-General UN (2021) Population, food security, nutrition and sustainable development: report of the Secretary-General. In: Oxford handbook. United Nations, pp 1–20. https://www.un.org/development/desa/pd/sites/www.un.org.development.desa.pd/files/undesa_pd_2021_e_cn.9_2021_2_advanceunedited.pdf. Accessed 1 Feb 2024.8.31

Senyolo MP, Long TB, Blok V, Omta O (2018) How the characteristics of innovations impact their adoption: an exploration of climate-smart agricultural innovations in South Africa. J Clean Prod 172:3825–3840

Shahid SA, Ahmed M (2014) Environmental cost and face of agriculture in the Gulf Cooperation Council countries: fostering agriculture in the context of climate change. Springer, pp 29–54

Sharma R, Kamble S, Gunasekaran A, Kumar V, Kumar A (2020) A systematic literature review on machine learning applications for sustainable agriculture supply chain performance. Comput Oper Res 119:104926

Sikka AK, Islam A, Rao KV (2018) Climate-smart land and water management for sustainable agriculture. Irrig Drain 67:72–81. https://doi.org/10.1002/ird.2162

Singh M, Karada MS, Rai RK, Pratap D, Agnihotri D, Singh AK, Singh BK (2023) A review on remote sensing as a tool for irrigation monitoring and management. Int J Environ Climate Change 13(6):203–211

Stafford JV (2000) Implementing precision agriculture in the 21st century. J Agric Eng Res 76(3):267–275

Stafford-Smith M, Griggs D, Gaffney O, Ullah F, Reyers B, Kanie N et al (2017) Integration: the key to implementing the sustainable development goals. Sustain Sci 12:911–919

Subeesh A, Mehta CR (2021) Automation and digitization of agriculture using artificial intelligence and internet of things. Artif Intell Agric 5:278–291

Swanson B, Davis KE (2014) Status of agricultural extension and rural advisory services worldwide: summary report. Global Forum for Rural Advisory Services (GFRAS), Lindau

Taylor M (2018) Climate-smart agriculture: what is it good for? J Peasant Stud 45(1):89–107

Teklewold H, Gebrehiwot T, Bezabih M (2019) Climate smart agricultural practices and gender diferentiated nutrition outcome: an empirical evidence from Ethiopia. World Dev 122:38–53. https://doi.org/10.1016/j.worlddev.2019.05.010

Tenzin S, Siyang S, Pobkrut T, Kerdcharoen T (2017) Low cost weather station for climate-smart agriculture. In: Proceedings of the 2017 9th international conference on Knowledge and Smart Technology (KST), Chonburi, Thailand, 1–4 February 2017, pp 172–177

Thottadi BP, Singh SP (2024) Climate-smart agriculture (CSA) adaptation, adaptation determinants and extension services synergies: a systematic review. Mitig Adapt Strateg Glob Chang 29(3):22

Trendov NM, Varas S, Zeng M (2019) Digital technologies in agriculture and rural areas – status report. Rome. http://www.fao.org/3/ca4985en/ca4985en.pdf

Vallejo L (2017) Insights from national adaptation monitoring and evaluation systems. In: OECD/IEA Climate Change Expert Group Papers, No. 2017/03. OECD Publishing, Paris. https://doi.org/10.1787/da48ce17-en

Vermeulen S, Zougmore R, Wollenberg E, Thornton P, Nelson G, Kristjanson P et al (2012) Climate change, agriculture and food security: a global partnership to link research and action for low-income agricultural producers and consumers. Curr Opin Environ Sustain 4(1):128–133

Vincent A, Balasubramani N (2021) Climate-smart agriculture (CSA) and extension advisory service (EAS) stakeholders' prioritisation: a case study of Anantapur district, Andhra Pradesh, India. J Water Clim change 12(8):3915–3931

Walker H, Sanz P (2024) Community-driven approaches to impact assessment. In: Handbook of public participation in impact assessment. Edward Elgar Publishing, pp 203–220
WB (2009) World development report 2010: development and climate change. The World Bank, Washington, DC
Weraikat D, Šoriˇc K, Žagar M, Sokăc M (2024) Data analytics in agriculture: enhancing decision making for crop yield optimization and sustainable practices. Sustain For 16:7331. https://doi.org/10.3390/su16177331
Westermann O, Förch W, Thornton P, Körner J, Cramer L, Campbell B (2018) Scaling up agricultural interventions: case studies of climate-smart agriculture. Agric Syst 165:283–293. https://doi.org/10.1016/j.agsy.2018.07.007
Witherspoon T, Harris E (2011) Avoiding the 30-pound paperweight: success via contextually appropriate technologies. In: *2011 IEEE global humanitarian technology conference*. IEEE, pp 402–408
WMO (World Meteorological Organization) (2020) State of the climate in Africa 2019. https://library.wmo.int/doc_num.php?explnum_id=10421
World Bank (2012) World Bank report: helping the poor through localized development is tough, but essential, 2012). Available online on: https://www.worldbank.org/en/news/feature/2012/11/15/world-bank-report-helping-the-poor-through-localized-development-is-tough-but-essential. Retrieved on: August 13, 2024
World Bank (2015a) Accelerating climate-resilient and low-carbon development: the Africa climate business plan.
World Bank (2015b) Climate-smart agriculture in Kenya. CSA Country Profiles for Africa, Asia, and Latin America and the Caribbean Series
World Bank (2016) Strengthening Community-Driven Development in East Asia, South Asia, and the Pacific. Available Online on:https://www.worldbank.org/en/news/feature/2016/08/24/strengthening-community-driven-development-in-east-asia-south-asia-and-the-pacific. Retrieved on: August 14, 2024
World Bank (2019) Mali climate-smart agriculture investment plan. Washington, DC
World Bank Group (2024) Climate-smart agriculture. Climate-smart agriculture (CSA) is an integrated approach to managing landscapes—cropland, livestock, forests and fisheries—that address the interlinked challenges of food security and climate change. Available online on: https://www.worldbank.org/en/topic/climate-smart-agriculture
Wreford A, Ignaciuk A, Gruère G (2017) Overcoming barriers to the adoption of climate-friendly practices in agriculture. (OECD Food, Agriculture and Fisheries Papers 101; OECD Food, Agriculture and Fisheries Papers, Vol. 101). https://doi.org/10.1787/97767de8-en.
Zaharia A, Diaconeasa MC, Maehle N, Szolnoki G, Capitello R (2021) Developing sustainable food systems in Europe: national policies and stakeholder perspectives in a four-country analysis. Int J Environ Res Public Health 18(14):7701. https://doi.org/10.3390/ijerph18147701
Zhang X, Huang Y, Li M (2011) Collaboration on disaster prevention education curriculum development project and its implications for social work. China J Soc Work 4(2):165–173
Zhao J, Liu D, Huang R (2023) A review of climate-smart agriculture: recent advancements, challenges, and future directions. Sustain For 15(4):3404
Zheng H, Ma W, He Q (2024) Climate-smart agricultural practices for enhanced farm productivity, income, resilience, and greenhouse gas mitigation: a comprehensive review. Mitig Adapt Strateg Glob Chang 29(4):28. https://doi.org/10.1007/s11027-024-10124-6
Zizinga A, Mwanjalolo JGM, Tietjen B, Bedadi B, Pathak H, Gabiri G, Beesigamukama D (2022) Climate change and maize productivity in Uganda: simulating the impacts and alleviation with climate smart agriculture practices. Agric Syst 199:103407. https://doi.org/10.1016/j.agsy.2022.103407
Zougmoré R, Partey S, Ouédraogo M, Omitoyin B, Thomas T, Ayantunde A et al (2016) Toward climate-smart agriculture in West Africa: a review of climate change impacts, adaptation strategies and policy developments for the livestock, fishery and crop production sectors. Agric Food Secur 5:1–16

Plant Growth Promoting Rhizobacteria for Mitigation of Environmental Stresses

10

Ayesha T. Tahir, Ammarah Bint e Sarwar, Farzana Kausar, Tayyaba Yasmin, and Zahra Jabeen

Abstract

One of the progressively growing global threats to agriculture is climate change, which leads to enhanced environmental stresses and consequently lower crop yield. Among the number of available approaches, use of PGPR is regarded as a viable, economical, and eco-friendly solution for minimizing the effect of environmental stressors and enhancing crop production. They can work directly by synthesizing the compounds with the ability to mitigate the effect of stress, or indirectly by triggering the responses related to the plant immune system. The current chapter will describe the PGPR strains effectively in use for various abiotic and biotic stresses and their mechanism of action for helping plants to thrive under changing climatic conditions.

Keywords

Microbial inoculants · Bacterial taxonomies · Environmental stress · Plant resistance

10.1 Introduction

The world's population is expected to reach a peak in the mid-2080s with 10.3 billion people, up from 8.2 billion in 2024 (World Population Prospectus 2024). Consequently, the total global food demand is also to increase from 35 to 56% for

A. T. Tahir (✉) · A. Bint e Sarwar · T. Yasmin · Z. Jabeen
Department of Biosciences, COMSATS University Islamabad, Islamabad, Pakistan
e-mail: ayesha.tahir@comsats.edu.pk

F. Kausar
Department of Environmental Science, PMAS Agriculture University, Rawalpindi, Pakistan

© The Author(s), under exclusive license to Springer Nature Singapore Pte Ltd. 2025

S. Faiz et al. (eds.), *Climate Smart Agriculture for Future Food Security*,
https://doi.org/10.1007/978-981-96-4499-5_10

the period 2010–2050 (Van Dijk et al. 2021). Plants have been a source of human food for ages and are frequently subjected to a range of abiotic and biotic stresses, in their natural habitat. Climate change intensifies such stresses, due to the plant's ineptness to handle them (Kumar et al. 2023). All such stressors consequently result in lower yield and, as a result, increased hunger around the globe.

To mitigate the effect of stresses on plant development, several strategies have been adopted by the agriculture community (Srivastava et al. 2020). Physical methods, such as composting, cover cropping, deep plowing, bed planting, and chemical methods (foliar spray of chemicals like Indole Acetic acid (IAA), Abscisic acid (ABA), Zeatin as well as osmo protectants like silicon, silver, betaine, glycine, salicylic acid, and jasmonic acid), have been adopted to mitigate the effects of environmental stresses (Aslam et al. 2022). The development of resistant varieties, lowering the impact of stressors through suitable cultivation and cropping practices, and other measures can all be used as strategies to reduce environmental stresses (Egamberdieva et al. 2019; Mahmood et al. 2022; Tahir et al. 2022a). All of these techniques have a low rate of success and are either labor-intensive and costly or are time-consuming and are not environmentally friendly; hence, it is difficult to use these techniques on a broad scale (Saddiq et al. 2021). One of the sustainable methods to boost agricultural productivity while protecting the environment is the use of microbial inoculants mainly comprised of Plant Growth Promoting Rhizobacteria (PGPR): bacteria that inhabit the rhizosphere of the plants (Danish et al. 2020). Plant growth-promoting rhizobacteria can act as symbionts or live freely, depending on their relationship with plants (Ullah et al. 2017). In symbiotic association, plant root exudates are absorbed by microbes, which in turn help plants to uptake nutrients, develop resistance to abiotic and biotic stresses, and to remove dangerous contaminants from the soil, ultimately leading to increased plant growth and improved plant nutrition (Dutta and Podile 2010). Scientists, worldwide are interested in the use of microbes in agriculture, commonly described as "biofertilizers or bioinoculants" for several extended benefits beyond enhanced plant growth (Sammauria et al. 2020; Tahir et al. 2022b). The main objective of this chapter is to explore PGPR that enhance stress resistance in plants and their mechanisms that are crucial for mitigating the impacts of climate change on crop plants. This understanding can help provide a paradigm for sustainable agriculture.

10.2 PGPR and Environmental Stresses

Environmental stresses negatively influence the plant nourishment and growth process. Abiotic stresses are the main factors reducing plant output, and they can vary in intensity. These stresses have the potential to permanently harm plants, resulting in altered performance, decreased yield, impaired development, disrupted metabolic processes, and abnormalities in the upcoming generation (Abulfaraj and Jalal 2021). Using PGPR to enhance plant resilience to environmental challenges is an effective strategy to combat the above-mentioned stresses. Initial research on PGPR featured bacteria antagonistic to soil-borne plant pathogens, such as fluorescent *Pseudomonas* and *Bacillus subtilis*, and was primarily concerned with biological management of plant diseases rather than with promoting growth (Kumar et al.

2022). Recent investigations encompassing a greater variety of plant species, advancements in bacterial taxonomy and an improved understanding of PGPR regulation have led to a rise in the variety of bacterial strains that are considered PGPR. Currently, PGPR comprise members of a wide range of bacterial taxa with roles in management of diverse abiotic and biotic stresses in addition to plant growth promotion (Morcillo and Manzanera, 2021). The impact of environmental stresses on crop plants and PGPR interventions to minimize these effects is discussed in upcoming sections.

10.3 Drought Stress

Drought is one of the main factors restricting sustainable agricultural production, which imposes serious threats to the world's food security (Gowtham et al. 2022). Approximately one-third of the world has been turned into an arid zone (4.21% hyper-arid, 14.62% desert and 12.3% semi-arid zones), in which there is drought or insufficient precipitation for agricultural production (Zhang et al. 2022; Adil et al. 2022). The persistent drought in various regions of the world is predicted to affect 40% of the global population by the year 2030, putting 700 million people, animals, and crops in danger (Seleiman et al. 2021). According to estimates, drought stress has reduced the global yield of cereals by up to 10% over the past 50 years, and is expected to effect the production of more than 50% of agricultural land by 2050 (Shah et al. 2021; Mohamed et al. 2021). In wheat, drought has resulted in upto 21% decline in productivity, while in maize it has caused a 40% reduction in production from 1980 to 2015 (Cao et al. 2022). Moreover, nearly 60% of common beans grown under rain-fed conditions are affected by drought with potential grain yield reductions of upto 80% in some areas (Chen et al. 2023; Cao et al. 2022).

PGPR have the ability to protect plants against drought stress. Numerous investigations have shown that PGPR reduce plant drought stress through various mechanisms as detailed in Table 10.1.

10.4 Salinity Stress

Salt stress poses a serious threat to agricultural production, crop yield, and ultimately food security. Seawater intrusion, inappropriate irrigation, soil salinization has become a significant global issue due to climate change. Nearly, 10% of all cropland and more than 833 million hectares of soil is at risk of salinization worldwide (Ali et al. 2022; Canton 2021). Salt stress results in annual losses of more than $27.3 billion crop production (Qadir et al. 2014), and farmlands are projected to become 50% more saline by 2050 (Shrivastava and Kumar 2015).

The application of salinity-alleviating microbes is a green approach for boosting crop output in saline agricultural areas (Shahid et al. 2021; Islam et al. 2022). Soil microbes such as *Bacillus, Pseudomonas, Agrobacterium, Enterobacter, Klebsiella, Streptomyces, Azospirillum, Azotobacter*, and *Serratia* are the few examples of PGPRs to mitigate the impact of salinity on plants (Table 10.2).

Table 10.1 Rhizobacterial strains effective in reducing Drought stress along with their control mechanisms in various crops

PGPR strain	Crops	Mechanism/Impact on plant	References
***Bacillus paranthracis* DT-97**	Wheat	Generates proline, catalase and peroxidase enhancing root biomass and water absorption.	Yadav et al. (2022)
***Bacillus amyloliquefaciens*QST713**	Alfalfa varieties: Galalxie Max' (drought-tolerant) and 'Saidi 7' (drought-sensitive)	Improves iron and potassium content, and overall morphology of plant root and shoot system. Decreases soil pH and water stress deficit (WSD).	Han et al. (2024)
***Bacillus rugosus* (MMH101)**	Wheat	Boosts chlorophyll and proline levels and reduces malondialdehyde (MDA) levels that enhances root and shoot biomass of plants.	Salem et al. (2024)
***Aeromonas caviae* A1–2, *A. veronii* C7_8**	Maize	Increase nutrient absorption and water-holding capacity enhancing root and shoot biomass.	Agunbiade et al. (2024)
***Streptomyces pactum* Act12**	Wheat	Elevates chlorophyll, osmolyte, and antioxidant enzyme levels improving grain hardness.	Yang et al. (2024)
***Pseudomonas brassicacearum* X123**	Walnut	Enhances the production of protease and IAA, improves physiological and biochemical properties like stomatal activity, and increases phytohormone production.	Jing et al. (2023)
Bacillus cereus L90	Walnut	Elevates cytokinin and IAA production, improves antioxidant activity and plant development.	Liu et al. (2023)
B. cereus LC514122 and B. haynesii LC 514123	Black gram and Bean	Enhance the production of IAA,1-aminocyclopropane-1-carboxylate (ACC) deaminase, and exopolysaccharides (EPS).	Andy et al. (2023)
***Bacillus* sp. *strains* DLGB2 and DKB26 *Pantoea* sp. *strains* DKB63, DKB70, DKB68, DKB64, and DKB65**	Strawberry	Elevate siderophore production, improve phosphate movement for better absorption and enhance -aminocyclopropane-1-carboxylate deaminase (ACCD).	Paliwoda et al. (2022)
***Pseudomonas* sp. MUS04, *Rhodococcus* sp. BAR03, *and Variovorax* sp. BAR04**	Tomato	Boost physiological activities and yield	Nemeskéri et al. (2022)

Table 10.2 Mechanisms of Rhizobacterial strains reducing salinity stress in various crops

PGPR strain	Crops	Mechanism/ Impact of Plant	References
Bacillus altitudinis AD13–4	Alfalfa	Regulates plant endogenous hormones and secondary metabolites. Improves photosynthesis and antioxidant activity.	Khoso et al. (2024)
***S. niveoruber* ATMLC132021 *and S. lienomycini* ATMLC122021**	Lettuce	Improve proline content, enhance biochemical processes like stomatal conductance and photosynthesis and promote efficient water usage.	González et al. (2023)
***Lysinibacillus* spp. *and Bacillus* spp.**	Tomato	Produce extracellular enzymes and phytohormones, regulate ion equilibrium and improve plant development.	Damodaran et al. (2023)
***Bacillus velezensis* UTB96**	Wheat	Increases proline content, antioxidant enzyme activity and plant development.	Mehrabi et al. (2024)
***Oceanobacillus jordanicus* GSFE11**	Wheat	Boosts genotype-specific responses and nutrient uptake.	Albdaiwi et al. (2024)
Microbacterium azadirachtae CNUC13	Maize	Improves biomass, enhances photosynthetic pigments, and boosts antioxidant enzyme activity.	Luo et al. (2024)
Microbacterium ginsengiterrae S4	Rice	Regulates gene expression, boosts osmotic balance and increases antioxidant enzyme activity.	Ji et al. (2024)
Enterobacter asburiae D2	Rice	Improves IAA production, siderophore synthesis, phosphate solubility, and enhances plant biomass and development.	Ning et al. (2024)
Bacillus paralicheniformis LBEndo1	Pecan	Increases proline content and improves plant morphology and biomass.	Palacio-Rodríguez et al. (2023)
Achromobacter xylosoxidans BOA4	Tomato	Regulate glutamic and proline pathways. Improve biomass.	Rai et al. (2023)
Priestia megaterium ZS-3	Camphor-Tree	Improves plant development and increases resistance to volatile organic compounds (VOCs).	Shi et al. (2023)
Pseudomonas putida, Alcaligenes* sp., *Klebsiella* sp., *and Pseudomonas cedrina	Alfalfa	Enhance plant growth by improving proline and chlorophyll content and reducing oxidative stress.	Tirry et al. (2022)
Bacillus velezensis QA2	Quinoa	Enhances plant development through production of IAA and antioxidant enzymes.	Mahdi et al. (2022)

10.5 Heat Stress

The surface temperatures of the Earth may rise by 1.5 to 11 °C by 2100 due to greenhouse gas emissions, which may significantly lower agricultural yield. Temperature stress (Heat shock or heat wave) has a major effect on plant growth phases. It decreases grain size and seed growth, causes leaf decay, disrupts cell stability and reduces water utilization efficiency (Michaletz et al. 2016; Ahmad et al. 2024). High temperatures and drought can cause production of reactive oxygen species (ROS), degradation of lipids and proteins, disrupted membrane integrity, and reduce photosynthesis and crop yield (Hein et al. 2021).

Elevated temperatures at any of the plant growth cycle (from germination until their maturity) can be lethal. Bacteria that can withstand high temperatures are vital for the survival and grwoth of the plants, as mentioned in Table 10.3.

10.6 Impact of Climate Change on Biotic Stresses

Extreme weather events are direct outcomes of climate change and are accountable not only for enhanced abiotic stresses but also for increasing the burden of biotic stressors. Changes in precipitation patterns, humidity, and temperature fluctuations lead to an expansion of the geographic range of plant pathogens and exposure to new regions and plant species. Such alternating patterns significantly influence the natural ecosystems, human well-being, and the agriculture sector (Lahlali et al. 2024; Leisner et al. 2022). In this section, we will discuss how climate change can influence the spread of biotic diseases and the role of PGPR in their mitigation.

Temperature ranks as one of the factors highly correlated with plant growth, development, and yield along with the spread of diverse plant pathogens, such as the late blight of potatoes (caused by *Phytophthora infestans*). It is considered the most damaging disease of plants due to its association with the Irish Potato Famine, which occurred in the 1840s. As the temperature increased due to climate change, it led to a 15–20% higher severity of late blight of potatoes, particularly in temperate zones (Chacón-Cerdas et al. 2020). Similarly, under high temperatures, the spread of Citrus Greening caused by *Candidatus Liberibacter* spp. expanded up to 40–60% in Southeast Asia (Wei et al. 2021).

Under high humidity and warmer temperatures, wheat rust disease (by *Puccinia striiformis)* also increases (Soroye et al. 2020). High precipitation also promotes the spread of downy (by *Plasmopara viticola)* and powdery mildew (by *Erysiphe necater)* (Sanghavi et al. 2021). In addition to pathogens, under hot and humid conditions as well as high CO_2, aggressive weeds like *Chenopodium album* (common lambsquarters) and *Amaranthus palmeri* (palmer amaranth) develop more biomass and seeds, making them more competitive against crops (Kumar et al. 2023).

Table 10.3 Mechanisms of Rhizobacterial strains reducing heat stress in various crops

PGPR strain	Crops	Mechanism	References
Bacillus subtilis strains, BsCP1	*Arabidopsis* seedlings	Improves overall growth of plants. Boosts chlorophyll levels and chances of survival.	Chang et al. (2023)
Bacillus subtilis strains AP-3 and AP-12	Common bean	Improve vegetative and reproductive traits, boosting plant biomass.	de Lima et al. (2022)
Bacillus* sp. *IHBT-705	Saffron	Produces siderophores and enhances phosphate solubility. Exhibits enhanced root colonization.	Ali et al. (2023)
Bacillus paramycoides Ph-04	Wheat	Enhances membrane structure stability and antioxidant enzymes activity. Improves plant development and morphological traits.	Dubey et al. (2022)
P. megaterium SH-19	Soybean	Produces IAA, siderophores, and EPS. Effectively regulates antioxidant enzymes enhancing defensive mechanisms.	Shaffique et al. (2024)
Bacillus* spp. *AH-08, AH-67, SH-16* and *Pseudomonas* spp. *SH-29 consortia	Maize	Maintain low levels of MDA, Enhance chlorophyll content and improve plant architecture and morphology.	Ahmad et al. (2023)
Bacillus zanthoxyli HS1	Cabbage and Cucumber	Enhances germination rate, increases enzymatic activities and reduces oxidation stressors by regulation of osmotic networks.	Barghi and Jung (2024)
Bacillus subtilis PM32	Potato	Produces IAA, ACC and EPS, boosting crop yield.	Mehmood et al. (2021)
B. licheniformis BE-L60	Spinach	Produces amylases and proteases. Enhances production of antioxidant enzymes and proline. Helps in maintaining ion equilibrium.	Li et al. (2023)
Enterobacter cloacae S23	Groundnut	Produces EPS, siderophores, and IAA, forms Biofilm and enhances phosphate solubility.	Ramakrishnan et al. (2023)

The impact of climate change on plant diseases is different in each region. For example, Juroszek and von Tiedemann (2013) predicted that rising temperatures and humidity in Northern Europe could increase the severity of wheat diseases like Septoria leaf blotch (*Zymoseptoria tritici*) by 18–40%. However, under moderate to high climate change scenarios, diseases like powdery mildew and downy mildew in grapevines could increase by 30–50% in the Mediterranean region due to a combination of higher temperatures and greater relative humidity (Caffarra et al. 2012).

In addition to the increasing prevalence and severity of plant diseases, climate change also influences the efficacy of disease management strategies, creating a dire need to adopt environment-friendly approaches to agricultural practices. As discussed above for abiotic stressors, PGPR can also help plants under certain biotic stresses by enhancing their growth both directly and indirectly (Shahid et al. 2021).

The most widespread bacterial genera with plant growth-promoting abilities include *Pseudomonas, Serratia, Streptomyces, Bacillus, Paenibacillus, Enterobacter, Pantoea, Burkholderia, Azospirillum, Azotobacter,* and *Paraburkholderia*, which have been reported to significantly inhibit pathogens and improve plant health (Santhanam et al. 2015; Gupta et al. 2021). Recent evidencees are summarized in Table 10.4.

10.7 PGPR Mechanism in Response to Environmental Stresses

The ability of plants to grow in environments with low water availability, excessive salinity, and high temperature depends on multiple factors such as their genetic makeup, interaction with endophytic/exophytic PGPR, and agronomic practices. PGPR interaction is complex as it can influence exogenous exudates and endogenous hormone levels in plants, as part of defense system against the damaging effects of environmental stress. Hence, PGPR can mitigate stress through direct or indirect mechanisms (Fig. 10.1), as detailed in recent reviews (Shah et al. 2021; De Andrade et al. 2023)

PGPR secrete secondary metabolites, including volatile organic compounds, exopolysaccharides (EPS), and phytohormones such as abscisic acid (ABA), auxins, IAA, and many more (Kaur et al. 2022; Das et al. 2022; Tahir et al. 2022b). *Bacillus subtilis* and *Azospirillum brasilense* can produce ABA, that increases biomass and chlorophyll content by decreasing water loss and enabling its efficient use under certain environmental stresses (Chang et al. 2023).

Similarly, the production of phytohormone indole-3-acetic acid (IAA) is crucial in mitigating environmental stresses. Certain strains of *Bacillus, Pseudomonas,* and *Enterobacter* produce IAA that encourages the development of lateral and adventitious roots, improving root architecture and enabling plants to explore a greater volume of soil and uptake more nutrients and water during salinity, drought, and heat stress, leading to higher crop yield (Tables 10.1, 10.2, and 10.3).

Numerous PGPR strains also generate ACC deaminase, an enzyme that lowers ethylene concentrations in stressed plants, alongside IAA (Mehmood et al. 2021).

Table 10.4 Impact of climate change on biotic stress and the PGPR efficacy in mitigating its effects on crops

Name of pathogen	Disease	Climate factors enhancing the pathogen spread	Crop affected by pathogen	PGPR used to mitigate the effect of pathogen	Reduction in disease severity	References
Fusarium equiseti	Wilt	High temperature	Lentil	*Rhizobium pusense*	44.4%	Ayub et al. (2024)
Fusarium equiseti	Wilt	High temperature	Lentil	*Burkholderia contaminans*	91.67%	Ayub et al. (2024)
Phytophthora infestans	Late blight of potato	High temperature	Potato	*Pseudomonas* sp. and *Bacillus velezensis*	50–75%	Wang and Long (2023)
Puccinia striiformis	Rust	High temperature	Wheat	*Bacillus subtilis*	8.18%	Khan et al. (2023)
Fusarium oxysporum	Vascular wilt, root rot	High temperature	Bananas, potatoes and fruits	*Pseudomonas syringae*	40–60%	Zakaria (2023)
Botrytis cinerea	Gray mold	High temperature	Strawberry	*Streptomyses* sp.	50–90%	Yong et al. (2022)
Xanthomons oryzae	Leaf spot and blight	Higher temperature	Rice	*Bacillus* spp. Rh219	Upto 80%	Thakur (2024)
Pythium ultimum	Damping off	Moisture	Corn, tomato	*Bacillus halotolerans, Massilia alkalitolerans, and Bacillus aryabhattai*	50–56%	Abdelaziz et al. (2023)
Colletotrichum gloesporoides	Anthracnose	Moisture	Avocado	*Bacillus thuringiensis*	63–65%	Magallón-Andalón (2025)
Zymoseptoria tritici	Septoria tritici leaf blotch	High temperature	Wheat	*Bacillus subtilis and B. simplex*	above 50%	Allioui et al. (2021)
Aspergillus flavus	Ear rot and yellow mold	High temperature	Maize	*Bacillus megaterium*	above 50%	Saleh et al. (2021)

(continued)

Table 10.4 (continued)

Name of pathogen	Disease	Climate factors enhancing the pathogen spread	Crop affected by pathogen	PGPR used to mitigate the effect of pathogen	Reduction in disease severity	References
Xanthomonas spp.	Bacterial spot	Increased temperatures and humidity	Tomato	*Bacillus subtilis (QST 713)*	50–80%	Marin et al. (2019)
Botrytis cinerea	Gray mold	Wind direction and extreme weather events	Grapes, strawberry	*Bacillus velezensis XT1*	50–60%	Toral et al. (2020)
Fusarium spp.	Fusarium ear blight	Increased precipitation and humidity	Wheat	*Consortia*: *Four strains of Pseudomonas and two strains of Serratia*	96%	Thuraga et al. (2024)
Pythium spp.	Root rot, damping off	Excessive rainfall leading to waterlogged conditions	Chilli	*Pseudomonas spp*	50–80%	Hyder et al. (2021)
Erysiphe spp	Powdery mildew	Increased solar radiation	Grapevines	*Bacillus subtilis*	50–70%	Hashem et al. (2019)
Fusarium equiseti	Seedling rot disease	Increased solar radiations	Wheat	*Bacillus strain FJ3*	39%	Jan et al. (2023)

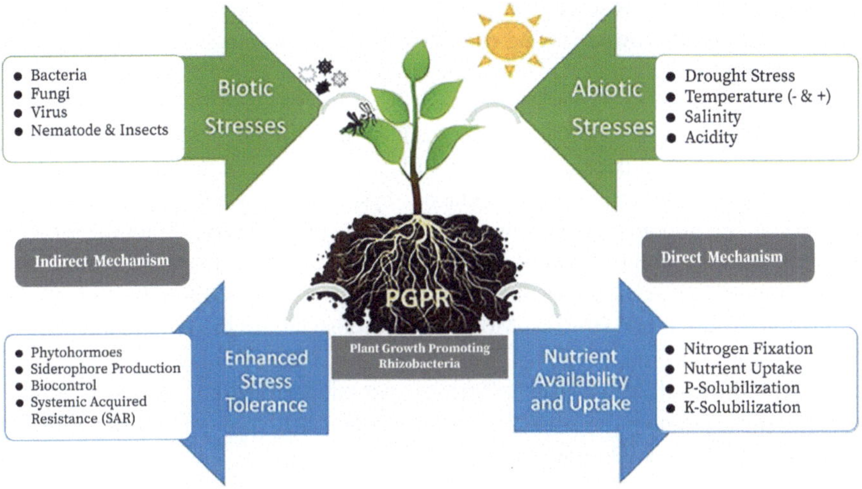

Fig. 10.1 Direct and indirect mechanisms of PGPR for stress mitigation and plant development. (Modified from Shah et al. 2021)

When plants are exposed to abiotic stresses, ethylene, a hormone associated with stress, usually builds up and causes early leaf mortality and decreased development. Plants can withstand environmental stresses better when ACC deaminase lowers ethylene levels, allowing plants to continue growing and delay senescence caused by environmental stresses.

The synthesis of EPS is another strategy utilized by PGPR to strengthen the host plant during stress. EPS creates a protective biofilm that encloses plant roots, trapping moisture and preventing the passage of toxic ions. By lowering osmotic pressure during drought and shielding plants from the toxic effects of salt during saline conditions, this barrier aids in the maintenance of a healthy soil microenvironment. Thus, biofilm formation helps maintain moisture and osmotic balance that enables plants to combat environmental stresses. In addition, biofilm production alst protect plant roots from pathogen attack under biotic stress incidences. *Pseudomonas, Bacillus, and Enterobacter* are among such PGPR to employ this mechanism (Ramakrishnan et al. 2023; Shaffique et al. 2024).

Nutrient uptake by plants is aided by PGPR. In soil ecosystems, nitrogen-fixing bacteria transform atmospheric nitrogen into usable form for various plant biological processes. Moreover, soil PGPR have the capacity to solubilize phosphorous (P) and potassium (K) and release advantageous compounds. These processes work together to prevent nitrogen loss from the soil, promote P and K solubilization, regulate soil pH, enhance soil integrity, and support the development of plant root systems (Table 10.1).

Among micronutrients, iron (Fe) is another element that is necessary for plant growth. Siderophores are iron-chelating substances generated by advantageous soil microbes that are important when the Fe level inside plant cells is insufficient (De Serrano et al. 2016). Siderophores generated by PGPR have a variety of uses, such as transport of nutrients to activate plant resistance and improving the biomass of plants under abiotic stresses (Ning et al. 2024).

PGPR substantially mitigate the detrimental effects of drought, salinity, and heat stresses by combining all of these processes: IAA production, ACC deaminase activity, EPS secretion, and the enhancement of the antioxidant system. This eventually encourages the development and survival of plants under stressful circumstances. These phytohormones, Induced systemic resistance, production of antimicrobial compounds (siderophores, antibiotics etc.,), nutrient mobilization by PGPRs, helps the plant growth development in a way to combat both abiotic and biotic stresses. Several adaptive mechanisms are reported by PGPR to help plants for mitigation of with biotic stresses. Moreover, they also induce the production of various antioxidant enzymes in plants, including superoxide dismutase (SOD), peroxidase (POD), catalase (CAT), ascorbate peroxidase (APX), glucanases and phenylalanine ammonia lyase (PAL) to scavenge reactive oxygen species under both types of stresses (De Andrade et al. 2023).

10.8 Conclusion and Future Perspectives

Climate change is the most unpredictable and challenging factor affecting agriculture in the current era. Utilizing the potential of beneficial microbes is becoming an essential approach for boosting plant stress resilience in this context. Such microbes are not only cost-effective and sustainable but are also beneficial for plant growth owing to their production of secondary metabolites and volatile organic compounds, enhanced availability of nutrients/minerals for the plants, biocontrol, and the ability of transcriptional regulation for minimizing the adverse impacts of environmental stresses. More accurate PGPR monitoring and applications are possible through emerging technologies and tools like remote sensing and advanced meta-OMICS approaches. Hence, in the current era of climate change, plant growth must be supported by PGPR screened for multi-stress tolerance and multiple plant growth-promoting traits to overcome food shortage and hunger in terms of both quality as well as quantity. Furthermore, there is a dire need to explore avenues for their field applications while maintaining their viability and effectiveness for their host.

References

Abdelaziz S, Belal EE, Al-Quwaie DA, Ashkan MF, Alqahtani FS, El-Tarabily KA, Hemeda NF (2023) Extremophilic bacterial strains as plant growth promoters and biocontrol agents against Pythium ultimum and Rhizocotnia solani. J Plant Pathol 105(4):1347–1369

Abulfaraj AA, Jalal RS (2021) Use of plant growth-promoting bacteria to enhance salinity stress in soybean (Glycine max L.) plants. Saudi J Biol Sci 28(7):3823–3834

Adil M, Wang J, Shah AN, Tanveer M, Fiaz S (2022) Effect of fallow management practices on soil water, crop yield and water use efficiency in winter wheat mono-culture systems: a meta-analysis. Front Plant Sci 13

Agunbiade VF, Fadiji AE, Agbodjato NA, Babalola OO (2024) Isolation and characterization of Plant-Growth-Promoting, Drought-Tolerant rhizobacteria for improved maize productivity. Plan Theory 13(10):1298. https://doi.org/10.3390/plants13101298

Ahmad M, Imtiaz M, Nawaz MS, Mubeen F, Sarwar Y, Hayat M, Asif M, Naqvi RZ, Ahmad M, Imran A (2023) Thermotolerant PGPR consortium B3P modulates physio-biochemical and molecular machinery for enhanced heat tolerance in maize during early vegetative growth. Ann Microbiol 73(1). https://doi.org/10.1186/s13213-023-01736-5

Ahmad Y, Haider S, Iqbal J, Naseer S, Attia KA, Mohammed AA, Fiaz S, Mahmood T (2024) In-silico analysis and transformation of OsMYB48 transcription factor driven by CaMV35S promoter in model plant–Nicotiana tabacum L. conferring abiotic stress tolerance. GM Crops & Food 15(1):130–149

Albdaiwi R, Al-Sayaydeh R, Al-Rifaee MK, Alhindi T, Ashraf M, Al-Abdallat AM (2024) Halotolerant endophytic bacteria regulate growth and field performance of two durum wheat genotypes with contrasting salinity tolerance potential. Plan Theory 13(9):1179. https://doi.org/10.3390/plants13091179

Ali B, Wang X, Saleem MH, Azeem MA, Afridi MS, Nadeem M et al (2022) Bacillus mycoides PM35 reinforces photosynthetic efficiency, antioxidant defense, expression of stress-responsive genes, and ameliorates the effects of salinity stress in maize. Life 12(2):219

Ali N, Swarnkar MK, Veer R, Kaushal P, Pati AM (2023) Temperature-induced modulation of stress-tolerant PGP genes bioprospected from Bacillus sp. IHBT-705 associated with saffron (Crocus sativus) rhizosphere: a natural-treasure trove of microbial biostimulants. *Frontiers*. Plant Sci 14:1141538

Allioui N, Driss F, Dhouib H, Jlail L, Tounsi S, Frikha-Gargouri O (2021) Two novel Bacillus strains (subtilis and simplex species) with promising potential for the biocontrol of Zymoseptoria tritici, the causal agent of septoria tritici blotch of wheat. BioMed Res Int 2021(1):6611657

Andy AK, Rajput VD, Burachevskaya M, Gour VS (2023) Exploring the identity and properties of two bacilli strains and their potential to alleviate drought and heavy metal stress. Horticulturae 9(1):46. https://doi.org/10.3390/horticulturae9010046

Aslam H, Mushtaq S, Maalik S, Bano N, Eed EM, Bibi A et al (2022) Exploring the effect of Jasmonic Acid for Aphids control for improving the yield of Triticum aestivum varieties. PeerJ 10:e14018

Ayub J, Tahir A, Iqbal U, Ayub H, Hyder MZ, Kiyani A et al (2024) Effective biological control of Lentil (Lens culinaris) Fusarium wilt and plant growth promotion through native Rhizobacteria. Physiol Mol Plant Pathol 129:102203

Barghi A, Jung HW (2024) Insights into Bacillus zanthoxyli HS1-mediated systemic tolerance: multifunctional implications for enhanced plant tolerance to abiotic stresses. Physiol Plant 176(4):e14458

Caffarra A, Rinaldi M, Eccel E, Rossi V, Pertot I (2012) Modelling the impact of climate change on the interaction between grapevine and its pests and pathogens: European grapevine moth and powdery mildew. Agric Ecosyst Environ 148:89–101

Canton H (2021) Food and agriculture organization of the United Nations—FAO. In: The Europa directory of international organizations 2021. Routledge, pp 297–305

Cao D, Zhang J, Han J, Zhang T, Yang S, Wang J, Prodhan FA, Yao F (2022) Projected increases in global terrestrial net primary productivity loss caused by drought under climate change. Earth S Future 10(7). https://doi.org/10.1029/2022ef002681

Chacón-Cerdas R, Barboza-Barquero L, Albertazzi FJ, Rivera-Méndez W (2020) Transcription factors controlling biotic stress response in potato plants. Physiol Mol Plant Pathol 112:101527. https://doi.org/10.1016/j.pmpp.2020.101527

Chang PE, Wu YH, Tai CY, Lin IH, Wang WD, Tseng TS, Chuang HW (2023) Examining the transcriptomic and biochemical signatures of Bacillus subtilis Strains: impacts on plant growth and abiotic stress tolerance. Int J Mol Sci 24(18):13720

Chen Q, Timmermans J, Wen W, Van Bodegom PM (2023) Ecosystems threatened by intensified drought with divergent vulnerability. Remote Sens Environ 289:113512. https://doi.org/10.1016/j.rse.2023.113512

Damodaran T, Jha SK, Kumari S, Gupta G, Mishra VK, Sharma PC, Gopal R, Singh A, Jat HS (2023) Development of halotolerant microbial consortia for salt stress mitigation and sustainable tomato production in Sodic soils: an enzyme mechanism approach. Sustain For 15(6):5186. https://doi.org/10.3390/su15065186

Danish S, Zafar-ul-Hye M, Mohsin F, Hussain M (2020) ACC-deaminase producing plant growth promoting rhizobacteria and biochar mitigate adverse effects of drought stress on maize growth. PLoS One 15(4):e0230615

Das J, Dangar TK, Adhya TK, Panigrahy M (2022) Plant growth promoting and antimicrobial functionalities of rhizosphere Pseudomonas species in flooded rice soils of coastal Odisha. Environ Qual Manag 32(2):125–137. https://doi.org/10.1002/tqem.21926

De Andrade LA, Santos CHB, Frezarin ET, Sales LR, Rigobelo EC (2023) Plant growth-promoting rhizobacteria for sustainable agricultural production. Microorganisms 11(4):1088

De Serrano LO, Camper AK, Richards AM (2016) An overview of siderophores for iron acquisition in microorganisms living in the extreme. BioMetals 29(4):551–571. https://doi.org/10.1007/s10534-016-9949-x

de Lima BC, Bonifacio A, de Alcantara Neto F, Araujo FF, Araujo ASF (2022) Bacillus subtilis rhizobacteria ameliorate heat stress in the common bean. Rhizosphere 21:100472

Dubey A, Kumar K, Srinivasan T, Kondreddy A, Kumar KRR (2022) An invasive weed-associated bacterium confers enhanced heat stress tolerance in wheat. Heliyon 8(7):e09893

Dutta S, Podile AR (2010) Plant Growth Promoting Rhizobacteria (PGPR): the bugs to debug the root zone. Crit Rev Microbiol 36(3):232–244. https://doi.org/10.3109/10408411003766806

Egamberdieva D, Wirth S, Bellingrath-Kimura SD, Mishra J, Arora NK (2019) Salt-tolerant plant growth promoting Rhizobacteria for enhancing crop productivity of saline soils. Front Microbiol 10:2791. https://doi.org/10.3389/fmicb.2019.02791

González F, Santander C, Ruiz A, Pérez R, Moreira J, Vidal G et al (2023) Inoculation with Actinobacteria spp. isolated from a hyper-arid environment enhances tolerance to salinity in lettuce plants (Lactuca sativa L.). Plants 12(10):2018

Gowtham HG, Singh SB, Shilpa N, Aiyaz M, Nataraj K, Udayashankar AC, Amruthesh KN, Murali M, Poczai P, Gafur A, Almalki WH, Sayyed RZ (2022) Insight into recent progress and perspectives in improvement of antioxidant machinery upon PGPR augmentation in plants under drought stress: a review. Antioxidants 11(9):1763. https://doi.org/10.3390/antiox11091763

Gupta A, Bano A, Rai S, Dubey P, Khan F, Pathak N, Sharma S (2021) Plant Growth Promoting Rhizobacteria (PGPR): a sustainable agriculture to rescue the vegetation from the effect of biotic stress: a review. Lett Appl NanoBioSci 10(3):2459–2465. https://doi.org/10.33263/lianbs103.24592465

Han L, Hu L, Lv Y, Li Y, Ma Z, Li B, Gao P, Liang Y, Zhao X (2024) Effects of Bacillus amyloliquefaciens QST713 on mineral nutrient utilization of Alfalfa (Medicago sativa L.) under drought stress. Agronomy 14(8):1793. https://doi.org/10.3390/agronomy14081793

Hashem A, Tabassum B, Allah FA, E. (2019) *Bacillus subtilis*: a plant-growth promoting rhizobacterium that also impacts biotic stress. Saudi J Biol Sci 26(6):1291–1297. https://doi.org/10.1016/j.sjbs.2019.05.004

Hein NT, Ciampitti IA, Jagadish SK (2021) Bottlenecks and opportunities in field-based high-throughput phenotyping for heat and drought stress. J Exp Bot 72(14):5102–5116

Hyder S, Gondal AS, Rizvi ZF, Atiq R, Haider MIS, Fatima N, Inam-Ul-Haq M (2021) Biological control of chili damping-off disease, caused by *Pythium myriotylum*. Front Microbiol 12:587431. https://doi.org/10.3389/fmicb.2021.587431

Islam MR, Naveed SA, Li Z, Zhao X, Fiaz S, Zhang F, Wu Z, Hu Z, Fu B, Shi Y, Shah SM, Xu J, Wang W (2022) Identification of candidate genes for salinity and anaerobic tolerance at the germination stage in rice by genome-wide association analyses. Front Genet 13

Jan F, Arshad H, Ahad M, Jamal A, Smith DL (2023) In vitro assessment of Bacillus subtilis FJ3 affirms its biocontrol and plant growth promoting potential. Front Plant Sci 14. https://doi.org/10.3389/fpls.2023.1205894

Ji H, Qi Y, Zhang X, Yang G (2024) Transcriptomics and metabolomics analysis revealed the ability of microbacterium ginsengiterrae S4 to enhance the saline-alkali tolerance of Rice (Oryza sativa L.) seedlings. Agronomy 14(4):649. https://doi.org/10.3390/agronomy14040649

Jing D, Liu B, Ma H, Liu F, Liu X, Ren L (2023) Effects of inoculation with different plant growth-promoting Rhizobacteria on the eco-physiological and stomatal characteristics of walnut seedlings under drought stress. Agronomy 13(6):1486. https://doi.org/10.3390/agronomy13061486

Juroszek P, von Tiedemann A (2013) Climate change and potential future risks through wheat diseases: a review. Eur J Plant Pathol 136:21–33

Kaur R, Sharma N, Tikoria R, Ali M, Kour S, Kumar D, Ohri P (2022) Insights into biosynthesis and signaling of Cytokinins during plant growth, development and stress tolerance. In: Signaling and communication in plants, pp 153–187. https://doi.org/10.1007/978-3-031-05427-3_7

Khan MA, Raheel M, Khan SA, Abid AD, Shahzad S, Siddiqui HZ, Hanif A (2023) Eco-friendly management of wheat stripe rust through application of Bacillus subtilis in combination with plant defense activators. J King Saud Univ Sci 35(4):102587

Khoso MA, Wang M, Zhou Z, Huang Y, Li S, Zhang Y, Qian G, Ko SN, Pang Q, Liu C, Li L (2024) Bacillus altitudinis AD13–4 enhances saline–alkali stress tolerance of alfalfa and affects composition of rhizosphere soil microbial community. Int J Mol Sci 25(11):5785. https://doi.org/10.3390/ijms25115785

Kumar A, Bhattacharya T, Mukherjee S, Sarkar B (2022) A perspective on biochar for repairing damages in the soil–plant system caused by climate change-driven extreme weather events. Biochar 4(1). https://doi.org/10.1007/s42773-022-00148-z

Kumar U, Kaviraj M, Kundu S, Rout S, Priya H, Nayak AK (2023) Microbial alleviation of abiotic and biotic stresses in rice. In: Sustainable agriculture reviews, pp 243–268. https://doi.org/10.1007/978-3-031-24181-9_10

Lahlali R, Taoussi M, Laasli SE, Gachara G, Ezzouggari R, Belabess Z, Barka EA (2024) Effects of climate change on plant pathogens and host-pathogen interactions. Crop Environ 3(3):159–170

Leisner CP, Potnis N, Sanz-Saez A (2022) Crosstalk and trade-offs: plant responses to climate change-associated abiotic and biotic stresses. Plant Cell Environ 46(10):2946–2963. https://doi.org/10.1111/pce.14532

Li SS, Li SY, Zhang ML, Wang D, Li S, Zhai Y (2023) Bacillus licheniformis (BE-L60) improved tolerance of spinach seedlings against heat stress. Russ J Plant Physiol 70(4):65

Liu F, Ma H, Liu B, Du Z, Ma B, Jing D (2023) Effects of plant growth-promoting rhizobacteria on the physioecological characteristics and growth of walnut seedlings under drought stress. Agronomy 13(2):290. https://doi.org/10.3390/agronomy13020290

Luo H, Win CS, Lee DH, He L, Yu JM (2024) Microbacterium azadirachtae CNUC13 enhances salt tolerance in maize by modulating osmotic and oxidative stress. Biology 13(4):244. https://doi.org/10.3390/biology13040244

Magallón-Andalón CG, Calderón-Santoyo M, Balois-Morales R, Ochoa-Jiménez VA, Casas-Junco PP, López-Guzmán GG, Bautista-Rosales PU (2025) Exploring the biocontrol potential of Bacillus thuringiensis against Colletotrichum gloeosporioides in'Hass' avocado fruits. Physiol Mol Plant Pathol 102724

Mahdi I, Allaoui A, Fahsi N, Biskri L (2022) Bacillus velezensis QA2 potentially induced salt stress tolerance and enhanced phosphate uptake in quinoa plants. Microorganisms 10(9):1836

Mahmood S, Ahmad W, Ali Z, Eed EM, Khalifa AS, Naeem M et al (2022) Exploring the potential of Moringa leaf extract for mitigation of cadmium stress in Triticum aestivum L. Appl Sci 12(16):8199

Marin VR, Ferrarezi JH, Vieira G, Sass DC (2019) Recent advances in the biocontrol of Xanthomonas spp. World J Microbiol Biotechnol 35(5):72

Mehmood S, Muneer MA, Tahir M, Javed MT, Mahmood T, Afridi MS et al (2021) Deciphering distinct biological control and growth promoting potential of multi-stress tolerant Bacillus subtilis PM32 for potato stem canker. Physiol Mol Biol Plants 27:2101–2114

Mehrabi SS, Sabokdast M, Bihamta MR, Dedičová B (2024) The coupling effects of PGPR inoculation and foliar spraying of strigolactone in mitigating the negative effect of salt stress in wheat plants: insights from phytochemical, growth, and yield attributes. Agriculture 14(5):732. https://doi.org/10.3390/agriculture14050732

Michaletz ST, Weiser MD, McDowell NG, Zhou J, Kaspari M, Helliker BR, Enquist BJ (2016) The energetic and carbon economic origins of leaf thermoregulation. Nat Plants 2(9):1–9

Mohamed AAE, Fiaz S, Ali M, Abdallah AA, Ahmed SI (2021) Performance of some Rice (Oryza sativa L.) cultivars under water shortage and high temperature stress. Sanis Malysia 50(3):617–628. http://journalarticle.ukm.my/16903/

Morcillo RJL, Manzanera M (2021) The effects of Plant-Associated bacterial exopolysaccharides on plant abiotic stress tolerance. Meta 11(6):337. https://doi.org/10.3390/metabo11060337

Nemeskéri E, Horváth KZ, Andryei B, Ilahy R, Takács S, Neményi A, Pék Z, Helyes L (2022) Impact of plant growth-promoting Rhizobacteria inoculation on the physiological response and productivity traits of field-grown tomatoes in Hungary. Horticulturae 8(7):641. https://doi.org/10.3390/horticulturae8070641

Ning Z, Lin K, Gao M, Han X, Guan Q, Ji X, Yu S, Lu L (2024) Mitigation of salt stress in rice by the halotolerant plant growth-promoting bacterium Enterobacter asburiae D2. Journal of Xenobiotics 14(1):333–349. https://doi.org/10.3390/jox14010021

Palacio-Rodríguez R, Sáenz-Mata J, Trejo-Calzada R, Ochoa-García PP, Arreola-Ávila JG (2023) Halotolerant Rhizobacteria promote plant growth and decrease salt stress in Carya illinoinensis (Wangenh.) K. Koch. Agronomy 13(12):3045. https://doi.org/10.3390/agronomy13123045

Paliwoda D, Mikiciuk G, Mikiciuk M, Kisiel A, Sas-Paszt L, Miller T (2022) Effects of rhizosphere bacteria on strawberry plants (Fragaria × ananassa Duch.) under water deficit. Int J Mol Sci 23(18):10449. https://doi.org/10.3390/ijms231810449

Qadir M, Quillérou E, Nangia V, Murtaza G, Singh M, Thomas RJ et al (2014) Economics of salt-induced land degradation and restoration. In: Natural resources forum, vol 38, No. 4. Wiley, pp 282–295

Rai A, Santana MM, Maia RN, Tavares J, Nabti E, Cruz C (2023) Bacterial inoculation and extracts of opuntia rackets or marine algae trigger distinct proline balances in tomato salt Stress alleviation. Agronomy 13(12):2921. https://doi.org/10.3390/agronomy13122921

Ramakrishnan P, Ariyan M, Rangasamy A, Rajasekaran R, Ramasamy K, Murugaiyan S, Janahiraman V (2023) Draft genome sequence of Enterobacter cloacae S23 a plant growth-promoting passenger endophytic bacterium isolated from groundnut nodule possesses stress tolerance traits. Curr Genomics 24(1):36

Saddiq MS, Wang X, Iqbal S, Hafeez MB, Khan S, Raza A, Iqbal J, Maqbool MM, Faiz S, Qazi MA, Bakhsh A, Zaib J, Gulshan AB (2021) Effect of water stress on grain yield and physiology characters of Quinoa (Chenopodium quinoa) genotypes. Agronomy (Basel) 11(10):1934. https://doi.org/10.3390/agronomy11101934

Saleh AE, Ul-Hassan Z, Zeidan R, Al-Shamary N, Al-Yafei T, Alnaimi H, Jaoua S (2021) Biocontrol activity of Bacillus megaterium BM344-1 against toxigenic fungi. ACS Omega, 6(16):10984–10990

Salem MA, Ismail MA, Radwan KH, Abd-Elhalim HM (2024) Unlocking the potential of plant growth-promoting Rhizobacteria to enhance drought tolerance in Egyptian Wheat (Triticum aestivum). Sustain For 16(11):4605. https://doi.org/10.3390/su16114605

Sammauria R, Kumawat S, Kumawat P, Singh J, Jatwa TK (2020) Microbial inoculants: potential tool for sustainability of agricultural production systems. Arch Microbiol 202(4):677–693

Sanghavi K, Sanghavi M, Rajurkar AM (2021) Early stage detection of Downey and Powdery Mildew grape disease using atmospheric parameters through sensor nodes. Artificial Intelligence in Agriculture 5:223–232

Santhanam R, Luu VT, Weinhold A, Goldberg J, Oh Y, Baldwin IT (2015) Native root-associated bacteria rescue a plant from a sudden-wilt disease that emerged during continuous cropping. Proceedings of the National Academy of Sciences 112(36):E5013–E5020

Seleiman MF, Al-Suhaibani N, Ali N, Akmal M, Alotaibi M, Refay Y, Dindaroglu T, Abdul-Wajid HH, Battaglia ML (2021) Drought stress impacts on plants and different approaches to alleviate its adverse effects. Plan Theory 10(2):259. https://doi.org/10.3390/plants10020259

Shaffique S, Shah AA, Peter O, Injamum-Ul-Hoque M, Elansary HO, Al Azzawi TNI et al (2024) The rhizobacterial Priestia megaterium strain SH-19 mitigates the hazardous effects of heat stress via an endogenous secondary metabolite elucidation network and molecular regulation signalling. BMC Plant Biol 24:827

Shah A, Nazari M, Antar M, Msimbira LA, Naamala J, Lyu D, Rabileh M, Zajonc J, Smith DL (2021) PGPR in agriculture: a sustainable approach to increasing climate change resilience. Front Sustain Food Sys 5. https://doi.org/10.3389/fsufs.2021.667546

Shahid M, Ameen F, Maheshwari HS, Ahmed B, AlNadhari S, Khan MS (2021) Colonization of Vigna radiata by a halotolerant bacterium Kosakonia sacchari improves the ionic balance, stressor metabolites, antioxidant status and yield under NaCl stress. Appl Soil Ecol 158:103809

Shi L, Zhu X, Qian T, Du J, Du Y, Ye J (2023) Mechanism of salt tolerance and plant growth promotion in Priestia megaterium ZS-3 revealed by cellular metabolism and whole-genome studies. Int J Mol Sci 24(21):15751. https://doi.org/10.3390/ijms242115751

Shrivastava P, Kumar R (2015) Soil salinity: a serious environmental issue and plant growth promoting bacteria as one of the tools for its alleviation. Saudi journal of biological sciences 22(2):123–131

Soroye P, Newbold T, Kerr J (2020) Climate change contributes to widespread declines among bumble bees across continents. Science 367(6478):685–688

Srivastava R, Panda R, Chakraborty A (2020) Assessment of climate change impact on maize yield and yield attributes under different climate change scenarios in eastern India. Ecol Indic 120:106881. https://doi.org/10.1016/j.ecolind.2020.106881

Tahir A, Kang J, Ali J, Bibi A, Abbas S (2022a) Nanotechnology for future sustainable plant production under changing environmental conditions. In: Bioresource technology: concept, tools and experiences, pp 466–492

Tahir AT, Kang J, Bint-e-Mansoor M, Ayub J, Naureen Z, Hafeez FY (2022b) Recent trends in characterization of endophytic microorganisms. In: Biocontrol mechanisms of endophytic microorganisms. Academic Press, pp 31–53

Thakur R (2024) Management of bacterial leaf blight of rice caused by Xanthomonas oryzae using organinc approaches (Doctoral dissertation, Kashmir University)

Thuraga V, Ghadamgahi F, Dadi FA, Vetukuri RR, Chawade A (2024) A new bacterial consortia for management of Fusarium head blight in wheat. Sci Rep 14(1):10131

Tirry N, Ferioun M, Kouchou A, Laghmari G, Bahafid W, El Ghachtouli N (2022) Enhanced Salinity tolerance of Medicago sativa, roots AM colonization and soil enzyme activities by PGPR. Environmental Sciences Proceedings 16(1):14

Toral L, Rodríguez M, Béjar V, Sampedro I (2020) Crop protection against *Botrytis cinerea* by Rhizhosphere biological control agent *Bacillus velezensis* XT1. Microorganisms 8(7):992. https://doi.org/10.3390/microorganisms8070992

Ullah MA, Hyder SI, Mahmood IA, Sultan T, Badshah L (2017) Inducing salt tolerance in wheat through inoculation with rhizobacteria containing 1-aminocyclopropane-1-carboxylate deaminase activity. J Agri Sci Pract 2(4):54–57. https://doi.org/10.31248/jasp2017.055

Van Dijk M, Morley T, Rau ML, Saghai Y (2021) A meta-analysis of projected global food demand and population at risk of hunger for the period 2010-2050. Nature food 2(7):494–501. https://doi.org/10.1038/s43016-021-00322-9

Wei X, Mira A, Yu Q, Gmitter FG (2021) The mechanism of citrus host defense response repression at early stages of infection by feeding of Diaphorina citri transmitting Candidatus Liberibacter asiaticus. Front Plant Sci 12. https://doi.org/10.3389/fpls.2021.635153

Yadav V, Yadav RC, Choudhary P, Sharma SK, Bhagat N (2022) Mitigation of drought stress in wheat (Triticum aestivum L.) by inoculation of drought tolerant Bacillus paramycoides DT-85 and Bacillus paranthracis DT-97. J Appl Biol Biotechnol 10(1):59–69

Yang B, Wen H, Wang S, Zhang J, Wang Y, Zhang T, Yuan K, Lu L, Liu Y, Xue Q, Shan H (2024) Enhancing drought resistance and yield of wheat through inoculation with Streptomyces pactum Act12 in drought field environments. Agronomy 14(4):692. https://doi.org/10.3390/agronomy14040692

Yong D, Li Y, Gong K, Yu Y, Zhao S, Duan Q, Li R (2022) Biocontrol of strawberry gray mold caused by Botrytis cinerea with the termite associated Streptomyces sp. sdu1201 and actinomycin D. Front Microbiol 13:1051730

Zakaria L (2023) Fusarium species associated with diseases of major tropical fruit crops. Horticulturae 9(3):322

Zhang X, Liu W, Lv Y et al (2022) Effects of drought stress during critical periods on the photosynthetic characteristics and production performance of Naked oat (Avena nuda L.). Sci Rep 12:11199. https://doi.org/10.1038/s41598-022-15322-3

Slow Release Nitrogen Fertilizer for Better Crop Yield: A Tool of Food Security

11

Amir Aziz, Mukkram Ali Tahir, and Noor-Us-Sabah

Abstract

Chemical fertilizers are widely used in modern agriculture. Fertilizers rich in nitrogen (N) are of special attention. However, numerous issues arise when nitrogen fertilizer is introduced into the soil. Significant losses occur due to soil-climate environment as volatilization, nitrification, and subsequent denitrification processes. Different strategies have been employed in the past to cope with the issues. Several organic and inorganic materials are used for the management of N dynamics in the environment with high efficiency. With the increase in temperature and moisture levels, the ammonium and nitrate release was also increased. However, ammonium and nitrate release from coated urea was significantly lower under different temperatures and moisture levels. Coated urea enhanced the growth and yield of crops significantly. Thus, using organic material as coating material had the potential to not only enhance the growth, yield, and nitrogen use efficiencies but also control negative effects on the environment. Issues related to urea utilization efficiency could be managed very effectively by using environmentally notorious materials like neem oil. This will not only lead to more fertilizer use efficiency but also help to manage the environmental problems originating due to ammonia emission and nitrate leaching into groundwater. This approach has the potential to increase the farmer's income with minimum loss to the environment.

A. Aziz (✉) · M. A. Tahir
Department of Soil and Environmental Sciences, College of Agriculture, University of Sargodha, Sargodha, Pakistan

Noor-Us-Sabah
Department of Soil and Environmental Sciences, College of Agriculture, University of Sargodha, Sargodha, Pakistan

CFAES Rattan Lal Center for Carbon Management and Sequestration, The Ohio State University, Columbus, OH, USA

© The Author(s), under exclusive license to Springer Nature Singapore Pte Ltd. 2025

S. Faiz et al. (eds.), *Climate Smart Agriculture for Future Food Security*,
https://doi.org/10.1007/978-981-96-4499-5_11

Keywords

Slow-release fertilizer · Coated urea · Crop yield · Food security

11.1 Role of Nitrogen on Plant Productivity

Nitrogen is one of the vital nutrients needed by the plant's growth, development, structural and functional systems. It comprises 1–5% of plant dry weight and most plant growth depends upon the supply of nitrogen (Shahid et al. 2020; Aziz et al. 2024). When inorganic nitrogen is assimilated, the first products that are formed are called amides and amino acids. The protoplasm is the place where chemical reactions take place. The majority of these reactions are responsible for improvements in the expansion and maturation of plant life (Schuler et al. 2012). Protein synthesis relies on a group of chemicals called amino acids. Nitrogen is a fundamental element of amino acids, which are the primary structural units of proteins and enzymes in plants. Enzymes catalyze the myriad biochemical events occurring within a plant, and proteins serve as the building blocks of all living things. Chlorophyll requires nitrogen, and when plants don't have enough of it, they turn yellow (McCauley et al. 2009). N is important in sucrose synthesis and maintenance (Guendouz et al. 2012).

Nitrogen (N) is used by all plant types, namely as NO_3^- and NH_4^+. It plays a crucial role in the biochemical processes of plants, allowing for their optimal growth and development, which in turn increases yield and improves the quality of that produce (Fiaz et al. 2021). As a result of its high demand, pivotal N must be provided to plants at a rate of around 1000 ug kg^{-1} dry matter. The Pakistani soils having low organic content is the primary cause of this deficiency, but there are also other issues at play, including losses due to volatilization, leaching, denitrification, and crop removals. About 78–79% of the atmospheric N is in a form (N_2) that plants cannot utilize; therefore, this portion is not taken up directly. Fixation of N from the atmosphere, biology, and organic sources all contribute to the global supply. Therefore, the best nutrient management practices will be technically advanced and provide maximum benefit. The importance of nitrogen to plants is illuminated in this review article, which will be of great use to farmers, students, and scientists. In addition, this exhaustive analysis drew the conclusion that nitrogen is of tremendous value to farmers. Due to incorrect and non-judicious usage of N, limited yield with lower-than-average quality of crops occurs globally. This is caused by a lack of understanding. (Leghari et al. 2016).

11.2 Urea: A Novel and Widespread Fertilizer

Ullah et al. (2019) conducted field trials in 2014, 2015, and 2016 at Cotton Research Station, Dera Ismail Khan, Pakistan to discover N management practices for ZT cotton grown into standing wheat straw. The experiment used a randomized complete block (RCB) design with three repetitions. We must find a technique for increasing the effectiveness of nitrogenous fertilizers that have been administered. There have been a variety of methods tested to increase efficiency, including split application, deep placement, organic fertilizers, and liquid fertilizers. Slow and controlled-release fertilizers, for example, are currently in use in industrialized countries as a modern strategy to improve nitrogen use efficiency.

Shahzad et al. (2019) conducted a long-term fertilizer experiment from 1961 to 2014 in Pakistan, nitrogen fertilizer uses and N excess grew at a much faster pace than mean yields; however, NUE and PFP continued to decline to much lower levels. As in many other places, increasing NUE in Pakistan can be achieved through other agronomic practices, such as better management of N, the use of high-efficiency fertilizers, and a more balanced approach to crop nutrition. Overuse of nitrogen (N) fertilizer, a crucial productivity factor in world cropping systems, is detrimental to the sustainability of crop production from an economic and environmental perspective. Mean yield, N fertilizer use, partial factor productivity (PFP) of N, N use efficiency (NUE), and N surplus in wheat, cotton, and rice production in Pakistan, which account for 75% of total national fertilizer N use, were compared with those for the top producer countries for 2014. Pakistan was among the top four in terms of N use but had low mean yields, with the lowest PFP and NUE and the highest N surplus.

11.3 Issue of Urea in Soil-Plant System

Urea fertilizer is produced commercially by heating a mixture of highly compressed liquid ammonia and gaseous carbon dioxide. Following dehydration, ammonium carbamate is processed in a granulator or prilling tower to form granules or prills. When compared to other solid nitrogenous fertilizers, urea artificial fertilizer has the highest nitrogen content (46%). In addition, it has the maximum solubility in water. Urea is superior to other nitrogenous fertilizers in terms of the amount of nitrogen it supplies to plants and soil due to its high nitrogen concentration. When nitrogen is obtained from living organisms like plants and animals, it is considered a natural fertilizer and can be broken down by the environment. Artificial or synthetic substances are produced by the use of industrial methods (Olson and Kurtz 1982).

How well urea fertilizer performs over a given time period is based on how much nitrogen is taken up by the crop. Recent years have seen a sustained interest in research into methods to improve the effectiveness of urea application to soils and crops (Dobermann 2005). Recent years have seen the development of a number of new approaches to improving urea performance (Grant and Wu 2008). Slow-release or controlled-release mechanisms can help with the problem of fast release and losses in conventional urea fertilizer. The best technique to give crops nitrogen on a consistent basis to fulfill the iron-going requirements is through the delayed release of nutrients via fertilizers. This procedure improves the efficiency of the applied urea and leads to reasonable crop development (Bröckel and Hahn 2004; Shaviv 2005; Aziz et al. 2024).

11.4 Coating of Urea to Make Intelligent Fertilizer

Nitrogen in the soil is released slowly because urea's covering does not break down completely and contains water-repellent elements. The covering material becomes more porous over time, allowing the applied fertilizer's nitrogen to be released rapidly into the soil. Coating materials have been shown to have qualities between those of hydrophobic and hydrophilic in a number of investigations. This leads to nitrogen being released from the fertilizer as water penetrates the covered surface. The coating substance is water-repellent and water-resistant, which slows down nutrient release (Choi and Meisen 1997; Chen et al. 2010; Aziz et al. 2024).

To ensure that only one application provides all of the minerals required for optimal crop growth, the ideal delayed-release fertilizer is coated with a biodegradable organic or inorganic polymer that slows down its release. The phrase "Enhanced Efficiency Fertilizers" (EEF) could be used interchangeably with the words "Controlled Release Fertilizers" (CRF) and "Slow Release Fertilizers" (SRF). However, ME-Trenkel (2021) and Shaviv (2005) explained the distinction between them. The soil type, the character of the soil, and the climate in the area all play a role in the mechanism by which a mineral from a slow-release fertilizer is released. The SRF mechanism operates based on chance alone. The process, amount, and rate of release in the soil are completely predictable for a controlled-release fertilizer (CRF) (Blouin et al. 1971; Aziz et al. 2024).

11.5 Working of Slow Release Fertilizer

In an effort to lessen the buildup of inorganic N in the soil solution and the risk of N losses, nitrogen fertilizers are developed to discharge N into the soil solution at a rate that more closely matches nutrient absorption by the crop (Shoji et al. 2001). Polymer-coated urea solutions have become more affordable recently, prompting serious consideration of their use in a larger variety of grain and oilseed cropping systems (Chen et al. 2008; Grant and Wu 2008; Nelson and Spaner 2010; Wang et al. 2015). The effectiveness of controlled-release N fertilizers is debated, despite the fact that they have been proven to boost crop productivity and nitrogen accumulation. Experiments conducted in the Mixed Grass ecoregion of the Canadian prairies found that using a polymer-coated CRU in winter wheat production increased seedling safety with seed-placed fertilizer applications, but did not increase crop yield when the fertilizers were side-banded away from the seed row at the time of seeding or broadcast on the soil surface (McKenzie et al. 2007). Polymer-coated urea (PCU) and other slow-release fertilizers are examples of better N management practices that have been investigated for their potential to boost agricultural production while reducing environmental N losses.

In comparison to urea, Shoji et al. (2001) found that using an SRF source improved N usage efficiency in maize, barley, and potato, and reduced soil N_2O emissions. Decreased N_2O emissions (Vallejo et al. 2001; McTaggart and Tsuruta 2003) and reductions in NO_3 leaching losses (Wang and Alva 1996) switching from fast-acting fertilizers like urea to SRF. It is possible to choose an enhanced efficiency fertilizer that is most suited to the growing conditions and cropping method at hand from among several distinct options. By using conventional N, environmental N loss in climates with high variability may be mitigated through the use of these fertilizers. If these fertilizers work as advertised, environmental N losses could be reduced by reducing the amount of N fertilizer used (Motavalli et al. 2008; Aziz et al. 2024). However, there are a number of obstacles that prevent the widespread use of these fertilizers in agronomic applications. These include the fertilizer's higher price compared to conventional options, the potential need for specialized handling and storage methods, and the absence of recommendations regarding whether fertilizers would be most effective.

Natural organic extract also contains different kinds of chemicals that act as nitrification inhibitors including Neem oil, Parthenium oil, Citronella oil, and many others (Afzal et al. 2021). Neem oil is one of the most common examples of trees that reduces nitrification and is excessively used in India (Malla et al. 2005). Weeds such as Parthenium and Palmarosa also contain compounds that successfully decline nitrogen losses by stopping the nitrification process (Malla et al. 2005).

Urease inhibitors (UI) slow urea hydrolysis, which reduces the pool of NH_4 or the potential for volatilization loss, thus reducing NH_3 emissions (Turner et al. 2010). Urease inhibitors had different effects depending upon the edaphic conditions of the plant (Suter et al. 2013). Nitrification inhibitors slow down the activities of Nitrobacter and, as a result, limit nitrate formation (Kim et al. 2012). In Asian countries like Pakistan and India nitrification inhibitors remain unacceptable mainly because of high cost and restricted availability to the common farmers (Kumar et al. 2017).

A 2006 study by Cui and coauthors suggests that water and soil pollution can be mitigated by using nanotechnology to boost plant uptake of fertilizer nutrients and decrease the pace at which they are lost via leaching (Johnson 2006). Naderi and Danesh Shahraki (2011) state that Nano-fertilizers reduce soil pollution, shorten the amount of time it takes to apply fertilizer and boost nutrient utilization efficiency. Nanoparticles and Nanocapsules can be used to create effective fertilizers. DeRosa et al. (2010) and Rahimpour et al. (2010) reported that the use of nano-fertilizers greatly enhances both NUE and crop productivity.

11.6 Means of Controlling N Losses

Non-leguminous crops may one day be able to fix their own nitrogen from the air (Dent and Cocking 2017; Hopkin et al. 2006; Beatty and Good 2011), as a result, they won't need Haber-Bosch N. However, the most important objective for the time being is to identify strategies to increase N use efficiency while lowering losses (Erisman et al. 2008; Bodirsky et al. 2014). Implementing innovative fertilizer use methods, which can be characterized by the 4R principle (right time (application), right placement (e.g., incorporation below soil surface), right amount (rate), and right product (e.g., urea, ammonia, or nitrate) is one strategy to pursue greater NUE, which has long been supported and driven by management (Flis 2017; Davidson et al. 2018). Another strategy that has gotten a lot of attention is the creation of enhanced efficiency nitrogen fertilizers (EEFs) (Smith et al. 2007; Sher et al. 2021). The main idea behind EEFs is to use inhibitors or coating materials to decrease the release rate of N or to slow down the processes of N transformation, resulting in greater synchrony between crop uptake and N release, lowering losses and increasing NUE (Aziz et al. 2024). There have been hundreds of EEF products since their beginning in the 1960s. It's possible to categorize these products into four groups: A partially permeable coating is used to limit nitrogen release in PCFs (polymer-coated fertilizers). Nitrous oxide emission and nitrate leaching decreased by inhibiting the activity of nitrifying bacteria (Chien et al. 2009; Akiyama et al. 2013; Aziz et al. 2024). Urea hydrolysis is delayed by urease inhibitors, reducing the ammonia emission potential. There are double inhibitors that are vital to reduce both N_2O and NH_3 emissions.

1. **Slow/controlled Release Fertilizers**

Granules of CRFs are inserted with excipients to restrict nutrient release, eventually increasing nutrient availability for crops while decreasing health, ecological, and environmental concerns. It is a more sophisticated approach to feeding nutrients to crops (compared to previous ways) because of the progressive pattern of nutrient release, which enhances fertilizer usage efficiency (FUE) in that way (Subbarao et al. 2013). CRFs enable nutrients to be released over a longer length of time, allowing for more control over the pace and pattern of release, relating to coating thickness, by controlling the time it takes nutrients to be released into the soil, excipients reduce the need for frequent fertilization while increasing the overall effectiveness of the fertilizer (Dubey and Mailapalli 2019; Aziz et al. 2024).

Guo et al. (2006) devised that three processes (Fig. 11.1) are used to release nitrogen from urea-formaldehyde (UF). *Step one:* The covering materials absorb

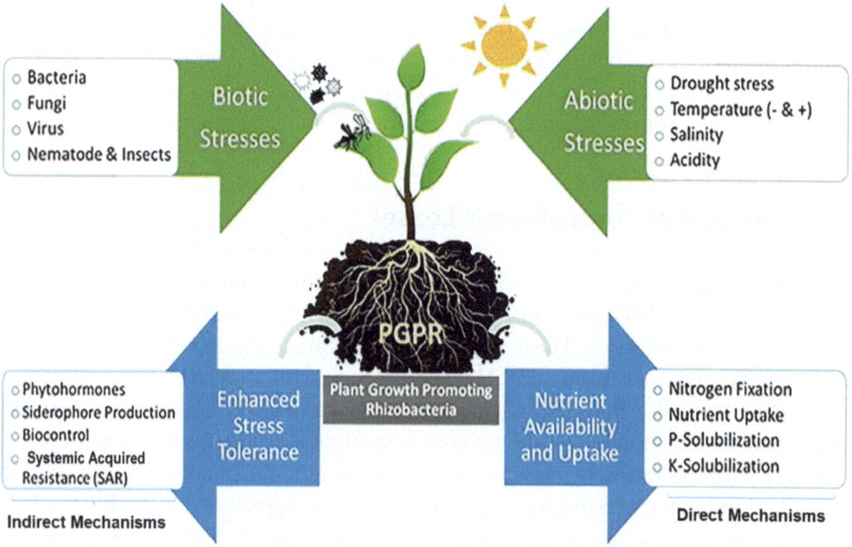

Fig. 11.1 Steps involved in the release of nutrients from coated fertilizer (Sempeho et al. 2014)

water from the soil and become hydrogels that expand the orifice size of 3D networks, allowing more fertilizer to be absorbed into the gel network's core. A water film develops between the inflated coats and the center of the fertilizer granules. *Step two:* Weakly cross-linked polymer networks allow soluble fertilizer to gently diffuse into the soil. This occurs because of a dynamic exchange between the hydrogel and soil waters, which causes fertilizer to slowly permeate into the soil. *Step three:* It is via this dynamic interchange that the fertilizer granules' insoluble nitrogen is broken down into urea and ammonia, which are then progressively released into the soil. The lag time, linear stage, and decay period have been alluded to in other research. (Du et al. 2006).

2. **Nitrification Inhibitors**

A huge proportion of nitrogen is in the form of urea in animal urine, which is promptly hydrolyzed into ammonium $\left(NH_4^+\right)$ when it enters the soil:

$$\left(NH_2\right)_2CO + 2H_2O\,\text{Urease} \rightarrow \left(NH_4\right)_2CO_3 \rightarrow NH_4 + NH_3(\text{gas}) + CO_2 + OH^-\left(\text{Urea}\right)$$

The negatively charged soil cation exchange complex, notably soil clays, and organic matter adsorbs and holds most of the ammonium $\left(NH_4^+\right)$ cation, which is lost by volatilization. As a consequence, NH_4^+ leaching is seldom an issue in soils with sufficient cation exchange capacity (CEC). However, the nitrification process in soil quickly oxidizes NH_4^+ to hydroxylamine, NO_2, and eventually NO_3: It is via this dynamic interchange that the fertilizer granules' insoluble nitrogen is broken down into urea and ammonia, which are then progressively released into the soil. The lag time, linear stage, and decay period have been eluded.

$$NH_3 + O_2 + 2H^+ + 2e- \rightarrow NH_2OH + H_2O$$

$$NH_2OH + H_2O \rightarrow NO_2^- + 5H^+ + 4e-$$

$$2NO_2^- - + O_2 \rightarrow 2NO_3^-$$

Negatively charged soil colloids do not absorb nitrate since it is an anion. Contaminated water is discharged into nearby bodies of water while drainage takes place. Nitrous oxide is also created via nitrification, which occurs when intermediates like nitrite or hydroxylamine decompose, by Nitrifier denitrification, which is derived from autotrophic ammonia oxidizers (Wrage et al. 2001). NO_3 generated by nitrification is also subject to denitrification in which NO_3 is reduced sequentially to N_2O, NO, and N_2 as an intermediate product of denitrification:

$$NO_3^- \rightarrow NO_2^- \rightarrow NO \rightarrow N_2O \rightarrow N_2$$

As a result, nitrification produces NO_3, which can be leached into surface waters or groundwater and produces N_2O as a potent greenhouse gas, either directly or indirectly (via denitrification). NO_3 concentrations and N_2O emissions from nitrification and denitrification may be lowered by adding a nitrification inhibitor (NI) to the soil, which prevents ammonia oxidation, the primary step in nitrification. Ammonium (NH_4^+) adsorbed on soil colloid then used for plant uptake. Successful NI results in reducing environmental implications of NO_3 leaching and N_2O emissions, in addition, it would also improve plant N use efficiency.

3. **Urease Inhibitors**

One strategy for improving urea performance in agriculture and reducing urea-related pollution emissions is to utilize urease inhibitors (Prasad et al. 2017; Modolo et al. 2015; Mira et al. 2017; Li et al. 2017). Ammonia and carbon dioxide are produced in the presence of the nickel-dependent enzyme urease. This hydrolase is widely distributed in yeasts, plants, fungi, and algae, being found in bacteria and animal feces as a crucial enzyme for the world N cycle (Follmer 2008). Numerous urea analogs have been shown to inhibit urease catalytic activity. A significant quantity of N may be lost to the environment as NH_3 if urea hydrolysis creates NH_3, which is subsequently transformed into ammonium (NH_4^+) in soil solution prior to plant uptake (Bock and Kissel 1988; Cameron et al. 2013). The use of NH_3 inhibitors in the context of covering fertilization is especially noteworthy, since it reduces urea-derived NH_3 production on the soil surface, promoting urea transfer to deeper soil layers via programmed irrigation or rain episodes (Artola et al. 2011). Controlling urease activity in soil, on the other hand, could be an environmentally beneficial way to increase N concentration in soil (Rawluk et al. 2001; Aziz et al. 2024).

11.7 Response of Crops to N-Fertilizer

Winter wheat is a widely grown crop not only in Pakistan but all over the world. This type of wheat requires substantial amounts of nitrogen (N) fertilization input, which is frequently associated with N losses. It is possible to boost the amount of nitrogen that is available to plants through the use of fertilizers that contain nitrification and urease inhibitors. This would also prevent the possibility of nitrogen being lost (Aziz et al. 2024).

An experiment was conducted by Jan et al. (2010) in 2000–01. The experiment was conducted in a random complete block design with 23 treatment combinations and four replicates for each (RCBD). Fertilizers, plant residues (legume and Stover), their combination, as well as when to apply them, were examined in a study. Research found that fertilized plots had a significant impact on grain yield and size. Inorganic nitrogen sources increased 1000 grain weight, spikes m^{-2}, and grain yield in comparison to organic nitrogen sources. As a result of applying fertilizers at the time of sowing, spikes m^{-2}, and grain yields increased. The 1000 grain weight, spike m^{-2}, and grain spike^{-1} were all improved as a result of applying 120 kg N ha^{-1}. Split application of N fertilizer results in increased grain weight. Grain yield, spike m^{-2}, and 1000 grain weight were all higher when cereal residue and nitrogen fertilizers were used together rather than cereal residue alone. One pound of legume and nitrogen fertilizer mixtures increased grain weight by 1000 grains over leguminous fertilizer alone. When only inorganic fertilizers were used, grain yield and spike m^{-2} increased, as opposed to when both inorganic and cereal residue were used. Crops may be fertilized with inorganic fertilizers. There was an increase in grain yield and spike m^{-2}, compared to the use of an inorganic N/legume residue mixture. Aziz et al. (2024) showed that the neem oil-coated urea fertilizer improved crop management and reduced N losses, which are usually caused by conventional fertilizers.

11.8 Improve Nitrogen Use Efficiency

In order to avoid environmental damage, nitrogen rates must be controlled to the amount necessary for maximum yield while considering the environmental impact. When the plants are actively growing and receiving an adequate supply of N for uptake, N fertilization must be applied on schedule. Side dress application, split fertilizers application, genetic approaches, and slow and controlled release fertilizers can all help reduce nitrogen loss.

Meng et al. (2017) conducted an experiment on the effect of controlled released fertilizers at different depths. This experiment consisted of five treatments as conventional urea application at 5 and 10 depths, and controlled released urea at 5, 10, and 15 cm depth. Data regarding dry matter accumulation, 1000-grain weight, grain-filling rate, N and water use efficiencies, and economic benefits were evaluated. Deep placement (15 cm) applied controlled released urea had increased yield, water and N use efficiencies compared to other treatments.

Yang et al. (2014) studied the effect of controlled-release fertilizer on CH_4 emission from paddy fields for 4 years in China. They estimated the grain yield, CH_4, and other agronomic parameters. It was concluded that during a 4-year study, average CH_4 emission was 65.78 and 64.9 kg ha^{-1} from conventional urea and coated urea (thermoplastic resin coated urea) respectively. CH_4 emissions were significantly reduced by up to 14.4% and 15.6% with urea and controlled-release fertilizer.

Geng and his coworkers investigate the separate and interactive effects of polymer-coated urea and Sulphur fertilizer on yield, nitrogen use efficiency, and physiological parameters of cotton under field conditions during 2014 and 2015. They use two N fertilizer sources including polymer-coated urea and conventional urea and three Sulphur levels i.e., 0, 60, and 120 kg ha^{-1}. The results indicated that the N release pattern and requirement of cotton were highly correlated to each other in both NO_3^- and NH_4^+. With polymer-coated urea application number of balls and lint yield about 8.91% and 11.17%, respectively. Yield was also increased when S + N fertilizers were applied in combination up to 9.26% but there was no significant difference between 60 and 120 kg S ha^{-1} (Geng et al. 2016).

Pan et al. (2016) conducted the experiment to determine the impact of amended urea with (pyrite, zeolite, organic acids), urease inhibitors, and polymer-coated urea on ammonia (NH_3) volatilization. They found that urea amended with pyrite significantly reduced ammonia volatilization losses by up to 75% compared to urease inhibitors and polymer-coated urea that inhibited volatilization by up to 54% and 65%, respectively.

Kiran et al. (2010) conducted an experiment to evaluate the effect of different CRU (controlled release urea) products. They used six CRU products CDU Uber-10, Meister-20, Meister-27, humate-coated urea, Duration type V, and sulfur-coated urea (gold–N). Fertilizers were applied once throughout the growing season of rice. CRU significantly increased rice yield (6 t ha^{-1}) compared to conventional urea. When comparing the CRU-treated plot to the control plot, the N accumulation in rice straw and rice grains was significantly higher in the treatment plot. It can be concluded that CRU outperforms granular urea significantly.

Geng et al. (2015) examined the efficiency of controlled-release nitrogen fertilizer (CRNF) including polymer and Sulphur-coated urea on yield and nutrient uptake of cotton under field conditions. The result indicated that there was a strong linear correlation between N release through controlled release fertilizer and N uptake of the crop. Crop yield (seed cotton) significantly increased up to 14.88% and 18.15% with polymer-coated and sulphur-coated urea respectively compared to control. Inorganic N contents were also improved in the soil as well as in plants compared to the control treatment.

Kumar et al. (2010) and Aziz et al. (2024) reported that neem oil-coated urea (NOCU) improved N recovery and agronomic efficiency of rice under field conditions compared to conventional urea. Dinesh et al. (2020) conducted a field experiment in India to investigate the effect of different thicknesses of neem oil coated on urea in low-land rice during 2004–2005. The experiment consisted of four thickness levels of neem oil (0, 500, 1000, and 2000 mg kg^{-1}) with three N levels (50, 100, and 150 N kg ha^{-1}). Neem oil-coated urea (with 1000 mg kg^{-1}) significantly improved growth, yield, and chemical parameters compared to the control treatment.

Effect of neem oil as nitrification inhibitor (Aziz et al. 2024). Kumar et al. (2011) studied to investigate the effect of neem cake and neem oil components under field conditions of rice in 2005 and 2006. Sixteen treatment combinations were used with five major neem oil constituents (fatty acids, pure oil, maliacins, saturated and unsaturated fractions) with three thickness levels (500, 1000, and 5000 mg kg^{-1} of urea). It was observed that maliacins proved most effective from all the neem oil components with 500 mg kg^{-1} thickness.

Kumar et al. (2016) conducted a field experiment to evaluate the effect of oil and three N levels (50, 100, and 150 kg ha^{-1}). The experiment was laid out according to a complete randomized block design (RCBD) with three replications. The result indicated that all the coating materials reduced N losses and improved the yield of rice except palmarosa oil over control. The highest yield and nitrogen use efficiency were obtained with citronella oil.

Sustainable agriculture is a demand of recent times for food security which is an alarming issue over the entire world especially in developing countries. Most farmers use urea as a source of N fertilizer due to the highest N contents. In our climate (arid and semiarid) N use efficiency is very low due to volatilization losses, denitrification, nitrate leaching, and surface runoff. Fast urea hydrolysis and nitrification losses can be stopped or reduced by using synthetic urease nitrification inhibitors and biodegradable polymers. So, we must have developed natural compounds and chemical-based nitrification and urease inhibitors that minimize N losses and are economical to the farmers.

11.9 Conclusion

Based on the above discussion, it is noted that nitrogen losses from applied N fertilizers can be minimized by coating the fertilizer granules with appropriate coating material. By sole application of urease inhibitors, nitrification inhibitors, and polymer as a coating material to be coated on fertilizer granules was found useful. While there are many waste products existing in nature that are notorious for living beings at the same time these are very efficient nitrification and urease inhibitors, thus extracts of such waste organic material could be used for the management of nitrogen. Coating of these waste materials like neem oil on the N fertilizer granules could be very effective from a crop growth perspective Some inorganic coating materials are also used to reduce the N losses in the soil system.

References

Afzal J, Wang X, Saleem HM, Sun X, Hussain S, Khan I, Rana MS, Awan SA, Fiaz S, Aziz O, Ali K, Ali S, Hu CX (2021) Application of Ferrous sulphate alleviates negative impact of cadmium in rice (Oryza sativa L.). Biocell 45(6):1631–1649. https://doi.org/10.32604/biocell.2021.014934

Akiyama H, Morimoto S, Hayatsu M, Hayakawa A, Sudo S, Yagi K (2013) Nitrification, ammonia-oxidizing communities, and N_2O and CH_4 fluxes in an imperfectly drained agricultural field fertilized with coated urea with and without dicyandiamide. Biol Fertil Soils 49:213–223

Artola E, Cruchaga S, Ariz I, Moran JF, Garnica M, Houdusse F, Aparicio-Tejo PM (2011) Effect of N-(n-butyl) thiophosphoric triamide on urea metabolism and the assimilation of ammonium by Triticum aestivum L. Plant Growth Regul 63:73–79

Aziz A, Tahir MA, Noor-Us-Sabah AR, Ahmad Z, Hashem A, Avila-Guezada GD, Abd Allah EF (2024) Optimizing neem oil coating thickness for the development of slow-release urea fertilizer to enhance wheat growth and yield. Pak. J. Bot 57:1

Beatty PH, Good AG (2011) Future prospects for cereals that fix nitrogen. Science 333(6041):416–417

Blouin GM, Rindt DW, Moore OE (1971) Sulfur-coated fertilizers for controlled release. Pilot-plant production. J Agric Food Chem 19(5):801–808

Bock BR, Kissel DE (1988) Ammonia volatilization from urea fertilizers, vol 25. National Fertilizer Development Center, Tennessee Valley Authority

Bodirsky BL, Popp A, Lotze-Campen H, Dietrich JP, Rolinski S, Weindl I, Stevanovic M (2014) Reactive nitrogen requirements to feed the world in 2050 and potential to mitigate nitrogen pollution. Nat Commun 5(1):3858

Bröckel U, Hahn C (2004) Product design of solid fertilizers. Chem Eng Res Des 82(11):1453–1457

Cameron KC, Di HJ, Moir JL (2013) Nitrogen losses from the soil/plant system: a review. Ann Appl Biol 162(2):145–173

Chen H, Bai Y, Wang Q, Chen F, Li H, Tullberg JN, Gong Y (2008) Traffic and tillage effects on wheat production on the Loess Plateau of China: 1. Crop yield and SOM. Soil Res 46(8):645–651

Chen C, Wang E, Yu Q (2010) Modelling the effects of climate variability and water management on crop water productivity and water balance in the North China plain. Agric Water Manag 97(8):1175–1184

Chien SH, Prochnow LI, Cantarella AH (2009) Recent developments of fertilizer production and use to improve nutrient efficiency and minimize environmental impacts. Adv Agron 102:267–322

Choi MM, Meisen A (1997) Sulfur coating of urea in shallow spouted beds. Chem Eng Sci 52(7):1073–1086

Davidson EA, Markewitz D, de Figueiredo OR, de Camargo PB (2018) Nitrogen fixation inputs in pasture and early successional forest in the Brazilian Amazon region: evidence from a claybox mesocosm study. J Geophys Res Biogeo 123(2):712–721

Dent D, Cocking E (2017) Establishing symbiotic nitrogen fixation in cereals and other non-legume crops: the greener nitrogen revolution. Agric Food Secur 6(1):1–9

DeRosa MC, Monreal C, Schnitzer M, Walsh R, Sultan Y (2010) Nanotechnology in fertilizers. Nat Nanotechnol 5(2):91–91

Dinesh A, Benson C, Holborn P, Sampath S, Xiong Y (2020) Performance evaluation of nitrogen for fire safety application in aircraft. Reliab Eng Sys Saf 202:107044

Dobermann AR (2005) Nitrogen use efficiency-state of the art. Agronomy-Faculty Publications, p 316

Du XW, Fu YS, Sun J, Han X, Liu J (2006) Complete UV emission of ZnO nanoparticles in a PMMA matrix. Semicond Sci Technol 21(8):1202

Dubey A, Mailapalli DR (2019) Zeolite coated urea fertilizer using different binders: fabrication, material properties and nitrogen release studies. Environmental Technology & Innovation 16:100452

Erisman JW, Sutton MA, Galloway J, Klimont Z, Winiwarter W (2008) How a century of ammonia synthesis changed the world. Nat Geosci 1(10):636–639

Fiaz S, Khan SA, Noor MA, Ali K, Wang X (2021) Novel plant breeding techniques to advance nitrogen use efficiency in rice: A review. GM Crop and Food 12:627. https://doi.org/10.1080/21645698.2021.1921545

Flis S (2017) The 4Rs in crop nitrogen research. Crops & Soils 50(2):18–20

Follmer C (2008) Insights into the role and structure of plant ureases. Phytochemistry 69(1):18–28

Geng J, Ma Q, Zhang M, Li C, Liu Z, Lyu X, Zheng W (2015) Synchronized relationships between nitrogen release of controlled release nitrogen fertilizers and nitrogen requirements of cotton. Field Crop Res 184:9–16

Geng J, Ma Q, Chen J, Zhang M, Li C, Yang Y, Liu Z (2016) Effects of polymer coated urea and sulfur fertilization on yield, nitrogen use efficiency and leaf senescence of cotton. Field Crop Res 187:87–95

Grant C, Wu R (2008) Enhanced-efficiency fertilizers for use on the Canadian prairies. Crop Manag 7(1):1–15

Guendouz A, Guessoum S, Maamri K, Benidir M, Hafsi M (2012) Canopy temperature efficiency as indicators for drought tolerance in durum wheat (Triticum durum Desf.) in semi-arid conditions. J Agri Sustain 1(1)

Guo M, Liu M, Liang R, Niu A (2006) Granular urea-formaldehyde slow-release fertilizer with superabsorbent and moisture preservation. J Appl Polym Sci 99(6):3230–3235

Hopkins DW, Sparrow AD, Elberling B, Gregorich EG, Novis PM, Greenfield LG, Tilston EL (2006) Carbon,nitrogen and temperature controls on microbial activity in soils from an Antarctic dry valley. Soil Biol Biochem 38(10):3130–3140

Jan MT, Khan JM, Khan A, Arif M, Shafi M, Nullah N (2010) Wheat nitrogen indices response to nitrogen source and application time. Pak J Bot 42(6):4267–4279

Johnson RB (2006) Sustainable agriculture: competing visions and policy avenues. Int J Sustain Dev World Ecol 13(6):469–480

Kim DJ, Seo DW, Lee SH, Shipin O (2012) Free nitrous acid selectively inhibits and eliminates nitrite oxidizers from nitrifying sequencing batch reactor. Bioprocess Biosyst Eng 35:441–448

Kiran JK, Khanif YM, Amminuddin H, Anuar AR (2010) Effects of controlled release urea on the yield and nitrogen nutrition of flooded rice. Commun Soil Sci Plant Anal 41(7):811–819

Kumar SP, Kumar CV, Bandana B (2010) Effects of salicylic acid on seedling growth and nitrogen metabolism in cucumber (Cucumis sativus L.). Journal of Stress Physiology & Biochemistry 6(3):102–113

Kumar D, Devakumar C, Kumar R, Panneerselvam P, Das A, Shivay YS (2011) Relative efficiency of prilled urea coated with major neem (Azadirachta indica A. Juss) oil components in lowland irrigated rice of the Indo-Gangetic plains. Arch Agron Soil Sci 57(1):61–74

Kumar A, Sharma A, Upadhyaya C, K. (2016) Vegetable oil: nutritional and industrial perspective. Curr Genomics 17(3):230–240

Kumar R, Kumawat N, Sahu YK (2017) Role of biofertilizers in agriculture. Popular kheti 5(4):63–66

Leghari SJ, Wahocho NA, Laghari GM, HafeezLaghari A, MustafaBhabhan G, HussainTalpur K, Lashari AA (2016) Role of nitrogen for plant growth and development: A review. Adv Environ Biol 10(9):209–219

Li Y, Sun Y, Liao S, Zou G, Zhao T, Chen Y, Zhang L (2017) Effects of two slow-release nitrogen fertilizers and irrigation on yield, quality, and water-fertilizer productivity of greenhouse tomato. Agric Water Manag 186:139–146

Malla G, Bhatia A, Pathak H, Prasad S, Jain N, Singh J (2005) Mitigating nitrous oxide and methane emissions from soil in rice–wheat system of the Indo-Gangetic plain with nitrification and urease inhibitors. Chemosphere 58(2):141–147

McCauley A, Jones C, Jacobsen J (2009) Plant nutrient functions and deficiency and toxicity symptoms. Nutrient management module 9:1–16

McKenzie RH, Bremer E, Middleton AB, Pfiffner PG, Dowbenko RE (2007) Controlled-release urea for winter wheat in southern Alberta. Can J Soil Sci 87(1):85–91

McTaggart IP, Tsuruta H (2003) The influence of controlled release fertilisers and the form of applied fertiliser nitrogen on nitrous oxide emissions from an andosol. Nutr Cycl Agroecosyst 67:47–54

Meng F, Dungait JA, Xu X, Bol R, Zhang X, Wu W (2017) Coupled incorporation of maize (Zea mays L.) straw with nitrogen fertilizer increased soil organic carbon in Fluvic Cambisol. Geoderma 304:19–27

ME-Trenkel T (2021) Slow-and controlled-release and Stabilized Fertilizers: an option for enhancing nutrient use effiiency in agriculture. International Fertilizer Industry Association (IFA)

Mira AB, Cantarella H, Souza-Netto GJM, Moreira LA, Kamogawa MY, Otto R (2017) Optimizing urease inhibitor usage to reduce ammonia emission following urea application over crop residues. Agric Ecosyst Environ 248:105–112

Modolo LV, de Souza AX, Horta LP, Araujo DP, de Fátima Â (2015) An overview on the potential of natural products as ureases inhibitors: A review. J Adv Res 6(1):35–44

Motavalli PP, Goyne KW, Udawatta RP (2008) Environmental impacts of enhanced-efficiency nitrogen fertilizers. Crop Manag 7(1):1–15

Naderi MR, Danesh Shahraki A (2011) Nano-fertilizers and their role in agriculture. Takta J 86:66–73

Nelson AG, Spaner D (2010) Cropping systems management, soil microbial communities, and soil biological fertility. In: Genetic engineering, biofertilisation, soil quality and organic farming, pp 217–242

Olson RA, Kurtz LT (1982) Crop nitrogen requirements, utilization, and fertilization. Nitrogen Agri Soils 22:567–604

Pan J, Yuan F, Yu L, Huang L, Fei H, Cheng F, Zhang Q (2016) Performance of organics and nitrogen removal in subsurface wastewater infiltration systems by intermittent aeration and shunt distributing wastewater. Bioresour Technol 211:774–778

Prasad R, Shivay YS, Kumar D (2017) Current status, challenges, and opportunities in rice production. In: Rice production worldwide. Springer, pp 1–32

Rahimpour MR, Barmaki MM, Mottaghi HR (2010) A comparative study for simultaneous removal of urea, ammonia and carbon dioxide from industrial wastewater using a thermal hydrolyser. Chem Eng J 164(1):155–167

Rawluk CDL, Grant CA, Racz GJ (2001) Ammonia volatilization from soils fertilized with urea and varying rates of urease inhibitor NBPT. Can J Soil Sci 81(2):239–246

Schuler M, Rellán-Álvarez R, Fink-Straube C, Abadía J, Bauer P (2012) Nicotianamine functions in the phloem-based transport of iron to sink organs, in pollen development and pollen tube growth in Arabidopsis. Plant Cell 24(6):2380–2400

Sempeho SI, Kim HT, Mubofu E, Hilonga A (2014) Meticulous overview on the controlled release fertilizers. Adv Chem 2014(1):363071

Shahid A, Malik S, Zhu H, Xu J, Nawaz MZ, Nawaz S, Mehmood MA (2020) Cultivating microalgae in wastewater for biomass production, pollutant removal, and atmospheric carbon mitigation; a review. Sci Total Environ 704:135303

Shahzad AN, Qureshi MK, Wakeel A, Misselbrook T (2019) Crop production in Pakistan and low nitrogen use efficiencies. Nat Sustain 2(12):1106–1114

Shaviv A (2005) Environmental friendly nitrogen fertilization. Sci China Ser C Life Sci 48:937–947

Sher A, Wang X, Sattar A, Ijaz M, Manaf A, Fiaz S (2021) Exogenous applied thiourea for improving the productivity and nutritional quality of bread wheat (Triticum aestivum L.). Agronomy-Basel 11(7):1432. https://doi.org/10.3390/agronomy11071432

Shoji S, Delgado J, Mosier A, Miura Y (2001) Use of controlled release fertilizers and nitrification inhibitors to increase nitrogen use efficiency and to conserve air andwater quality. Commun Soil Sci Plant Anal 32(7–8):1051–1070

Smith DR, Owens PR, Leytem AB, Warnemuende EA (2007) Nutrient losses from manure and fertilizer applications as impacted by time to first runoff event. Environ Pollut 147(1):131–137

Subbarao GV, Nakahara K, Ishikawa T, Ono H, Yoshida M, Yoshihashi T, Sahrawat KL (2013) Biological nitrification inhibition (BNI) activity in sorghum and its characterization. Plant Soil 366:243–259

Suter H, Sultana H, Turner D, Davies R, Walker C, Chen D (2013) Influence of urea fertilizer formulation, urease inhibitor and season on ammonia loss from ryegrass. Nutr Cycl Agroecosyst 95:175–185

Turner JC, Hattey JA, Warren JG, Penn CJ (2010) Electrical conductivity and sodium adsorption ratio changes following annual applications of animal manure amendments. Commun Soil Sci Plant Anal 41(9):1043–1060

Ullah S, Ai C, Ding W, Jiang R, Zhao S, Zhang J, He P (2019) The response of soil fungal diversity and community composition to long-term fertilization. Appl Soil Ecol 140:35–41

Vallejo A, Diez JA, López-Valdivia LM, Gascó A, Jiménez C (2001) Nitrous oxide emission and denitrification nitrogen losses from soils treated with isobutylenediurea and urea plus dicyandiamide. Biol Fertil Soils 34:248–257

Wang FL, Alva AK (1996) Leaching of nitrogen from slow-release urea sources in sandy soils. Soil Sci Soc Am J 60(5):1454–1458

Wang S, Zhao X, Xing G, Yang Y, Zhang M, Chen H (2015) Improving grain yield and reducing N loss using polymer-coated urea in Southeast China. Agron Sustain Dev 35:1103–1115

Wrage N, Velthof GL, Van Beusichem ML, Oenema O (2001) Role of nitrifier denitrification in the production of nitrous oxide. Soil Biol Biochem 33(12–13):1723–1732

Yang Z, Xu M, Liu Y, He F, Gao F, Su Y, Zhang Y (2014) Nitrogen-doped, carbon-rich, highly photoluminescent carbon dots from ammonium citrate. Nanoscale 6(3):1890–1895

Genome Editing of Biotic and Abiotic Stress-Resistant Crops

12

Bilal Ahmad Khan, Ijaz Rasool Noorka, J. S. Pat Heslop-Harrison, Muhammad Ather Nadeem, Tehreem Zehra, Hussam F. Najeeb Alawadi, Sheza Ayaz, Sabir Jamil, Muhammad Mubashar Zafar, F. N. U. Abdullah, Abdul Wahid, and Bismark Anokye

Abstract

Agricultural productivity is constantly challenged by biotic and abiotic stresses. However, conventional plant breeding methods have previously been utilized to address these setbacks. With the advent of genome editing technologies, such as TALENs, ZFN, and CRISPR/Cas9, scientists can now precisely modify the genetic makeup of crops to introduce stress-resistance traits. Apart from the diverse site-directed nucleases that were accessible over a decade ago, the CRISPR/Cas (clustered regularly interspaced short palindromic repeats/CRISPR-associated protein) method of genome editing has garnered attention due to its cost-effectiveness, ease of use, versatility, and simplicity. Genetic engineering and biotechnology are poised for a true revolution thanks to the development of genome editing techniques. In addition to enabling research into the basic prin-

B. A. Khan
Departnment of Soil and Crop Sciences, Texas A&M University, College Station, TX, USA

Department of Agronomy, College of Agriculture, University of Sargodha, Sargodha, Pakistan

I. R. Noorka (✉)
Department of Plant Breading and Genetics, College of Agriculture, University of Sargodha, Sargodha, Pakistan

J. S. P. Heslop-Harrison
Department of Genetics and Genome Biology, Institute for Environmental Futures, University of Leicester, Leicester, UK

South China National Botanical Garden, Chinese Academy of Sciences, Guangzhou, China

M. A. Nadeem
Department of Agronomy, College of Agriculture, University of Sargodha, Sargodha, Pakistan

© The Author(s), under exclusive license to Springer Nature Singapore Pte Ltd. 2025

S. Faiz et al. (eds.), *Climate Smart Agriculture for Future Food Security*,
https://doi.org/10.1007/978-981-96-4499-5_12

ciples of biological systems, targeted editing of living organisms' genomes enables the pursuit of numerous objectives aimed at enhancing crop quality and yield. This involves breeding plants to have desirable compositional features and characteristics that provide resistance to a range of biotic and abiotic stressors.

Keywords

Genome editing · Biotechnology · Development · CRISPR/Cas9 · Stressors

12.1 Genome Editing Technology

Genome editing is a super cool way to make precise changes to an organism's DNA. It involves using special tools called nucleases to make targeted breaks in the DNA. Several repair processes, including homologous recombination (HR) or non-homologous end-joining (NHEJ), can be used to repair these breaks. NHEJ is quick but can be a bit messy, while HR is more precise and can even replace genes. Zinc finger nucleases (ZFNs), transcription activator-like effector nucleases (TALENs), and the CRISPR/Cas system are the three primary types of nucleases that are utilized. (Yin and Qiu 2019).

Genome editing has been booming in recent years and has incredible potential to be the go-to technology for managing abiotic and biotic stress in plants. The CRISPR-Cas system, specifically the Cas9 protein, is a super efficient and convenient method for precisely manipulating targeted genes. It works by creating specific breaks in the DNA, which triggers the plant's natural DNA repair mechanisms and allows for precise genome modifications (Fiaz et al. 2021). Despite having defense systems such as Effectors-triggered immunity (ETI) and PAMP-triggered immunity (PTI), plants are susceptible to pests such as fungi, bacteria, viruses, and insects. Genome editing offers a promising way to enhance plant resistance and better protect our crops (Table 12.1).

T. Zehra · S. Ayaz · S. Jamil
Department of Botany, Division of Science and Technology University of Education, Lahore, Pakistan

H. F. Najeeb Alawadi
College of Agriculture, University of Al-Qadisiyah, Al Diwaniyah, Iraq

M. M. Zafar
Department of Botany, Minhaj University Lahore, Lahore, Pakistan

F. N. U. Abdullah · A. Wahid · B. Anokye
Departnment of Soil and Crop Sciences, Texas A&M University, College Station, TX, USA

Table 12.1 Various Genome editing tools, their Applications and Mechanism used in crop development

Technology	Applications	Mechanism
TALENs	Effective in precise genome modifications for stress tolerance (Voytas 2013).	Transcription ActivatorLike Effector Nucleases that recognize specific DNA sequences through repeat
		Domains and induce targeted mutations.
ZFN	Used for genetic modification but is less popular due to design complexity (Urnov et al. 2010).	Zinc Finger Nucleases that target specific DNA sequences using engineered zinc-finger domains fused with a nuclease domain.
CRISPR/ Cas9	Widely used for both biotic and abiotic stress resistance due to its versatility (Jiang and Doudna 2017).	Introduces double-strand breaks (DSBs) using guide RNA and Cas9 nuclease to induce site-specific mutations.

12.2 Enhancing Crops and Reducing Biotic and Abiotic Stress

Plant genome editing technologies have become incredibly valuable for developing crop varieties with exceptional qualities and the ability to withstand challenging environmental conditions. The plant's DNA can be altered more precisely and specifically with the use of contemporary genome editing tools than with conventional breeding techniques like mutation and selective breeding. This focused strategy greatly quickens the process of creating crops with desirable characteristics, such as higher yield and improved resistance to environmental changes. The authors of this article examine the potential use of genome editing technologies in the future to mitigate the effects of biotic and abiotic pressures on agriculture. They concentrate on the CRISPR/Cas system in particular, which has drawn a lot of interest as an innovative genome-editing tool for a variety of species (Ahmad et al. 2021) (Fig. 12.1).

Crop plants face many challenges from both biotic and abiotic stresses, which can really reduce their yield. It's crucial to address these issues to ensure we can produce enough food for our growing global population and avoid food shortages in the future. We can't just keep expanding the area of land dedicated to food crops indefinitely. That's where biotechnological interventions come in as a game-changer. With the advancements in genomics and genome editing technologies like TALENs, ZFNs, and CRISPR/Cas9, we have more tools than ever to overcome these stresses in various food crops.

12.3 Biotic Stress Resistance

Biotic stresses include pathogens (fungi, bacteria, and viruses), insects, and pests, which result in significant yield losses. Genome editing has been applied to enhance resistance by modifying host genes that either regulate immunity or are targets of pathogen effectors (Dangl et al. 2013).

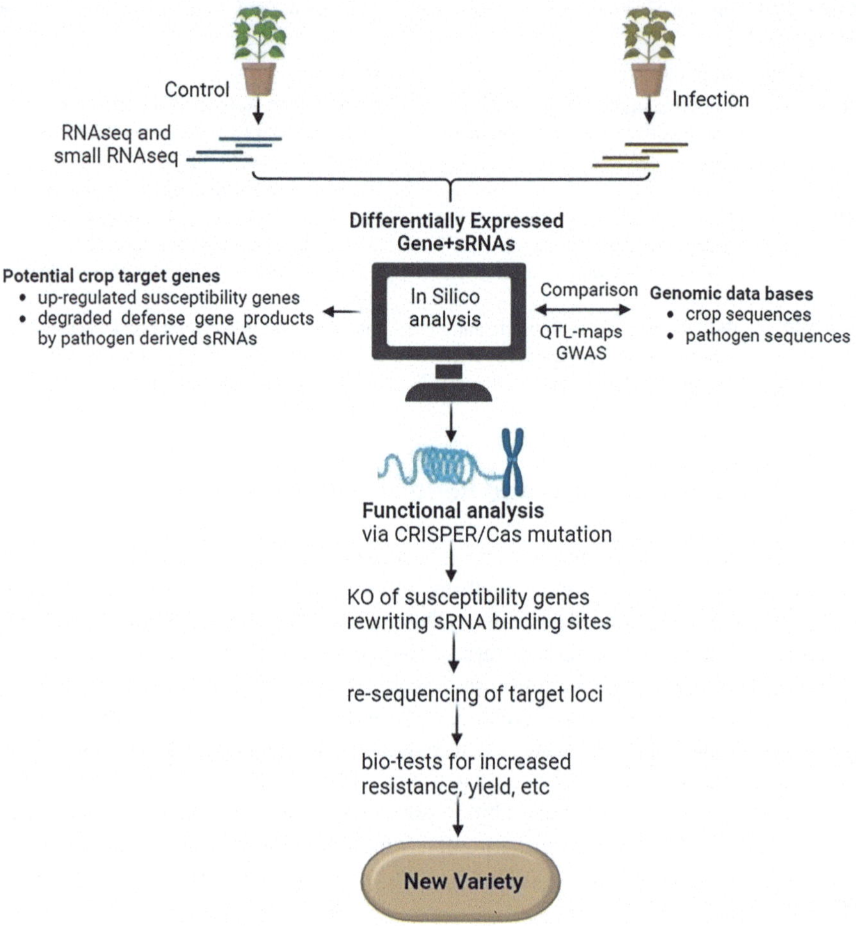

Fig. 12.1 "De novo" pipeline for identifying targets to create resistance to biotic stress

12.4 Applications of Genome Editing for Biotic Stress Resistance

12.4.1 Editing Plants for Virus, Fungal, and Insect Resistance

CRISPR/Cas9 technology has been used to knock out susceptibility (S) genes in crops such as rice, wheat, and tomato. For example, editing the OsSWEET genes in rice conferred resistance to bacterial leaf blight caused by Xanthomonas oryzae. Targeting the CYP71A1 gene in maize via CRISPR/Cas9 enhanced resistance to the maize stalk borer, a major pest. In cassava, genome editing has been used to induce resistance against cassava mosaic virus by altering the susceptibility genes using CRISPR/Cas9 (Odipio et al. 2017: Li et al. 2020; Kim et al. 2021) (Table 12.2).

Table 12.2 Targeted traits improved through genome editing technique

Crop	Stress type	Targeted gene	Outcome	Reference
Rice	Bacterial blight	OsSWEET	Increased resistance to *Xanthomonas*	(Li et al. 2020)
Wheat	Powdery mildew	TaMLO	Reduced susceptibility	(Wang et al. 2019)
Maize	Stalk borer	CYP71A1	Enhanced insect resistance	(Kim et al. 2021)

12.4.2 Editing Plants for Disease Resistance

Breeding plants for disease resistance has been done using different methods. For self-pollinated crops like wheat, the pure line method was used to develop rust-resistant varieties. The pedigree method is commonly used to create disease-resistant varieties controlled by major genes (Hamdan et al. 2022).

12.5 Genome Editing Towards Enhancing Abiotic Stress Tolerance in Plants

Abiotic stress factors like cold, drought, and saline-alkali stress, along with biotic stressors such as diseases and insect pests, really take a toll on plant growth and agricultural productivity. However, because molecular biology is developing so quickly, genome editing methods are already very common in the domains of botany and agronomy. These techniques are known for their high efficiency, controllability, and precise editing capabilities (Barman et al. 2019). They have enormous potential for breeding resistant crop varieties. They've already achieved some incredible results in developing resilient cereal crops like rice, wheat, and maize, as well as fruits and vegetables like tomatoes and cucumbers. One of the superstar genome editing systems is CRISPR/Cas, which plays a major role in ensuring stable crop yields worldwide. It's mind-blowing how genome editing is revolutionizing agriculture (Nascimento et al. 2023).

Twenty major agricultural crops were taken into consideration to generate a string search, and of these, rice has been examined the most, with 50 papers, according to FAOSTAT. Abiotic stressors in rice that have been reported the most include salt, water deficiency, salinity plus water deficit, and cold tolerance. With 24 papers, Arabidopsis thaliana—a common model plant in biological experiments—was the second most investigated. The two most common themes in A. thaliana were low temperatures and water deprivation. The majority of studies on tomatoes, maize, and beans concentrated on salinity and water deficiency tolerance. Osmotic stress, salinity, water deficiency, and high temperatures were among the other subjects with lower article frequency (Ghosh and Dey 2022).

12.5.1 Drought Tolerance

Drought stress triggers complex physiological responses in plants. Genome editing can enhance drought resistance by targeting genes involved in water uptake, transpiration, and osmotic regulation. For example, editing the DREB1A gene in rice has resulted in improved drought tolerance by enhancing the expression of drought-responsive genes (Wei et al. 2021).

12.5.2 Salinity Tolerance

Salinity is a significant abiotic stress in arid and semi-arid regions. Genes such as **NHX1** (which regulates sodium ion transport) have been targeted using genome editing techniques to develop salt-tolerant varieties. In wheat, CRISPR/Cas9-mediated knockouts of the **TaHKT1;5** genes have shown improved salt tolerance (Zhang et al. 2020).

12.5.3 Heat Tolerance

Heat stress affects crop growth, particularly in regions prone to high temperatures. Genome editing approaches have targeted genes such as **HsfA1** (a heat-shock factor) in tomatoes, which regulates heat response pathways, to improve heat tolerance (Sun et al. 2019) (Table 12.3).

The backcross method and recurrent selection are used to improve resistance in elite rice varieties. Interspecific hybrids have been used to transfer resistance genes from one cotton species to another. And the mutation breeding method has been successful in creating rice lines resistant to blast. However, these traditional methods can be time-consuming and sometimes affect plant development.

12.6 Transgene-Free Genome Editing for Biotic and Abiotic Stress

12.6.1 Resistance in Sugarcane

An immensely valuable crop for both industrial and food use is sugarcane. However, several biotic stressors, including fungi, bacteria, viruses, and insect pests, as well as abiotic stressors, such as drought, salt, extreme heat or cold, waterlogging, and heavy metals, pose serious obstacles to its development. Quickly improving sugarcane genetics is essential because of the growing need for biofuel and sugar, as well as the introduction of novel pest and disease variations. Conventional breeding techniques might demand a lot of time and resources. Herein lies the utility of the CRISPR/Cas genome editing technology. It enables the development of novel sugarcane cultivars that are more resilient and tolerant of various biotic and abiotic

Table 12.3 Application of the CRISPR-based genome editing approach in plants for improvement of drought and salinity stress tolerance

Application of CRISPR-based genome editing	Plant species	Target gene	Gene ID	Method of delivery	Reference
Drought Tolerance: Enhanced stomatal regulation to minimize water loss	*Arabidopsis thaliana*	*AtOST2*	NM_001335616	Agrobacterium- mediated	(Osakabe et al. 2016).
Increased drought tolerance through improved ABA-responsive gene expression	*A. thaliana*	*AtAREB1*	AT1G45249.3	Agrobactrium- mediated	(Roca Paixão et al. 2019)
Improved vacuolar proton transport to enhance water use efficiency	*A. thaliana*	*AtAVP1*	NM_101437	Agrobacterium- mediated	(Park et al. 2017; Chen et al. 2019)
Enhanced transcriptional regulation under drought stress	*A. thaliana*	*AtmiR169a*	–	Agrobacterium- mediated	(Zhao et al. 2016)
Improved root architecture and water uptake capacity	*Brassica napus*	*BnaA6.RGA*	LOC106445425	Agrobacterium- mediated	(Wu et al. 2020)
Enhanced lignin biosynthesis to improve structural integrity	*Cicer arietinum*	*At4CL, AtRVE7*	LOC101502718, LOC101509066	PEG-mediated	(Badhan et al. 2021)
Enhanced transcription factors activity to improve drought response	*Glycine max*	*GmMYB118*	GLYMA_17G094400	Agrobacterium- mediated	Du et al. 2018)
Improved reproductive drought tolerance	*Oryza sativa*	*OsERA1*	LOC_Os01g53600	Agrobacterium- mediated	(Ogata et al. 2020)

Source: Kumar et al. (2023)

stressors. With the use of this technology, sugarcane production might be revolutionized and assist in satisfying the needs of the (Gong and Liu 2013).

The CRISPR/Cas genome editing technique has created new and intriguing opportunities for cultivar development with enhanced stress tolerance and resistance. The genome editing cassette that is present in the modified plants, however, is a problem and may result in regulatory obstacles for commercial application. Don't worry, though; transgene-free genome editing methods exist to get around this restriction. Among these strategies are Agrobacterium-mediated transformation for the temporary expression of CRISPR/Cas reagents, virus-induced genome editing (VIGE), and delivery of CRISPR/Cas components through biolistics or protoplast fusion, among others. Advanced screening technologies including PCR-based assays, visual marker systems, and the

Transgene killer CRISPR systems (TKC) are employed for fast identification in order to guarantee transgene-free modifications.

12.6.2 Resistance in Citrus

The most significant tree fruit crop in the world is citrus. But it encounters numerous difficulties, including viruses, illnesses, dryness, and extremely high and low temperatures. Fortunately, developments in genetics and plant biotechnology have made it easier to understand how citrus can withstand these pressures. Researchers have made strides at both genetically modifying citrus trees and identifying and cloning genes that confer tolerance. This creates opportunities for further advancements in citrus crop improvement in the future (Sicilia et al. 2022). Traditional breeding methods have played a crucial role in developing new and improved citrus cultivars. These methods have utilized natural mutations and explored the tolerance of naturally occurring or induced polyploids. However, when it comes to the cross-hybridization of citrus, there have been some challenges. Reproductive habits like polyembryony, incompatibility, and sterility, as well as a long juvenile period, have limited progress in this area. New developments in computational analysis and high-throughput sequencing have produced invaluable genomic data for the Citrus genus. Citrus breeding now offers more options thanks to the application of biotechnological techniques such as genomic selection, marker-assisted selection, genome-wide association studies, and the identification of quantitative trait locus (QTL) areas.

12.6.3 Resistance in Potato

Potato (Solanum tuberosum L.) is indeed the third most important food crop worldwide, after rice and wheat. In portions of the Americas and Europe, it is a staple diet. Global potato production in 2018 was 368.17 million tonnes, with 90.26 million tonnes coming from China and 48.53 million tonnes from India. As the world's population is predicted to reach 9.7 billion people by 2050, securing food supply will be extremely difficult. Regrettably, diseases, insect pests, and environmental abiotic stresses all of which are getting worse because of climate change—affect potatoes. For instance, it is predicted that in India, the yield of potatoes in the main states where they are grown will decrease by 2.0% in the 2050s and 6.4% in the 2080s.To address these problems, traditional breeding techniques have been essential in creating new potato varieties. Additionally, marker-assisted selection has been used, especially (Tiwari et al. 2022).

A very effective method for making precise changes to the genome is CRISPR/Cas. It makes it simple to design and build single guide RNA (sgRNA) vectors that are unique to a gene. These vectors can be reprogrammed to direct the Cas9 (SpCas9) protein of Streptococcus pyogenes to

cause double-strand breaks (DSBs) in DNA. Then, the error-prone homology-directed repair (HR) or non-homologous end joining (NHEJ) pathways fix the fractures. Two classes apply to CRISPR/Cas: Class 2 (types II, V, and VI) uses a single effector protein, whereas Class 1 (types I, III, and IV) uses a multi-protein effector complex. Type VI targets RNA, whereas types II and V target DNA. The Class 2, type-II system is the most widely utilized CRISPR/Cas9 version for DNA targeting. Another important (Chincinska et al. 2023).

12.7 Challenges in Genome Editing for Stress Resistance

Despite the success of genome editing in developing stress-resistant crops, several challenges.

12.7.1 Off-Target Effects

Unintended modifications can occur at non-target sites, potentially causing harmful mutations in an organism (Zhang et al. 2019).

12.7.2 Public Acceptance

Public perception and acceptance of genome-edited crops continue to be a significant hurdle. Education and outreach are necessary to communicate the benefits of genome editing technologies (Jinek et al. 2012).

12.7.3 Regulatory Concerns

The legal and regulatory landscape for genome-edited crops varies widely across countries. While some countries, like the USA, are more tolerant, others, such as the European Union, have stricter regulations (Voytas 2013).

12.7.4 Gene Redundancy

Some plants have multiple copies of genes with similar functions, complicating efforts to knock out or modify specific genes effectively (Jiang and Doudna 2017).

12.8 Future Directions toward Genome Editing Technologies

Future research should focus on different genome editing technologies that are.

12.8.1 Improving Precision

Developing more refined genome editing tools with reduced off-target effects (Zhu et al. 2016).

12.8.2 Integration with Other Breeding Techniques

Combining genome editing with traditional breeding or marker-assisted selection to create more resilient crop varieties (Li et al. 2020).

12.8.3 Enhancing Multi-Stress Resistance

Targeting pathways that confer resistance to multiple stresses.

12.8.4 Crop-Specific Applications

More work is needed to adapt genome editing technologies to a wider range of crops, including orphan crops important in developing countries (Sun et al. 2019).

12.9 Conclusion

Genome editing has emerged as a powerful tool for developing crop resistance to both biotic stresses e.g., pest disease and abiotic stress like salinity, drought, salt, etc. Genome editing technologies have opened new frontiers in the field of crop improvement. Different genome editing technologies are CRISPR/Cas9, TALENs, and ZFN. The precision and flexibility of techniques like CRISPR/Cas9, TALENs, and ZFN have made it possible to develop crop varieties that can withstand the increasing challenges posed by climate change and the growing global population. However, challenges related to public perception, technical hurdles, and regulatory acceptance must be addressed to fully realize the potential of genome editing in the field of crop improvement.

Author Contribution Statement Bilal Ahmad Khan contributed to the overall conceptualization, led the research design, and coordinated the manuscript's development. Ijaz Rasool Noorka provided expertise in data analysis, and statistical methods, and played a key role in writing the methodology and results sections. J.S. (Pat) Heslop-Harrison contributed to the theoretical framework, integrating advanced scientific perspectives, and reviewed the final manuscript. Muhammad Ather Nadeem assisted with research implementation and co-wrote parts of the discussion and conclusion. Tehreem Zehra supported data collection, analyzed key findings, and contributed to the writing of the introduction. Hussam F. Najeeb Alawadi contributed to the research design and provided valuable feedback on manuscript revisions. Sheza Ayaz supported the writing process and assisted in reviewing and editing the manuscript. Sabir Jamil worked on the literature review, providing critical insights and analysis. Muhammad Mubashar Zafar assisted with data analysis, co-wrote the methodology section, and helped in finalizing the draft. FNU Abdullah contributed to research execution and provided support in data visualization and manuscript revision. Abdul Wahid played a key role in the manuscript's final editing and proofreading to ensure clarity and consistency. Bismark Anokye provided guidance on research methods, contributed to writing and editing, and ensured the chapter's scientific rigor.

References

Ahmad M, Ali Q, Hafeez MM, Malik A (2021) Improvement for biotic and abiotic stress tolerance in crop plants. Biol Clin Sci Res J 2021(1)

Badhan S, Ball AS, Mantri N (2021) First report of CRISPR/Cas9 mediated DNA-free editing of 4CL and RVE7 genes in chickpea protoplasts. Int J Mol Sci 22:396. https://doi.org/10.3390/ijms22010396

Barman HN, Sheng Z, Fiaz S, Zhong M, Wu Y, Cai Y, Wang W, Jiao G, Tang S, Wei X, Peisong H (2019) Generation of a new thermo-sensitive genic male sterile rice line by targeted mutagenesis of TMS5 gene through CRISPR/Cas9 system. BMC Plant Biol 19:109. https://doi.org/10.1186/s12870-019-1715-0

Chen K, Wang Y, Zhang R, Zhang H, Gao C (2019) Genome editing for crop improvement: challenges and opportunities. Front Plant Sci 10:360. https://doi.org/10.3389/fpls.2019.00360

Chincinska IA, Miklaszewska M, Sołtys-Kalina D (2023) Recent advances and challenges in potato improvement using CRISPR/Cas genome editing. Planta 257(1):25

Dangl JL, Horvath DM, Staskawicz BJ (2013) Pivoting the plant immune system from dissection to deployment. Science 341(6147):746–751. https://doi.org/10.1126/science.1236011

Du YT, Zhao MJ, Wang CT, Gao Y, Wang YX, Liu YW et al (2018) Identification and characterization of GmMYB118 responses to drought and salt stress. BMC Plant Biol 18:1–9. https://doi.org/10.1186/s12870-018-1551-7

Fiaz S, Khan SA, Noor MA, Ali K, Wang X (2021) Novel plant breeding techniques to advance nitrogen use efficiency in rice: a review. GM Crop and Food 12:627. https://doi.org/10.1080/21645698.2021.1921545

Ghosh S, Dey G (2022) Biotic and abiotic stress tolerance through CRISPR-Cas mediated genome editing. J Plant Biochem Biotechnol 31(2):227–238

Gong XQ, Liu JH (2013) Genetic transformation and genes for resistance to abiotic and biotic stresses in Citrus and its related genera. Plant Cell Tissue Organ Cult (PCTOC) 113:137–147

Hamdan MF, Karlson CKS, Teoh EY, Lau SE, Tan BC (2022) Genome editing for sustainable crop improvement and mitigation of biotic and abiotic stresses. Plan Theory 11(19):2625

Jiang F, Doudna JA (2017) CRISPR–Cas9 structures and mechanisms. Annu Rev Biophys 46:505–529. https://doi.org/10.1146/annurev-biophys-062215-010822

Jinek M, Chylinski K, Fonfara I, Hauer M, Doudna JA, Charpentier E (2012) A programmable dual-RNA–guided DNA endonuclease in adaptive bacterial immunity. Science 337(6096):816–821. https://doi.org/10.1126/science.1225829

Kim H, Lee K, Hwang H, Kim S (2021) Insect resistance in maize conferred by CRISPR/Cas9-based editing of the CYP71A1 gene. Plant Biotechnol J 19(1):85–98. https://doi.org/10.1111/pbi.13431

Kumar M, Prusty MR, Pandey MK, Singh PK, Bohra A, Guo B, Varshney RK (2023) Application of CRISPR/Cas9-mediated gene editing for abiotic stress management in crop plants. Front Plant Sci 14:1157678. https://doi.org/10.3389/fpls.2023.1157678

Li X, Zhou Y, Wang Q, Li J (2020) CRISPR/Cas9 editing of OsSWEET genes confers resistance to bacterial blight in rice. Nat Biotechnol 38(4):471–475. https://doi.org/10.1038/s41587-020-0400-5

Nascimento FDS, Rocha ADJ, Soares JMDS, Mascarenhas MS, Ferreira MDS, Morais Lino LS, Amorim EP (2023) Gene editing for plant resistance to abiotic factors: a systematic review. Plan Theory 12(2):305

Odipio J, Alicai T, Ingelbrecht I, Nusinow DA, Bart R, Taylor NJ (2017) Efficient CRISPR/Cas9 genome editing of phytoene desaturase in cassava. Front Plant Sci 8:1780. https://doi.org/10.3389/fpls.2017.01780

Ogata T, Ishizaki T, Fujita M, Fujita Y (2020) CRISPR/Cas9-targeted mutagenesis of OsERA1 confers enhanced responses to abscisic acid and drought stress and increased primary root growth under nonstressed conditions in rice. PLoS One 15:e0243376. https://doi.org/10.1371/journal.pone.0243376

Osakabe Y, Watanabe T, Sugano SS, Ueta R, Ishihara R, Shinozaki K et al (2016) Optimization of CRISPR/Cas9 genome editing to modify abiotic stress responses in plants. Sci Rep 6:26685. https://doi.org/10.1038/srep26685

Park CS, Helmbrecht A, Htoo JK, Adeola O (2017) Comparison of amino acid digestibility in full-fat soybean, two soybean meals, and peanut flour between broiler chickens and growing pigs. J Anim Sci 95(7):3110–3119. https://doi.org/10.2527/jas.2017.1404

Roca Paixão JF, Gillet FX, Ribeiro TP, Bournaud C, Lourenço-Tessutti IT, Noriega DD et al (2019) Improved drought stress tolerance in arabidopsis by CRISPR/dCas9 fusion with a histone AcetylTransferase. Sci Rep 9:8080. https://doi.org/10.1038/s41598-019-44571-y

Sicilia A, Basu S, Lo Piero AR (2022) Genomic Design for Abiotic Stress Resistant Citrus. In: Genomic designing for abiotic stress resistant fruit crops. Springer International Publishing, Cham, pp 41–59

Sun X, Zhang Z, Deng X, Zhang Y (2019) Heat tolerance in tomato plants engineered using CRISPR technology. Plant Mol Biol 99(1–2):111–123. https://doi.org/10.1007/s11103-018-0804-9

Tiwari JK, Buckseth T, Challam C, Zinta R, Bhatia N, Dalamu D, Kumar M (2022) CRISPR/Cas genome editing in potato: current status and future perspectives. Front Genet 13:827808

Urnov FD, Rebar EJ, Holmes MC, Zhang HS, Gregory PD (2010) Genome editing with engineered zinc finger nucleases. Nat Rev Genet 11(9):636–646. https://doi.org/10.1038/nrg2842

Voytas DF (2013) TALENs: customizable DNA-binding proteins for genome engineering. Plant Physiol 163(1):3–6. https://doi.org/10.1104/pp.113.225813

Wang Y, Cheng X, Shan Q, Zhang Y, Liu J, Gao C, Qiu J (2019) Genome editing in wheat: application of CRISPR/Cas9 to develop powdery mildew-resistant wheat. Plant Biotechnol J 17(10):1812–1820. https://doi.org/10.1111/pbi.13171

Wei J, Li X, Chen M, Zheng J (2021) Improved drought tolerance in rice through CRISPR/Cas9-mediated editing of DREB1A. Plant Cell Rep 40(8):1367–1378. https://doi.org/10.1007/s00299-021-02669-z

Wu J, Yan G, Duan Z, Wang Z, Kang C, Guo L et al (2020) Roles of the brassica napus DELLA protein BnaA6.RGA, in modulating drought tolerance by interacting with the ABA signaling component BnaA10.ABF2. Front Plant Sci 11. https://doi.org/10.3389/fpls.2020.00577

Yin K, Qiu JL (2019) Genome editing for plant disease resistance: applications and perspectives. Philos Trans R Soc B 374(1767):20180322

Zhang H, Si X, Ji X, Fan R, Liu J, Chen K, Gao C (2019) Genome editing of multistress resistant rice using CRISPR/Cas9. Nat Biotechnol 37(2):236–238. https://doi.org/10.1038/s41587-019-0342-2

Zhang F, Guo Z, Zhang H, Wang Q (2020) Enhancing salinity tolerance in wheat through CRISPR/Cas9 editing of TaHKT1;5. Plant J 103(5):1282–1292. https://doi.org/10.1111/tpj.14800

Zhao Y, Zhang C, Liu W, Gao W, Liu C, Song G et al (2016) An alternative strategy for targeted gene replacement in plants using a dual-sgRNA/Cas9 design. Sci Rep 6:1–11. https://doi.org/10.1038/srep23890

Zhu JK, Liu J, Xiong L (2016) Abiotic stress signaling and responses in plants. Cell 167(2):313–324. https://doi.org/10.1016/j.cell.2016.08.029

Biotechnology and Digital Revolution for Climate Smart Crops

13

Ijaz Rasool Noorka, Bilal Ahmad Khan, Zafar Iqbal Khan, Muhammad Ather Nadeem, Tehreem Zehra, Abdul Wahid, Bismark Anokye, and J. S. Pat Heslop-Harrison

Abstract

Biotechnology and digital technologies are the viable drivers of change to be part of revolution in research and development to combat changing climatic conditions by climate smart crops. Climate smart crops are resilient to extreme meteorological conditions like fluctuating temperature, rainfall, and increased CO_2 level. These crops maintain the soil health and increase productivity.

I. R. Noorka (✉)
Department of Plant Breading and Genetics, College of Agriculture, University of Sargodha, Sargodha, Pakistan

B. A. Khan
Departnment of Soil and Crop Sciences, Texas A&M University, College Station, TX, USA

Department of Agronomy, College of Agriculture, University of Sargodha, Sargodha, Pakistan

Z. I. Khan
Department of Botany, University of Sargodha, Sargodha, Pakistan

M. A. Nadeem
Department of Agronomy, College of Agriculture, University of Sargodha, Sargodha, Pakistan

T. Zehra
Department of Botany, Division of Science and Technology, University of Education, Lahore, Pakistan

A. Wahid · B. Anokye
Departnment of Soil and Crop Sciences, Texas A&M University, College Station, TX, USA

J. S. Pat Heslop-Harrison
Department of Genetics and Genome Biology, Institute for Environmental Futures, University of Leicester, Leicester, UK

South China National Botanical Garden, Chinese Academy of Sciences, Guangzhou, China

© The Author(s), under exclusive license to Springer Nature Singapore Pte Ltd. 2025

S. Faiz et al. (eds.), *Climate Smart Agriculture for Future Food Security*,
https://doi.org/10.1007/978-981-96-4499-5_13

Biotechnology itself plays a pivotal role through diverse and sustainable techniques to overcome the vagaries of biotic and abiotic stresses, e.g. genetic engineering, tissue culture, genomics, and genome editing. New and resistant genotypes are evolved through genetic engineering technology. Similarly, CRISPR and other gene-editing tools developed crop varieties which are resistant against intense climatic conditions. Marker-Assisted selection is another technique which selects the desirable traits by using molecular markers and speed up the breeding process. Digital technology is the most recent era of farming that is very helpful for farmers in agricultural practices. Various digital technologies have been introduced for proficient farming like artificial intelligence (AI) that provides information on all aspects of farming, plant growth-related issues and ultimately recommends essential actions to improve productivity and sustainability. Drones and satellite monitor the growth parameters and field condition and provide relevant data. Along with these, there are many other devices and systems being used to record the real time data. Precision agriculture, robotics, AI, and mobile applications are the latest digital technologies, ICT are the pre-requisites of modern agriculture to sustain growth and production quantum.

Keywords

Artificial intelligence · Genomics · ICT · Mobile apps · Genome-editing

13.1 Introduction

Under the changing climatic conditions, the smart techniques to mitigate the onset of vagaries in agriculture can be overcome by the followings.

13.2 Biotechnology and Digital Revolution

Biotechnology and the digital revolution are transforming agriculture by increasing productivity, improving sustainability, and addressing universal food safety challenges. Innovative genomic techniques, such as cis-genesis and genome editing, combined with in vitro regeneration technologies, enabled the creation of innovative, value-added products. Simultaneously, the integration of digital tools and data analytics optimized the agricultural practices, leading to more efficient resource management and better decision-making. Together, these advancements may play a key role in meeting the demands of a growing population and certifying the sustainability of agricultural systems in the face of climate change (Noorka and Heslop-Harrison 2014; Taranto et al. 2018).

13.3 Impact of Climate Change on Agriculture

Climate change, characterized by global warming, extreme weather conditions, and an increase in weeds, pests, and pathogens, significantly impacts major cropping systems. Change in climate effects the agriculture and other components of food system. In this challenging environment, diverse strategies are essential to accelerate genetic gains and develop new crop varieties. Key components of next-generation breeding include large plant breeding populations, efficient high-throughput technologies, big data management tools and advanced biotechnology, and molecular techniques (Noorka 2014).

Biotechnology has performed its pivotal role in climate-resilient crops through (Sarfraz et al. 2023)

- Genome editing.
- Genomics.
- Genetic engineering.

13.4 Plant Breeding Techniques and Smart Climate Approaches

Plant breeding has always been a crucial tool for adapting to changing climates by developing new cultivars with resistance to stresses such as drought, temperature extremes, salinity, waterlogging, pests, and diseases. Plant breeders are leveraging agricultural biotechnology to create drought-tolerant varieties of crops such as wheat, rice, tomatoes, soybeans, and cotton (Noorka and Heslop-Harrison. (2019)). In the United States, drought-tolerant corn, including genetically engineered varieties, is already being cultivated in drought-prone regions. Additionally, drought-tolerant wheat has been approved for use in Argentina and Brazil. To ensure food security under challenging climatic conditions, various plant breeding techniques (Jamil et al. 2023), such as:

13.4.1 Crop Wild Relatives

Crop wild relatives are wild plant species that are closely related to domesticated crops. They possess genetic diversity that can be invaluable for breeding programs, particularly in developing new crop varieties with improved traits such as disease resistance, drought tolerance, and adaptability to changing climatic conditions. By incorporating genes from CWR into cultivated crops, plant breeders can enhance the resilience and productivity of agricultural systems (Fig. 13.1).

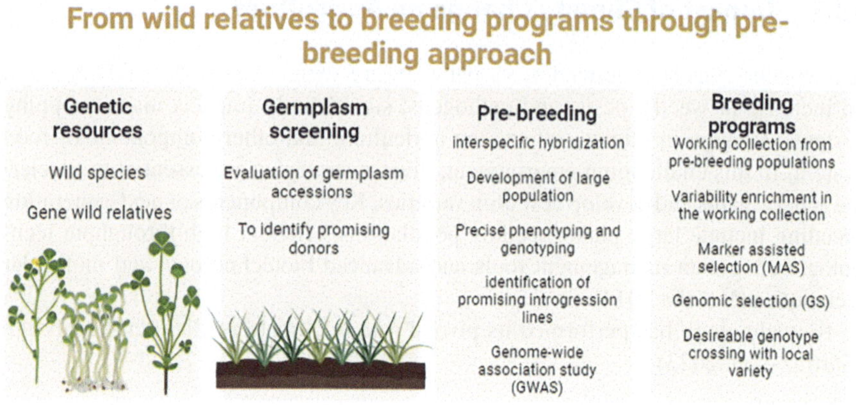

Fig. 13.1 Impact of plants wild relatives in breeding programs

13.4.2 Shuttle Breeding

Shuttle breeding is a plant breeding technique that involves growing plants in multiple locations or environments within the same year. This method accelerates the selection process for desirable traits, such as resistance to various stresses or adaptability to different climatic conditions. By exposing plants to diverse environments, breeders can quickly identify and develop cultivars that are more resilient and perform better under a range of conditions (Bakala et al. 2020).

13.4.3 Genomic-Assisted Breeding

It is a modern plant breeding technique that uses genetic information to guide the selection of desirable traits in crops. By analyzing the genome of plants, breeders can identify specific markers associated with beneficial traits such as disease resistance, drought tolerance, or higher yield (Noorka (2019)). This information allows for more precise and efficient selection of parent plants to develop new varieties, speeding up the breeding process and improving the likelihood of success in creating crops that are better adapted to various environmental conditions (Mahood et al. 2020).

13.4.4 Genome-Wide Association Studies

Genome-wide association studies are a powerful tool used in plant breeding to identify genetic variants associated with specific traits across the entire genome. By analyzing the DNA of a large number of plants, these studies can detect associations between genetic markers and traits such as disease resistance, drought tolerance, or

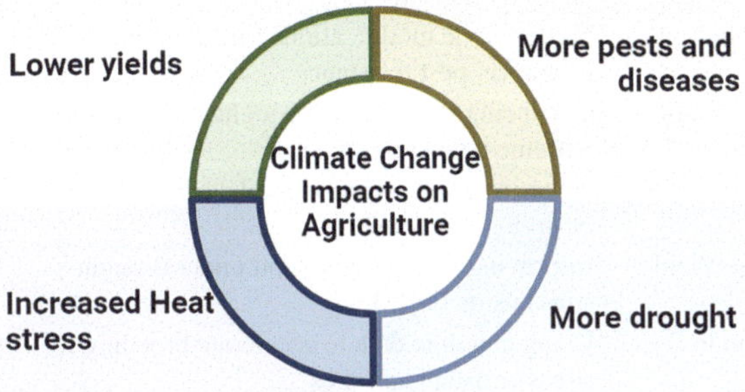

Fig. 13.2 Smart farming for climate-resilient crops

yield. This approach helps breeders understand the genetic basis of complex traits and can inform the selection of parent plants for breeding programs (Fig. 13.2). GWAS accelerates the development of improved crop varieties by pinpointing the genetic factors that contribute to desirable characteristics, enabling more targeted and efficient breeding strategies (Li et al. 2018).

13.4.5 Double Haploid Technology

Double haploid technology is a plant breeding technique used to produce homozygous plants in a single generation. This method involves generating plants that have two sets of chromosomes from a single parent, resulting in plants with uniform genetic material. Induction of Haploids and Doubling of Chromosomes are 2 stages of this technique. These methods accelerate the evolution of crop varieties, leading to the development of climate-resilient crops that perform better under both biotic and abiotic stress conditions (Gosal and Wani 2020).

13.5 Biotechnology Role in Climate Smart Agriculture

Agricultural systems face significant threats from climate change, leading to reduced productivity and impacting food security. To address and mitigate these challenges, climate-smart agriculture has emerged as an effective strategy. This approach aims to enhance agricultural productivity, build resilience, and reduce greenhouse gas emissions (Kole 2020) (Table 13.1).

Table 13.1 Biotechnological tools for climate-smart crops

Technology	Area of Application	Reference
CRISPR-Cas9 Gene Editing	Precise gene modifications for drought tolerance, pest resistance	Chen et al. (2019)
Gene Editing for Abiotic Stress	Enhancing tolerance to drought, salinity, and extreme temperatures	Alok et al. (2020)
Marker-Assisted Selection (MAS)	Selecting plants with desirable traits for heat or drought tolerance	Varshney et al. (2018)
CRISPR for Maize Yield Improvement	Improving maize grain yield under drought stress using CRISPR	Shi et al. (2018)
Genomic Selection in Plant Breeding	Using genomic data to accelerate breeding for stress-resistant varieties	Crossa et al. (2017)

13.5.1 Climate-Smart Molecular Plant Breeding

Climate-smart plant molecular breeding is an advanced method that integrates climate resilience into crop development by applying molecular techniques. This technique focuses on developing crop varieties that can withstand climate-related stresses such as extreme temperature conditions, drought, and pests. By leveraging genetic information and molecular tools, climate-smart breeding aims to enhance crop productivity and sustainability in the face of climate change (Munawar et al. 2020).

13.5.2 QTL Mapping and Marker-Assisted Selection (MAS)

MAS is a key biotechnological tool in breeding programs used to select plants with desired traits based on genetic markers. In the context of climate-smart crops, MAS can identify and breed plants that carry genes for drought tolerance, heat resistance, and pest resistance. This technology accelerates the breeding process, ensuring that only plants with the most beneficial traits are selected for further cultivation (Varshney et al. 2018). MAS emphasizes the utility of the technology in developing climate-resilient crops like chickpeas, which show enhanced resistance to high temperatures and drought, highlighting MAS's role in future-proofing agriculture (Varshney et al. 2020).

QTL mapping involves in detecting specific regions of genome (loci) which are associated with variation in quantitative traits, such as drought tolerance, disease resistance, or its yield. By evaluating the genetic makeup and phenotypic data of a population, researchers can locate QTLs and understand the genetic basis of important traits.

Marker-assisted selection uses molecular markers linked to desirable traits to guide breeding decisions. Once QTLs are identified, these markers can be used to screen and select plants that carry the desired traits, even at the seedling stage. This phenomenon accelerates the breeding process and increases the precision of developing new crop varieties with specific characteristics (Massruhá et al. 2023).

13.5.3 Genetic Modification of Crops

Genetic modification of crops involves altering the DNA of plants to achieve desirable traits such as increased yield, resistance to pests and diseases, improved nutritional content, and tolerance to environmental stresses.

13.5.4 Genome Editing

Genome-editing technologies fall under broader category of new plant breeding techniques and can be divided into two main types: "oligonucleotide directed mutagenesis (ODM)" and "site directed nucleases (SDNs)." These technologies enable gene transfer, precise mutagenesis, and gene expression control.

Genetic Engineering and CRISPR-Cas9: On the other hand, Genetic engineering has long been used to introduce traits such as herbicide resistance and improved yield in crops. However, with CRISPR-Cas 9, gene editing has become more precise and efficient, allowing scientists to make targeted modifications to crop genomes. CRISPR enables the editing of genes responsible for traits like drought tolerance and pest resistance. By editing these genes, crops can be made more resilient to climate-induced stresses without the need to introduce foreign DNA. CRISPR-Cas 9 was used to engineer rice varieties with improved drought resistance, reducing yield loss under water-scarce conditions by 25% (Chen et al. 2019; Mao et al. 2019).

Oligonucleotide-Directed Mutagenesis (ODM): In ODM, chemically synthesized DNA fragments, 20 to 100 nucleotides long, are introduced into plant cells using common methods such as PEG transfection or particle delivery. These fragments induce mutations at target sites with low efficiency.

Site-Directed Nucleases (SDNs): SDNs are enzymes that specifically bind to short target DNA sequences (9 to 40 nucleotides) and perform various biochemical reactions in situ, such as introducing double-strand breaks (DSBs) methylation, demethylation, acetylation, and deamination. These reactions can alter biological activities such as gene expression, gene silencing, and base editing.

Gene Editing for Abiotic Stress Tolerance: Abiotic stresses, including drought, salinity, and extreme temperatures, are exacerbated by climate change. Biotechnological interventions can improve crops' resilience to these stresses. For example, gene-editing techniques have been employed to enhance water-use efficiency in crops, making them more drought-tolerant. The modification of transcription factors like DREB proteins has been instrumental in developing drought-resistant crop varieties (Alok et al. 2020).

Moreover, gene-editing technology was used to modify abscisic acid (ABA) pathways, plants exhibited enhanced resilience to drought conditions, demonstrating significant potential for agricultural sustainability (Shi et al. 2018).

13.5.5 Integration of AI and Biotechnology for Climate-Smart Crop Development

The integration of AI and biotechnology offers a powerful approach for developing climate-smart crops. AI models can analyze genetic data to identify key genes responsible for stress tolerance, while biotechnological tools can modify or enhance these genes. For example, AI can predict how genetic modifications will impact a plant's response to environmental stresses, enabling researchers to focus on the most promising traits for genetic enhancement (Leng and Hall 2020). Moreover, AI algorithms can simulate the performance of genetically modified crops across various environments, further optimizing their adaptation to climate change. This synergistic approach is already being used in projects such as the Genomic Selection program initiated by the International Maize and Wheat Improvement Center (CIMMYT), where AI and genomic tools are combined to accelerate the breeding of heat- and drought-tolerant maize varieties This integration represents a new frontier in crop science, with the potential to significantly enhance the efficiency of breeding programs and improve crop resilience (Crossa et al. 2017) (Table 13.1).

The convergence of AI and biotechnology marks a paradigm shift in the development of climate-smart crops. AI-driven tools enable real-time monitoring, precision management, and predictive modeling, while biotechnological advancements such as gene editing and marker-assisted selection offer the molecular innovations needed to enhance crop resilience. Together, these technologies address the challenges posed by climate change, ensuring the development of resilient crops capable of securing food production in an uncertain future. As AI and biotechnology continue to advance, their integration will be crucial for creating sustainable agricultural systems that can adapt to and mitigate the impacts of climate change.

13.6 Digital Farming: A Smart Solution for Sustainable Agriculture

Agriculture has undergone numerous revolutions that have significantly boosted yield, productivity, and profitability to previously incredible elevations. According to market predictions for the next decade indicate that "digital agricultural revolution" will be the latest transformation ensuring that agriculture continues to meet the demands of the global population. Digital farming focuses on extracting value from data (Sahoo 2020). Smart farming & precision farming both combined into digital farming. AI (Artificial Intelligence) revealed as digital technology used by farmers to enhance crop production. Agriculture with AI is still in its beginning (Khan et al. 2022).

Farmers, particularly those reliant on timely rainfall for their crops, may be significantly affected by changing weather patterns such as rising temperatures and variations in precipitation. The entire agriculture food chain will transform with digitalization, enabling customized, optimal, and anticipatory resource management across the system. This transformation will operate in real-time, driven by data and interconnected networks. According to a report by the Direct Line Group, digital farming involves the "consistent application of precision farming and smart farming methods, both internally and externally networking the farm and utilizing web-based data platforms along with Big Data analysis"(Klerkx et al. 2019).

Digital farming is empowering farmers with the tools and knowledge which is needed for the development of climate-smart crops.

13.6.1 Precision Farming

13.6.1.1 Irrigation Management by Sensors and IoT

Various agriculture devices are located in the fields to collect data on soil moisture content and nutrient levels. This data aids the farmers to apply fertilizers and water precisely and enhancing crop yields by reducing waste.

13.6.1.2 Satellite Imagery and Drones

Drones and satellite imagery are digital technological tools examine the crop conditions from above and detecting the areas that require attention and after that according to field conditions treatments are applied (Ayre et al. 2019).

13.6.2 Big Data and Analytics

In this technology massive amount of data (about planting, harvesting, irrigation) is collected from various sources including weather stations, oil sensors, and satellite imagery to build comprehensive datasets. Predictive analysis is part of digital technology we use machine learning algorithms to analyze the data and forecast consequences, e.g. pest infestations, yield of crops, and optimal planting times. This data supports the farmers in making proactive decisions (Sarker et al. 2020).

13.6.3 Artificial Intelligence (AI) and Machine Learning in Climate-Smart Agriculture

AI, particularly machine learning (ML) and deep learning (DL), is revolutionizing agricultural systems through data-driven approaches. These technologies are integral to precision agriculture, which optimizes inputs such as water, fertilizers, and pesticides to reduce waste and environmental degradation (Fig. 13.3; Table 13.2).

13.6.3.1 Decision Support Systems

AI-powered platforms provide recommendations on irrigation schedules, fertilization, and pest control based on data analysis. In this way it is improving efficiency and reducing resource use.

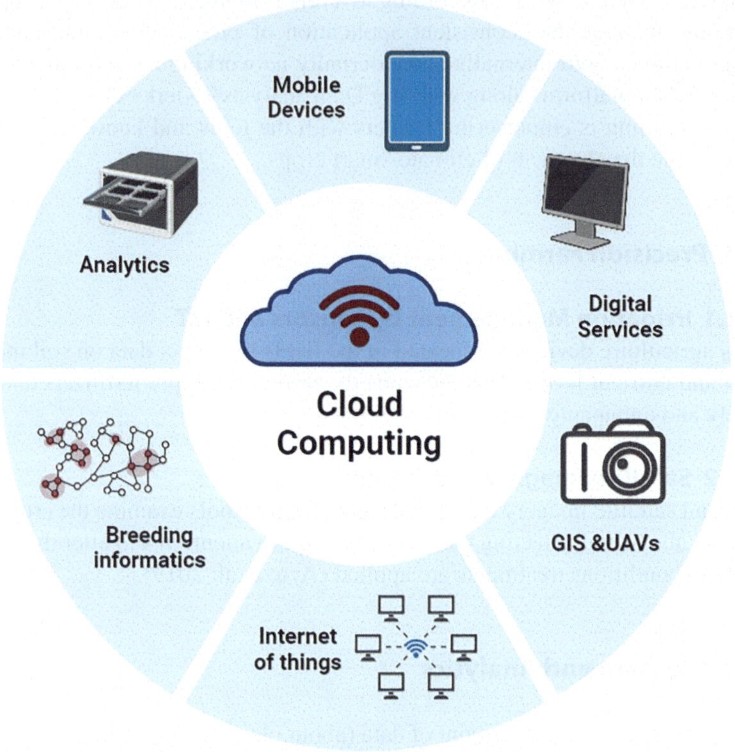

Fig. 13.3 Digital farming through various digital technologies

Table 13.2 Artificial intelligence in climate-smart agriculture

Technology	Area of application	Reference
AI-Powered Crop Monitoring	Real-time monitoring of crop health, detecting diseases and stress	Kamilaris and Prenafeta-Boldú (2018)
AI for Climate Prediction	Climate modeling, forecasting climate patterns, and crop adaptation	Leng and Hall (2020)
AI-Enhanced Breeding Programs	Genomic selection for breeding climate-resilient crops	Montesinos-López et al. (2023)
Machine Learning for Crop Yield	Predicting yield outcomes based on environmental conditions	Xu et al. (2021)
Deep Learning for Disease Detection	Identifying plant diseases using image-based detection systems	Ramcharan et al. (2017)

13.6.3.2 Climate Modeling

AI models forecast future climate circumstances and their impact on agriculture. Artificial intelligence is helping farmers to implement their practices according to changing climatic conditions. AI offers an innovative benchmark summary for analytical and computational methods used in the integrated analysis of vast datasets across the big data spectrum.

Utilizing AI to deduce interrelations between candidate genes and CREs represents a novel approach for identifying and categorizing previously unknown genes critical for significant crop improvements.

Additionally, AI strategies hold immense potential for interpreting crop yield, assessing climatic variations, high-throughput crop stress phenotyping, and evaluating climate factors such as temperature, UV radiation, wind, and hail.

The role of AI is increasingly vital in acquiring, analyzing, integrating, and managing genomic and phenomic data to enhance agricultural climate resilience.

13.6.4 AI-Powered Crop Monitoring and Management

AI-driven tools enable real-time monitoring of crop health using satellite imagery, drones, and ground-based sensors. These technologies, supported by remote sensing data, allow AI models to analyze plant health, detect early signs of stress, and predict yields with high accuracy. For instance, Convolutional Neural Networks (CNNs) can process multispectral images of crops to identify disease outbreaks or nutrient deficiencies (Kamilaris and Prenafeta-Boldú 2018).

This capability facilitates timely interventions, reducing crop losses and optimizing resource use. AI-based models were applied to cassava crop monitoring, demonstrating over 98% accuracy in detecting disease, showcasing AI's potential to revolutionize crop management practices (Ramcharan et al. 2017).

13.6.4.1 AI for Climate Prediction and Crop Adaptation

AI's capacity to analyze vast datasets makes it invaluable for climate modeling and forecasting. By integrating historical climate data with real-time weather information, AI can predict future climate patterns, allowing farmers to adjust planting schedules and select climate-resilient crops. For example, ML algorithms can assess the impact of climatic factors such as rainfall and temperature on crop growth, helping breeders develop climate-resilient varieties. A significant application of AI in this area is the use of recurrent neural networks (RNNs) for predicting seasonal climate variability, which has been instrumental in optimizing irrigation practices and reducing crop failures (Leng and Hall 2020).

13.6.4.2 AI-Enhanced Breeding Programs

AI is transforming breeding programs by rapidly analyzing genetic data to identify desirable traits such as drought tolerance, heat resistance, and pest resilience. Traditional breeding methods are often time-consuming, but AI-driven genomic selection accelerates this process by predicting the performance of new crop varieties based on genetic markers. This allows for the rapid development of crops that can withstand climate stressors, improving agricultural sustainability. AI's role in breeding is further highlighted by studies like Xu and colleagues, where ML models improved the prediction of maize yields by up to 15%, demonstrating the efficacy of AI in accelerating crop improvement (Montesinos-López et al. 2023; Xu et al. 2021).

13.6.5 Digital Platforms and Mobile Application

Digital farming is now very easy way farming. Farmers have easy access to information by using mobile apps. Farmers use mobile apps to access weather forecasts, market prices of sale and purchase of products, and agricultural guidance enables them to make informed decisions (Dayıoğlu and Turker 2021).

13.6.6 Robotics

Robotics and automation reduces the manual labor and production is increased in sufficient time with accuracy. Now robots can be used for some difficult tasks like weeding & spraying, gathering nuts and fruits, and also used for crop monitoring (Shamshiri et al. 2018).

13.6.7 Implementation of ICTs in the Agriculture Sector

In the development and growth of the agriculture sector, information and communication technology (ICT) contributes a lot. ICT helps in socio-economic development. The nations who implement ICT into the agriculture sector can greatly improve efficiency and productivity of the agricultural value chain (Porwal and Verma 2023).

13.7 Advantages of Digital Technology for Climate-Smart Crop

13.7.1 Water and Nutrient Management

Precise irrigation reduces water usage by watering plant in precise amount at the right time. Fertilizers are applied to field according to data driven by digital techniques to provide optimal nutrition to plant (Abioye et al. 2022).

13.7.2 Risk Management

By using digital tool, weather predictions are monitored which help the farmers design their farming activities. Detection of pest and disease is also possible through his technique which helps in minimizing crop loss.

13.7.3 Improved Productivity

Data generated by digital technology field management practices and crop rotation lead to higher yields. Manual labor is reduced due to robotics and automation.

13.7.4 Sustainability

Use of efficient resource decreases the environmental footprint of agriculture practices. These Digital tools help farmers to adapt extreme weather events and ensure sustainable crop production.

13.8 Summary

The convergence of biotechnology and digital revolution plays a vital role in developing climate-smart crops. This technology enables the farmers to accept the challenges of climate change. Biotechnology aids in development of resilient crops through different techniques like genetic engineering, genome editing, biotech traits, and precision breeding which helps in improving yield, increases water efficiency and making crops resistant against diseases. Digital revolution includes smart and precision farming which leverages precision agriculture, artificial intelligence, data analytics, and the Internet of Things (IoT) to optimize the crop adaptation and management, monitor the climatic conditions and predict the amount of yield. Due to combination of these technologies, farmers can access the information services, enhance the crop resilience, improved the productivity of crops and reduced chemical and water usage. Monitoring of crops can be done in a better way through genomic selection and digital crop monitoring. This leads to enhanced crop resilience, improved productivity, and reduced water and chemical usage. This integrated approach helpful for farmers to increase their income and livelihood security while addressing about climate change. Ultimately, biotechnology and digital revolution technologies help in ensuring sustainable agriculture, global food security, and improved livelihood.

Author Contribution Statement Ijaz Rasool Noorka conceptualized the chapter, led the overall coordination, and contributed significantly to the methodology and data analysis sections. Bilal Ahmad Khan assisted with the research design, contributed to the data analysis, and co-wrote key sections of the manuscript. Zafar Iqbal Khan provided expert insights into the theoretical framework and contributed to the discussion and conclusion sections. Muhammad Ather Nadeem supported the writing process and provided critical input on data interpretation and manuscript revisions. Tehreem Zehra assisted with data collection and helped structure the introduction and literature review. Abdul Wahid contributed to data analysis and played a key role in reviewing and editing the manuscript. Bismark Anokye provided support in the research design, assisted with data visualization, and helped finalize the draft. J.S. (Pat) Heslop-Harrison offered critical expertise in the overall theoretical framework, contributed to the manuscript's scientific rigor, and participated in the final revisions.

References

Abioye EA, Hensel O, Esau TJ, Elijah O, Abidin MSZ, Ayobami AS et al (2022) Precision irrigation management using machine learning and digital farming solutions. Agric Eng 4(1):70–103
Alok A, Sandhya D, Jogawat A, Bhati KK (2020) CRISPR/Cas9 and its applications for crop improvement in the era of climate change. In: Biotechnology for sustainable agriculture. Springer, pp 23–39
Ayre M, Mc Collum V, Waters W, Samson P, Curro A, Nettle R et al (2019) Supporting and practising digital innovation with advisers in smart farming. NJAS-Wageningen J Life Sci 90:100302
Bakala HS, Singh G, Srivastava P (2020) Smart breeding for climate resilient agriculture. In: Plant breeding-current and future views. IntechOpen
Chen K, Wang Y, Zhang R, Zhang H, Gao C (2019) CRISPR/Cas genome editing and precision plant breeding in agriculture. Annu Rev Plant Biol 70:667–697
Crossa J, Pérez-Rodríguez P, Cuevas J, Montesinos-López O, Jarquín D, de Los Campos G (2017) Genomic selection in plant breeding: methods, models, and perspectives. Trends Plant Sci 22(11):961–975
Dayıoğlu MA, Turker U (2021) Digital transformation for sustainable futureagriculture 4.0: a review. J Agric Sci 27(4):373–399
Gosal SS, Wani SH (2020) Accelerated plant breeding, volume 1: cereal crops. Springer
Jamil S, Kanwal S, Kanwal R, Razzaq H, Shahzad R, Afza N et al (2023) Climate-smart agriculture: a way to ensure food security. Pak J Bot 55(3):1157–1167
Kamilaris A, Prenafeta-Boldú FX (2018) Deep learning in agriculture: a survey. Comput Electron Agric 147:70–90
Khan MHU, Wang S, Wang J, Ahmar S, Saeed S, Khan SU et al (2022) Applications of artificial intelligence in climate-resilient smart-crop breeding. Int J Mol Sci 23(19):11156
Klerkx L, Jakku E, Labarthe P (2019) A review of social science on digital agriculture, smart farming and agriculture 4.0: new contributions and a future research agenda. NJAS-Wageningen J Life Sci 90:100315
Kole C (ed) (2020) Genomic designing of climate-smart cereal crops. Springer, Amsterdam, The Netherlands
Leng G, Hall J (2020) Crop yield sensitivity of global major agricultural countries to droughts and the projected changes in the future. Sci Total Environ 715:13696
Li Y, Shi W, Wasserman WW (2018) Genome-wide prediction of cis-regulatory regions using supervised deep learning methods. BMC Bioinformatics 19:202. https://doi.org/10.1186/s12859-018-2187-1
Mahood EH, Kruse LH, Moghe GD (2020) Machine learning: a powerful tool for gene function prediction in plants. Appl Plant Sci 8:e11376
Mao Y, Zhang H, Xu N, Zhang B, Gou F, Zhu JK (2019) Application of the CRISPR/Cas9 system for efficient genome engineering in plants. Mol Plant 6(6):2008–2020
Massruhá SMFS, Leite MDA, Luchiari Junior A, Evangelista SRM (2023). Digital transformation in the field towards sustainable and smart agriculture
Montesinos-López OA, Montesinos-López A, Juliana P, Singh RP, Barrón-López JA (2023) A review of deep learning applications for genomic selection in crops. Plant Sci 310:110972
Munawar S, ul Qamar MT, Mustafa G, Khan MS, Joyia FA (2020) Role of biotechnology in climate resilient agriculture. Environ Clim Plant Vegetation Growth:339–365
Noorka IR (2014) Effect of drought/water stress and adaptation of unintended consequences on wheat growth and development in pakistan. Hand book of Plant and Crop Physiology. 441–452. https://taylorandfrancis.com
Noorka IR (2019) Climate risks and adaptation to crop yield in Pakistan: toward water stress tolerance for food security. In: Leal FW, Azul A, Brandli L, Özuyar P, Wall T (eds) Climate action. Encyclopedia of the UN sustainable development goals. Springer, Cham. https://doi.org/10.1007/978-3-319-71063-1_113-1. www.springer.com

Noorka IR, Heslop-Harrison JS(P) (2014) Water and crops: molecular biologists, physiologists, and plant breeders approach in the context of evergreen revolution. In: Hand book of plant and crop physiology. CRC Press, Taylor and Francis, pp 967–978. https://taylorandfrancis.com
Noorka IR, Heslop-Harrison. JS(P) (2019) Cross-disciplinary drivers to benefit smallholder farmer communities and to achieve the SDGs by various means. In: Leal FW (ed) Handbook of climate change resilience. Springer, Cham. https://doi.org/10.1007/978-3-319-71025-9_40-1. www.springer.com. 978-3-319-71025-9 Page 1-12
Porwal M, Verma B (2023) Agronomic interventions for the mitigation of climate change. Emerg Trnd Clim Change 2(1):27–39. https://doi.org/10.18782/2583-4770.122
Ramcharan A, Baranowski K, McCloskey P, Ahmed B, Legg J, Hughes DP (2017) Deep learning for image-based cassava disease detection. Front Plant Sci 8:1852
Sahoo JP (2020) Digital farming–a new era of Indian agriculture. Biotica Research Today 2(7):567–569
Sarfraz S, Ali F, Hameed A, Ahmad Z, Riaz K (2023) Sustainable agriculture through technological innovations. In: Sustainable agriculture in the era of the OMICs revolution. Springer International Publishing, Cham, pp 223–239
Sarker MNI, Islam MS, Murmu H, Rozario E (2020) Role of big data on digital farming. Int J Sci Technol Res 9(4):1222–1225
Shamshiri RR, Weltzien C, Hameed IA, Yule IJ, Grift TE, Balasundram SK et al (2018) Research and development in agricultural robotics: a perspective of digital farming. Int J Agric Biol Eng 11(4):1–14
Shi J, Gao H, Wang H, Lafitte HR, Archibald RL, Yang M, Li Z (2018) ARGOS8 variants generated by CRISPR-Cas9 improve maize grain yield under field drought stress conditions. Plant Biotechnol J 15(2):207–216
Taranto F, Nicolia A, Pavan S, De Vita P, D'Agostino N (2018) Biotechnological and digital revolution for climate-smart plant breeding. Agronomy 8(12):277
Varshney RK, Pandey MK, Bohra A (2018) Towards the next generation of crop genomics. Curr Opin Plant Biol 42:142–153
Varshney RK, Roorkiwal M, Sorrells ME (2020) Genomic selection for crop improvement: challenges and opportunities. Plant Breed 139(4):596–610
Xu Y, Liu X, Fu J, Wang H, Wang J, Huang C (2021) Improvement of maize yield prediction by integrating machine learning with crop modeling. Field Crop Res 264:108071

Climate Change and Changing Pests Scenario

14

Vyshnavi Sunil, K. Logeswaran, Akanksha Adhikari, Shreya Singh, Alagesan Keerthana, Abhibandana Das, Badavat Sujatha, Habib Ali, and Vinod Kumar Dubey

Abstract

Climate change has disturbed ecological balance, food security, sustainability, and human lives. Effects of climate change can adversely affect the pest life cycle, alter pest status, and increase their population which can pose a huge threat to humans around the world. Climate change impacts in urban areas are amplified compared to rural areas since it creates urban warm heat islands which make them warmer. This phenomenon increases pest activity by making pests move to warmer areas. The rise in carbon dioxide (CO_2) concentration causes an increase in temperature, which is responsible for critical changes in insect life patterns like behavior, diapauses, and an increase in generation numbers per year. CO_2 concentration and temperature are interrelated to each other. CO_2 increase leads to an increase in temperature. Also, climate change can lead to the ineffectiveness of natural enemies i.e., an asynchrony in the interaction of pests and natural enemies which can potentially contribute to pest outbreaks. These circumstances drastically reduce crop yields. Combination of climate forecasting and environmental changes can be used to develop models which predict the future pest infestation. Continuous pest monitoring and revised integrated pest management tactics can help to deal this problem.

V. Sunil
Department of Entomology, College of Agriculture, Kerala Agricultural University, Thiruvananthapuram, Kerala, India

K. Logeswaran
Division of Entomology and Nematology, Indian Institute of Horticultural Research, Bengaluru, Karnataka, India

A. Adhikari
Department of Zoology, Radhey Hari Government Post Graduate College, Kashipur, Udham Singh Nagar, Uttarakhand, India

© The Author(s), under exclusive license to Springer Nature Singapore Pte Ltd. 2025

S. Faiz et al. (eds.), *Climate Smart Agriculture for Future Food Security*,
https://doi.org/10.1007/978-981-96-4499-5_14

Keywords

Climate change · Crops · Food security · Pests · Environmental threat

14.1 Introduction

Climate change is explained by variation through time in the statistical properties of the climate system, irrespective of its nature (ENSAA 2011; IPCC 2001). India is more vulnerable to climate change because of its geographical location (Yadav et al. 2021). The increase in global temperature during the previous century was 0.8 °C, and by the end of the next century, it is predicted to be between 1.1 and 5.5 °C (Skendzic et al. 2021). The key elements that affect the growth and development of organisms are temperature, light, and water (Harvell et al. 2002; Rosenzweig et al. 2001; Stanaszek-Tomal 2020). Insect pests cause an estimated annual loss of 13.6% globally, and the extent of losses in India is 17.5% (Dhaliwal et al. 2010) and annual loss of US$36 billion (Dhaliwal et al. 2015). As per Food and Agriculture Organization (FAO), the global economy lost $290 billion to pests in yields. Life cycle and fecundity of pests depend on several factors like temperature, moisture, relative humidity, precipitation, etc. Insects are cold-blooded organisms. Low temperatures are core impactful for the geographical distribution of insect pests than higher ones. Changes in cropping patterns due to climate change also affect the geographical distribution of certain species (Maiorano et al. 2008). Minor pests are likely to become more problematic with global warming (Sharma 2013). Climate change develops new ecological niches for insects, creating opportunities to explore more diverse areas, causing them to migrate from one region to another (Skendzic et al. 2021).

S. Singh
Ramlalit Singh Mahavidyalaya, Kailhat, Chunar, Mirzapur, Uttar Pradesh, India

A. Keerthana
Department of Entomology, Tamil Nadu Agricultural University, Coimbatore, Tamil Nadu, India

A. Das
Department of Entomology, College of Agriculture, Assam Agricultural University, Jorhat, Assam, India

B. Sujatha
Department of Entomology, Dr. YSRHU-College of Horticulture, Parvathipuram, Vizianagaram, Andhra Pradesh, India

H. Ali
Department of Agricultural Engineering, Khwaja Fareed University of Engineering and Information Technology, Rahim Yar Khan, Pakistan

V. K. Dubey (✉)
Department of Genetics and Plant Breeding, Acharya Narendra Deva University of Agriculture and Technology, Ayodhya, Uttar Pradesh, India

14.2 Elevated CO_2 Levels

Reduced transpiration and stomatal conductance, increased water and light usage capacity, and an elevation in photosynthetic efficiency are the principal effects of high carbon dioxide (CO_2) on plants (Skendzic et al. 2021). Changing CO_2 levels affect plants' nutritional qualities by altering protein and amino acid content. The reduction in protein profile is noticed and insects feed more to compensate for that. The magnitude of the effect of changes in CO_2 levels differs from plant to plant i.e., C3 plants are more sensitive to CO_2 changes than C4 plants. According to Rötter and Van De Geijn (1999), an increase in atmospheric CO_2 content will be advantageous for C3 and C4 plants.

In the middle of the twenty-first century, climate change models predict a high concentration of O_2 and CO_2 in the atmosphere using free air gas concentration enrichment technology. When soybeans, *Glycine max* (Linn.) Merr, were cultivated at these elevated CO_2 levels, their sugar content increased by 31%. This caused 57% more insect damage than when they were planted at ambient temperature. Hence, enhanced sugar content in leaves grown at elevated CO_2 acted as a phagostimulant for soybean-damaging insect pests (Hamilton et al. 2005). Researchers reported that an increase in CO_2 levels increased the growth of stem and leaves in plants which increased the crop canopy which resulted in a higher humidity favorable to pathogens (Elad and Pertot 2014; Chowdhury et al. 2023).

Elevated levels of temperature and CO_2 due to global warming affect the plant pathogenic nematode population by altering host physiology. A number of studies have shown that CO_2 has a positive or neutral effect on plant parasitic nematodes, while global warming increases the geographical distribution of these parasites. Analysis of different cropping systems on the effect of CO_2 enhancement on nematode populations was analyzed. It was found that the nematode channel ratio (NCR) for sugar beet and wheat exhibited higher CO_2 levels (Sticht et al. 2009). It was reported that plant parasitic nematodes from the genera needle nematode, *Longidorus* spp. and citrus nematode, *Tylenchlus* spp. This could be attributed to an increase in root production due to elevated CO_2 conditions (Yeates et al. 2003).

CO_2 may expand plant canopies, producing more biomass with higher nutritional value and considerably higher relative humidity in the microclimate. This would undoubtedly attract plant diseases including fungal diseases. Increased overwintering crop waste would also boost inoculum potential. Transitions in the type of pathogen and the host plant can reshape the rates of infection (Manning and Tiedemann 1995). Research revealed on Asian rice (*Oryza sativa* Linn.) diseases which include leaf blast. In Japan, rising amount of carbon dioxide leads to an increased risk of the leaf blast infections, but once merged with ozone, it exponentially weakens the infection rate of downy mildew in soybean (Kobayashi et al. 2006). When compared to atmospheric CO_2, remarkably increases the incidence limits of the amount of soyabean brown spot of mildew-infected leaf area grown at 400 ppm (Eastburn et al. 2010). As a result of shrubby stylus grown under elevated emission profiles, more conidia of anthracnose, *Colletotrichum gloeosporioides* (Penz & Sacc.) were trapped under these conditions. That one, following higher

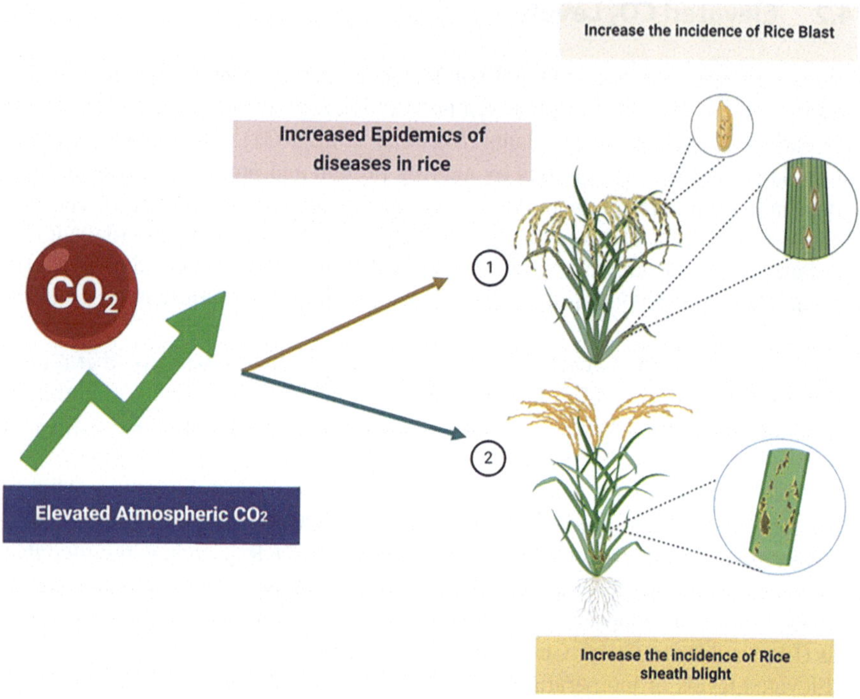

Fig. 14.1 Graphical representation of CO_2 on the incidence of pests

humidity in the dense canopy, stimulated more intense in tear stain than in seedling in lower emission (Chakraborty et al. 2000) (Fig. 14.1).

14.3 Rising Temperatures

Temperature greatly affects insect development. An increase in metabolism and physiological activity increases metabolism, hence the insect feeds more which ultimately results in huge damage to agricultural crops (González-Tokman et al. 2020). For plenty of pest species, including corn lice, climate can radically change the fertility rates, which could have an inflationary effect. Aphid alarm pheromone which is generated when predators strike is less sensitive for detecting at extreme temps (Awmack et al. 1997). Due to juvenile stages growing faster in warmer climates, many insects belong to order lepidoptera have altered their phenology, with adults emerging earlier in recent years. For *instance,* a change in phenology of butterflies has been reported in Europe and for every 1 °C increment, by 2–10 days along with prolonged flight periods (Roy and Sparks 2000; Stefanescu et al. 2003).

Temperature changes can also cause phenotypic changes in plants, such as flowering, grain formation, etc. (Luo 2011). In temperate areas, increased temperature results in elevated pest attack, hence a higher number of insecticidal sprays.

Repetitive use of insecticides in such areas is responsible for insecticide resistance development in pest species and pest resurgence issues (Shelton et al. 2001; Georghiou and Taylor 1986). Pests have wide flight range, such as the gram pod borer, can adapt to climatic changes conditions from different geographical areas (Sharma 2005).

Disease transmission and outbreak intensity were correlated with temperature as per laboratory experiments conducted by Elderd and Reilly (2014). With a combined effect of host immune responses and rising temperatures, crown rot of wheat and barley crown rot will indeed be reduced (Sabburg et al. 2015). It was reported that increased temperature increased oat and wheat susceptibility to rust disease and made them less susceptible to forage species (Coakley et al. 1999). Models for tomato late blight and potato late blight showed that temperatures between 45 °F (7.2 °C) and 80 °F (26.8 °C) at high moisture levels favored the most successful infection and reproduction of the fungus (Wallin and Waggoner 1950). UV radiation, sunlight, and extreme temperatures inactivate the virus (Thompson et al. 2006). For example, Stewart's wilt caused by *Xanthomonas stewarti* (Smith) Dowson is spread by the vector corn flea beetle, *Chaetocnema pulicaria* F. E. Melsheimer. The severity of this disease is due to pathogen overwintering during winter season. A forecast model is designed to predict disease severity based on winter temperatures. The prototype acknowledges the fact that clover flea beetle survival is higher in warmer than in colder ones (Castor et al. 1975). It was reported that a warm dry climate increases the plant's resistance to disease infection, hence less fungicidal application is required (Patterson et al. 1999). Their dependence on temperature for essential physiological functions is well established. Temperature influences the migration, development, and capacity of the root-knot nematode to detect the roots of its host. Studies show 28 °C is the ideal temperature for root knot nematode development. Above that point, development is reduced, and at 36.5 °C, no development takes place (Fig. 14.2).

14.4 Drought, Flooding, and Moisture Levels

An increase in global temperature can result in more precipitation and floods i.e., globally, humid areas would be diminishing by 3%, while dry and semi-arid areas would grow by 10.3% and 9.9% as per the experiment conducted by Tabari (2020). When we include the thermodynamic Clausius–Clapeyron interaction, moisture intensity rises with temperature at a rate of about 5% per degree increase that time only few insects thrive mostly in humid conditions (Tabari 2020). For instance, the desert locust, *Schistocerca gregaria* Forsskal, emerges after rains and population increases hugely. Insects are washed away in heavy rainfall whereas some insects overwinter in soils and crop residues (Pathak et al. 2012). Since climate patterns is the best way to maximize of crop production worldwide, variations in seasonal have a substantial impact (Van Doren Jr and Allmaras 1978). The soils dwelling wireworm was studied in plots, response to increased summer rainfall and drought. Crops such as potato (*Solanum tuberosum* Linn.), corn (*Zea mays* Linn.), sugar

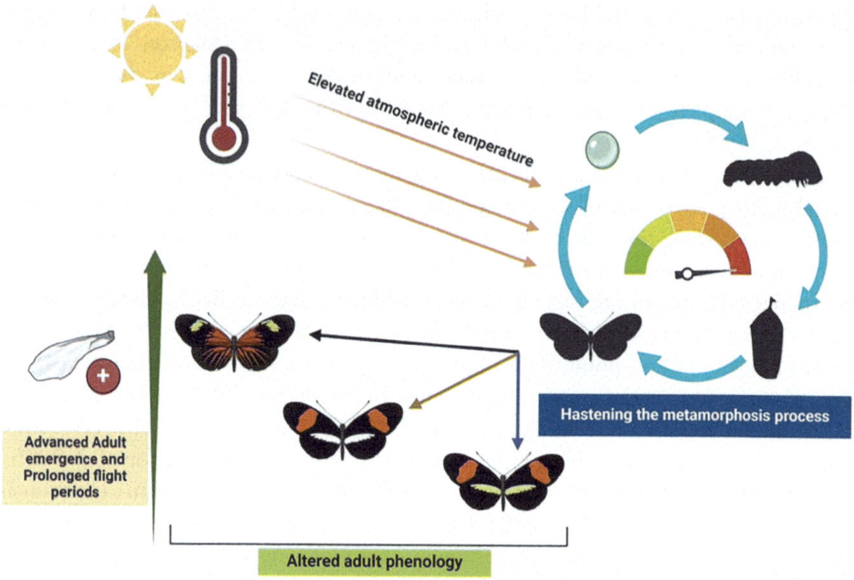

Fig. 14.2 Pictorial representation of temperature and how the temperature is affecting the insect pests population

beets, etc., are extremely vulnerable to damage from wireworms, especially if they are grown in grassland plots. Climate change will likely make wireworms a more serious problem. In direct contradiction to atmospheric and water restrictions, wireworm species in the topsoil area increased efficiently in terms of steadily increasing summer rainfall (Staley et al. 2007; Johnson et al. 2008).

It affects insects by favoring their growth and development like spider and mites. Also, stressed plants can activate the signaling pathway and release signals that attract some insect species. For example, when transpiration takes place, some gaps are created in the xylem which makes an acoustic sound that attracts certain beetles. The secondary metabolites that protect plants from herbivores are not produced under drought stress, resulting in more pest attacks (Skendzic et al. 2021).

According to latest studies, the international allocation of potentially pathogenic microbes is tremendously triggered by environmental attributes (Větrovský et al. 2019; Delgado-Baquerizo et al. 2020). For fungal spores to germinate and for the germ tube to enter the host, humidity is highly desirable. Solvent and precipitation are integral for the transmission of pathogens on crops, as they make them more vulnerable to specific pathogens and alter the transmission path. Pathogens causing diseases like rot, late blight, and apple scab favor high moisture for their growth and development. Pathogen causing powdery mildew favors low temperatures (Hunjan and Lore 2020). The composition of plant parasitic nematodes can be directly influenced by various parameters (Verschoor et al. 2001) or may not have an effect. In stunt nematode and dagger nematode in terms of humidity and temperature, densities are more affected than lesion nematode (Reddy 2013).

14.5 Reduced Potency of Predators

Natural enemies can be highly sensitive to climate change (Thomson et al. 2010a, b). A depreciation in the natural enemy can result in pest build-up. In this case, the pest continues damaging the crops since the natural enemy is not available in the newly created niche. Here, natural enemies also decline since they do not have their prey (Thomson et al. 2010a, b; Gonzalez et al. 2016). Drought followed by heavy rain increases the army worm species, which harms natural enemies (Sharma 2005). Cropping landscape has evolved as an impacts of climate change, alarming the rhythm between pests and their predators. Higher trophic levels more affected since they depend on lower trophic level for their growth and development. Increase in functional diversity in ecosystems improves system resilience which reduces insect loss (Newton et al. 2009).

14.6 Range Expansions with Climate Change

In terms of climate, customize the time frame of sexual maturity (Maurer et al. 2018), transition species, aggravate species compliment or experimentation with new species (Wellenreuther et al. 2010; Ryan et al. 2018), and species evolving their ranges to higher coordinates (Hill et al. 2011). In North America, the black hills beetle is transforming its zone, moving north and east towards higher coordinates with climate variability (Allan et al. 2003).

14.7 Climate Change and Fate of Pesticides

Increased degradation and volatilization are affected greatly by direct exposure to light, elevated temperatures which can reduce pesticide concentration (Noyes et al. 2009; Zhang et al. 2006; Johnson et al. 1995; Otieno et al. 2013). Inorganic chemicals are deteriorating more quickly as a result of climate, the chemical and microbial reaction rates and reduces the pesticide concentration in the environment (Chen and McCarl 2001; Ismail and Azlizan 2002; Ahmad et al. 2003; Singh et al. 2003; Athanasopoulos et al. 2004; Stenrød et al. 2008; Kookana 2010; Delcour et al. 2015). Higher volatilization rates of organochlorines have been shown in studies conducted in America (Yeo et al. 2003).

An increase in ultra violet (UV) radiation and temperature can affect biopesticide effectiveness on insects since they are highly temperature sensitive. Mishandling of synthetic pesticides prompts to depletion of natural resources, and health implications to humans (Sharma 2013). Climate change also affects parasitoids' sex ratio and fecundity (Dhillon and Sharma 2009). Organophosphates, organochlorines, and carbamates are positively correlated with temperature whereas synthetic pyrethroids showed a negative effect. The temperature feedback of phosphorus insecticides is feature of interest, meaning that they work better in extreme heat. Hence, in contrast to organophosphates, pyrethroids exhibit a negative connection with temperature, suggesting a possible justification for their lower toxicity at elevated temperature, as has been documented by a number of academics (Lydy et al. 1999; Chen and McCarl 2001; Wang and Shen 2007).

Generally, temperature and ecotoxicity are positively correlated, except for some pyrethrins (Noyes et al. 2009). The crucial piece that climate tends to play in insecticide exposure is underlined by the profound effect that has on the adequacy of pesticides against divergent vectors of malaria genus (*Anopheles arabiensis* and *An funestus*). With dichloro-diphenyl-trichloroethane (DDT) indicating a positive correlation, which illustrates elevated lethality and applicability at lower pressures, the temperature coefficient interprets the bond with both pyrethroid and heat (Miller and Adams 1982).

The relationship between temperature and nematodes effectiveness is nuanced and poorly understood. Temperature increases may enhance the target nematode's metabolic activity, change the chemical's solubility at which the nematicide is broken down by microbes or chemicals (Chen et al. 2020). Low soil temperature may be an issue in some situations as insecticides are frequently sprayed from early season. The anti-mortality and anti-infectivity effects of ethylene dibromide (EDB). At 5 °C compared to 15 °C temperature, the root knot nematodes, *Meloidogyne javanica* (Treub) in fumigation chambers was significantly lower. The dagger nematode and *M. incognita* (Kofoid & White) both showed stronger resistance to methyl bromide at 30 °C than at 15 °C in soils in sealed cans. In controlled temperature experiments, the increase in methyl bromide mechanism against citrus nematodes at maximum temperatures suggested that soil solarization might increase nematodes' efficacy (Miller and Adams 1982).

Climate change impacts chemical control. Two factors impact the fate of fungicide residue on crop foliage. One of these is the relationship between temperature and precipitation; when rainfall frequency rises, it washes the fungicide off and reduces control (Gatto et al. 2016). Dynamics of fungicides are intricate and rely on precipitation intensity and frequency. For a select few fungicides, post-precipitation application results in improved efficacy. At two intensities, stimulated rain was applied to the potato crop (Agbeniyi et al. 2017). When there is high-intensity rainfall, there is less fungicide residue on the leaf, but the disease intensity is unaffected. A morphological and physiological alteration in crops grown in high CO_2 environments that affects uptake is the second element in metabolism and translocation of systemic fungicides. For instance, chemicals uptake and spray coverage are negatively correlated with increased cuticle wax thickness and crop canopy size, respectively. This results in the dilution of the chemical in host tissue. According to research, a rise in temperature causes pesticides to degrade, which changes the morphology and physiology of plants, increases fungicide uptake, high metabolic rates, and target organism toxicity (Neuhaus et al. 1974) (Fig. 14.3).

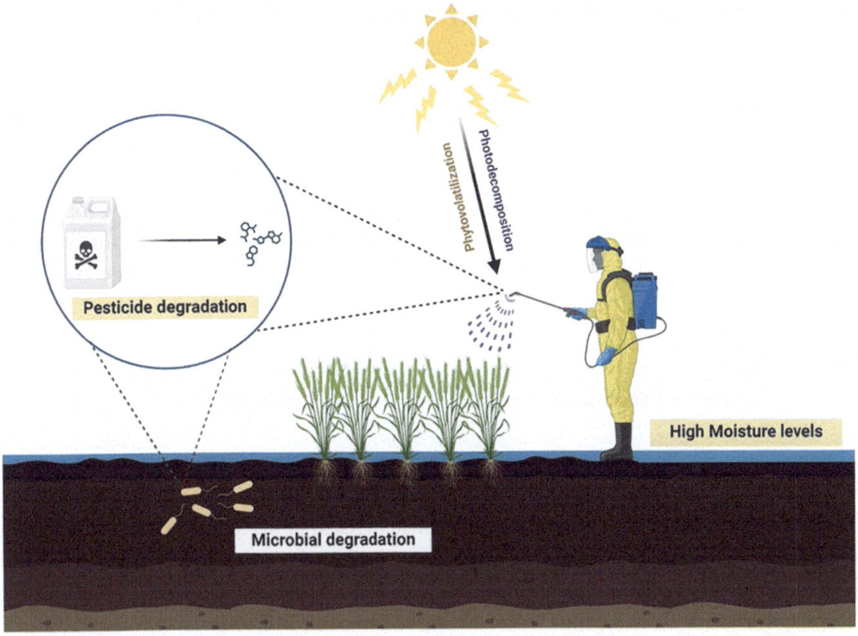

Fig. 14.3 Climate change and fate of pesticides

14.8 Coupling and Signals

It is still mostly questionable how insect fertility is shaped by changes in weather pattern (Gibert et al. 2016). Acoustic communication is the process by which insect songs and sexual signals are used by insects for courtship and mate location. Such transmission has a radiative specificity (Conrad et al. 2017). The parameters of songs are influenced by environmental conditions, especially the pulse rate, which is proven to increase uniformly in cricket songs (Walker and Cade 2003).

14.9 Host Plant Shifts

The quality and availability of hosts is affected by climatic conditions like precipitation, temperature, etc. As plant populations deteriorate, insect deepen their bands and undergo seasonal adjustments to seeking new hosts, as affirmed by *Ragoletis pomonella* (Bush 1969; Forbes et al. 2017).

14.10 Strategies to Mitigate Effects of Climate Change

One important attribute of innovative insect control is host plant resistance where we can develop stable resistant plants against the pests. Combining host plant resistance with trans gene expression act effectively in variable environmental conditions. The sole purpose of crop yields is to optimize control strategies by reinforcing the spectrum of possible pests (Canto et al. 2009; Dash et al. 2021). Develop pesticide formulations and spray schedules which are least affected by the climate change (Chen and McCarl 2001). The major impact of climate change on crops pests viz., insect pests, nematodes, and diseases is given in Table 14.1.

14.11 Climate Forecasting and Model Development

By stimulating pathogen risk and host defense explanations, temperature data and regulatory issues can feature an international viewpoint of changes made, enhancing the hypothesis of insect infestation metrics (Raza et al. 2014). Ecological niche frameworks give a better understanding of agricultural pest's species, reshaping the correlations among these organisms and their ecosystems, pointing out different territory for these to examine (Jiménez-Valverde et al. 2021).

14.12 Approaches for Managing Pests Influence by Weather

Many species, especially pests, are experiencing changes in populations, and life histories because of weather change. It affects a variety of insects' emergence and arrival times, as well as phenology. The way people live is being impeded by the contamination and pests brought on by progressions in agriculture. Following list highlights how to manage pests in changing climate scenarios (Pareek et al. 2017; Malhi et al. 2021).

14.12.1 Developing Climate-Resistant Plants

Breeding novel varieties with increased tolerance to abiotic and biotic challenges will reduce variations in the weather patterns. There is a risk that *Rabi*/cold season crops will grow later and/or shorter. This is due to winter's later onset and/or shorter length. Therefore, we should focus on developing varieties suited to planting late. These varieties can withstand challenging weather conditions as well as pests and diseases (Li and Yan 2020; Rivero et al. 2022).

Table 14.1 Effects of climate change on agricultural insect pests, nematodes, and diseases

Sr. no.	Effect	Description	References
1.	Increased insect populations	Warmer temperatures lead to quicker multiplication and higher endurance paces of bugs, bringing about bigger vermin populations	Deutsch et al. (2018)
2.	Expansion of pest range	Pests, like bugs and nematodes, are venturing into new geographic regions beforehand excessively cold for their survival	Bebber et al. (2013)
3.	Modified pest life cycles	Climate change upsets the planning of life cycles, prompting more successive irritation ages each year, expanding their effect on crops	Parmesan (2006)
4.	Increased pesticides resistance	Frequent utilization of pesticides because of rising irritation populaces prompts speedier improvement of obstruction in pests	Pimentel and Peshin (2014)
5.	Disruption of natural predators	Beneficial bugs that control bug populaces are adversely impacted by environmental change, diminishing natural control	Thomson et al. (2010a, b)
6.	Increased crop damage	Higher temperatures and changing precipitation designs intensify bother harm to crops, prompting huge losses	Rosenzweig and Parry (1994)
7.	Emergence of new pests species	Climate change works with the rise and foundation of new nuisance species in agribusiness, causing extra difficulties for crop management	Bebber et al. (2014)
8.	Enhanced pathogens virulence	Warmer and wetter circumstances increment the destructiveness of plant microorganisms, prompting more extreme episodes of diseases	Chakraborty and Newton (2011)
9.	Increased nematode populations	Warmer soil temperatures speed up the propagation of nematodes, prompting higher populace densities and expanded crop damage	Nicol et al. (2011)
10.	Longer growing seasons for pests	Extended developing seasons because of milder winters give bothers additional opportunity to take care of and imitate, expanding their general impact	Jepsen et al. (2008)
11.	Reduced efficacy of control methods	Changes in environment conditions might decrease the viability of conventional nuisance control techniques, requiring new methodologies for bug management	Ziska and Dukes (2011)
12.	Increased resistance to chemical controls	Pests are creating protection from substance pesticides all the more rapidly because of their sped up life cycles under changing environment conditions	Palumbi (2001)
13.	Increased incidence of soil-borne diseases	Warmer and wetter soil conditions favor the multiplication of soil-borne microorganisms, prompting more continuous illness episodes in crops	Gahukar (2009)

(continued)

Table 15.1 (continued)

Sr. no.	Effect	Description	References
14.	Increased nematode virulence	Climate change is prompting more harmful nematode strains, which make more noteworthy harm crops	Jones et al. (2013)
15.	Changing in disease prevalence	Changing environment conditions are causing a change in the commonness of illnesses, for certain sicknesses turning out to be more normal in specific regions	Anderson et al. (2004)
16.	Increased pest overwintering	Milder winters are empowering more vermin to endure the colder time of year, prompting higher populace levels in the accompanying developing season	Bale and Hayward (2010)
17.	Modified host–pest dynamics	Climate change is upsetting the connection among bothers and their host plants, in some cases prompting more serious infestations.	Roux and Robert (2019)
18.	Increased disease spread	Warmer temperatures and changed precipitation designs are working with the spread of plant infections, influencing a more extensive scope of crops	Garrett et al. (2006)
19.	Changes in pest behavior	Climate change is causing modifications in bother conduct, for example, taking care of examples, which can prompt surprising harvest damage	**Harrington et al.** (2007)
20.	Emergence of resistant nematode strains	Hotter temperatures are best for nematode strains that are more impervious to conventional control strategies	Nicol et al. (2011)

14.12.2 Crops Planting Dates Fluctuating

Crop sowing dates would shift due to global climate change, changing host–pest synchronization. Through an interpretation of transitions in host plant engagement, the study seeks possible ways to lessen identify the planting time to generate at high output. Therefore, it is necessary to adjust crop calendars to the shifting crop environment (Ghaffar et al. 2022; Sunil et al. 2023; Raj et al. 2024; Das et al. 2024). Crop growers must take into account crops losses and predicted climate changes in pest incidence when adjusting insect management tactics. Global warming is also a factor affecting host-induced resistance to plant pathogenic nematodes. *Mi*-1 gene confers resistance to *Meloidogyne javanica* (Treub) and *M. arenaria* (Neal) in most crops of the Solanaceae family but in case of diurnal fluctuation and soil temperatures beyond 28 °C *Mi*-induced resistance so *Mi*-mediated resistance is limited to sub-tropical and tropical crops (Jablonska et al. 2007; Dubey et al. 2021, 2022).

14.12.3 Crop Pest Risk Mapping Using GIS

It is recommended to use geographic information systems (GIS) in pest control programs because they enable entomologists to relate pest outbreaks to the biographical and physiographical characteristics of the terrain. These feasible shifts by documenting patterns in the agriculture hotspots' spatial range and subsequent invasive species threat zones. It is a great help for contemplating how rising temperatures changes the progression, and trends of pest populations (Rafoss 2003; Sporleder et al. 2009).

14.12.4 Testing Insecticides with New Mechanisms of Action

Moderate moisture and temperature are essential for carbamates and organo phosphates to induce their toxicity against plant pathogenic nematodes; fluctuation in precipitation and temperature costs the reachability with persistence in soil (Delcour et al. 2015). In addition to its insecticidal activity, salicylic acid-associated plant defense responses have been shown to increase plant vigor and abiotic stress tolerance when neonicotinoid pesticides are used to combat sucking pests. This provides information about the impact insecticides may have in improving plant's ability to withstand stress. It is necessary to identify additional chemicals for use in crops pests' management in the future. In initiatives to mitigate to any such mechanisms, we will also have to make the call on future works and devising pest infestation in climate crisis (Ford et al. 2010; Guedes et al. 2023).

14.13 Conclusions

The expected consequence of climate variability is a massive shift in the variety of pests, and abundance as well as the magnitude of pest-related losses, which will affect crop productivity and food security. It is possible to optimize pest treatment plans and alleviate the detrimental climatic changes of agricultural yields by predicting transitions in the host range and dynamics of pests. Continuous monitoring can help us mitigate climate change's adverse effects on crops. Long-term extreme weather events are followed by heavy downpours likely to result in pest outbreaks. An increase in global temperature leads to a change in the life cycle with a distribution pattern of pests. Tropical pests may move to temperate regions and increase their distribution pattern which damages crops in those areas. Likewise, CO_2 increases photosynthetic efficiency and increase in moisture levels resulted in more pest attack. Climate change will shift pest–host interactions. Current pest management sessions must follow environment's potential to optimize disease-resistant plants into account, as it may also have an implication on the intensity of predators. According to the study, there is indeed a tiny possibility that climate will alter how far natural pest control methods, host-induced resistance, plant–pest interactions, pest population dynamics, pest composition work, etc. It is important to note that elevated CO_2, moisture, and temperature have a neutral or positive effect on pest populations, but they do not negatively affect them. Proper monitoring of the pest's population can help to control the occurrence and spread of pests to an extent.

Author Contribution Statement Vyshnavi Sunil led the overall conceptualization of the chapter, coordinated the research, and contributed to writing and editing key sections. Logeswaran K provided valuable insights into the theoretical framework, co-wrote sections, and assisted with data analysis. Akanksha Adhikari contributed to the research design and methodology, as well as helped draft the introduction and conclusion. Shreya Singh supported data collection, analyzed key findings, and co-wrote portions of the results and discussion sections. Alagesan Keerthana assisted in structuring the chapter, reviewed relevant literature, and contributed to the final editing. Abhibandana Das supported the writing process and contributed to the analysis and interpretation of data. Avantika Pandey helped in the data visualization and provided critical feedback for the final draft of this chapter. Badavat Sujatha helped in data visualization, provided critical feedback, and participated in final manuscript revisions. Habib Ali contributed to the methodology and co-wrote parts of the discussion, ensuring accuracy and clarity throughout. Vinod Kumar Dubey provided technical expertise, assisted with the manuscript revisions, and ensured scientific rigor in the final draft.

References

Agbeniyi SO, Adeniyi DO, Ogundeji D (2017) Impact of climate change on timing and frequency of fungicide application for the control of phytophthora pod rot of cacao in Nigeria. In International Symposium on Cocoa Research (ISCR). pp. 13–17

Ahmad R, James TK, Rahman A, Holland PT (2003) Dissipation of the herbicide clopyralid in an allophanic soil: laboratory and field studies. J Environ Sci Health B 38(6):683–695. https://doi.org/10.1081/PFC-120025553

Allan LC, Steve WT, Jacques R, Les S (2003) Effect of climate change on range expansion by the mountain pine beetle in British Columbia. In: Mountain pine beetle symposium: challenges and solutions. Canadian Forest Service, Victoria, BC, Canada, pp 223–232

Anderson PK, Cunningham AA, Patel NG, Morales FJ, Epstein PR, Daszak P (2004) Emerging infectious diseases of plants: pathogen pollution, climate change, and agrotechnology drivers. Trends Ecol Evol 19(10):535–544

Athanasopoulos PE, Kyriakidis NV, Stavropoulos P (2004) A study on the environmental degradation of pesticides azinphos methyl and parathion methyl. J Environ Sci Health B 39(2):297–309. https://doi.org/10.1081/PFC-120030243

Awmack CS, Woodcock CM, Harrington R (1997) Climate change may increase vulnerability of aphids to natural enemies. Ecol Entomol 22:366–368

Bale JS, Hayward SAL (2010) Insect overwintering in a changing climate. J Exp Biol 213(6):980–994. https://doi.org/10.1242/jeb.037911

Bebber DP, Ramotowski MA, Gurr SJ (2013) Crop pests and pathogens move polewards in a warming world. Nat Clim Change 3(11):985–988. https://doi.org/10.1038/nclimate1990

Bebber DP, Holmes T, Gurr SJ (2014) The global spread of crop pests and pathogens. Glob Ecol Biogeoraph 23(12):1398–1407. https://doi.org/10.1111/geb.12214

Bush GL (1969) Sympatric host race formation and speciation in frugivorous flies of the genus *Rhagoletis* (Diptera, Tephritidae). Evolution 3:237–251. https://doi.org/10.2307/2406788

Canto T, Aranda MA, Fereres A (2009) Climate change effects on physiology and population processes of hosts and vectors that influence the spread of hemipteran-borne plant viruses. Glob Change Biol 15(8):1884–1894. https://doi.org/10.1111/j.1365-2486.2008.01820.x

Castor LL, Ayers JE, McNabb A, Krause RA (1975) Computerized forecast system for Stewart's bacterial disease on corn. Plant Disease Repot 59:533–536

Chakraborty S, Newton AC (2011) Climate change, plant diseases, and food security: an overview. Plant Pathol 60(1):2–14. https://doi.org/10.1111/j.1365-3059.2010.02411.x
Chakraborty S, Pangga IB, Lupton J, Hart L, Room PM, Yates D (2000) Production and dis pers al of *Colletotrichum gloeosporioides* spores on Stylosanthes scabra under elevated CO_2. Environ Pollut 108:317–326. https://doi.org/10.1016/S0269-7491(99)00217-1
Chen CC, McCarl BA (2001) An investigation of the relationship between pesticide usage and climate change. Clim Chang 50(4):475–487. https://doi.org/10.1023/A:1010655503471
Chen J, Li QX, Song B (2020) Chemical nematicides: recent research progress and outlook. J Agric Food Chem 68(44):12175–12188. https://doi.org/10.1021/acs.jafc.0c02871
Chowdhury S, Dubey VK, Choudhury S, Das A, Jeengar D, Sujatha B, Kumar A, Kumar N, Semwal A, Kumar V (2023) Insects as bioindicator: a hidden gem for environmental monitoring. Front Environ Sci 11:273. https://doi.org/10.3389/fenvs.2023.1146052
Coakley S, Scherm MH, Chakraborty S (1999) Climate change and disease management. Annu Rev Phytopathol 37:399–426. https://doi.org/10.1146/annurev.phyto.37.1.399
Conrad T, Stöcker C, Ayasse M (2017) The effect of temperature on male mating signals and female choice in the red mason bee, *Osmia bicornis* (L.). Ecol Evol 7:8966–8975. https://doi.org/10.1002/ece3.3331
Das A, Kadawla K, Nath H, Chakraborty S, Ali H, Singh S, Dubey VK (2024) Drone-based intelligent spraying of pesticides: current challenges and its future prospects. In: Applications of computer vision and drone Technology in Agriculture 4.0. Springer Nature Singapore, pp 199–223. https://doi.org/10.1007/978-981-99-8684-2-12
Dash SP, Dipankar P, Burange PS, Rouse BT, Sarangi PP (2021) Climate change: how it impacts the emergence, transmission, resistance and consequences of viral infections in animals and plants. Crit Rev Microbiol 47(3):307–322. https://doi.org/10.1080/1040841X.2021.1879006
Delcour I, Spanoghe P, Uyttendaele M (2015) Literature review: impact of climate change on pesticide use. Food Res Int 68:7–15. https://doi.org/10.1016/j.foodres.2014.09.030
Delgado-Baquerizo M, Guerra CA, Cano-Díaz C, Egidi E, Wang JT, Eisenhauer N, Maestre FT (2020) The proportion of soil-borne pathogens increases with warming at the global scale. Nat Clim Chang 10(6):550–554. https://doi.org/10.1038/s41558-020-0759-3
Deutsch CA, Tewksbury JJ, Tigchelaar M, Battisti DS, Merrill SC, Huey RB, Naylor RL (2018) Increase in crop losses to insect pests in a warming climate. Science 361(6405):916–919. https://doi.org/10.1126/science.aat3466
Dhaliwal GS, Jindal V, Dhawan AK (2010) Insect pest problems and crop losses: changing trends. Indian J Ecol 37(1):1–7
Dhaliwal GS, Jindal V, Mohindru B (2015) Crop losses due to insect pests: global and Indian scenario. Indian J Entomol 77(2):165–168. https://doi.org/10.5958/0974-8172.2015.00033.4
Dhillon MK, Sharma HC (2009) Temperature influences the performance and effectiveness of field and laboratory strains of the ichneumonid parasitoid, *Campoletis chlorideae*. Biol Control 54:743–750. https://doi.org/10.1007/s10526-009-9225-x
Dubey VK, Kalleshwaraswamy CM, Shivanna BK (2021) Seasonal incidence of major sternorrhynchan insect pests infesting arecanut in South India. Indian J Agric Res 47(5):436–440. https://doi.org/10.33451/florafauna.v27i1pp101-108
Dubey VK, Kalleshwaraswamy CM, Joshi S, Shivanna BK (2022) Diversity and diagnostics of Sternorrhynchan insect pests infesting Arecanut. Indian J Entomol 84(3):509–515. https://doi.org/10.55446/IJE.2021.56
Eastburn DM, Degennaro MM, Delucia EH, Dermody O, Mcelrone AJ (2010) Elevated atmospheric carbon dioxide and ozone alter soybean diseases at SoyFACE. Glob Change Biol 16(1):320–330. https://doi.org/10.1111/j.1365-2486.2009.01978.x
Elad Y, Pertot I (2014) Climate change impacts on plant pathogens and plant diseases. J Crop Improve 28(1):99–139. https://doi.org/10.1080/15427528.2014.865412
Elderd BD, Reilly JR (2014) Warmer temperatures increase disease transmission and outbreak intensity in a host-pathogen system. J Anim Ecol 83(4):838–849. https://doi.org/10.1111/1365-2656.12180
ENSAA (2011) Defining climate change. (City)

Forbes AA, Devine SN, Hippee AC, Tvedte ES, Ward AKG, Widmayer HA (2017) Revisiting the particular role of host shifts in initiating insect speciation. Evolution 71:1126–1137. https://doi.org/10.1111/evo.13164

Ford KA, Casida JE, Chandran D, Gulevich AG, Okrent RA, Durkin KA, Wildermuth MC (2010) Neonicotinoid insecticides induce salicylate-associated plant defense responses. Proc Natl Acad Sci 107(41):17527–17532

Gahukar RT (2009) Sustainable agriculture in India: Current situation and future needs. Int J Agric Sci 5:1–7

Garrett KA, Dendy SP, Frank EE, Rouse MN, Travers SE (2006) Climate change effects on plant disease: genomes to ecosystems. Annu Rev Phytopathol 44:489–509. https://doi.org/10.1146/annurev.phyto.44.070505.143420

Gatto MP, Cabella R, Gherardi M (2016) Climate change: the potential impact on occupational exposure to pesticides. Annali dell'Istituto superiore di sanita 52(3):374–385. https://doi.org/10.4415/ANN_16_03_09

Georghiou GP, Taylor CE (1986) Factors influencing the evolution of resistance. In: Pesticide resistance: strategies and tactics for management. National Research Council, National Academy Press, pp 157–169

Ghaffar A, Rahman MHU, Ahmed S, Haider G, Ahmad I, Khan MA, Ahmed A (2022) Adaptations in cropping system and pattern for sustainable crops production under climate change scenarios. In: Improvement of plant production in the era of climate change. CRC Press, pp 1–34. https://doi.org/10.1201/9781003286417

Gibert JP, Chelini MC, Rosenthal MF, DeLong JP (2016) Crossing regimes of temperature dependence in animal movement. Glob Change Biol 22:1722–1736. https://doi.org/10.1111/gcb.13245

Gonzalez F, Tkaczuk C, Dinu MM, Fiedler Ż, Vidal S, Zchori-Fein E, Messelink GJ (2016) New opportunities for the integration of microorganisms into biological pest control systems in greenhouse crops. J Pest Sci 89:295–311. https://doi.org/10.1007/s10340-016-0751-x

González-Tokman D, Córdoba-Aguilar A, Dáttilo W, Lira-Noriega A, Sánchez-Guillén RA, Villalobos F (2020) Insect responses to heat: physiological mechanisms, evolution and ecological implications in a warming world. Biol Rev 95(3):802–821. https://doi.org/10.1111/brv.12588

Guedes RNC, Biondi A, Agathokleous E, Nunes-Nesi A (2023) (systemic) insecticides in plants: Phytotoxicity, bioactivation, or Hormesis? Agric Commun:100002

Hamilton JGO, Dermody M, Aldea AR, Zangerl A, Rogers MR, Berenbaum ED (2005) Anthropogenic changes in tropospheric composition increase susceptibility of soybean to insect herbivory. Environ Entomol 34:2479–2485. https://doi.org/10.1603/0046-225X-34.2.479

Harrington R, Fleming RA, Woiwod IP (2007) Insect movement and climate change. Agric Forest Entomol 9(2):69–84

Harvell CD, Mitchell CE, Ward JR, Altizer S, Dobson AP, Ostfeld RS (2002) Climate warming and disease risks for terrestrial and marine biota. Science 296(5576):2158–2162. https://doi.org/10.1126/science.1063699

Hill JK, Griffiths HM, Thomas CD (2011) Climate change and evolutionary adaptations at species' range margins. Annu Rev Entomol 56:143–159. https://doi.org/10.1146/annurev-ento-120709-144746

Hunjan MS, Lore JS (2020) Climate change: impact on plant pathogens, diseases, and their management. In: Jabran K, Florentine S, Chauhan B (eds) Crop protection under changing climate. Springer, Cham, pp 85–100. https://doi.org/10.1007/978-3-030-46111-9_4

IPCC (2001) Climate change 2001: The scientific basis. https://www.ipcc.ch/2001/ Accessed 11/06/2024

Ismail BS, Azlizan BA (2002) Persistence and bioactivity of metsulfuron-methyl in three soils. J Environ Sci Health B 37(4):345–353. https://doi.org/10.1081/PFC-120004475

Jablonska B, Ammiraju J, Bhattarai K, Mantelin S, Ilarduya O, Robert P, Kaloshian I (2007) The *Mi-9* gene from *Solanum arcanum* conferring heat-stable resistance to root-knot nematodes is homolog of *Mi-1*. Plant Physiol 143:1044–1054. https://doi.org/10.1104/pp.106.089615

Jepsen JU, Hagen SB, Ims RA, Yoccoz NG (2008) Climate change and outbreaks of the geometrid moths in the subarctic birch forest: evidence of a recent outbreak range expansion. Glob Change Biol 14(8):1556–1565. https://doi.org/10.1111/j.1365-2656.2007.01339.x

Jiménez-Valverde A, Rodriguez-Rey M, Pena-Aguilera P (2021) Climate data source matters in species distribution modelling: the case of the Iberian Peninsula. Biodivers Conserv 30:67–84. https://doi.org/10.1007/s10531-020-02075-6

Johnson AW, Wauchope RD, Burgoa B (1995) Effect of simulated rainfall on leaching and efficacy of fenamiphos. J Nematol 27(4):555–562

Johnson SN, Anderson EA, Dawson G, Griffiths DW (2008) Varietal susceptibility of potatoes to wireworm herbivory. Agric Forest Entomol 10:167–174. https://doi.org/10.1111/j.1461-9563.2008.00372.x

Jones JT, Haegeman A, Danchin EG, Gaur HS, Helder J, Jones MG, Perry RN (2013) Top 10 plant-parasitic nematodes in molecular plant pathology. J Nematol 45(2):89–107. https://doi.org/10.1111/mpp.12057

Kobayashi T, Ishiguro K, Nakajima T, Kim HY, Okada M, Kobayas K (2006) Effects of elevated atmospheric CO2 concentration on the infection of rice blast and sheath blight. Phytopathology 96:425–431. https://doi.org/10.1094/PHYTO-96-0425

Kookana RS (2010) The role of biochar in modifying the environmental fate, bioavailability, and efficacy of pesticides in soils: a review. Soil Res 48(7):627–637. https://doi.org/10.1071/SR10007

Li Q, Yan J (2020) Sustainable agriculture in the era of omics: knowledge-driven crop breeding. Genome Biol 21:1–5. https://doi.org/10.1186/s13059-020-02073-5

Luo Q (2011) Temperature thresholds and crop production: a review. Clim Change 109:583–598. https://doi.org/10.1007/s10584-011-0028-6

Lydy MJ, Belden JB, Ternes MA (1999) Effects of temperature on the toxicity of M-parathion, chlorpyrifos, and pentachlorobenzene to *Chironomus tentans*. Arch Environ Contam Toxicol 37:542–547. https://doi.org/10.1007/s002449900550

Maiorano AM, Blandino Reyneri A, Vanara F (2008) Effects of maize residues on the *Fusarium* spp. infection and deoxynivalenol (DON) contamination of wheat grain. Crop Prot 27:182–188. https://doi.org/10.1016/j.cropro.2007.05.004

Malhi GS, Kaur M, Kaushik P (2021) Impact of climate change on agriculture and its mitigation strategies: a review. Sustain For 13(3):1318. https://doi.org/10.3390/su13031318

Manning WJ, Tiedemann AV (1995) Climate change: potential effects of increased atmospheric carbon dioxide (CO2), ozone (O3), and ultraviolet-B (UV-B) radiation on plant diseases. Environ Pollut 88(2):219–245. https://doi.org/10.1016/0269-7491(95)91446-R

Maurer EP, Kayser G, Doyle L, Wood AW (2018) Adjusting flood peak frequency changes to account for climate change impacts in the western United States. J Water Res Plan Manag 144(3):05017025. https://doi.org/10.1061/(ASCE)WR.1943-5452.0000903

Miller TA, Adams ME (1982) Mode of action of pyrethroids. In: Insecticide mode of action. Academic Press (Elsevier), pp 3–27. https://doi.org/10.1016/b978-0-12-177120-1.50006-0

Neuhaus W, Stachewicz H, Dunsing M (1974) Uber den Einfluss von Nieder-schl¨agen auf die biologische Wirkung von Fungiziden zur Phytophthora-Bek¨ampfung. Nachrichtenbl Pflanzenschutz 28:149–153

Newton AC, Begg GS, Swanston JS (2009) Deployment of diversity for enhanced crop function. Ann Appl Biol 154:309–322. https://doi.org/10.1111/j.1744-7348.2008.00303.x

Nicol JM, Turner SJ, Coyne DL, Nijs LD, Hockland S, Maafi ZT (2011) Current nematode threats to world agriculture. Field Crop Res 126:18–31. https://doi.org/10.1007/978-94-007-0434-3_2

Noyes PD, McElwee MK, Miller HD, Clark BW, Van Tiem LA, Walcott KC (2009) The toxicology of climate change: environmental contaminants in a warming world. Environ Int 35(6):971–986. https://doi.org/10.1016/j.envint.2009.02.006

Otieno PO, Owuor PO, Lalah JO, Pfister G, Schramm KW (2013) Impacts of climate-induced changes on the distribution of pesticides residues in water and sediment of Lake Naivasha, Kenya. Environ Monitor Assess 185(3):2723–2733. https://doi.org/10.1007/s10661-012-2743-5

Palumbi SR (2001) Humans as the world's greatest evolutionary force. Science 293(5536):1786–1790. https://doi.org/10.1126/science.293.5536.1786

Pareek A, Meena BM, Sharma S, Tetarwal ML, Kalyan RK, Meena BL (2017) Impact of climate change on insect pests and their management strategies. Clim Change Sustain Agric:253–286

Parmesan C (2006) Ecological and evolutionary responses to recent climate change. Annu Rev Ecol Evol Syst 37:637–669. https://doi.org/10.1146/annurev.ecolsys.37.091305.110100

Pathak H, Aggarwal PK, Singh SD (2012) Climate change impact, adaptation and mitigation in agriculture: methodology for assessment and applications. Indian Agricultural Research Institute, New Delhi, India, ISBN 978-81-88708-82-6., pp 1–322

Patterson DT, Westbrook JK, Joyce RJV, Lingren PD, Rogasik J (1999) Weeds, insects, and diseases. Clim Chang 43(4):711–727. https://doi.org/10.1023/A:1005549400875

Pimentel D, Peshin R (2014) Integrated pest management: pesticide problems. Springer, pp 23–38

Rafoss T (2003) Spatial stochastic simulation offers potential as a quantitative method for pest risk analysis. Risk Anal 23(4):651–661. https://doi.org/10.1111/1539-6924.00344

Raj M, Lal K, Satdev KP, Kumari S, Dubey VK, Kumar S (2024) Potential nutrient cycling and management in agroforestry. In: Agroforestry to combat global challenges: current prospects and future challenges. Springer Nature Singapore, Singapore, pp 71–92. https://doi.org/10.1007/978-981-99-7282-1-4

Raza MM, Khan MA, Arshad M, Sagheer M, Sattar Z, Shafi J, Haq EU, Ali A, Aslam U, Mushtaq A (2014) Impact of global warming on insects. Arcc Phytopathol Plant Prot 48:84–94. https://doi.org/10.1080/03235408.2014.882132

Reddy PP (2013) Impact of climate change on insect pests, pathogens and nematodes. Pest Manag Hortic Ecosyst 19(2):225–233

Rivero RM, Mittler R, Blumwald E, Zandalinas SI (2022) Developing climate-resilient crops: improving plant tolerance to stress combination. Plant J 109(2):373–389. https://doi.org/10.1111/tpj.15483

Rosenzweig C, Parry ML (1994) Potential impact of climate change on world food supply. Nature 367(6459):133–138. https://doi.org/10.1038/367133a0

Rosenzweig C, Iglesias A, Yang X, Epstein PR, Chivian E (2001) Climate change and extreme weather events; implications for food production, plant diseases, and pests. Glob Change Hum Health 2(2):90–104. https://digitalcommons.unl.edu/nasapub/24

Rötter R, Van De Geijn S (1999) Climate change effects on plant growth, crop yield and livestock. Clim Chang 43:651–681. https://doi.org/10.1023/A:1005541132734

Roux O, Robert A (2019) The impact of climate change on host-parasitoid interactions. Ecol Lett 22(10):1641–1654

Roy DB, Sparks TH (2000) Phenology of British butterflies and climate change. Glob Chang Biol 6:407–416. https://doi.org/10.1046/j.1365-2486.2000.00322.x

Ryan SF, Deines JM, Scriber JM, Pfrender ME, Jones SE, Emrich SJ (2018) Climate-mediated hybrid zone movement revealed with genomics, museum collection, and simulation modeling. Proc Natl Acad Sci USA 16:201714950. https://doi.org/10.1073/pnas.1714950115

Sabburg R, Obanor F, Aitken E (2015) Effect of increasing temperature on the pathogenic fitness of *Fusarium pseudograminearum*. Procedia Environ Sci 29:182–183. https://doi.org/10.1016/j.proenv.2015.10.066

Sharma HC (2005) *Heliothis/Helicoverpa* management: emerging trends and strategies for future research. CRC Press, pp 283–308. https://doi.org/10.1201/9781482280340

Sharma HC (2013) Climate change effects on insects: implications for crop protection and food security. In: Combating climate change: an agricultural perspective. CRC Press, pp 213–236

Shelton AM, Wilsey WR, Soderlund DM (2001) Classification of insecticides and acaricides for resistance management. Department of Entomology, NYSAES, Geneva, NY 14456. pp. 315–787

Singh N, Singh B, Dureja P, Sethunathan N (2003) Persistence of phorate in soils: role of moisture, temperature, preexposure and microorganisms. J Environ Sci Health B 38(6):723–735. https://doi.org/10.1081/PFC-120025556

Skendzic S, Zovko M, Zivkovic IP, Lesic V, Lemic D (2021) The impact of climate change on agricultural insect pests. Insects 12(5):440. https://doi.org/10.3390/insects12050440

Sporleder M, Chavez D, Gonzales JC, Juarez H, Simon R, Kroschel J (2009) ILCYM-insect life cycle modeling: software for developing temperature-based insect phenology models with applications for regional and global pest risk assessments and mapping. In: Proceedings of the 15th Triennial ISTRC Symposium of the International Society for Tropical Root Crops (ISTRC). Lima, Peru, pp 216–223

Staley JT, Hodgson CJ, Mortimer SR, Morecroft MD, Masters GJ, Brown VK, Taylor ME (2007) Effects of summer rainfall manipulations on the abundance and vertical distribution of herbivorous soil macro-invertebrates. Eur J Soil Biol 43:189–198. https://doi.org/10.1016/j.ejsobi.2007.02.010

Stanaszek-Tomal E (2020) Environmental factors causing the development of microorganisms on the surfaces of national cultural monuments made of mineral building materials-review. Coat 10:1203–1209. https://doi.org/10.3390/coatings10121203

Stefanescu C, Peñuelas J, Filella I (2003) Effects of climatic change on the phenology of butterflies in the Northwest Mediterranean Basin. Glob Chang Biol 9:1494–1506. https://doi.org/10.1046/j.1365-2486.2003.00682.x

Stenrød M, Heggen HE, Bolli RI, Eklo OM (2008) Testing and comparison of three pesticide risk indicator models under Norwegian conditions-a case study in the Skuterud and Heiabekken catchments. Agric Ecosyst Environ 123(1–3):15–29. https://doi.org/10.1016/j.agee.2007.03.003

Sticht C, Schrader S, Giesemann A, Wiegel HJ (2009) Sensitivity of nematode feeding types in arable soil to free-air CO_2 enrichment (FACE) is crop specific. Pedobiologia 52:337–349. https://doi.org/10.1038/s41598-020-70816-2

Sunil V, Majeed W, Chowdhury S, Riaz A, Shakoori FR, Tahir M, Dubey VK (2023) Insect population dynamics and climate change. In: Climate change and insect biodiversity. CRC Press, pp 121–146. https://doi.org/10.1201/9781003382089-8

Tabari H (2020) Climate change impact on flood and extreme precipitation increases with water availability. Sci Rep 10:13768. https://doi.org/10.1038/s41598-020-70816-2

Thompson SR, Brandenburg RI, Arends JJ (2006) Impact of moisture and UV degradation on *Beauveria bassiana* (Balsamo) Vuillemin conidial viability in turfgrass. Biol Control 39:401–407

Thomson LJ, Macfadyen S, Hoffmann AA (2010a) Predicting the effects of climate change on natural enemies of agricultural pests. Glob Chang Biol 16(11):3170–3184. https://doi.org/10.1016/j.biocontrol.2009.01.022

Thomson LJ, Macfadyen S, Hoffmann AA (2010b) Predicting the effects of climate change on natural enemies of agricultural pests. Biol Control 52:296–306. https://doi.org/10.1016/j.biocontrol.2009.01.022

Van Doren Jr DM, Allmaras RR (1978) Effect of residue management practices on the soil physical environment, microclimate, and plant growth. Crop Resid Manag Syst 31:49–83. https://doi.org/10.2134/asaspecpub31.c4

Verschoor B, De Goede R, De Vries F, Brussaard L (2001) Changes in the composition of the plant-feeding nematode community in grasslands after cessation of fertiliser application. Appl Soil Ecol 17:1–17. https://doi.org/10.1016/S0929-1393(00)00135-9

Větrovský T, Kohout P, Kopecký M, Machac A, Man M, Bahnmann BD, Baldrian P (2019) A meta-analysis of global fungal distribution reveals climate-driven patterns. Nat Commun 10(1):5142. https://doi.org/10.1038/s41467-019-13164-8

Walker SE, Cade WH (2003) The effects of temperature and age on calling song in a field cricket with a complex calling song, *Teleogryllus oceanicus* (*Orthoptera: Gryllidae*). Can J Zool 81:1414–1420. https://doi.org/10.1139/z03-106

Wallin JR, Waggoner PE (1950) The influence of climate on the development and spread of Phytophthora infestans in artificially inoculated potato plots. Plant Dis Rep Suppl:19–33

Wang XY, Shen ZR (2007) Potency of some novel insecticides at various environmental temperatures on *Myzus persicae*. Phytoparasitica 35:414–422. https://doi.org/10.1007/BF02980705

Wellenreuther M, Tynkkynen K, Svensson EI (2010) Simulating range expansion: male species recognition and loss of premating isolation in damselflies. Evolution 64:242–252. https://doi.org/10.1111/j.1558-5646.2009.00815.x

Yadav S, Bhattacharya P, Areendran G, Sahana M, Raj K, Sajjad H (2021) Predicting impact of climate change on geographical distribution of major NTFP species in the Central India region. Model Earth Syst Environ 6:1–20. https://doi.org/10.1007/s40808-020-01074-4

Yeates G, Newton PCD, Ross DJ (2003) Significant changes in soil macrofauna in grazed pasture under elevated carbon dioxide. Biol Fertil Soils 38:319–326. https://doi.org/10.1007/s00374-003-0659-5

Yeo HG, Choi M, Chun MY, Sunwoo Y (2003) Concentration distribution of polychlorinated biphenyls and organochlorine pesticides and their relationship with temperature in rural air of Korea. Atmos Environ 37(27):3831–3839. https://doi.org/10.1016/S1352-2310(03)00456-4

Zhang ZY, Zhang CZ, Liu XJ, Hong XY (2006) Dynamics of pesticide residues in the autumn Chinese cabbage (*Brassica chinensis* L.) grown in open fields. Pest Manag Sci 62(4):350–355. https://doi.org/10.1002/ps.1174

Ziska LH, Dukes JS (2011) Weed biology and climate change. Wiley-Blackwell. (ISBN: 978-0-813-81417-9), p 246

Nanomaterials as Nanofertilizers for Climate-Smart Agriculture

15

Umair Ashraf, Fatima Batool, Rabia Ghaffar, Muhammad Imran, Ayesha Riaz, Muhammad Hussaan, Muhammad Ansar Farooq, and Fahd Rasul

Abstract

Sustainable agricultural practices are inevitable for long-term benefits under changing climate scenarios. The utilization of nanomaterials as nanofertilizers is important to enhance nutrient use efficiency, reduce environmental impacts, and support the resilience of agricultural systems in the face of climate change. Nanofertilizers possess unique properties, such as enhanced solubility, controlled release of nutrients, and increased surface area that enable precise nutrient delivery to plants with minimum loss. This chapter focuses on the potential use of nanomaterials in synthesizing nanofertilizers (physically, chemically, and biologically) and the effects of nanoparticles/nanofertilizers on plant growth and

U. Ashraf (✉)
Department of Botany, Division of Science and Technology, University of Education, Lahore, Punjab, Pakistan

Upper Peninsula Research and Extension Center, Michigan State University, Chatham, Michigan, United States
e-mail: umair.ashraf@ue.edu.pk; ashrafum@msu.edu

F. Batool · R. Ghaffar
Department of Botany, Division of Science and Technology, University of Education, Lahore, Punjab, Pakistan

M. Imran
The Hong Kong University of Science and Technology, Jiangmen Laboratory of Carbon Science and Technology, Jiangmen, Guangdong, China

Guangdong-Hong Kong Joint Laboratory for Carbon Neutrality, Jiangmen, China

A. Riaz
Department of Botany, Division of Science and Technology, University of Education, Lahore, Punjab, Pakistan

Department of Botany, Government College University, Faisalabad, Pakistan

© The Author(s), under exclusive license to Springer Nature Singapore Pte Ltd. 2025

S. Faiz et al. (eds.), *Climate Smart Agriculture for Future Food Security*,
https://doi.org/10.1007/978-981-96-4499-5_15

development with emphasis on how nanofertilizers can optimize nutrient uptake, reduce nutrient loss, and minimize the excessive use of conventional fertilizers, in order to address the key challenges of modern day agriculture. The role of nanofertilizers in mitigating abiotic stresses is also emphasized by showing their potential to improve crop resilience under sub-optimal conditions. Either via soil or foliar application, the nanofertilizers were found beneficial regarding crop growth and yield improvement than conventional fertilizers. Additionally, the short- and long-term benefits of nanofertilizers include improvement in nutrient use efficiency, plant and soil health, less environmental pollution, an alternative to conventional fertilizers, and a step toward climate-smart agriculture. Overall, nanomaterials as nanofertilizers present a promising strategy for advancing climate-smart agriculture, offering a pathway to more sustainable, resilient, and efficient agricultural systems globally.

Keywords

Nanofertilizer · Nutrient use efficiency · Crop yield · Greenhouse gases · Soil health

15.1 Introduction

The agricultural sector serves as fundamental pillar of global food security but faces significant challenges due to climate change and environmental stresses. According to Pouratashi and Iravani (2012), food demand in Pakistan is projected to increase by up to 70% by 2050, driven by rapid population growth. This escalating demand underscores the necessity of adopting of innovative technologies to enhance both food production and nutritional quality (Smith and Gilbertson 2018). Conventional fertilizers, while essential for enhancing crop yields, present several challenges such as low nutrient use efficiency (NUE) and adverse environmental impacts (Rajput et al. 2021a, b). Nutrient use efficiency is a crucial factor in agricultural productivity, influencing both plant growth, and nutrient availability. Fertilizers can be applied through various methods e.g., soil incorporation, fertigation, and hydroponic techniques (Bindraban et al. 2018). However, these traditional methods

M. Hussaan
Department of Botany, Government College University, Faisalabad, Pakistan

M. A. Farooq
Institute of Environmental Sciences and Engineering (IESE), School of Civil and Environmental Engineering (SCEE), National University of Sciences and Technology (NUST), Islamabad, Pakistan

CFAES Rattan Lal Center for Carbon Management and Sequestration, School of Environment and Natural Resources, The Ohio State University, Columbus, United States

F. Rasul
Department of Agronomy, University of Agriculture, Faisalabad, Pakistan

of fertilizer application often results in nutrient losses, environmental pollution, and negative health impacts (Foley et al. 2011; Smith and Gilbertson 2018). To mitigate these challenges, researchers are increasingly exploring environmentally friendly alternatives, such as nanotechnology-based fertilizers, to enhance nutrient delivery efficiency, and minimize environmental contamination (Sharifi et al. 2020). These advancements could play a crucial role in addressing the pressing needs of climate-smart agriculture.

In the agricultural sector, nano-enabled fertilizers are increasingly being utilized to boost crop yields across diverse environmental conditions (Gogos et al. 2012). Recent advancements in engineered nanomaterials (ENMs) have particularly focused on nanofertilizers and nanopesticides, offering innovative and efficient nutrient delivery systems. Nanofertilizers have gained significant traction in agriculture due to their advanced formulations and delivery mechanisms, which enhance nutrient uptake and utilization in plants (Adisa et al. 2019; Fraceto et al. 2016; Verma et al. 2022). The application of nanofertilizers increases the nutrient delivery efficiency in plants by ensuring targeted nutrient release, minimizing nutrient losses, and reducing fertilizer application rates thgrough improved NUE (Kumar et al. 2021a, b). These fertilizers contribute to enhanced crop productivity while mitigating excessive reactive nutrient emissions into the environment (Kah et al. 2019; Kumar et al. 2021a, b). However, their widespread use has also raised concerns regarding potential environmental degradation and pollution, which may contribute to climate change. Nanotechnology, particularly the use of nanomaterials smaller than 100 nm, presents a cost-effective and efficient alternative due to their high absorption rates, superior utilization potential, and reduced nutrient losses (Lahari et al. 2021). These nanoparticles efficiently penetrate plant leaves, activate nutrient pathways, and enhance overall NUE, offering a promising approach to sustainable agriculture.

Nanofertilizers effectively meet crop nutrient demands by increasing nutrient bioavailability within the plant system and the root rhizosphere (Liu and Lal 2015). Exogenous applications of nanofertilizers improve NUE and nutritional quality through bio-fortification, employing mechanisms such as direct uptake, stomatal absorption, and surface adsorption (Sun et al. 2023; Mai et al. 2024). The ultra-small particle size of nanofertilizers facilitates rapid distribution within plant tissues, reducing nutrient losses and maximizing nutrient utilization. Nanotechnology thus presents a promising strategy for optimizing plant productivity (Preetha and Balakrishnan 2017). A wide range of nanomaterials has been applied to various plant species, demonstrating significant positive effects on growth, development, and overall yield (Li et al. 2021; Ashraf et al. 2022). The impact of nanoparticles on plant physiology is influenced by their intrinsic physicochemical properties (e.g., size, shape, surface area, and surface charge) as well as their extrinsic interactions with the biological environment (Raliya et al. 2018; Rawat et al. 2017; Chen et al. 2021). Additionally, nanoparticles play a vital role in enhanicng plant resilience under environmental stresses, such as elevated temperatures, mineral deficiencies, and metal toxicity (Zulfiqar et al. 2019; Sharifi et al. 2020).

The use of nanofertilizers presents an effective strategy for addressing crucial agricultural challenges while minimizing environmental pollution. The development of nanofertilizers marks a significant advancement in modern agriculture, enhancing both the quality and quantity of produce (Ashraf et al. 2023). Ongoing innovations in nanofertilizer technology continue to explore novel approaches to achieving higher productivity levels. Recent studies indicate that nanofertilizers coated with nanomaterials can sustain nutrient release by 40–50 days. For instance, nanohybrid urea has demonstrated a nitrogen release rate that is 12 times slower than conventional synthetic urea, potentially increasing crop yields by up to 50%. However, before commercialization, it is essential to conduct comprehensive studies on the mechanisms of action of these fertilizers. Future research should prioritize safety assessments, bioavailability, and toxicity toxicity to ensure their sustainable application in agriculture. Moreover, the use of green-synthesized nanobiofertilizers offers a promising approach to improving crop yields with in a sustainable agricultural framework. The global challenge of meeting rising food demand is already a pressing concern, and climate change is expected to intensify this issue further. Addressing these challenges requires the continued development and responsible implementation of advanced agricultural technologies.

Factors determining food security, such as availability of water, production of rain-fed crops, and crop distribution may greatly be affected by rising temperatures, shortage of water, frequent weather events, and increasing sea levels (Raza et al. 2024). In this context, Climate Smart Agriculture (CSA) is a key approach by the Food and Agriculture Organization of the United Nations (UN-FAO), providing restructuring and realigning strategies to enhance the resiliance of the agriculture system to climate change (Campbell et al. 2014). Moreover, the integration of various farming/agronomic systems and practices, together with the improvement of input utilization, such as seeds, pesticides, water, etc., all fall within the concept of CSA (Lipper et al. 2014; Khatri-Chhetri et al. 2019).

In general, the CSA aims to achieve three primary objectives via site-dependent activities, i.e., (a) to increase agricultural output and revenue in a sustainable manner, (b) to increase resilience to climate change, and (c) to reduce greenhouse gas emissions. The so-called "triple win" scenario is achieved when these goals are successfully accomplished. The CSA approach relies on both traditional agricultural methodologies and new cutting-edge technology that needs to be incorporated into the agricultural industry as a whole (Budaev et al. 2019; Mazzetto et al. 2020). Conventional farming is a broad term that is increasingly being replaced by digital farming, which incorporates the latest technologies and innovative farm practices. However, the integration of these practices, along with their short- and long-term environmental impacts, highlights the need for sustainable and climate-smart agriculture. This approach focuses on minimizing environmental degradation and promoting eco-friendly farm practices. The interlinking of different farming practices has been described in Fig. 15.1.

Fig. 15.1 Simplified link between different agricultural systems

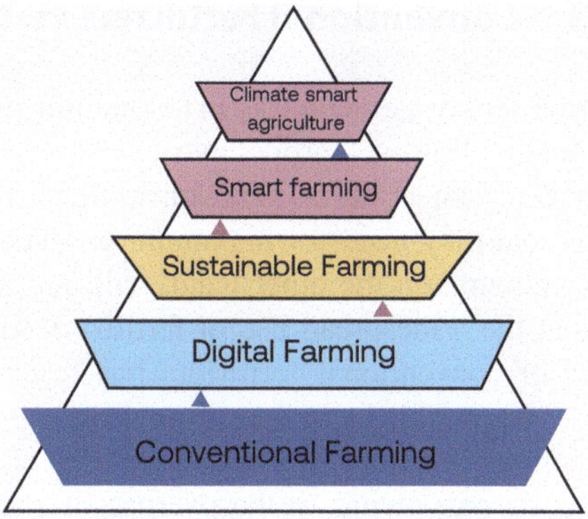

15.2 Role of Nanofertilizers for Climate-Smart Agriculture

Considering the socio-ecological and economic viewpoints, climate change has emerged as one of the most controversial and pressing issues of the twenty-first century. It exerts significant impacts on several sensitive systems, with agriculture being one of the most vulnerable. Climate change resulting in biotic and abiotic stresses (Robroek et al. 2017; Hussein and Abou-Baker 2018) and disrupting ecosystems, potentially leading to severe damage to crop production (Mittal et al. 2020). Fertilizers are g applied to the soil either by broadcasting or by mixing them with water for irrigation. However, traditional chemical fertilizers can negatively impact the environment, contaminate water bodies, and cause air pollution (Adisa et al. 2019).

In recent years, reducing the use of synthetic fertilizers has been encouraged by the farming community aiming to preserve agroecosystems, reduce pollutants, and minimize residual effects on crops. Therefore, nanotechnology might provide long-term solutions for the critical issues regarding ecosystems, biodiversity, and climate change (Rajput et al. 2021a, b) faced by modern intensive agriculture. Nanofertilizers possess the potential to enhance the efficiency of crop production systems and satisfy plant nutritional demands without compromising crop yields (Janmohammadi et al. 2016). Tarafder et al. (2020) conducted a field experiment on Okra , in which copper (Cu^{2+} (Fe^{2+} and zinc (Zn^{2+}) nanoparticles were added to urea-modified hydroxyapatite and reported that the application of nanofertilizer—which was 100 times less potent than traditional fertilizer—led to a notable increase in the absorption of the nutrients such as Cu^{2+}, Fe^{2+}, and Zn^{2+}. The use of a mixture of zinc oxide, manganese dioxide, and molybdenum trioxide nanoparticles in common bean plants by foliar spraying was also found to enhance vegetative growth, flower number per-plant, and yield (Salama et al. 2022). In addition to crop yield benefits, using nanofertilizers can lower greenhouse gas emissions into the atmosphere (Mohanraj et al. 2017).

15.3 Conventional Fertilizers vs. Nanofertilizers

Fertilizers play a crucial role in maintaining soil fertility while enhancing crop yield and quality; however, proper nutrient management is essential to boost crop productivity (Zulfiqar et al. 2019). Conventional fertilizers supply essential nutrients to plants that they need for maximum productivity and growth, but existing agricultural systems, on the other hand, will not be able to fulfill growing food demand without the widespread use of fertilizers (Babu et al. 2022). However, long-term usage of conventional fertilizers has resulted in major environmental problems across the world, including groundwater pollution, eutrophication, air pollution, and soil degradation (Mahapatra et al. 2022; Babu et al. 2022). Moreover, production costs rise owing to the overuse of chemical fertilizers with lower producer profit margins (Fatima et al. 2021). In traditional nutrient application methods, most of the applied nutrients are lost by leaching, gaseous emissions, drift, runoff, evaporation, microbial degradation, or volatilization, or become inaccessible to the plant due to high availability, obstructing root absorption and occasionally generating harmful consequences (Al-Mamun et al. 2021).

Therefore, efforts have been dedicated to develop novel fertilizers to improve NUE, and other strategies including precision fertigation, targeted fertilizer delivery, and the application of nanofertilizers (Beig et al. 2022). Thus, one of the possible possibilities to improve nutrient use efficiencies is to use nanotechnology to synthesize new forms of fertilizers (Jogaiah et al. 2021). Nanofertilizers are nanostructured formulations of various fertilizers with a smaller size (almost 1–100 nm) and have more surface area with a higher volume ratio (Seleiman et al. 2021). Nanostructured fertilizers can better upake by plant roots owing to their excellent physicochemical properties, such as smaller size, higher reactivity, more surface area, higher ionizing power, more pH tolerance, higher thermal stability, and absorbability than conventional fertilizers (Fatima et al. 2021). Furthermore, nanofertilizers release nutrients into the agricultural soil steadily and in a precise way. Nanofertilizers have nano-dimensional adsorbents that discharge nutrients extremely gradually parallel to traditional fertilizers, whether applied alone or in combination with other fertilizer types.

Moreover, nanofertilizers provide a number of advantages, e.g., low application rates, low transportation costs, and ease of application (Babu et al. 2022), which not only improves fertilizer utilization but also reduces nutrient loss. The combination of nanofertilizers with microbes, i.e., "nano-biofertilizers," could be more beneficial than conventional and nano-fertilizers (Zulfiqar et al. 2019). The benefits of nanofertilizers over conventional fertilizers have been demonstrated in Fig. 15.2.

Prasad et al. (2017) reported that nanofertilizers are more effective than conventional fertilizers owing to enhanced solubility, bioavailability, and ultimately improved accessibility of nutrients that in turn, lower the nutrient losses. Furthermore, these are easily entered through the plant cell wall, which has a very small pore size (about 5–20 nm). Therefore, nanofertilizer utilizes different ion channels through advanced mechanisms and successfully reduces issues like premature conversion, nutrient leaching, and soil fixation (Cieschi et al. 2019). Moreover, the functionality of nanofertilizers is flexible regarding root and foliar

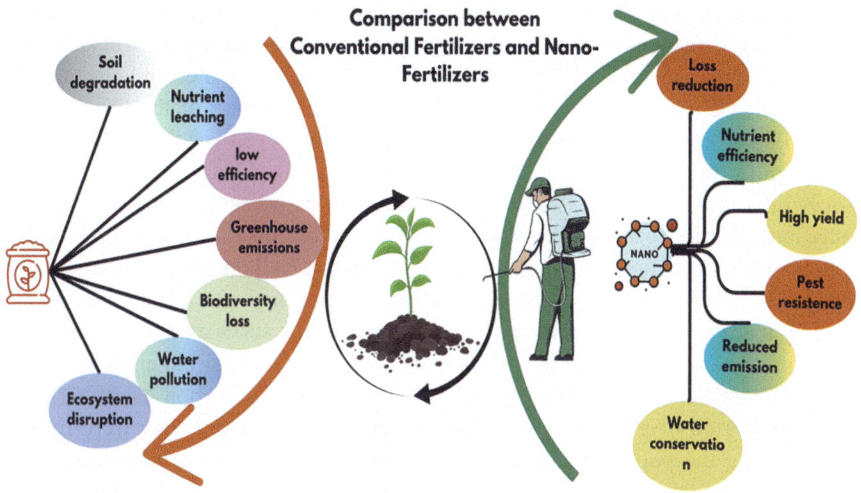

Fig. 15.2 Benefits of applying nanofertilizers over conventional fertilizers

entry in plants as compared to chemical fertilizers (Zulfiqar et al. 2019). Consequently, nanofertilizers play an important role in countering several abiotic stresses like heavy metal stress (Salam et al. 2022), drought (Kandhol et al. 2022), salinity (Sarkar and Kalita 2022), etc. Recently, several studies have found that various nanofertilizers, such as zinc (Zafar et al. 2022), iron (Waqas et al. 2022), titanium (Silva et al. 2022), copper (Saffan et al. 2022), selenium (Sarkar and Kalita 2022), silicon (Bansal et al. 2022), and magnesium (Ahmed et al. 2021) oxides, have revealed that nanofertilizers effectively improve the agricultural yield by boosting seed germination, photosynthesis, seedling development, proteins and carbohydrate synthesis, and nitrogen metabolism, as well as improving the plant defense system against various abiotic stresses.

15.4 Synthesis and Potential Use of Nanofertilizers in Agriculture

Synthesis of nanopartilces includes two methods i.e., top-down synthesis involves physical and biological synthesis, while bottom-up synthesis involves chemical synthesis. Bottom-up nanotechnology uses self-assembly and self-organization to make nanoscale materials. Nanoscale fertilizer delivers high-production, nutritious food, saves natural resources, improves soil biological health, and mobilizes soil nutrients (Raliya et al. 2017). There are numerous reports on nonmaterials being synthesized and used as nanofertilizers (Abbasifar et al. 2020; Cota-Ruiz et al. 2020; de França Bettencourt et al. 2020; Ramírez-Rodríguez et al. 2020). The broad categories of nonmaterial synthesis routes, such as top-down and bottom-up, hold well for nanofertilizers. Shebl et al. (2019) produced manganese (Mn), Zn, and Fe nanofertilizers with a particle size ranging from 20 to 60 nm by using a green microwave-aided hydrothermal approach and starting with nitrate precursors and further reported that the identical synthesis process was utilized in the production of manganese–zinc ferrite nanofertilizers for the growth of squash plants (Shebl et al. 2020).

Nanofertilizer manufacturing is to consider novel dosage mechanisms such as gradual release and controlled release. The ionotropic gelation method for silicon-encapsulating chitosan nanofertilizers resulted in improved seedling vigor index and overall crop growth of maize (Kumaraswamy et al. 2021). The leaves of the *Morus alba* plant were used in the biogenic synthesis process to manufacture the potassium nanofertilizer, which resulted in substantial improvement in growth and yield attributes of wheat as compared to conventional KCl fertilizer (Sheoran et al. 2021). The nanocomposites of zeolite and iron oxide nanoparticles were synthesized by loading iron oxide nanoparticle phases onto the surface of zeolite by rapid green precipitation, resulting in a slow release of Fe^{2+} (Jahangirian et al. 2020). In order to construct a slow-release nanofertilizer method, the carbon nanofibers that were transported over Cu–Zn micronutrient on PVA–starch substrate were formulated. During the polymerization, the micronutrients and carbon nanofibers were disseminated in the substrate in situ, and the synthesis included acid and heat treatment, followed by chemical vapor deposition, and ball milling to synthesize the product (Kumar et al. 2018). The three most common methods used to synthesize the nanocomposites are briefed below:

15.4.1 Chemical Synthesis of Nanofertilizers

Chemical synthesis methods include the synthesis of nanofertilizers as bottom-up approaches since the synthesis begins at the atomic size and builds up nanoparticles through chemical reactions. These approaches improve particle size control, minimize contaminants, entail a low-temperature procedure, and enable controlled synthesis (Belal and El-Ramady 2016). Chemical synthesis methods such as chemical vapor deposition (CVD), chemical precipitation, and the sol-gel technique are used to produce various metal oxides, borides, sulfides, and other nanoparticles (Tarafder et al. 2020; Rajonee et al. 2016; Qureshi et al. 2018).

15.4.2 Physical Synthesis of Nanofertilizers

Physical methods for synthesizing nanomaterials are based on principles that involve the application of thermal energy, radiation, or mechanical pressure, leading to processes like condensation, dissolution, evaporation, or abrasion. Unlike chemical methods, these techniques are considered eco-friendly as these do not involve chemical solvents (Sudha et al. 2018). Physical methods include the disintegration of bulk materials through top-down methods (Baig et al. 2021). On the other hand, the physical synthesis method is primarily hampered by problems such as source-precursor incompatibility, lower control over size distribution, a greater quantity of contaminants, induced stresses, faults, and imperfections, and the high cost of equipment. Physical synthesis methods are commonly employed as the primary approach to synthesize nanofertilizers due to their ease of use, less time required for nanoparticle synthesis, and potential for large-scale production (Subramanian and Thirunavukkarasu 2017). There are several physical synthesis procedures, such as the gas condensation method (Uhm et al. 2007), attrition ball mill (Qazi and Javaid 2016), planetary ball mill (Nieman et al. 1989), low energy tumbling mill (Garrigue et al. 2004), and high energy ball mill (Konrad et al. 2001) have been employed by researchers for the synthesis of nanomaterial.

15.4.3 Biological Synthesis of Nanofertilizers

The biological synthesis of nanofertilizers includes various plant extracts or microorganisms, whereas plant waste could be an ideal material as a reducing agent for the biological synthesis of nanofertilizers (Makarov et al. 2014). The present focus of nano-biotechnology is on environmentally friendly biosynthetic methods (Saratale et al. 2018), whereby microorganisms like bacteria and fungi are used as prospective bio-manufacturing units for nano-synthesis (Kalimuthu et al. 2008). Biosynthesized nanoparticles have applications beyond nanofertilizers, including antimicrobial coatings, medical imaging and drug delivery, catalytic water treatment, and environmental sensors. Plant extracts and plant components such as leaves, roots, stems, and fruits are employed to synthesize nanoparticles or nanomaterials to manufacture nanofertilizers. The biological synthesis approach is far superior to physical and chemical synthesis methods, as well as a more cost-effective and eco-friendly. With the help of effective enzymes and diverse microbial activities, it is also possible to produce nanoparticles on a large scale using plant extracts and microorganisms that are readily available. In biological systems, biosynthesis alters inorganic crystals with nano-fractional phase, orientation, and topography. These biosynthesis processes are biomimetic approaches, which have several advantages over other chemical and physical methods (Hussain et al. 2017). There have been instances of silver, gold, titanium, lead, copper, cadmium sulfide, zinc, and iron being manufactured from plants (Saravanan et al. 2018; Pugazhendhi et al. 2018; Chellapandian et al. 2019; Srinivasan et al. 2019). These methods rely on low-cost, ecologically friendly bacteria, fungi, and microalgae. Controlling pH, temperature, stirring rate, temperature, and reaction environment improves the form and size of nanoparticles in biosynthetic processes. The biosynthesis approaches have significant advantages over other alternative routes of nanofertilizer manufacturing at small and large scales (Buitrago-Vásquez and Ossa-Orozco 2018; Sathiyavimal et al. 2019; Mostaghaci et al. 2009).

15.5 Potential Use of Nanofertilizers in Agriculture

Nanotechnology is expanding globally and offers a lot of potential applications in the agriculture sector (Chhipa and Joshi 2016). Different kinds of nanoparticles have been transformed into nanofertilizers, including nanotubes of C, Cu, Si, Mo,

Zn, Fe, Ag, Ti, and their oxides, as well as nanoformulations containing typical crop inputs like phosphorus and sulfur (Taha et al. 2016; Pradhan et al. 2013; Mahajan et al. 2011). The application of nanoparticles at optimum levels has shown positive impacts on seed germination, plant growth, nutrient uptake, crop yields, and grain enrichment with essential micronutrients. Moreover, it is considered important in offering the potential for sustainable agriculture by replacing conventional methods of farming with precision farming (Chen and Yada 2011). Owing to effective uptake and utilization by plants, the nanoscale fertilizers limit nutrient losses owing to less or no leaching and are effective for agricultural applications. Nanoscale particles are smaller in size and have different dynamics than bulk particles or ionic salts (Dimkpa et al. 2017a, b; Chugh et al. 2021). Utilizing fertilizers that are equipped with nanotechnology may result in increased nutrient delivery efficiency in plants (Chhipa 2017), owing to the fact that nanofertilizers have been shown to increase crop yield by facilitating the targeted delivery and slow release of nutrients, as well as cutting down the amount of fertilizer applied while simultaneously boosting nutrient use efficiencies (Kah et al. 2019). Nanofertilizers may replace the application of conventional fertilizers by providing an alternative option for farmers in future (El-Kereti et al. 2013). The impacts of nanoparticles/nanomaterials at different concentrations on different plants have been summarized in Table 15.1.

In general, nanofertilizers act as a source and carrier in terms of on-site delivery of nutrients in precise quantities, which improves the overall growth and development of plants. Overall, the effect of nanofertilizers undeniably varies depending on plant characteristics, with several key factors influencing their performance, such as application techniques, size, shape, morphology, and concentration.

15.6 Role of Nanofertilizers in Crop Improvement under Changing Climate

Nanotechnology is of great importance in agriculture to improve crop production and grain quality under multiple abiotic stresses (Tanveer et al. 2020). The chemical combination and physical adsorption between nanocomposites and nutrient elements are valuable due to the small size and surface reactions of nanocomposites. Nanofertilizers release nutrients in a regulated way and supply the right amount of nutrients in the right proportion to crops, thus maintaining environmental safety and improving crop production (DeRosa et al. 2010). These multifunctional fertilizers also reduce soil leaching and fertilizer fixation. Yang et al. (2009) reported that the nanomaterials efficiently controlled the penetrability and structure as well as improved the mineral granule of soil, increased water-holding capacity and fertilizer storage, strengthened the activity of microorganisms, maintained the carbon and nitrogen ratio, and improved the soil fertility.

The soil application of inorganic fertilizers ultimately results in decline of nutrients that eventually become inaccessible to plants. For example, 80–90% of phosphorus, 50–90% of potassium, and 40–70% of nitrogen fertilizers are lost, leading to economic losses (Ombódi and Saigusa 2000). Hence, more demand for fertilizer

Table 15.1 Impact of nanoparticles/nanofertilizers on different crops

Nanoparticles	Plant	Application	Concentration	Impact	Reference
nFe_2O_3	*Glycine max* L.	Foliar	0.25–1 M	Enhanced leaf biomass with seed weight as compared to normal plants	Sheykhbaglou et al. (2010)
	Solanum lycopersicum L.	Hydroponic	50–800 ppm	Improved germination of seeds, morphological traits, dry weight, and Fe uptake as compared to control	Shankramma et al. (2016)
$nSiO_2$	*Zea mays* L.	Hydroponic	20–40 nm	Enhanced germination (%) rate, biomass, Si uptake, and nutrient uptake	Suriyaprabha et al. (2012)
	Glycine max L.	Soil	30–50 nm	Reduced the toxic effects on plant under Hg stress and reduced Hg uptake in the epidermis and pericycle of the plant roots and leaves. Increase leaf gas exchange and enzymatic responses.	Li et al. (2020)
	Triticum aestivum L.	Hydroponic	10 μM	Alleviated harmful effects of UV radiation on plants	Tripathi et al. (2017)
	Hordeum vulgare L.	Soil	12–250 ppm	Significantly enhanced plant growth performance, chlorophyll content, leaf gas exchange, osmolytes, antioxidative enzyme activities, cell membrane efficiency, and profile of metabolites	Ghorbanpour et al. (2020)
nCuO	*Solanum lycopersicum* L.	Foliar	50–500 ppm	Enhanced vitamin C, lycopene, ABTS, catalase, and superoxide dismutase and reduced the ascorbate peroxidase and glutathione peroxidase activities. Increased Cu accumulation in tomato fruits	López-Vargas et al. (2018)
$nCeO_2$	*Triticum aestivum* L.	Soil	0–500 mg kg^{-1} of soil	Enhanced overall plant fitness and productivity as compared to normal plants—Increased Ce uptake in roots but no change in leaves, hull, and seeds	Rico et al. (2014)

(continued)

Table 15.1 (continued)

Nanoparticles	Plant	Application	Concentration	Impact	Reference
$nTiO_2$	*Hordeum vulgare* L.	Soil	500–1000 mg kg^{-1} of soil	Applied NPs found to stimulate plant performance by enhancing germination (%) as compared to normal and treated plants	Nhan et al. (2015)
nZnO	*Oryza sativa* L.	Soil	0–750 mg kg^{-1} of soil	Enhanced plant performance, P level in roots to grains. Upregulated the level of metabolites, i.e., amino acids, palmitic acids, and glycerol levels in rice seeds	Zahra et al. (2017)
	Triticum vulgare L.	Hydroponic	5–40 ppm	No significant effects on plant performance. Leaf photosynthetic pigments were reduced with increasing NP levels. Increased nutrient uptake and accumulation except for K level	Dağhan et al. (2020)
	Zea mays L.	Foliar	150–300 ppm	Enhanced maximum growth characteristics, physiological and biochemical activities during high pH treatment	Raddy et al. (2018)
	Sorghum bicolor L.	Soil and foliar	6 mg kg^{-1} soil	Enhanced plant performance and yield component, uptake of N and K elements, improved grain nutrient profile and NUE as compared to normal plants	Dimkpa et al. (2017a, b)
Nano chelated iron fertilizer	*Oryza sativa* L.	Foliar	2.5 g L^{-1}	Improved grain yield and grain protein as well as NPK, Zn, and Fe contents	Fakharzadeh et al. (2020)
Nano-Urea	*Oryza sativa* L.	Foliar spray	10–50% and 100% nano-urea	Urea and nano-urea (50% each) substantially improved the growth and yield attributes of rice	Midde et al. (2022)

to overcome the loss negatively influences the nutrient balance of soil (Baruah and Dutta 2009). Nanofertilizers are used to minimize the frequent applications of traditional fertilizers and may be an alternative option for dissolved inorganic fertilizers due to the slow and steady release of nutrients during crop production that results in efficient absorption of nutrients with minimal loss. According to Vega-Vásquez et al. (2020), urea nanoencapsulation can minimize the N-associated environmental effect of fertilizer application by reducing its frequency and enhancing NUE.

Nanocomposites help to reduce the reliance on conventional fossil fuels, thereby reducing greenhouse gas emissions and mitigating global warming (Subramanian et al. 2020). Plants applied with nanomaterials have the ability for greater absorption of CO_2 (Hayat et al. 2022) and are able to carry out photosynthesis more effectively by means of genetic and nitrogen fertilization modifications, leading to a greater uptake of CO_2 from the environment. Likewise, N_2O emissions are also significantly decreased by the use of nanomaterials in soil amendments (Wu et al. 2020).

15.7 Short and Long-Term Benefits of Nanofertilizers

Nanoparticles have unique physical, biological, and chemical properties that can be manipulated to get certeain short and long-term benefits due to their high surface area-to-volume ratio (El-Saadony et al. 2021). Based on the nutritional requirements of plants, nanofertilizers are categorized into three types: micro nanofertilizers, nanoparticulate fertilizers, and macro nanofertilizers (Chhipa and Joshi 2016). The nanomaterials can be applied as liquids or powders, delivering nutrients in an easily accessible form, leading to improved plant growth and yield. Furthermore, nanofertilizers play a key role in preventing fertilizer-related environmental contamination, contributing to sustainable agricultural practices (Naik et al. 2021). Nanomaterials also enhance plant stress tolerance, promoting overall plant health (Tiwari et al. 2012). The key characteristics of nanofertilizers include efficient nutrient delivery, low-cost nutrition, high fertilization efficiency, and their ability to prevent pollution, especially in the context of water contamination (Guru et al. 2015).

Moreover, multi-walled carbon nanotubes (MWCNTs) can stimulate seed germination and growth in tomato plants and promote cell development in tobacco (Khodakovskaya et al. 2009, 2012). The MWCNTs have also been successfully used in mustard plant development (Mondal et al. 2011). Additionally, silica nanoparticles have been shown to boost seedling development and stress tolerance by stimulating antioxidant enzymes under both biotic and abiotic stress conditions (Adisa et al. 2019; Li et al. 2020; Rajput et al. 2021a, b). These nanoparticles can also enhance potassium (K) uptake while reducing sodium (Na) absorption, aiding in stress tolerance (Rizwan et al. 2015; Liang et al. 2015). Silicon nanoparticles are particularly favorable for improving plant growth and mitigating the adverse effects of climate change, such as erratic rainfall, temperature fluctuations, and increased evaporation (Younis et al. 2020). By facilitating controlled nutrient release over time, nanofertilizers minimize nutrient volatility and leaching, reducing environmental impact. Additionaly, nanoparticles possess a high absorption capacity and large surface area, enabling them to target specific plant sites for enhanced and efficient nutrient uptake (Rameshaiah et al. 2015).

In the short term, nanofertilizers enhance nutrient use efficiency by targeting nutrient supply through roots or leaves, in case of soil or foliar application, in a controlled manner. The targeted delivery of nanofertilizers reduces nutrient loss through leaching or volatilization, leading to immediate improvements in crop growth, yield, and overall plant health. Additionally, nanofertilizers promote faster germination and development by ensuring that essential nutrients are readily available at critical growth stages and are also helpful in grain enrichment with micronutrients.

Regarding long-term benefits, the use of nanofertilizers can reduce the overall use of conventional chemical fertilizers, decreasing the environmental impact associated with over-fertilization, such as soil degradation and water contamination. Better nutrient absorption through the application of nanofertilizers ensures a substantial reduction in the cost of production (in general) and fertilizer costs (in specific) for farmers and enhances soil health over time. Furthermore, the sustained release of nutrients provided by nanofertilizers helps maintain long-term soil fertility and improve crop productivity. Overall, nanofertilizers could be promising solutions for enhancing nutrient use efficiencies and crop productivity in short and the long run (Fig. 15.3).

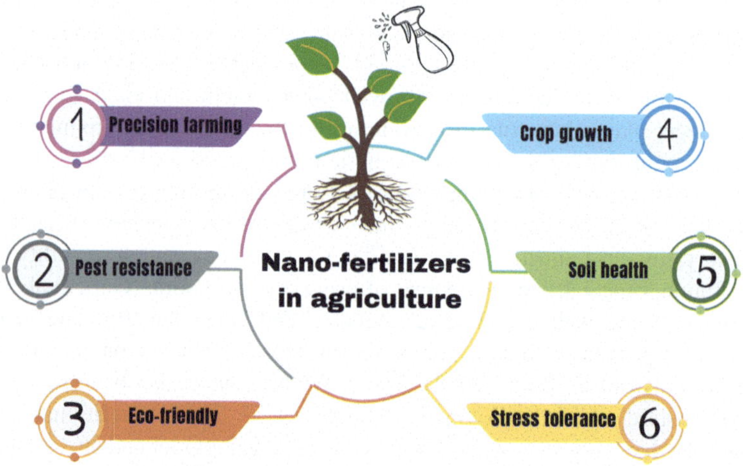

Fig. 15.3 Multi-facet benefits of nanofertilizers in agriculture and crop production

15.8 Conclusions and Future Perspectives

Nanomaterials, specifically in the form of nanofertilizers, have emerged as a promising approach to improve NUE and crop productivity. The unique properties of nanomaterials, such as high surface area and controlled release, allow for precise nutrient delivery to plants, minimizing nutrient losses and environmental pollution. The nanofertilizer application was found effective in either applied to soil or foliar spray as compared to conventional fertilizer. The synthesis of nanofertilizers and their potential application to agricultural crops could be a promising step toward climate-smart agriculture, offering solutions that promote both productivity and sustainability and provide an avenue for enhancing resource use efficiency while mitigating the negative impacts on the environment.

Future research should be focused on understanding the long-term effects of nanofertilizers on soil health, microbial ecosystems, and overall crop productivity. Additionally, the development of eco-friendly and biodegradable nanomaterials would be crucial in ensuring environmental sustainability and biosafety purposes. Before large-scale commercialization and widespread adoption, scientists, farmers, and policymakers need to work together to formulate regulations and frameworks for safe and effective use.

Author Contribution Statement Umair Ashraf led the conceptualization of this chapter, done with theortical framework, wrote the origina draft, and was responsible for writing and revising key sections. Fatima Batool, Rabia Ghaffar, and Muhammad Imran co-wrote significant portions of the chapter. Muhammad Hussaan and Muhammad Ansar Farooq assisted with the literature review, collection of previous studies/research and contributed to the overall chapter writeup. Ayesha Riaz helped to collect tabulated data and preperation of figures. Fahd Rasul helped in reviewing, revising, and improving the chapter. All authors approved the final draft.

References

Abbasifar A, Shahrabadi F, ValizadehKaji B (2020) Effects of green synthesized zinc and copper nano-fertilizers on the morphological and biochemical attributes of basil plant. J Plant Nutr 43(8):1104–1118

Adisa IO, Pullagurala VLR, Peralta-Videa JR, Dimkpa CO, Elmer WH, Gardea-Torresdey JL, White JC (2019) Recent advances in nano-enabled fertilizers and pesticides: a critical review of mechanisms of action. Environ Sci Nano 6(7):2002–2030

Ahmed T, Noman M, Manzoor N, Shahid M, Hussaini KM, Rizwan M, Li B (2021) Green magnesium oxide nanoparticles-based modulation of cellular oxidative repair mechanisms to reduce arsenic uptake and translocation in rice (*Oryza sativa* L.) plants. Environ Pollut 288:117785

Al-Mamun MR, Hasan MR, Ahommed MS, Bacchu MS, Ali MR, Khan MZH (2021) Nanofertilizers towards sustainable agriculture and environment. Environ Technol Innov 23:101658

Ashraf U, Zafar S, Ghaffar R, Sher A, Mahmood S, Noreen Z et al (2022) Impact of nano chitosan-NPK fertilizer on field crops. In: Role of chitosan and chitosan-based nanomaterials in plant sciences. Academic Press, pp 165–183

Ashraf U, Shahid MN, Batool F, Mahmood S, Mustafa GM, Aqeel M et al (2023) Application of nanomaterials in agriculture. In: The impact of nanoparticles on agriculture and soil. Academic Press, pp 259–283

Babu S, Singh R, Yadav D, Rathore SS, Raj R, Avasthe R, Singh VK (2022) Nanofertilizers for agricultural and environmental sustainability. Chemosphere 292:133451

Baig N, Kammakakam I, Falath W (2021) Nanomaterials: a review of synthesis methods, properties, recent progress, and challenges. Mater Adv 2(6):1821–1871

Bansal K, Hooda V, Verma N, Kharewal T, Tehri N, Dhull V, Gahlaut A (2022) Stress alleviation and crop improvement using silicon nanoparticles in agriculture: a review. SILICON:1–14

Baruah S, Dutta J (2009) Nanotechnology applications in pollution sensing and degradation in agriculture: a review. Environ Chem Lett 7(3):191–204

Beig B, Niazi MBK, Sher F, Jahan Z, Malik US, Khan MD, Vo DVN (2022) Nanotechnology-based controlled release of sustainable fertilizers. A review. Environ Chem Lett:1–18

Belal ES, El-Ramady H (2016) Nanoparticles in water, soils and agriculture. Nanosci Food Agric 2:311–358

Bindraban PS, Dimkpa CO, Angle S, Rabbinge R (2018) Unlocking the multiple public good services from balanced fertilizers. Food Secur 10:273–285

Budaev D, Lada A, Simonova E, Skobelev P, Travin V, Yalovenko O, Voshchuk G, Zhilyaev A (2019) Conceptual design of smart farming solution for precise agriculture. Manag App Complex Syst 13:309–316

Buitrago-Vásquez M, Ossa-Orozco CP (2018) Hydrothermal synthesis of hydroxyapatite nanorods using a fruit extract template. Dyna 85(204):283–288

Campbell BM, Thornton P, Zougmoré R, Van Asten P, Lipper L (2014) Sustainable intensification: what is its role in climate smart agriculture? Curr Opin Environ Sustain 8:39–43

Chellapandian C, Ramkumar B, Puja P, Shanmuganathan R, Pugazhendhi A, Kumar P (2019) Gold nanoparticles using red seaweed *Gracilaria verrucosa*: green synthesis, characterization and biocompatibility studies. Process Biochem 80:58–63

Chen H, Yada R (2011) Nanotechnologies in agriculture: new tools for sustainable development. Trends Food Sci Technol 22(11):585–594

Chen J, Zeng X, Yang W, Xie H, Ashraf U, Mo Z et al (2021) Seed priming with multiwall carbon nanotubes (MWCNTs) modulates seed germination and early growth of maize under cadmium (Cd) toxicity. J Soil Sci Plant Nutr 21:1793–1805

Chhipa H (2017) Nanofertilizers and nanopesticides for agriculture. Environ Chem Lett 15:15–22

Chhipa H, Joshi P (2016) Nanofertilisers, nanopesticides and nanosensors in agriculture. Nanosci Food Agric 1:247–282

Chugh G, Siddique KH, Solaiman ZM (2021) Nanobiotechnology for agriculture: smart technology for combating nutrient deficiencies with nanotoxicity challenges. Sustain For 13(4):1781

Cieschi MT, Polyakov AY, Lebedev VA, Volkov DS, Pankratov DA, Veligzhanin AA, Lucena JJ (2019) Eco-friendly iron-humic nanofertilizers synthesis for the prevention of iron chlorosis in soybean (*Glycine max*) grown in calcareous soil. Front Plant Sci 10:413–421

Cota-Ruiz K, Ye Y, Valdes C, Deng C, Wang Y, Hernández-Viezcas JA, Gardea-Torresdey JL (2020) Copper nanowires as nanofertilizers for alfalfa plants: understanding nano-bio systems interactions from microbial genomics, plant molecular responses and spectroscopic studies. Sci Total Environ 742:140572

Dağhan H, Gülmezoğlu N, Köleli N, Karakaya B (2020) Impact of titanium dioxide nanoparticles (TiO2-NPs) on growth and mineral nutrient uptake of wheat (*Triticum vulgare* L.). Biotech Stud 29(2):69–76

de França Bettencourt GM, Degenhardt J, Torres LAZ, de Andrade Tanobe VO, Soccol CR (2020) Green biosynthesis of single and bimetallic nanoparticles of iron and manganese using bacterial auxin complex to act as plant bio-fertilizer. Biocatal Agric Biotechnol 30:101822

DeRosa MC, Monreal C, Schnitzer M, Walsh R, Sultan Y (2010) Nanotechnology in fertilizers. Nat Nanotechnol 5(2):91–91

Dimkpa CO, Bindraban PS, Fugice J, Agyin-Birikorang S, Singh U, Hellums D (2017a) Composite micronutrient nanoparticles and salts decrease drought stress in soybean. Agron Sustain Dev 37:1–13

Dimkpa CO, White JC, Elmer WH, Gardea-Torresdey J (2017b) Nanoparticle and ionic Zn promote nutrient loading of sorghum grain under low NPK fertilization. J Agric Food Chem 65(39):8552–8559
El-Kereti M, El-feky SA, Khater MS, Osman YA, El-sherbini AES (2013) ZnO nanofertilizer and he ne laser irradiation for promoting growth and yield of sweet basil plant. Recent Pat Food Nutr Agric 5(3):169–181
El-Saadony MT, Alkhatib FM, Alzahrani SO, Shafi ME, Abdel-Hamid SE, Taha TF, Ahmed NH (2021) Impact of mycogenic zinc nanoparticles on performance, behavior, immune response, and microbial load in *Oreochromis niloticus*. Saudi J Biol Sci 28(8):4592–4604
Fakharzadeh S, Hafizi M, Baghaei MA, Etesami M, Khayamzadeh M, Kalanaky S, Akbari ME, Nazaran MH (2020) Using nanochelating technology for biofortification and yield increase in rice. Sci Rep 10(1):4351
Fatima F, Hashim A, Anees S (2021) Efficacy of nanoparticles as nanofertilizer production: a review. Environ Sci Pollut Res 28(2):1292–1303
Foley JA, Ramankutty N, Brauman KA, Cassidy ES, Gerber JS, Johnston M, Mueller ND, O'Connell C, Ray DK, West PC (2011) Solutions for a cultivated planet. Nature 478:337–342
Fraceto LF, Grillo R, de Medeiros GA, Scognamiglio V, Rea G, Bartolucci C (2016) Nanotechnology in agriculture: which innovation potential does it have? Front Environ Sci 4:20
Garrigue P, Delville MH, Labrugère C, Cloutet E, Kulesza PJ, Morand JP, Kuhn A (2004) Top– down approach for the preparation of colloidal carbon nanoparticles. Chem Mater 16(16):2984–2986
Ghorbanpour M, Mohammadi H, Kariman K (2020) Nanosilicon-based recovery of barley (*Hordeum vulgare*) plants subjected to drought stress. Environ Sci Nano 7(2):443–461
Gogos A, Knauer K, Bucheli TD (2012) Nanomaterials in plant protection and fertilization: current state, foreseen applications, and research priorities. J Agric Food Chem 60:9781–9792
Guru T, Veronica N, Thatikunta R, Reddy SN (2015) Crop nutrition management with nano fertilizers. Int J Environ Sci Technol 1(1):4–6
Hayat MA, Chen Y, Bevilacqua M, Li L, Yang Y (2022) Characteristics and potential applications of nano-enhanced phase change materials: A critical review on recent developments. Sustain Energy Technol Assess 50:101799
Hussain I, Singh NB, Singh A, Singh H, Singh SC, Yadav V (2017) Exogenous application of phytosynthesized nanoceria to alleviate ferulic acid stress in *Solanum lycopersicum*. Sci Hortic 214:158–164
Hussein MM, Abou-Baker NH (2018) The contribution of nano-zinc to alleviate salinity stress on cotton plants. R Soc Open Sci 5(8):171809
Jahangirian H, Rafiee-Moghaddam R, Jahangirian N, Nikpey B, Jahangirian S, Bassous N, Webster TJ (2020) Green synthesis of zeolite/Fe_2O_3 nanocomposites: toxicity & cell proliferation assays and application as a smart iron nanofertilizer. Int J Nanomedicine 15:1005–1020
Janmohammadi M, Amanzadeh T, Sabaghnia N, Dashti S (2016) Impact of foliar application of nano micronutrient fertilizers and titanium dioxide nanoparticles on the growth and yield components of barley under supplemental irrigation. Acta Agriculturae Slovenica 107(2):265–276
Jogaiah S, Paidi MK, Venugopal K, Geetha N, Mujtaba M, Udikeri SS, Govarthanan M (2021) Phytotoxicological effects of engineered nanoparticles: an emerging nanotoxicology. Sci Total Environ 801:149809
Kah M, Tufenkji N, White JC (2019) Nano-enabled strategies to enhance crop nutrition and protection. Nat Nanotechnol 14(6):532–540
Kalimuthu K, Babu RS, Venkataraman D, Bilal M, Gurunathan S (2008) Biosynthesis of silver nanocrystals by bacillus licheniformis. Colloids Surf B: Biointerfaces 65(1):150–153
Kandhol N, Jain M, Tripathi DK (2022) Nanoparticles as potential hallmarks of drought stress tolerance in plants. Physiol Plant e13665
Khatri-Chhetri A, Pant A, Aggarwal PK, Vasireddy VV, Yadav A (2019) Stakeholders prioritization of climate-smart agriculture interventions: evaluation of a framework. Agric Syst 174:23–31

Khodakovskaya M, Dervishi E, Mahmood M, Xu Y, Li Z, Watanabe F, Biris AS (2009) Carbon nanotubes are able to penetrate plant seed coat and dramatically affect seed germination and plant growth. ACS Nano 3(10):3221–3227

Khodakovskaya MV, De Silva K, Biris AS, Dervishi E, Villagarcia H (2012) Carbon nanotubes induce growth enhancement of tobacco cells. ACS Nano 6(3):2128–2135

Konrad A, Herr U, Tidecks R, Kummer F, Samwer K (2001) Luminescence of bulk and nanocrystalline cubic yttria. J Appl Phys 90(7):3516–3523

Kumar R, Ashfaq M, Verma N (2018) Synthesis of novel PVA–starch formulation-supported Cu–Zn nanoparticle carrying carbon nanofibers as a nanofertilizer: controlled release of micronutrients. J Mater Sci 53(10):7150–7164

Kumar Y, Singh T, Raliya R, Tiwari KN (2021a) Nano fertilizers for sustainable crop production, higher nutrient use efficiency and enhanced profitability. Indian J Fertil 17(11):1206–1214

Kumar Y, Tiwari KN, Singh T, Raliya R (2021b) Nano fertilizers and their role in sustainable agriculture. Ann Plant Soil Res 23:238–255

Kumaraswamy RV, Saharan V, Kumari S, Choudhary RC, Pal A, Sharma SS, Biswas P (2021) Chitosan-silicon nanofertilizer to enhance plant growth and yield in maize (*Zea mays* L.). Plant Physiol Biochem 159:53–66

Lahari S, Hussain SA, Parameswari YS, Sharma KHS (2021) Grain yield and nutrient uptake of rice as influenced by the nano forms of nitrogen and zinc. Int J Environ Clim Chang 11(7):1–6

Li Y, Zhu N, Liang X, Bai X, Zheng L, Zhao J, Gao Y (2020) Silica nanoparticles alleviate mercury toxicity via immobilization and inactivation of Hg (II) in soybean (*Glycine max*). Environ Sci Nano 7(6):1807–1817

Li Y, Liang L, Li W, Ashraf U, Ma L, Tang X et al (2021) ZnO nanoparticle-based seed priming modulates early growth and enhances physio-biochemical and metabolic profiles of fragrant rice against cadmium toxicity. J Nanobiotechnol 19:1–19

Liang Y, Nikolic M, Bélanger R, Gong H, Song A (2015) Silicon in agriculture. In: Liang Y et al (eds) Silicon-mediated tolerance to salt stress. Springer Science, pp 123–142

Lipper L, Thornton P, Campbell BM, Baedeker T, Braimoh A, Bwalya M, Torquebiau EF (2014) Climate-smart agriculture for food security. Nat Clim Chang 4(12):1068–1072

Liu R, Lal R (2015) Potentials of engineered nanoparticles as fertilizers for increasing agronomic production. Sci Total Environ 514:131–139

López-Vargas ER, Ortega-Ortíz H, Cadenas-Pliego G, de Alba Romenus K, Cabrera de la Fuente M, Benavides-Mendoza A, Juárez-Maldonado A (2018) Foliar application of copper nanoparticles increases the fruit quality and the content of bioactive compounds in tomatoes. Appl Sci 8(7):1020

Mahajan P, Dhoke SK, Khanna AS (2011) Effect of nano-ZnO particle suspension on growth of mung (*Vigna radiata*) and gram (*Cicer arietinum*) seedlings using plant agar method. J Nanotechnol 2011(1):696535

Mahapatra DM, Satapathy KC, Panda B (2022) Biofertilizers and nanofertilizers for sustainable agriculture: Phycoprospects and challenges. Sci Total Environ 803:149990

Mai Y, Ren Y, Deng S, Ashraf U, Tang X, Duan M, Mo Z (2024) Influence of ZnO nanoparticles on early growth stage of fragrant Rice at low temperature (LT) stress. J Soil Sci Plant Nutr 24(1):1301–1317

Makarov VV, Love AJ, Sinitsyna OV, Makarova SS, Yaminsky IV, Taliansky ME, Kalinina NO (2014) "Green" nanotechnologies: synthesis of metal nanoparticles using plants. Acta Nat 6(1 (20)):35–44

Mazzetto F, Gallo R, Sacco P (2020) Reflections and methodological proposals to treat the concept of "information precision" in smart agriculture practices. Sensors 20(10):2847

Midde SK, Perumal MS, Murugan G, Sudhagar R, Mattepally VS, Bada MR (2022) Evaluation of nano urea on growth and yield attributes of rice (*Oryza sativa* L.). Chem Sci Rev Lett 11(42):211–214

Mittal D, Kaur G, Singh P, Yadav K, Ali SA (2020) Nanoparticle-based sustainable agriculture and food science: recent advances and future outlook. Front Nanotechnol 2:579954

Mohanraj J, Lakshmanan A, Subramanian K (2017) Nano-zeolite amendment to minimize greenhouse gas emission in rice soil. J Environ Nanotechnol 6(3):73–76

Mondal A, Basu R, Das S, Nandy P (2011) Beneficial role of carbon nanotubes on mustard plant growth: an agricultural prospect. J Nanopart Res 13:4519–4528

Mostaghaci B, Fathi MH, Sheikh-Zeinoddin M, Soleimanian-Zad S (2009) Bacterial synthesis of nanostructured hydroxyapatite using *Serratia marcescens* PTCC 1187. Int J Nanotechnol 6(10–11):1015–1030

Naik BSS, Mahawar N, Rupesh T, Dhegavath S, Singh Meena R (2021) Nanotechnology based nano-fertilizer: a sustainable approach for enhancing crop productivity under climate changing situations. Curr Res Agric Farm 2(1):21–29

Nhan LV, Ma C, Rui Y, Liu S, Li X, Xing B, Liu L (2015) Phytotoxic mechanism of nanoparticles: destruction of chloroplasts and vascular bundles and alteration of nutrient absorption. Sci Rep 5(1):11618

Nieman GW, Weertman JR, Siegel RW (1989) Micro-hardness of nanocrystalline palladium and copper produced by inert-gas condensation. *Scripta Metallurgica (USA)* 23:12

Ombódi A, Saigusa M (2000) Broadcast application versus band application of polyolefin-coated fertilizer on green peppers grown on andisol. J Plant Nutr 23(10):1485–1493

Pouratashi M, Iravani H (2012) Farmers' knowledge of integrated pest management and learning style preferences: implications for information delivery. Int J Pest Manag 58:347–353

Pradhan S, Patra P, Das S, Chandra S, Mitra S, Dey KK, Goswami A (2013) Photochemical modulation of biosafe manganese nanoparticles on *Vigna radiata*: a detailed molecular, biochemical, and biophysical study. Environ Sci Technol 47(22):13122–13131

Prasad R, Bhattacharyya A, Nguyen QD (2017) Nanotechnology in sustainable agriculture: recent developments, challenges, and perspectives. Front Microbiol 8:1014–1021

Preetha PS, Balakrishnan N (2017) A review of nano fertilizers and their use and functions in soil. Int J Curr Microbiol App Sci 6:3117–3133

Pugazhendhi A, Prabakar D, Jacob JM, Karuppusamy I, Saratale RG (2018) Synthesis and characterization of silver nanoparticles using *Gelidium amansii* and its antimicrobial property against various pathogenic bacteria. Microb Pathog 114:41–45

Qazi UY, Javaid R (2016) A review on metal nanostructures: preparation methods and their potential applications. Adv Nanoparticles 5(01):27–43

Qureshi A, Singh DK, Dwivedi S (2018) Nano-fertilizers: a novel way for enhancing nutrient use efficiency and crop productivity. Int J Curr Microbiol App Sci 7(2):3325–3335

Raddy R, Salimath M, Geetha K, Shankar A (2018) ZnO nanoparticle improves maize growth, yield and seed zinc under high soil pH condition. Int J Curr Microbiol Appl Sci 7:1593–1601

Rajonee AA, Nigar F, Ahmed S, Huq SI (2016) Synthesis of nitrogen nano fertilizer and its efficacy. Can J Pure Appl Sci 10(2):3913–3919

Rajput VD, Minkina T, Feizi M, Kumari A, Khan M, Mandzhieva S, Choudhary R (2021a) Effects of silicon and silicon-based nanoparticles on rhizosphere microbiome, plant stress and growth. Biology 10(8):791

Rajput VD, Singh A, Minkina T, Rawat S, Mandzhieva S, Sushkova S, Shuvaeva V, Nazarenko O, Rajput P, Komariah. (2021b) Nano-enabled products: challenges and opportunities for sustainable agriculture. Plan Theory 10:2727

Raliya R, Saharan V, Dimkpa C, Biswas P (2017) Nanofertilizer for precision and sustainable agriculture: current state and future perspectives. J Agric Food Chem 66(26):6487–6503

Raliya R, Saharan V, Dimkpa C, Biswas P (2018) Nano fertilizers for precision and sustainable agriculture: current status and future perspectives. J Agric Food Chem 66:6487–6503

Rameshaiah GN, Pallavi J, Shabnam S (2015) Nano fertilizers and nano sensors–an attempt for developing smart agriculture. Int J Eng Res Gen Sci 3(1):314–320

Ramírez-Rodríguez GB, Dal Sasso G, Carmona FJ, Miguel-Rojas C, Pérez-de-Luque A, Masciocchi N, Delgado-López JM (2020) Engineering biomimetic calcium phosphate nanoparticles: a green synthesis of slow-release multinutrient (NPK) nanofertilizers. ACS Appl Bio Mater 3(3):1344–1353

Rawat M, Nayan R, Negi B, Zaidi M, Arora S (2017) Physio-biochemical basis of iron sulfide nanoparticle induced growth and seed yield enhancement in *B. juncea*. Plant Physiol Biochem 118:274–284

Raza A, Safdar M, Adnan Shahid M, Shabir G, Khil A, Hussain S, … Akram HMB (2024) Climate change impacts on crop productivity and food security: an overview. Transforming Agricultural Management for a Sustainable Future: Climate Change and Machine Learning Perspectives, 163–186

Rico CM, Lee SC, Rubenecia R, Mukherjee A, Hong J, Peralta-Videa JR, Gardea-Torresdey JL (2014) Cerium oxide nanoparticles impact yield and modify nutritional parameters in wheat (*Triticum aestivum* L.). J Agric Food Chem 62(40):9669–9675

Rizwan M, Ali S, Ibrahim M, Farid M, Adrees M, Bharwana SA, Abbas F (2015) Mechanisms of silicon-mediated alleviation of drought and salt stress in plants: a review. Environ Sci Pollut Res 22:15416–15431

Robroek BJ, Jassey VE, Beltman B, Hefting MM (2017) Diverse fen plant communities enhance carbon-related multifunctionality, but do not mitigate negative effects of drought. R Soc Open Sci 4(10):170449

Saffan MM, Koriem MA, El-Henawy A, El-Mahdy S, El-Ramady H, Elbehiry F, Prokisch J (2022) Sustainable production of tomato plants (*Solanum lycopersicum* L.) under low-quality irrigation water as affected by bio-nanofertilizers of selenium and copper. Sustain For 14(6):3236

Salam A, Khan AR, Liu L, Yang S, Azhar W, Ulhassan Z, Gan Y (2022) Seed priming with zinc oxide nanoparticles downplayed ultrastructural damage and improved photosynthetic apparatus in maize under cobalt stress. J Hazard Mater 423:127021

Salama DM, Abd El-Aziz ME, Shaaban EA, Osman SA, Abd El-Wahed MS (2022) The impact of nanofertilizer on agro-morphological criteria, yield, and genomic stability of common bean (*Phaseolus vulgaris* L.). Sci Rep 12(1):18552

Saratale RG, Karuppusamy I, Saratale GD, Pugazhendhi A, Kumar G, Park Y, Shin HS (2018) A comprehensive review on green nanomaterials using biological systems: recent perception and their future applications. Colloids Surf B: Biointerfaces 170:20–35

Saravanan M, Barik SK, MubarakAli D, Prakash P, Pugazhendhi A (2018) Synthesis of silver nanoparticles from *Bacillus brevis* (NCIM 2533) and their antibacterial activity against pathogenic bacteria. Microb Pathog 116:221–226

Sarkar RD, Kalita MC (2022) Se nanoparticles stabilized with *Allamanda cathartica* L. flower extract inhibited phytopathogens and promoted mustard growth under salt stress. Heliyon 8(3):09076

Sathiyavimal S, Vasantharaj S, LewisOscar F, Pugazhendhi A, Subashkumar R (2019) Biosynthesis and characterization of hydroxyapatite and its composite (hydroxyapatite-gelatin-chitosan-fibrin-bone ash) for bone tissue engineering applications. Int J Biol Macromol 129:844–852

Seleiman MF, Almutairi KF, Alotaibi M, Shami A, Alhammad BA, Battaglia ML (2021) Nano-fertilization as an emerging fertilization technique: why can modern agriculture benefit from its use? Plan Theory 10(1):2

Shankramma K, Yallappa S, Shivanna MB, Manjanna J (2016) Fe_2O_3 magnetic nanoparticles to enhance *S. lycopersicum* (tomato) plant growth and their biomineralization. Appl Nanosci 6:983–990

Sharifi M, Faryabi K, Talaei AJ, Shekha MS, Ale-Ebrahim M, Salihi A, Nanakali NMQ, Aziz FM, Rasti B, Hasan A (2020) Antioxidant properties of gold nanozyme: A review. J Mol Liq 297:112004

Shebl A, Hassan AA, Salama DM, Abd El-Aziz ME, Abd Elwahed MS (2019) Green synthesis of nanofertilizers and their application as a foliar for *Cucurbita pepo* L. J Nanomater 2019(1):3476347

Shebl A, Hassan AA, Salama DM, Abd El-Aziz ME, Abd Elwahed MS (2020) Template-free microwave-assisted hydrothermal synthesis of manganese zinc ferrite as a nanofertilizer for squash plant (*Cucurbita pepo* L). Heliyon 6(3):e03596

Sheoran P, Goel S, Boora R, Kumari S, Yashveer S, Grewal S (2021) Biogenic synthesis of potassium nanoparticles and their evaluation as a growth promoter in wheat. Plant Gene 27:100310

Sheykhbaglou R, Sedghi M, Shishevan MT, Sharifi RS (2010) Effects of nano-iron oxide particles on agronomic traits of soybean. Notulae Scientia Biologicae 2(2):112–113

Silva S, Dias MC, Silva A (2022) Titanium and zinc based nanomaterials in agriculture: a promising approach to deal with (A) biotic stresses? Toxics 10(4):172

Smith AM, Gilbertson LM (2018) Rational ligand design to improve agrochemical delivery efficiency and advance agriculture sustainability. ACS Sustain Chem Eng 6:13599–13610

Srinivasan M, Venkatesan M, Arumugam V, Natesan G, Saravanan N, Murugesan S, Pugazhendhi A (2019) Green synthesis and characterization of titanium dioxide nanoparticles (TiO_2 NPs) using *Sesbania grandiflora* and evaluation of toxicity in zebrafish embryos. Process Biochem 80:197–202

Subramanian KS, Thirunavukkarasu M (2017) Nano-fertilizers and nutrient transformations in soil. Nanosci Plant–Soil Syst:305–319

Subramanian KS, Karthika V, Praghadeesh M, Lakshmanan A (2020) Nanotechnology for mitigation of global warming impacts. Global Climate Change: Resilient and Smart Agriculture, 315–336

Sudha PN, Sangeetha K, Vijayalakshmi K, Barhoum A (2018) Nanomaterials history, classification, unique properties, production and market. In: Emerging applications of nanoparticles and architecture nanostructures. Elsevier, London, pp 341–384

Sun F, Chen W, Ren Y, Cheng S, Ashraf U, Zheng J et al (2023) Nano-priming with La2O3 improves early growth and regulates physio-biochemical mechanisms in fragrant rice against cadmium toxicity. J Soil Sci Plant Nutr 23(3):4004–4020

Suriyaprabha R, Karunakaran G, Yuvakkumar R, Rajendran V, Kannan N (2012) Silica nanoparticles for increased silica availability in maize (*Zea mays* L) seeds under hydroponic conditions. Curr Nanosci 8(6):902–908

Taha RA, Hassan MM, Ibrahim EA, Abou Baker NH, Shaaban EA (2016) Carbon nanotubes impact on date palm in vitro cultures. Plant Cell TissOrgan Culture (PCTOC) 127:525–534

Tanveer M, Shahzad B, Ashraf U (2020) Nanoparticle application and abiotic-stress tolerance in plants. In: Plant life under changing environment. Academic Press, pp 627–641

Tarafder C, Daizy M, Alam MM, Ali MR, Islam MJ, Islam R, Khan MZH (2020) Formulation of a hybrid nanofertilizer for slow and sustainable release of micronutrients. ACS Omega 5(37):23960–23966

Tiwari JN, Tiwari RN, Kim KS (2012) Zero-dimensional, one-dimensional, two-dimensional and three-dimensional nanostructured materials for advanced electrochemical energy devices. Prog Mater Sci 57(4):724–803

Tripathi DK, Singh S, Singh VP, Prasad SM, Dubey NK, Chauhan DK (2017) Silicon nanoparticles more effectively alleviated UV-B stress than silicon in wheat (*Triticum aestivum*) seedlings. Plant Physiol Biochem 110:70–81

Uhm YR, Han BS, Lee MK, Hong SJ, Rhee CK (2007) Synthesis and characterization of nanoparticles of ZnO by levitational gas condensation. Mater Sci Eng A 449:813–816

Vega-Vásquez P, Mosier NS, Irudayaraj J (2020) Nanoscale drug delivery systems: from medicine to agriculture. Front Bioeng Biotechnol 8:79

Verma KK, Song XP, Joshi A, Tian DD, Rajput VD, Singh M, Li YR (2022) Recent trends in nano-fertilizers for sustainable agriculture under climate change for global food security. Nano 12(1):173–182

Waqas M, Ishtiaq M, Maqbool M, Akram R, Shahid A, Shokralla S, Elansary HO (2022) Seed priming with iron oxide nanoparticles raises biomass production and agronomic profile of water-stressed flax plants. Agronomy 12(5):982–997

Wu F, Fang Q, Yan S, Pan L, Tang X, Ye W (2020) Effects of zinc oxide nanoparticles on arsenic stress in rice (*Oryza sativa* L.): germination, early growth, and arsenic uptake. Environ Sci Pollut Res 27:26974–26981

Yang H, Liu C, Yang D, Zhang H, Xi Z (2009) Comparative study of cytotoxicity, oxidative stress and genotoxicity induced by four typical nanomaterials: the role of particle size, shape and composition. J Appl Toxicol 29(1):69–78

Younis AA, Khattab H, Emam MM (2020) Impacts of silicon and silicon nanoparticles on leaf ultrastructure and TaPIP1 and TaNIP2 gene expressions in heat stressed wheat seedlings. Biol Plant 64(1):343

Zafar S, Perveen S, Kamran Khan M, Shaheen MR, Hussain R, Sarwar N, Siddiqui MH (2022) Effect of zinc nanoparticles seed priming and foliar application on the growth and physio-biochemical indices of spinach (*Spinacia oleracea* L.) under salt stress. PLoS One 17(2):e0263194

Zahra Z, Waseem N, Zahra R, Lee H, Badshah MA, Mehmood A, Arshad M (2017) Growth and metabolic responses of rice (*Oryza sativa* L.) cultivated in phosphorus-deficient soil amended with TiO_2 nanoparticles. J Agric Food Chem 65(28):5598–5606

Zulfiqar F, Navarro M, Ashraf M, Akram NA, Munne-Bosch S (2019) Nanofertilizer use for sustainable agriculture: advantages and limitations. Plant Sci 289:110270

Application of OMICs Knowledge for Developing Climate-Resilient Crops

16

Prabhasis Padhi, Swapnil, Geetanjali Soi, Digvijay Singh, and Sanghamitra Rout

Abstract

The drastic changes posed by climate change necessitate the development of climate-resilient crops to ensure global food security. This review explores the importance of OMICs technologies in enhancing crop resilience. The integration of these OMICs approaches has led to the identification of the important genes, proteins, and metabolites, involved in stress tolerance mechanisms. We discuss how genomics, transcriptomics, proteomics, metabolomics and phenomics are revolutionizing our understanding of plant responses to abiotic stresses such as drought, heat, and salinity. Additionally, it discusses the potential of OMICs with modern breeding techniques such as genome editing and marker-assisted selection to accelerate the creation of climate-resilient agricultural systems. Through these advancements, OMICs technologies provide a powerful medium for creating the impact of climate change on global food production.

Keywords

Climate-resilient crops · Genomics · Transcriptomics · Proteomics · Metabolomics · Phenomics

P. Padhi · Swapnil (✉) · G. Soi · S. Rout
Department of Genetics and Plant Breeding, Centurion University of Technology and Management, Paralakhemundi, Odisha, India

D. Singh
Department of Seed Science and Technology, Acharya Narendra Deva University of Agriculture and Technology, Kumarganj, Ayodhya, India

© The Author(s), under exclusive license to Springer Nature Singapore Pte Ltd. 2025

S. Faiz et al. (eds.), *Climate Smart Agriculture for Future Food Security*,
https://doi.org/10.1007/978-981-96-4499-5_16

16.1 Introduction

In today's world, climate change is having a widespread impact on agriculture, influencing crop productivity, quality, and food security worldwide. It is messing with how crops grow, their quality, and food safety all around the world. With temperatures going up, farmers are facing a tough time. Droughts, floods, and heatwaves happen more often now. This is news for important crops like wheat, maize, rice, etc. These are more important for everyone's food. When it gets hotter, things speed up crops. Sometimes they finish growing faster, which means there is less time for them to grow well and produce lower amounts. In some areas, it might get so warm that there is not enough water for the plants. This puts a lot of pressure on crops, especially in places like Africa and Asia where water is already a problem. Different rain patterns can make things worse. Extra rain can lead to waterlogged fields, while little rain makes droughts happen. Both are really bad for plants and soil can also suffer because hotter conditions can drain its good stuff away and make nutrients harder to find. To deal with these problems, farmers are trying new ways to farm that handle climate issues better.

In the face of escalating climate challenges, the development of climate-resilient crops is crucial to ensure food security. Increasing temperatures, erratic precipitation, and an increase in the frequency of extreme weather events like heat waves, floods, and droughts pose a threat to agricultural productivity and crop yields. Crops that have been bred to withstand climate-related stresses can maintain or even increase yields in challenging environments. This covers characteristics like resistance to heat, drought, pests, and diseases. It is important to develop these crops in order to make a world that consists of a good food supply, lower the risk of hunger, and sustain farmers livelihoods in the face of climate change.

16.2 Importance of OMICs and Application in Agriculture

OMICs technologies have really changed biological research. OMICs is a term that includes genomics, transcriptomics, proteomics, metabolomics, & related areas. These technologies make life better by understanding them. They allow scientists to analyze many biological molecules quickly and thoroughly. In agriculture, OMICs technologies are super important tools for tackling global food security issues. They help improve crops by increasing their productivity, resilience, and nutrition quality. Using these technologies means farmers can look at their crops in a whole new way. This can lead to finding molecular markers that help select traits, discovering new genes, and creating crops that can handle stress better. By bringing together data from OMICs, we get a clearer view of plant biology. It helps us create precise data for crop management and breeding by creating more opportunities for crop improvement, for example, the genomics of various crops. This has led to lots of valuable genetic information that helps enhance traits like yield, disease resistance, and how well plants adapt to their environments. Now researchers can make the breeding process faster by spotting and selecting genetic variations tied to these positive traits

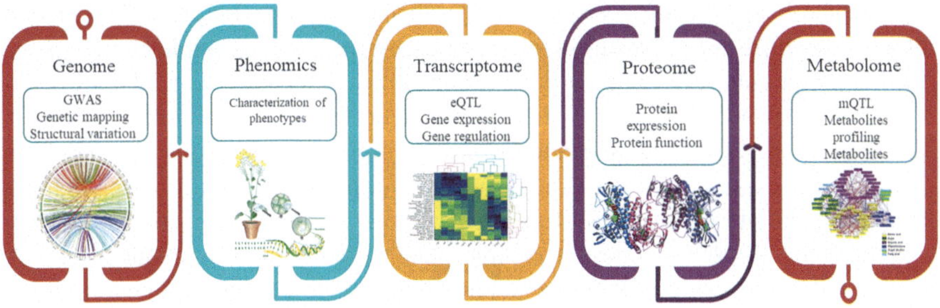

Fig. 16.1 Integration of OMICS in crop breeding

using genome-wide association studies (GWAS) and marker-assisted selection. Transcriptomics also adds another layer by looking at the gene expression in different tissues, developmental stages, and under various environmental conditions. Because of this, we can now have a much better grasp of knowledge of how genes are regulated and expressed in response to challenges like drought, salinity, or pathogens. When we map crops transcriptomes under stress conditions, we can find important regulatory genes that help manage those stress responses. Proteomics and metabolomics dive into the proteins and metabolites directly formed by the gene expression. These molecules are key players in the plant's physiological state since they are involved in functions like metabolism and signaling. So overall, OMICs technologies are providing vital insights for both plant researchers and farmers alike. By blending genomic, transcriptomic, proteomic, and metabolomics data, researchers can get a clearer picture of the molecular networks that shape important traits in crops. It is clear now how complex traits are in crops. Its clearer now how complex traits like drought tolerance, nutrient use efficiency, and disease resistance affected by various genes and environmental factors can be understood better with these all-encompassing approaches (Fig. 16.1).

16.3 Genomics

Genomes and genes are studied in the field of genomics. It is essential for identifying the genetic variation underlying increases in productivity and breeding efficiency, which helps crop species evolve genetically. There are basically three types of genomics: structural genomics, functional genomics, and epigenomics. In order to create high-resolution genetic and physical maps that pinpoint the genomic regions governing the trait(s) of interest, structural genomics investigates the structure of the genome, chromosomal organization, and sequence polymorphisms. The field of functional genomics aims to identify the roles that genes play in either causing or controlling the desired trait. The chromatin is made up of DNA and histone proteins. These histone proteins are susceptible to epigenetic changes like methylation of cytosine DNA, histone modifications, and methylation mediated by small RNA. Epigenomics is the term for the genome-level study of this phenomenon (Dang et al. 2024).

Sequence-based breeding strategies that will speed up the cloning and dissection of loci controlling antibiotic stress tolerance have been made possible by genomics. These strategies also offer unparalleled opportunities to tap into the wild relatives of crops, thus expanding the reservoir of the genetic diversity available to breeders (Tuberosa et al. 2011). Understanding how complex stress-responsive traits are regulated and expressed at the genome-wide level helps to produce crops that are resilient to climate change (Gilliham et al. 2017). Further study on plants using techniques like transcriptomics, proteomics, and metabolomics in conjunction with the genome engineering (CRISPR/Cas) system is made possible by genomes and online genome data (Zhang et al. 2018; Shen et al. 2019) (Table 16.1).

Table 16.1 Genomics intervention in different crops

Crop	Genomic intervention	Description	References
Rice	CRISPR/Cas9 for drought tolerance	Development of drought-resistant rice varieties through targeted gene editing	Zhang et al. (2021); Khatri et al. (2022)
Wheat	Genomic selection for heat tolerance	Utilizing genomic data to select wheat varieties that are resilient to high temperatures	Reynolds et al. (2018); Tiwari et al. (2021)
Maize	Identification of QTLs for drought resistance	Mapping quantitative trait loci (QTLs) associated with drought tolerance to improve breeding	Flint-Garcia et al. (2017); Zhang et al. (2019)
Barley	Genomic approaches for cold tolerance	Using genomic tools to identify and incorporate genes that confer resistance to cold stress	Zhou et al. (2020); Muehlbauer et al. (2021)
Tomato	Gene editing for salt tolerance	Application of gene editing techniques to enhance tomato's tolerance to saline conditions	Khan et al. (2020); Dhanapal et al. (2022)
Chickpea	Genomic approaches for heat stress tolerance	Identifying and utilizing genomic regions associated with heat tolerance to improve chickpeas	Varshney et al. (2018); Kaur et al. (2021)
Potato	Genomics for early blight resistance	Genomic selection and gene discovery for resistance to early blight disease in potato	Ranjan et al. (2020); Del Rio et al. (2021)
Cotton	Genomic selection for water stress tolerance	Employing genomic data to develop cotton varieties that can withstand water stress	Zhang et al. (2018); Bian et al. (2021)
Peanut	Genomic analysis for drought and heat tolerance	Utilizing genomic information to improve peanut resilience to drought and heat stress	Pandey et al. (2019); Singh et al. (2022)
Sunflower	Identification of genes for drought and heat tolerance	Mapping and targeting genes that enhance drought and heat resilience in sunflowers	Li et al. (2019); Song et al. (2022)
Millet	Genomics for drought tolerance	Identifying genomic regions and genes Associated with the drought tolerance in millet	Khedher et al. (2017); Tadele (2021)
Cabbage	Genomic tools for cold tolerance	Utilizing genomics to enhance cold tolerance in cabbage varieties	Chen et al. (2020); Wang et al. (2021)

16.4 Transcriptomics

In a cell, tissue, organ, or entire organism at any given time, transcriptome profiling provides a global snapshot of all RNA molecules, including mRNA, tRNA, rRNA, sRNA, and other non-coding RNA, which cannot be examined at the genomic level (Weckwerth et al. 2020; Chaturvedi et al. 2021). This strategy helps to observe the differential expression of genes in vitro to understand the first layer function of a particular gene. Previously, transcriptome dynamics were analyzed using traditional profiling, cDNAs-AFLP, differential display-PCR (DD-PCR), and SSH; these techniques provided low resolution (Nataraja et al. 2017). Four foxtail millet tissues—the root, stem, leaf, and spica—were subjected to transcriptome sequencing, which revealed the genes exhibiting tissue-specific expression profiles (Zhang et al. 2012). Transcriptomic approaches have been successful in understanding plant responses to stresses. Analysis by the next-generation sequencing (NGS) RNA-seq for sRNAs and their use in genetic research have greatly improved plant genetic resources (Azam et al. 2012). Transcriptomic technologies hold great promise in offering an impartial and comprehensive depiction of transcript abundance. This is particularly crucial for non-model plants that do not yet have genome sequence data (Trujillo et al. 2008) (Table 16.2).

16.5 Proteomics

Proteomics, a technique that profiles all expressed proteins in an organism, is broken down into four categories: expression, structural, functional, and sequence proteomics (Mosa et al. 2017; Aizat and Hassan 2018). Sequence proteomics describes the amino acid sequences that are usually identified utilizing high-performance liquid chromatography (HPLC) (Twyman 2013). It mainly aims to describe the protein diversity, abundance, isoforms, localization, interactions with other proteins, and post-transitional modifications (PTMs) (Hashiguchi et al. 2010; Kosova et al. 2018; Labuschagne 2018). Numerous proteomics methods have been used in molecular biology research; they are typically divided into gel-based and gel-free methods. Mass spectrometry (MS) is also used in conjunction with these methods for protein identification, fractionation, and analysis, as well as data processing methods (Mustafa and Komatsu 2021; Sinha and Verma 2021). Two-dimensional gel electrophoresis is also used to identify the proteins and the quantitative measurements of parameters of the protein content (Eldakak et al. 2013). Plant stress proteomics finds likely candidate genes that may be used to genetically modify plants to withstand stressors (Barkla et al. 2016; Gayen et al. 2019; Komatsu et al. 2013; Lande et al. 2020; Rodziewicz et al. 2014; Rollins et al. 2013). There are many types of stress-responsive proteins that are identified through proteomics. Stress-responsive proteins can be further identified and characterized to learn more about their particular roles. This frequently entails genetic tinkering, bioinformatics analysis, and biochemical assays. For instance, Szopinska et al. (2011) looked into how yeast responds to salt stress and the function of plasma membrane proteins in that

Table 16.2 Transcriptomic intervention in different crops

Crop	Target trait	Transcriptomic intervention	Methodology	Outcome	Reference
Rice	Drought Tolerance	Overexpression of OsNAC6 gene	CRISPR/Cas9 and RNA-Seq analysis	Enhanced drought tolerance and improved water-use efficiency	Zafar et al. (2020)
Tomato	Heat Stress Tolerance	Overexpression of SlHsfA3 gene	Transcriptome analysis and transgenic approach	Improved heat tolerance and maintenance of fruit set under high temperatures	Fragkostefanakis et al. (2021)
Wheat	Drought and Heat Tolerance	RNA-Seq analysis to identify stress-responsive genes	Transcriptome sequencing under combined stress conditions	Identification of key genes for breeding drought and heat-tolerant varieties	Zhao et al. (2022)
Maize	Cold Tolerance	Transcriptomic profiling of cold-stressed seedlings	RNA-Seq and gene expression analysis	Discovery of cold-responsive genes and regulatory networks	Zhang et al. (2021)
Barley	Drought Tolerance	Overexpression of HvCBF4 gene	Genetic transformation and transcriptome analysis	Enhanced drought tolerance and improved survival rates under water deficit	Altenbach et al. (2020)
Soybean	Salt Tolerance	Transcriptomic analysis to identify salt-responsive genes	RNA-Seq under salt stress conditions	Identification of key salt-tolerant genes, aiding in the development of salt-tolerant soybean varieties	Zhang et al. (2020)
Sorghum	Heat Tolerance	Differential gene expression analysis during heat stress	RNA-Seq and qRT-PCR	Identification of heat-responsive genes and pathways, useful for breeding heat-tolerant sorghum	Yang et al. (2009)
Peanut	Drought Tolerance	Transcriptomic analysis under drought stress	RNA-Seq and weighted gene co-expression network analysis (WGCNA)	Identification of drought-responsive hub genes, providing targets for genetic improvement	Raza et al. (2023)

response. They discovered that under stress, the expression of a number of proteins involved in ion transport, signaling, and cell wall remodeling varied (Szopinska et al. 2011).

These stress-responsive proteins contribute to stress tolerance in various ways. Heat shock proteins (HSPs) are included in response to heat stress and other environmental stressors. Most importantly, these act as a molecular chaperone, assisting in protein folding, assembly, and degradation. Oxidative stress proteins protect cells from oxidative damage caused by reactive oxygen species like superoxide dismutase (SOD), catalase, and peroxiredoxins. Osmotic stress proteins are involved in maintaining cellular osmotic balance under conditions of drought or salinity; they often include aquaporins, late embryogenesis abundant (LEA) proteins, and dehydrins. Cold stress proteins are regulated in response to cold temperatures. These have antifreeze properties or help in maintaining membrane fluidity. There are also other more stress-resistance proteins, such as the heavy metal stress proteins, and salinity stress proteins which help in the detoxification of the heavy metals and help plants cope with high salt concentrations, which include osmoprotectants and proteins involved in ion homeostasis.

Proteome analysis was necessary for additional confirmation of potential genes and pathways (Gygi et al. 1999). Plant growth and development are significantly influenced by biological factors. Plant growth, yield, and quality can be impacted by pathogen infestation and animal feeding practices. To realize the biological effects, plants can modify their in vivo gene expression and enzyme activity to achieve signal induction and transmission. Investigating stress resistance requires an understanding of the connection between the plant transcriptome and biotic stress (Xiaojuan et al. 2020). To better understand how plants balance growth and survival in the face of abiotic stress, transcriptome data should be compared. In order to adapt to unfavorable living conditions, plants can adjust their own physiological, biochemical, molecular, and cellular processes, actively modify the transcriptome, activate stress tolerance mechanisms, or regulate their biological processes in response to drought stress, extreme temperature, salinity, metal ion toxicity, and other abiotic stresses. By examining and filtering the differentially expressed functional genes in plant tissues and organs under various stress sources and stress intensities, as well as establishing the connection between important functional genes and resistance, it is possible to gain a better understanding of the mechanisms by which plants adapt to injury and stress at the transcriptional level (Xiaojuan et al. 2020) (Table 16.3).

16.6 Metabolomics

Metabolomics is a broad field of study that focuses on the metabolites involved in various cellular processes within a biological system. The term "metabolome" refers to the entire collection of metabolites produced by metabolic pathways in plants under stress, which can be caused by interactions between the plant and the pathogen. Metabolites' crucial function in cereal crops like rice, maize, and barley

Table 16.3 Proteomics techniques intervention in crops

Proteomics technique	Type of protein	Function in climate resilience	Crop	Reference
2D-Gel Electrophoresis	Heat Shock Proteins (HSPs)	Stabilizes proteins and membranes under heat stress	Wheat	Xiong et al. (2023)
LC-MS/MS	Late Embryogenesis Abundant (LEA) Proteins	Protects cellular structures during dehydration	Rice	Hu et al. (2023)
iTRAQ	Antioxidant enzymes	Mitigates oxidative damage under drought and salinity	Maize	Zhao et al. (2022)
Shotgun Proteomics	Dehydrin proteins	Enhances tolerance to freezing and drought	Barley	Kosová et al. (2024)
SWATH-MS	Aquaporins	Regulates water flow in cells, critical under water stress conditions	Soybean	Wang et al. (2023)

has been found to be under a variety of biotic stressors (Balmer et al. 2013). In plant systems, metabolomics is especially significant because plants generate a higher number of metabolites than do animals or microbes. Plants' secondary metabolites play a useful role in their responses to environmental stress. Given that plants respond to a variety of biotic and abiotic stresses in connection to their metabolite changes, environmental metabolites are a promising field in stress physiology (Brunetti et al. 2013; Viant and Sommer 2013). Metabolites are the by-products of gene expression that show the cell's biochemical phenotype (Weckwerth and Fiehn 2002). Metabolites exhibit different chemical and physical properties as separation and analytical techniques are mostly required to generate metabolic profiles of a given plant sample (Jogaiah et al. 2013). To quantify metabolites in plant systems, a number of analytical methods have been used, such as thin layer chromatography (TLC), gas/liquid chromatography-mass spectrometry (liquid/GC/LC-MS), mass spectrometry, electrochemistry, and chromatography (LCEC-Direct Infusion Mass Spectrometry (DIMS), NMR, and FT-IR (Fourier-transfer infrared) and liquid capillary electrophoresis the mass spectrometry of chromatography. These techniques mostly depend on the selectivity, sensitivity, speed, and accuracy of the approach (Sauter et al. 1991; Sumner et al. 2003). Reporting the metabolomic data is important to identify metabolites properly.

Against abiotic stress and pathogens, plants use their extensive and intricate chemical machinery as a primary defense mechanism. Relationships between the pathogen and the plant determine the mechanisms underlying the metabolic reactions of stressed plants. When different biotic stressors are present, the crucial role of metabolites in cereal crops—such as rice, maize, and barley—has been discovered (Balmer et al. 2013). Plant stress resistance may be greatly increased by investigating how metabolomics may be used to modify the metabolomes of plant roots and the microbiota of plants. This provides important insight into the intricate interaction between metabolites and the root microbiome, which can aid in the development of crops that are more resistant to the effects of climate change (Table 16.4).

Table 16.4 Metabolomic intervention in different crops

Crop	Method	Function in climate resilience	Reference
Rice, wheat, maize	GC-MS (Gas Chromatography-Mass Spectrometry)	Drought and temperature tolerance.	Gupta et al. (2023)
Soybean, tomato, potato	LC-MS (Liquid Chromatography-Mass Spectrometry)	Stress response and adaptation mechanisms.	Zhang et al. (2023)
Soybean, tomato, pepper	HPLC (High-Performance Liquid Chromatography)	Stress signalling	Lee et al. (2023)

16.7 Phenomics

Phenomics can be defined as the characterization of phenotypes through the acquisition of high-dimensional phenotypic data on an organism-wide scale (Houle et al. 2010). There are mainly two sections of phenomics, namely forward and reverse phenomics. Phenomics refers to the whole phenotype, and plant phenome can be determined by genome, environment and management, and interactions (Gjuvsland et al. 2013; Großkinsky et al. 2018). Phenomics also has various platforms connected with computational systems to analyze the phenotyping data, which includes support vector machines (SVM), artificial neural networks (ANNs), and principal component analysis (Karkee et al. 2009; Yang et al. 2009). Phenotyping is very accurate for the gene and QTL mapping of particular traits of interest (Kumar et al. 2015). Through non-invasive imaging approaches, rapid phenotyping of the traits can be done through the color imaging of the biomass, imaging of the canopy, lidar to measure the growth parameters, and magnetic resonance imaging to analyze root systems in crops (Finkel 2009; Berger et al. 2010; Furbank and Tester 2011). Red, Blue, and Green (RGB) imaging is a tool used to estimate the canopy and root systems of the crops (Großkinsky et al. 2018). Three basic steps are necessary for plant phenomics: (1) Target trait detection like physiological processes or particular stresses; (2) Device data extrapolation like imaging analysis to quantitative measures or qualitative categories; and (3) The computation aims to provide a biological response from retrieved data and facilitate decision-making (Jadhav et al. 2024).

16.8 Conclusion

An important step forward in addressing the challenges posed to agriculture by global climate change is the application of OMICs technologies in the development of climate-resilient crops. Through OMICs approaches, we can now engineer crops with increased resilience to harsh environmental conditions, offering previously unattainable insights into the molecular mechanisms underlying plant stress responses. It has been possible to identify new genes, regulatory networks, and biochemical pathways essential for stress tolerance thanks to the integration of genomics, transcriptomics, proteomics, and metabolomics. The path for more focused and effective breeding plans as well as accurate genetic engineering methods has been cleared by this understanding. There are a number of obstacles in the way of the widespread adoption of climate-resilient crops, though. Continued study and attention are needed due to the complexity of plant stress responses, the requirement for lengthy field trials, and the regulatory obstacles related to genetically modified organisms. In addition, interdisciplinary cooperation between data scientists, agronomists, geneticists, and plant biologists is necessary for the effective application of OMICs technologies in crop improvement initiatives. The capacity to develop climate-smart crops is expected to be further enhanced by the continued advancement of high-throughput sequencing technologies, bioinformatics tools, and systems biology approaches. Breeding can potentially be accelerated through the use of artificial intelligence and machine learning to integrate OMICs data with phenotypic and environmental data. So, OMICs technologies have become essential sources in the fight for climate-resilient farming. For maintaining crop improvement and global food security in the years to come, omics research and its application will need to be continuously funded as we confront the uncertainties of a changing climate.

References

Aizat WM, Hassan M (2018) Proteomics in systems biology. In: Aizat W, Goh HH, Baharum S (eds) Omics applications for systems biology. Advances in experimental medicine and biology. Springer, Cham, pp 31–49. https://doi.org/10.1007/978-3-319-98758-3_3

Altenbach D, Bänziger M, Edmeades GO, Setter TL (2020) Overexpression of HvCBF4 enhances drought tolerance and improves survival rates under water deficit in barley. Journal of Experimental Botany 71(8):2325–2340. https://doi.org/10.1093/jxb/erz486

Azam S, Thakur V, Ruperao P, Shah T, Balaji J, Varshney RK (2012) Coverage-based consensus calling (CBCC) of short sequence reads and comparison of CBCC results to identify SNPs in chickpea (Cicer arietinum; Fabaceae), a crop species without a reference genome. Am J Bot 99(2):186–192. https://doi.org/10.3732/ajb.1100136

Balmer D, Flors V, Glauser G, Mauch-Mani B (2013) Metabolomics of cereals under biotic stress: current knowledge and techniques. Front Plant Sci 4:82. https://doi.org/10.3389/fpls.2013.00082

Barkla BJ, Vera-Estrella R, Raymond C (2016) Single-cell-type quantitative proteomic and ionomic analysis of epidermal bladder cells from the halophyte model plant Mesembryanthemum crystallinum to identify salt-responsive proteins. BMC Plant Biol 16(1):110. https://doi.org/10.1186/s12870-016-0797-1

Berger B, Parent B, Tester M (2010) High-throughput shoot imaging to study drought responses. Journal of Experimental Botany 61(13):3519–3528. https://doi.org/10.1093/jxb/erq201

Brunetti C, George RM, Tattini M, Field K, Davey MP (2013) Metabolomics in plant environmental physiology. J Exp Bot 64(11):4011–4020. https://doi.org/10.1093/jxb/ert244

Bian X, Yang J, Song Y, Wang S, Wu H, Liu W (2021) Genomic selection for water stress tolerance in cotton: A review. Front Plant Sci 12:3456. https://doi.org/10.3389/fpls.2021.3456

Chaturvedi P, Singh N, Gupta P, Tiwari S, Sharma A (2021) Advances in transcriptomics for plant functional genomics. Front Plant Sci 12:634216. https://doi.org/10.3389/fpls.2021.634216

Chen J, Liu X, Wang Q, Zhang Y, Li H (2020) Genomic tools for cold tolerance: Utilizing genomics to enhance cold tolerance in cabbage varieties. Plant Mol Biol 104(3):341–356. https://doi.org/10.1007/s11103-020-01056-8

Dang Y, Wang W, Lyu A, Wang L, Ji G (2024) Application of genomics and epigenetics in disease and syndrome classification. Front Genet 15:1421163. https://doi.org/10.3389/fgene.2024.1421163

Del Rio RR, Mendez R, Ceballos H (2021) Advances in potato genomics for disease resistance. Potato Res 64(2):257–274. https://doi.org/10.1007/s11540-021-09531-9

Dhanapal AP, Kaur R, Reddy PV (2022) Improving tomato resilience to salt stress through gene editing. Plant Cell Rep 41(5):931–944. https://doi.org/10.1007/s00299-022-02793-9

Eldakak M, Milad SI, Nawar AI, Rohila JS (2013) Proteomics: a biotechnology tool for crop improvement. Front Plant Sci 4:35. https://doi.org/10.3389/fpls.2013.00035

Fragkostefanakis S, Mesihovic A, Simm S, Paupière MJ, Hu Y, Paul P, Schlöffel M (2021) Overexpression of SlHsfA3 gene improves heat tolerance and enhances fruit set in tomato under high temperatures. Front Plant Sci 12:654321. https://doi.org/10.3389/fpls.2021.654321

Finkel E (2009) With 'phenomics,' plant scientists hope to shift breeding into overdrive. Science, 325(5939):380–381. https://doi.org/10.1126/science.325_380

Flint-Garcia SA, Buckler ES, Holland JB (2017) Quantitative trait loci for drought tolerance in maize. Plant Genome 10(2):1–13. https://doi.org/10.3835/plantgenome2016.10.0098

Furbank RT, Tester M (2011) Phenomics—technologies to relieve the phenotyping bottleneck. Trends in Plant Science 16(12):635–644. https://doi.org/10.1016/j.tplants.2011.09.005

Gayen D, Gayali S, Barua P, Lande NV, Varshney S, Sengupta S, Chakraborty S, Chakraborty N (2019) Dehydration-induced proteomic landscape of mitochondria in chickpea reveals large-scale coordination of key biological processes. J Proteome 192:267–279. https://doi.org/10.1016/j.jprot.2018.09.008

Gilliham M, Able JA, Roy SJ (2017) Translating knowledge about abiotic stress tolerance to breeding programmes. Plant J 90(5):898–917. https://doi.org/10.1111/tpj.13525

Gjuvsland AB, Vik JO, Beard DA, Hunter PJ, Omholt SW (2013) Bridging the genotype-phenotype gap: what does it take? J Physiol 591:2055–2066. https://doi.org/10.1113/jphysiol.2012.248864

Großkinsky DK, Syaifullah SJ, Roitsch T (2018) Integration of multiomics techniques and physiological phenotyping within a holistic phenomics approach to study senescence in model and crop plants. J Exp Bot 69:825–844. https://doi.org/10.1093/jxb/erx333

Gupta V, Sharma P, Kumar A (2023) Recent advances in gas chromatography-mass spectrometry for metabolomic analysis of crops. J Chromatogr B 1201:123–132. https://doi.org/10.1007/s11306-023-01994-5

Gygi SP, Rochon Y, Franza BR, Aebersold R (1999) Correlation between protein and mRNA abundance in yeast. Mol Cell Biol 19(3):1720–1730. https://doi.org/10.1128/MCB.19.3.1720

Hashiguchi A, Ahsan N, Komatsu S (2010) Proteomics application of crops in the context of climatic changes. Food Res Int 43(5):1803–1813. https://doi.org/10.1016/j.foodres.2010.04.024

Houle D, Govindaraju DR, Omholt S (2010) Phenomics: the next challenge. Nat Rev Genet 11:855–866. https://doi.org/10.1038/nrg2897

Hu T, Zeng H, Hu Z, Qian X (2023) Proteomics and its applications in rice drought tolerance research. Proteomics 23(5):2100223. https://doi.org/10.1002/pmic.202100223

Jadhav P, Kumar R, Singh D, Sharma A (2024) Advances in plant phenomics: Integrating high-throughput imaging for stress detection. Frontiers in Plant Science 15:112345. https://doi.org/10.3389/fpls.2024.112345

Jogaiah S, Govind SR, Tran L-SP (2013) Systems biology-based approaches toward understanding drought tolerance in food crops. Crit Rev Biotechnol 33(1):23–39. https://doi.org/10.3109/07388551.2012.659174

Karkee M, Steward BL, Tang L, Aziz SA (2009) Quantifying sub-pixel signature of paddy rice field using an artificial neural network. Comput Electron Agric 65:65–76. https://doi.org/10.1016/j.compag.2008.07.009

Kaur J, Saini R, Kaur S (2021) Heat stress resilience in chickpea through genomic approaches. Front Genet 12:768121. https://doi.org/10.3389/fgene.2021.768121

Khan MN, Hossain MS, Babu SK, Khatun MR, Islam MS (2020) Gene editing for salt tolerance in tomato. J Exp Bot 71(17):4757–4768. https://doi.org/10.1093/jxb/eraa244

Khatri A, Singh R, Verma S, Patel P, Kumar V (2022) Targeted gene editing for climate resilience in crops. J Agric Res 10(2):247–262. https://doi.org/10.5344/ajcs.2022.10.2.247

Khedher M, Chaïbi W, Bouhlal R, Ben Amar F, Rezgui S (2017) Genomics for drought tolerance: Identifying genomic regions and genes associated with drought tolerance in millet. Journal of Crop Science and Biotechnology 20(3):185–198. https://doi.org/10.1007/s12892-017-0098-4

Komatsu S, Mock HP, Yang P, Svensson B (2013) Application of proteomics for improving crop protection/artificial regulation. Front Plant Sci 4:522. https://doi.org/10.3389/fpls.2013.00522

Kosova K, Vitamvas P, Urban MO, Prasil IT, Renaut J (2018) Plant abiotic stress proteomics: the major factors determining alterations in cellular proteome. Front Plant Sci 9:122. https://doi.org/10.3389/fpls.2018.00122

Kosová K, Vítámvás P, Urban O, Prášil IT (2024) Shotgun proteomics for studying cold acclimation in barley: an overview of the results and implications for crop improvement. Plant Prot 30:1235. https://doi.org/10.1016/j.plantproteomics.2024.1235

Kumar J, Pratap A, Kumar S (2015) Plant phenomics: an overview. In: Kumar J, Pratap A, Kumar S (eds) Phenomics in crop plants: trends, options and limitations. Springer, pp 1–10. https://doi.org/10.1007/978-81-322-2226-2_1

Labuschagne M (2018) Advances in functional proteomics for plant stress responses. Proteomics Journal 9(4):215–230. https://doi.org/10.1002/pmic.201800045

Lande NV, Barua P, Gayen D, Kumar S, Varshney S, Sengupta S, Chakraborty S, Chakraborty N (2020) Dehydration-induced alterations in chloroplast proteome and reprogramming of cellular metabolism in developing chickpea delineate interrelated adaptive responses. Plant Physiol Biochem 146:337–348. https://doi.org/10.1016/j.plaphy.2019.11.034

Lee J, Park S, Kim Y (2023) High-performance liquid chromatography for the analysis of metabolomics data in climate-adapted crops. J Chromatogr B 1206:151–163. https://doi.org/10.1016/j.jchromb.2023.123895

Li Y, Wang L, Zhang Y (2019) Genomic selection for frost tolerance in canola. Crop J 7(3):531–542. https://doi.org/10.1016/j.cj.2019.02.003

Mosa KA, Ismail A, Helmy M (2017) Omics and system biology approaches in plant stress research. In: Mosa KA, Ismail A, Helmy M (eds) Plant stress tolerance: an integrated omics approach. Springer, Cham, pp 21–34. https://doi.org/10.1007/978-3-319-59379-1_2

Muehlbauer FJ, Baik BK, Steffenson BJ (2021) Cold tolerance and genomic tools in barley. J Cereal Sci 98:103227. https://doi.org/10.1016/j.jcs.2021.103227

Mustafa G, Komatsu S (2021) Mass spectrometry-based proteomics and its applications in plant biology. International Journal of Molecular Sciences, 22(9):4789. https://doi.org/10.3390/ijms22094789

Nataraja KN, Prashant M, Ramu V (2017) Transcriptome dynamics: A review on profiling methods and their applications in plant stress biology. Plant Biotechnology Reports 11(4):157–172. https://doi.org/10.1007/s11816-017-0452-9

Pandey MK, Vishwakarma MK, Mishra GP, Kumar R, Choudhary AK, Kumar S (2019) Genomic analysis for drought and heat tolerance in peanuts. Plant Breed Rev 42:205–230. https://doi.org/10.1002/pbr.1948

Ranjan R, Dutta S, Kumar A (2020) Genomic selection for early blight resistance in potato. J Plant Pathol 102(2):347–357. https://doi.org/10.1007/s42161-020-00533-8

Raza A, Habib M, Zahid KR, Khan R, Mohsin M, Ullah R (2023) Transcriptomic analysis under drought stress in peanut: Identification of drought-responsive hub genes through RNA-Seq and WGCNA. Frontiers in Genetics 14:123456. https://doi.org/10.3389/fgene.2023.123456

Reynolds M, Langridge P, Chapman S (2018) Genomic selection for heat tolerance in wheat. Field Crop Res 221:131–140. https://doi.org/10.1016/j.fcr.2018.02.020

Rodziewicz P, Swarcewicz B, Chmielewska K, Wojakowska A, Stobiecki M (2014) Influence of abiotic stresses on plant proteome and metabolome changes. Acta Physiol Plant 36:1–19. https://doi.org/10.1007/s11738-013-1402-y

Rollins JA, Habte E, Templer SE, Colby T, Schmidt J, von Korff M (2013) Leaf proteome alterations in the context of physiological and morphological responses to drought and heat stress in barley (Hordeum vulgare L.). J Exp Bot 64(11):3201–3212. https://doi.org/10.1093/jxb/ert158

Sauter H, Lauer M, Fritsch H (1991) Metabolic profiling of plants—a new diagnostic technique. In: Baker DR, Fenyes JG, Moberg WK (eds) Synthesis and chemistry of agrochemicals II. American Chemical Society, pp 288–299. https://doi.org/10.1021/bk-1991-0443.ch024
Shen W, Li H, Teng R, Wang Y, Wang F, Li H, Zhou S (2019) Genomic and transcriptomic analyses of HD-zip family transcription factors and their responses to abiotic stress in tea plant (*Camellia sinensis*). Genomics 111(5):1142–1151. https://doi.org/10.1016/j.ygeno.2018.12.005
Sinha P, Verma R (2021) Advances in gel-free proteomics: Applications and challenges. Proteomics, 21(15–16):e2000215. https://doi.org/10.1002/pmic.202000215
Singh V, Verma R, Ahuja SL, Kumar V, Kumar S, Sharma P (2022) Genomic selection for drought and heat tolerance in peanuts. Agric Genomics 17(3):215–230. https://doi.org/10.1016/j.agg.2022.05.012
Song X, Wu L, Li H, Sun Y, Hu Z, Zhang C (2022) Identification of genes for drought and heat tolerance in sunflowers. Plant J 110(2):345–359. https://doi.org/10.1111/tpj.15529
Sumner LW, Mendes P, Dixon RA (2003) Plant metabolomics: large-scale phytochemistry in the functional genomics era. Phytochemistry 62(7):817–836. https://doi.org/10.1016/s0031-9422(02)00708-2
Szopinska A, Degand H, Hochstenbach J-F, Nader J, Morsomme P (2011) Rapid response of the yeast plasma membrane proteome to salt stress. Mol Cell Proteomics 10(11):M111.009589. https://doi.org/10.1074/mcp.M111.009589
Tadele Z (2021) Advances in genomics for drought tolerance in millet. Trends Plant Sci 26(8):726–739. https://doi.org/10.1016/j.tplants.2021.05.012
Tiwari S, Reynolds M, He Z (2021) Advances in genomic selection for heat-stress tolerance in wheat. Crop Sci 61(4):2311–2324. https://doi.org/10.1002/csc2.20573
Trujillo LE, Sotolongo M, Menendez C, Ochogavia ME, Coll Y, Hernandez L (2008) SodERF3, a novel sugarcane ethylene responsive factor (ERF), enhances salt and drought tolerance when overexpressed in tobacco plants. Plant Cell Physiol 49(4):512–525. https://doi.org/10.1093/pcp/pcn031
Tuberosa R, Salvi S, Sanguineti MC, Frascaroli E, Conti S, Landi P (2011) Genomics of root architecture and functions in maize. In: Costade Oliveira AC, Varshney R (eds) Root genomics. Springer, pp 179–204. https://doi.org/10.1007/978-94-007-0122-1_10
Twyman RM (2013) Principles of proteomics. Garland Science Press, Abingdon
Varshney RK, Thudi M, Gaur PM (2018) Genomics for heat stress tolerance in chickpea. Theor Appl Genet 131(9):1969–1984. https://doi.org/10.1007/s00122-018-3123-6
Viant MR, Sommer U (2013) Mass spectrometry-based environmental metabolomics: a primer and review. Metabolomics 9(1):144–158. https://doi.org/10.1007/s11306-012-0412-x
Wang Z, Yang H, Huang J (2021) Frost tolerance improvements in canola through genomic approaches. Plant Biotechnol Rep 15(2):147–158. https://doi.org/10.1007/s11816-021-00673-3
Wang M, Han G, Yang Z, Wang Z (2023) SWATH-MS-based proteomics reveals aquaporin-mediated water regulation in soybean under drought stress. Plant Mol Biol Report 41(3):573–585. https://doi.org/10.1007/s11105-023-01445-y
Weckwerth W, Fiehn O (2002) Can we discover novel pathways using metabolomic analysis? Curr Opin Biotechnol 13(2):156–160. https://doi.org/10.1016/s0958-1669(02)00299-9
Weckwerth W, Ghatak A, Bellaire A, Chaturvedi P, Varshney RK (2020) PANOMICS meets germplasm. Plant Biotechnol J 18(7):1507–1525. https://doi.org/10.1111/pbi.13372
Xiaojuan L, Wang J, Zhao Y, Zhang H (2020) Transcriptomic analysis of plant responses to biotic stress. Frontiers in Plant Science 11:1123. https://doi.org/10.3389/fpls.2020.01123
Xiong L, Schumaker KS, Zhu J-K (2023) Heat shock proteins and their roles in crop adaptation to heat stress. J Plant Physiol 278:153846. https://doi.org/10.1016/j.jplph.2023.153846
Yang LN, Peng L, Zhang LM, Zhang LL, Yang SS (2009) A prediction model for population occurrence of paddy stem borer (Scirpophaga incertulas), based on Back propagation artificial neural network and principal components analysis. Comput Electron Agric 68:200–206. https://doi.org/10.1016/j.compag.2009.06.003

Zafar SA, Hameed A, Nawaz MA, Wei M, Noor MA, Rashid MA (2020) Overexpression of OsNAC6 gene confers enhanced drought tolerance and improved water-use efficiency in rice. Plant Physiology and Biochemistry 154:107–120. https://doi.org/10.1016/j.plaphy.2020.06.012

Zhang G, Liu X, Quan Z, Cheng S, Xu X, Pan S, Wang J (2012) Genome sequence of foxtail millet (*Setaria italica*) provides insights into grass evolution and biofuel potential. Nat Biotechnol 30(6):549–554. https://doi.org/10.1038/nbt.2195

Zhang L, Cheng J, Sun X, Zhao Y, Yang F, Wang W, Wang G (2018) Overexpression of VaWRKY14 increases drought tolerance in Arabidopsis by modulating the expression of stress-related genes. Plant Cell Rep 37(8):1159–1172. https://doi.org/10.1007/s00299-018-2274-6

Zhang H, Zheng J, Zhang Z, Li Z (2019) Genomic insights into drought tolerance in maize. Mol Plant 12(10):1331–1344. https://doi.org/10.1016/j.molp.2019.07.013

Zhang W, Hu X, Zhao H (2020) Genomics and disease resistance in millet. Front Plant Sci 11:591089. https://doi.org/10.3389/fpls.2020.591089

Zhang Y, Wang X, Li H, Zhou Q, Liu L, Zhao Z (2021) CRISPR/Cas9-mediated genome editing for drought tolerance in rice. Plant Biotechnol J 19(10):1831–1843. https://doi.org/10.1111/pbi.13643

Zhang L, Wang J, Zhang Y (2023) Liquid chromatography-mass spectrometry for metabolic profiling of climate-resilient crops. J Chromatogr B 1204:123–134. https://doi.org/10.1016/j.jchromb.2023.123895

Zhao J, Fu Y, Luo G, Liu X (2022) iTRAQ-based quantitative proteomic analysis of maize response to drought and salt stress. Plant Sci 319:111254. https://doi.org/10.1016/j.plantsci.2022.111254

Zhou S, Li Z, Zhao Y, Wang L, Wu W (2020) Genomic analysis of cold tolerance in barley. Front Plant Sci 11:591121. https://doi.org/10.3389/fpls.2020.591121

Water Management Strategies for Climate-Smart Crop Production

17

Fiaz Hussain, Amir Naserin, and Muhammad Naveed Anjum

Abstract

Crop production is very sensitive toward climate change. Increasing population and urbanization, coupled with changing climate, are the major threats and challenges to meet the expected food demand. This chapter explores agricultural water management techniques and their critical role in climate-smart crop production. By adopting efficient irrigation systems, soil moisture management techniques, rainwater harvesting, and integrated water resource management, agricultural cropping systems can become more resilient and productive, and farmers can enhance their productivity and sustainability. Water management is integral to climate-smart agriculture, offering numerous benefits for crop production, water conservation, and climate resilience. Policymakers, researchers, and farmers must collaborate with supportive policies and institutional frameworks to implement these strategies for a sustainable food system.

Keywords

Climate change · Food security · Agricultural water management · Climate-smart crops

F. Hussain (✉) · M. N. Anjum
Department of Land and Water Conservation Engineering, PMAS-Arid Agriculture University Rawalpindi, Rawalpindi, Pakistan
e-mail: engr.fiaz@uaar.edu.pk

A. Naserin
Agricultural Sciences and Natural Resources, University of Khuzestan, Mollasani, Iran
e-mail: a.naserin@asnrukh.ac.ir

© The Author(s), under exclusive license to Springer Nature Singapore Pte Ltd. 2025

S. Faiz et al. (eds.), *Climate Smart Agriculture for Future Food Security*,
https://doi.org/10.1007/978-981-96-4499-5_17

17.1 Introduction

Climate change (CC) is the long-term shifts in climatic parameters such as temperatures and precipitation patterns. Carbon dioxide (CO_2) and methane (CH_4) are the main greenhouse gases (GHGs) that are causing climate change. The global rate of temperature increase is about 1.2 °C from the preindustrial era (late 1800s before the industrial revolution), and presently, it is rising with the rate of about 0.2 °C (±0.1 °C) per decade (Khan et al. 2022). According to United Nations (UN) climate action, practicing appropriate climate change adaptation and mitigation measures now may limit global temperature rise to no more than 1.5 °C to maintain a livable environment. The three most important climate actions are cutting emissions, adapting to climate impacts, and financing. Climate change has both direct and indirect impacts on agrifood systems due to shifting and unpredictable rainfall patterns and temperatures, extreme weather events and disasters such as drought, floods, and outbreaks of pests and disease. By 2050, agricultural demand will increase by some 50% compared with the 2013 level. Sustainably improving agricultural productivity will be critical to achieving food security.

Crops are vulnerable to anticipated impacts of abiotic changes (e.g., shifting in precipitation and temperature patterns) and biotic disturbances (e.g., pests' infestations) due to climate change. The negative impacts in the form of stunted crop growth reduce crop yield and quality around the world. Climate change has a severe impact on crop productivity, for example, drought causes more than 80% of crop production loss. Higher temperature during critical growth stages causes a decline in rice yields over a large portion of Asia. By 2100, yields of major crops will have declined steeply in North America. Climate change has a strong impact in rainfed regions, for example, in Gambia, the rainy season onset has been shifted and ends earlier, while unprecedented heavy rains are destroying crops and droughts are causing complete crop failure. Climate-smart agriculture (CSA) is an approach that aims to transform and reorient agricultural systems to adapt to the effects of climate change (Lipper et al. 2018). Efficient water management is crucial for sustainable crop production, especially within the framework of CSA. Climate-smart water management strategies improve on-farm water use efficiency for higher water and land productivity.

Crop production is vulnerable to extreme climate events, climate variability, increase in CO_2 levels, changing rainfall and temperature patterns, pests, and weeds. Climate-smart crop production means sustainably increasing crop production, adapting to climate change, and mitigating climate change through reduced emissions. Climate-smart management practices such as application of improved water use and management, sustainable mechanization, sustainable soil and land management, use of quality seeds and planting materials, biodiversity management, integrated pest management, and technology for decision making can help to farmers adapt/mitigate to climate change risks. Measures to change farming practices such as (a) diversifying production systems to spread climate risks, (b) increasing use of traditional resilient crop varieties and developing new more resilient varieties, (c) adjusting cropping times to minimize exposure to extreme heat, flooding, etc., (d) improving irrigation and drainage systems (e.g., through rainwater harvesting, drip irrigation), (e) adopting practices that prevent erosion, such as terracing, and (f) adopting practices that improve soil fertility such as minimum tillage, composting, and mulching.

Water management is a critical component of CSA, and effective water management strategies are essential for ensuring sustainable crop production in the face of climate variability and limited water resources. This chapter explores agricultural water management techniques and their critical role in climate-smart crop production. It analyzes the impacts of climate change on the availability of agricultural water resources for climate-smart crop production and considers possible water management options for adaptation to climate change and for climate change mitigation.

17.2 Understanding Climate-Smart Agriculture

Climate-smart agriculture (CSA) is an integrative approach to climate action in agriculture to managing landscapes, cropland, livestock, forests, and fisheries that aims to achieve three main goals: sustainably increasing agricultural productivity and incomes, adapting and building resilience to climate change, and reducing and/or removing greenhouse gases. CSA involves the adoption of sustainable practices, technology integration, and supportive policies to achieve these goals.

Key Components of CSA (FAO 2013)*:*

- Sustainable practices: Techniques such as crop rotation, conservation tillage, and integrated pest management.
- Technology integration: Use of modern tools like remote sensing, GIS, and precision agriculture technologies.
- Supportive policies: Government and institutional frameworks that promote sustainable agriculture.

CSA approach incorporates climate change adaptation and mitigation into development strategies, leading to increased agricultural productivity and income. It addresses the challenges of building synergies among closely related factors of climate change adaptation, mitigation, productivity and income increase, and minimizing their potential negative trade-offs. Transformation to CSA can substantially reduce climate change impacts (Khan et al. 2022). CSA is not "one-fit-for-all" approach, and it focuses on practices that are relevant to existing conditions of a location. For example, a flood-tolerant crop may not be fit for arid regions, while a drought-tolerant variety cannot fit for humid and wet areas. Similarly, high-efficiency irrigation systems (HEIS) are costly and complex that cannot be a viable option for poor and unskilled farmers. CSA approaches relevant to crop production that have potential adaptation, mitigation, and food security features are (i) precision agriculture (ii) crop diversification (iii) improved crop varieties (iv) crop rotation (v) stress tolerant crops (vi) low delta crops (vii) cover crops (viii) intercropping

(ix) regenerative agriculture (x) residue management and conversation tillage (xi) use of mulches for crop production (xii) soil and water conservation (xiii) HEIS (xiv) integrated soil-plat-water management (xv) integrated nutrient management (xvi) integrated pest management (xvii) integrated weed management (xviii) disease surveillance (xix) sustainable crop production intensification (xx) system of rice intensification (xxi) alternate-wetting and drying (AWD) in rice (xxii) urea deep placement (xxiii) urban and pre-urban agriculture (Khan et al. 2022).

17.2.1 Climate-Smart Crop Production

The exploration of the relationship between crop production and climate change has paramount importance in today's world where the population is increasing, the natural resources are getting scarce, and the climate is changing. Under these scenarios, the concept of climate-smart agriculture is a key for sustainable production of crops to meet the 3F-E (food, fiber, feed, and energy) needs. Due to anthropogenic activities, the concentration of greenhouse gases (GHGs, i.e., CO_2, CH_4, N_2O, CFCs) has been increased since the last century in the atmosphere that enhanced the greenhouse effect and heated the atmosphere and the earth surface. This phenomenon is called climate change indicating significant rise in global mean temperature, changes in precipitation patterns, and intense extreme weather events (FAO 2018). There is a two-way link between climate change and agriculture that is a severe threat to agricultural productivity and food security, while agriculture directly contributes about 20% of GHG emissions (Wu and Hussain 2024). Today, agriculture needs climate change adaptation measures to reduce vulnerability and mitigation measures to reduce GHG emissions. Climate-smart agriculture (CSA) is an approach to develop actions needed to transform agri-food systems toward green and climate resilient practices under climate change. CSA has three main objectives: sustainably increasing agricultural productivity and incomes; adapting and building resilience to climate change; and reducing and/or removing greenhouse gas emissions, where possible. The Save and Grow model of FAO for the sustainable intensification of crop production is the cornerstone of climate-smart agriculture (FAO 2013). The sustainable intensification options of crop production that support climate change adaptation and mitigation include:

- The use of good quality seeds and planting materials of well adapted varieties.
- The use of integrated pest management practices.
- The use of innovative tools and techniques.
- The use of conservation agriculture.
- The use of smart mechanization and smart irrigation water management.
- Integrated pest and nutrient management.
- Crop diversification, crop intercropping or rotations.

Crop production and climate change are closely interconnected, and there is a direct or indirect impact of climate change on crop production. Direct impacts mean changes in cropping calendars and crop phenology, soil loss and shifting of cultivation areas, changes in crop water demand and water supply, and direct effects of increased levels of CO_2 on crop growth. Indirect effects have further negative impacts on crop production, e.g., increase in crop diseases and pests. Each year soil erosion degrades 10 Mha of cropland, and 40% of this soil loss is due to tillage erosion. The loss of soil organic carbon and organic matter due to soil degradation increased chemical fertilization use to produce sufficient crop yield.

17.2.2 Adapting Crop Production to Impacts of Climate Change

Crop production is highly sensitive to climate variability and affected by extreme weather events, shifting of rainfall and temperature patterns, heat stress during different phenological stages, etc. Some crops are more vulnerable, and others are more tolerant to certain types of stress. Five-key environmental factors affect crop production, i.e., atmospheric CO_2 levels, temperature, precipitation, insects and pests, and weeds. The increased level of CO_2 concentration in the atmosphere affects crops according to the plants' metabolic pathways (C3 and C4 plants). The photosynthesis process of C3 plants (95% of the green plants are C3 plants) depends on CO_2 concentration in atmosphere only, while C4 plants (5% of the green plants are C4 plants include maize, sugarcane, pearl millet, sorghum, grasses) have the capacity to increase CO_2 concentration within their leaves before the photosynthesis begins (FAO 2013). A higher concentration of CO_2 may have a small fertilization effect on C3 crops if all other factors remain favorable. This effect may be offset due to adverse moisture conditions during growing season, insufficient nutrient availability, or temperatures above the optimum range. While for C4 plants, the increased concentrations of CO_2 may not provide benefits under normal conditions. The temperature variations mean either a change in average temperature, a change in maximum and minimum temperature, or a change in intensity, frequency, and duration of extremely hot/cold weather. These variations have a direct impact on crop production (FAO 2013). The crop phenological and morphological changes during development stage (reproductive and grain-filling) are more sensitive to high temperatures. The stress due to temperature change has a different effect on crop duration and productivity. For example, the increase of 1 °C in average temperatures can reduce the yields of major food and cash crops by 5–10% because plants use more energy for respiration rather than toward their growth (Allara et al. 2012). The vegetative growth of most crops remains optimum between 20 to 30 °C, while if daytime temperature exceeds 30 to 34 °C during the reproductive stage, the crop yield decreases. There may be permanent physical damage to plants above 30 °C persistent and above 37 °C can damage stored seeds. Due to climate change, the predicted temperature rise will remain above critical thresholds for wheat, rice, and maize.

Changes in rainfall pattern are critical to crop productivity, especially in rainfed regions (80% of the world's cropped area) where the cropping system depends on rainfall occurrence, distribution, and level. The impact of rainfall changes becomes strongest when combined with temperature changes that lead toward moisture stress. The projections of climate change already predicted that the wet area will be wetter, and the dry area will be drier in the near future, so there is a need to maintain water resources accordingly for sustainability. The intense rainfall variability makes it hard for farmers to shift cropping seasons accordingly, and they may not rely on the knowledge of the seasonality of climatic variables. The extreme weather events such as droughts, floods, and heat waves can physically damage crops, can delay sowing and harvesting, or complete crop failure. Plants are more vulnerable to indirect stress (e.g., pests and weeds) due to the direct effect of weather stresses. Climate change also modified the pests' attack on plants in space and over time (FAO 2013). For example, plants in waterlogged soil are less resilient to virus attack, and drought-affected plants are less able to outcompete weeds. Insect pests' distribution is influenced by temperature like pests (fall armyworm, aphids, pod borers, cereal stem borers) of vegetables, pulse, and cereals crops, likely to move to temperate regions. The crop loss due to insect pests depends on insect biotype, the population dynamics, the geographical distribution of insect pests, and the technological efficacy of crop protection. Climate change also favors perennial invasive C4 weed (Rottboellia cochinchinensis) germination and causes significant yield reduction of C3 crops (FAO 2013). Factors (temperature changes, droughts, weeds, insects, and pest attack, etc.) affecting crop production and their effects (low crop yield, low quality, and sugar contents, etc.) under climate change are given in Fig. 17.1.

The impacts of climate change on crop production are severe around the globe and wheat, maize, rice, and cotton productions significantly decreased 6%, 7.4%, 3.2%, and 10%, respectively worldwide (Table 17.1). In South Asia, the yield of staple crops including rice decreased up to 10%, while millet and maize production decreased more than 10% (Lobell et al. 2008). The crop productivity in high latitude areas showed the positive impact of climate change with an increase of 1 °C in average temperature, while in lower latitude areas, crop productivity decreased with a slight increase in the relative temperature range (IPCC 2007). Li et al. (2021a, b) research showed that 1 °C increase during the full growth cycle can reduce the growth period of rice, wheat, and corn crops by nearly 4, 10, and 7 days, resulting in a drop yields nearly 2, 4, and 3 percent, respectively, in China.

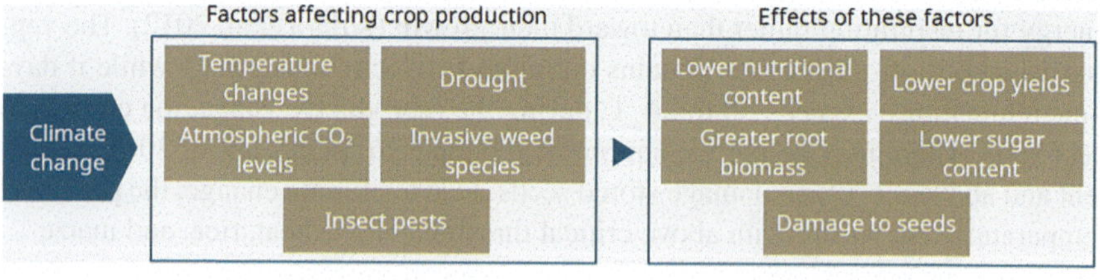

Fig. 17.1 Factors affecting crop production and their effects under climate change

Table 17.1 Impact of climate change on different crop yields around the globe

Crop type	Country/region	Yield reduction (%)	Reference
Wheat	Worldwide	6	Zhao et al. (2017)
	USA	5.5	
	France	6	
	India	9.1	
	Russia	7.8	
	China	2.6	
	Iran	37 29 to 42	Valizadeh et al. (2014) Javadi et al. (2024)
	India	7	Rao et al. (2015)
	Australia	27 50	Hochman et al. (2017) Asseng et al. (2011)
	Sudan	6	Asseng et al. (2017)
	Rainfed areas	5 to 7	Liu et al. (2016)
	North America	3.5 to 7.5	Zhang et al. (2022)
	Pakistan	6 14.7	Chaudhary (2017) Haq et al. (2021)
	China (Northern part)	2 to 4	Li et al. (2021a, b)
	North America	1 to 10	Wang et al. (2020)
Rice	Worldwide	3.2	Zhao et al. (2017)
	India	4.8	
	China	9.8	
	Indonesia	2.8	
	Vietnam	4.3	
	Bangladesh	5.2	
	Pakistan	15 to 18 20.5	Chaudhary (2017) Haq et al. (2021)
	Iran	21	Javadi et al. (2024)
	China	10.9	Jiang et al. (2021)
	Turkey	20	Özdoğan (2011)
	Italy	12	Bregaglio et al. (2017)
	Japan	11.3	Iizumi et al. (2011)
	Nepal	24	Khanal et al. (2018)
Maize	Worldwide	7.4	Zhao et al. (2017)
	USA	10.3	
	China	8	
	Brazil	5.5	
	India	5.2	
	France	2.6	
	Portugal	17	Yang et al. (2017)
	Pakistan	27	Ahmed et al. (2018)
	Africa	20	Abera et al. (2018)

(continued)

Table 17.1 (continued)

Crop type	Country/region	Yield reduction (%)	Reference
Cotton	Worldwide	10 to 17	Zafar et al. (2022)
	USA	19 17	FAO (2016) Adhikari et al. (2016)
	China	17 5.5	FAO (2016) Chen et al. (2015)
	Pakistan	5 8	FAO (2016) Rahman et al. (2018)
	Africa	7	Amouzou et al. (2018)
	Australia	17	Williams et al. (2015)
	Burkina Faso	13	Diarra et al. (2017)

17.2.3 Mitigating the Impacts of Crop Production on Climate Change

Crop production has both negative and positive impacts on climate change drivers such as GHG emissions. There are various practices involved in crop production, and it is difficult to measure GHG emissions accurately from each practice. The system is complex, and major GHG (CO_2, CH_4, and N_2O) emissions from crop production and agriculture have been described qualitatively, such as CO_2, CH_4, and N_2O emissions are 1%, 67%, and 32%, respectively (Kabange et al. 2023). Implementing GHG mitigation measures from agriculture is indeed a complex task. Chemical fertilizers such as nitrogen (N), phosphorous (P), and potassium (K) fertilizers significantly increased GHG emissions. Application of nitrogen fertilizer is the primary source of N_2O (Groenigen 2010). According to an estimate, the phosphorus net loss from crops is over 10.5 million metric tons per year (Liu et al. 2008). Potassium chloride (KCl) is the major source of potassium in agriculture (Prakash and Verma 2016). The use of organic amendments such as animal manure, composted manure, and biosolids is the cause of GHGs. The animal manure usage as organic fertilizer improves soil health and crop productivity, but also a source of 6% total CH_4 emissions (Yusuf et al. 2012). Composting is a biological process that generates different GHGs such as CO_2, CH_4, N_2O, and CO during the microbial decomposition of organic wastes (Andersen et al. 2010). Biosolids are leftovers from municipal wastewater treatment plants. It consists of 20 to 50% organic carbon and 2–5% total nitrogen (Haynes et al. 2009). Different cereal crops such as rice, wheat, maize, and oilseed crops such as soybean, sunflower, and peanuts are the source of GHG emissions. Rice has 3.75 times larger global warming potential than wheat and maize (Thangarajan et al. 2013). Globally, 30% CH_4 emission and 11% N_2O emission are from rice fields (Gupta et al. 2021). The contribution of cereal crops (68%), legumes and oilseeds (11%), vegetables (7%), other field crops (14%) in national total field crop's GHG emissions (Tongwane et al. 2016).

Mitigation measures for GHG emissions (Fig. 17.2) such as conservation tillage, smart water management, alternate cropping system, integrated nutrient management, and manure management are commonly used to reduce the GHGs emission in cropping system. Conservation tillage is minimum, or no-tillage method that

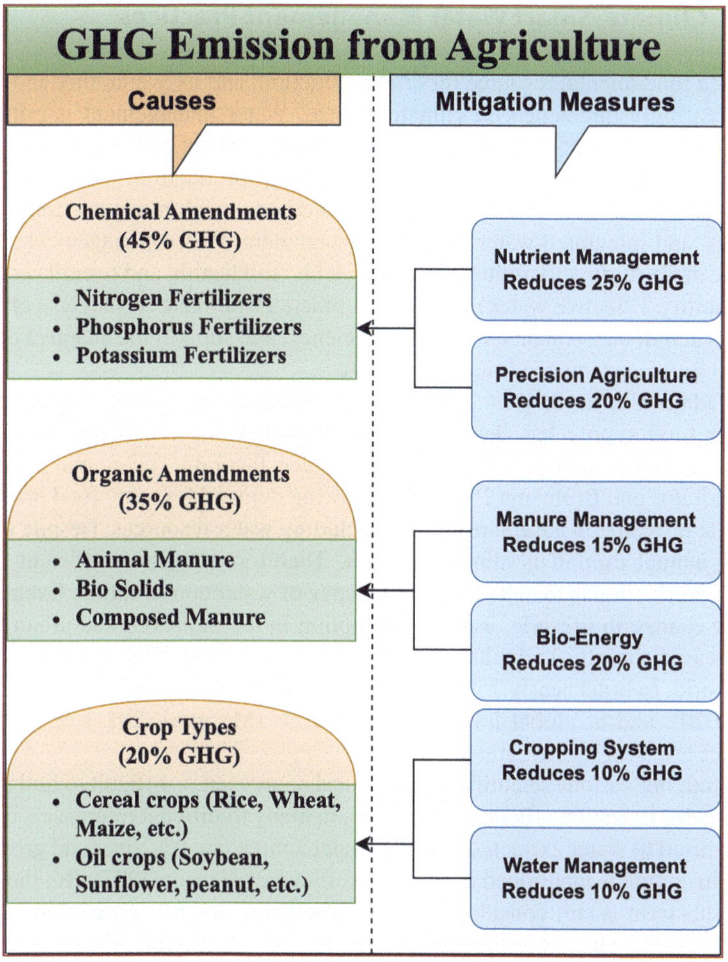

Fig. 17.2 Causes of GHG emission from agriculture and mitigation measures (Chataut et al. 2023)

improves soil physical conditions and lowers soil erosion and GHG emissions and enhances crop yield (Shakoor et al. 2021a, b). While another study showed adverse impact of no-tillage that GHG emissions CO_2 (7.1%), N_2O (12%), and CH_4 (20.8%) are more than conventional tillage (Shakoor et al. 2021a, b). Smart irrigation water management such as mulched drip irrigation (MDI) saves water efficiently and emits less GHGs than other irrigation methods (Ye et al. 2020). The sprinkler irrigation method reduces N_2O and CH_4 emissions by 40% and 99%, respectively, compared to flood irrigation (Fangueiro et al. 2017). Legume crops are an important source of nitrogen in root zone, and its rotation can reduce GHG emissions (Izaurralde et al. 2001). Ammonium sulfate fertilizer usage compared to urea reduces CH_4 emissions (Linquist et al. 2012). Agricultural leftovers and wastes can be used as bioenergy for GHGs' mitigation option, for example, its use can reduce 80 to 90% emission compared to fossil fuel usage (Chum et al. 2011).

17.3 Climate-Smart Water Management Practices

Water is a fundamental resource for crop production, and its availability and quality are increasingly threatened by climate change. Water management is critical for ensuring food security in the face of climate change. There are various water management strategies essential for climate-smart crop production, including technological advancements, efficient irrigation methods, soil moisture conservation practices, and integrated water resource management. Water management is a cornerstone of CSA, directly influencing crop yields, soil health, and overall ecosystem sustainability. Effective water management practices mitigate the adverse effects of droughts and floods, enhance water use efficiency, and support the adaptive capacity of agricultural systems to climate variability.

Nowadays, due to the promotion of living standards and growth population, the need for food supply has increased. It is predicted that food consumption will increase to 3000 kcal per day by 2050 under favorable economic conditions (Alexandratos and Bruinsma 2012). The growing demand for more food production puts great pressure on natural resources, including water resources. Despite this, the average annual rainfall is almost constant. Therefore, the most efficient way to increase production is to increase the efficiency of water consumption. Even though with the change in lifestyle, water consumption in the industrial and urban sectors has increased sharply. Agriculture is still the main consumer of water in many parts of the world. In India nearly 75% (Moin and Kamil 2018), in China about 62% (Han et al. 2020), and at global level more than 70% (Ma et al. 2021) of freshwater resources are used for irrigation of crops.

By studying various scientific and technical sources, it is difficult to find a single definition for irrigation efficiency. However, in many traditional references, the ratio of water used to water extracted from the water source (such as river and groundwater) for irrigation is mentioned as irrigation efficiency (Jensen 2007). But the definition of this term is still considered as a big challenge among experts. On the other hand, the term water use efficiency (WUE) is utilized, which is the amount of the produced yield in exchange for water consumption, and its unit is the ratio of biomass or grain produced to the amount of water consumed per hectare. WUE is definable from plant level to farm level. By moving up from the plant level to the farm level, due to the increase in factors affecting WUE, the complexity of its definition increases. However, by knowing these factors at higher levels, it is possible to improve WUE more (Vadez et al. 2023).

17.3.1 Efficient Irrigation Systems

Efficient irrigation systems, such as drip and sprinkler irrigation, play a vital role in climate-smart agriculture. These systems deliver water directly to the plant roots, reducing water loss through evaporation and runoff, increasing water use efficiency, reducing water consumption, and enhancing crop yield and quality. The "more crop per drop" is achievable via integrating irrigation scheduling (when, where, how, and how much) with innovative tools/ sensors and water conservation/ storage intervention called climate-smart irrigation options.

The literature review revealed that surface, sprinkler, and drip irrigation methods tend to have a WUE in the range of 70–40%, 90–50%, and 65–95%, respectively. Therefore, saving 35–65% water can be saved by changing from surface irrigation systems to pressurized systems (Anjum et al. 2023). For this reason, it is inevitable that efforts must be made to transform systems toward greater efficiency to uphold sustainability. Although a large portion of cultivated land in some regions is included under its coverage (e.g., above 60% in Europe and above 40% in Northern America (Pereira and Gonçalves 2018). But, because of technical and economic reasons, many arid and semi-arid regions have not received as much focus as others. To maximize farm yield, various measures such as fertilization, pest control, and use of different machines are necessary. In arid and semi-arid areas, precipitation is less than the crop water requirement of plants, and these precipitations are not uniformly distributed in terms of space and time. What is worse is that climate change has increased the incidence of extremes and has made precipitations more irregular than before (Tabari 2020). In these regions, the plants, at least in part of their growth and development period, are facing water deficiency. From this definition, the most efficient irrigation system is required to deliver water to the plant at the right time and precise amount (which is not supplied naturally).

Drip Irrigation This method delivers water directly to the plant root zone, reducing evaporation and runoff. Drip irrigation systems have demonstrated high water use efficiencies, particularly in arid and semi-arid regions. A study conducted in India demonstrated the effectiveness of drip irrigation in maize production. The system reduced water use by 40% and increased yield by 20%, showcasing the potential of efficient irrigation systems in enhancing water use efficiency and crop productivity. In conventional irrigation practices, water loss primarily arises from runoff, deep percolation, and soil surface evaporation. The introduction and advancement of drip irrigation systems aimed to address these three issues. Solving the issues not only increases the WUE, but it also has other agricultural and environmental advantages including: reducing the need for drainage, reducing weeds in the field, pests and weeds control, reducing the need for land leveling and loss of agricultural soil, increasing the efficiency of using fertilizers and pesticides through chemigation, and reducing their emissions to environment, the possibility of using low quality water.

In the field, water stored in the root development zone is lost in two ways: transpiration (by the plant) and evaporation (from the soil surface). Transpiration is considered beneficial due to its importance in the biological activities of the plant, while evaporation is regarded as a non-productive component of water loss because it does not contribute to the biological processes of the plant. By using a drip system that reduces the wetted surface of the soil and delivers enough water directly to the root zone, it is possible to significantly reduce evaporation, which is a non-productive loss. Accurate irrigation in the drip system effectively reduces deep percolation and runoff, and the use of pipes to deliver water to plants through a network of pipes, by reducing evaporation and infiltration along the way, significantly improves the efficiency of water distribution in the field. While this method offers numerous benefits, it does come with its limitations. These include high initial costs, salt accumulation in the root zone in saline conditions, as well as implementation and maintenance challenges. Despite these drawbacks, the method's high water consumption efficiency and significant increase in yield have made it a proper choice among farmers in low-water areas. However, there is still a challenge about supplying and costs of energy, caused by the energy need in the pressurized irrigation. Due to the expansion of automation in recent years, the use of night irrigation and renewable energy sources such as photovoltaic seems like proper options.

Sprinkler Systems Sprinkler systems are a modern irrigation system that operates under sufficient pressure and water sprays on the crop or soil in the manner of rain. This system is particularly effective for uneven terrains and can be used for a wide range of crops. Effective for a wide range of crops, sprinkler systems distribute water evenly and can be used to apply water at controlled rates, reducing water loss due to deep percolation and runoff. In this method, water is lost in three ways: through evaporating after exiting nozzle due to exposure to air, evaporating while resting on the canopy, and either and as runoff or deep percolation after reaching the ground. Apart from losses, uniformity of distribution is also important. Due to irrigation with high flow rate, lack of uniformity causes runoff and water deficit in shallow root plants. The losses are affected by climatic conditions such as wind speed and vapor pressure deficit, and these were affected by changing different design parameters. Redesigning system variables such as riser height and type of nozzles periodically can reduce losses. Sprinkler irrigation provides exclusive methods to enhance food security and environmental conditions that are not provided by other techniques: High-density crop irrigation made possible by this method results in greater yield per unit area and improves food security. The sprinkler method is the primary choice for irrigation in sandy and desert regions, making it ideal for transforming arid lands into lush green areas. Enabling the usage of degraded water in situations where purifying it is impossible or creates financial and environmental issues. It is very effective in the protection of plants from frost damage.

Furrow-Irrigated Bed Planting This technique involves planting crops on raised beds (40–130 cm wide and 10–20 cm tall) with furrows (20–50 cm wide) that facilitate efficient water distribution. Crops are grown on the top of raised beds and water is applied in furrows. This method ensures uniform water application and keeps soil at optimum moisture conditions. The design of furrow cum bed planting system depends upon field slope. If the field slope is less than 0.5%, the bed-furrow should be parallel with the slope, otherwise align it with the contours. In furrow cum bed planting system, the wetting of field area is small, and hydraulic resistance is high which needed less amount of water compared to flood irrigation. Proper land leveling is crucial for optimizing and uniform irrigation water use in this system. Water can be conserved more by beds mulch or plastic sheets covering. The bed planted-furrow irrigation system has numerous advantages over flood irrigation such as reducing cresting of soil surface and increasing germination rate. The WUE of bed planting with furrow irrigation improved by 21–30% with 17% savings in applied irrigation water, and nitrogen use efficiency (NUE) could improve by 10% and improve grain quality, and increase grain yield by more than 10% (Wang et al. 2004).

Subsurface Irrigation Subsurface irrigation involves burying irrigation pipes underground to deliver water directly to the root zone. This advanced technique reduces evaporation losses, reduces weeds and disease rate, and ensures efficient water use. Subsurface drip irrigation is a high-efficiency-low-pressure irrigation system to meet crop water requirement consisting of buried drip tubes or drip tape. A subsurface drip system requires higher initial cost depending upon filtration needs according to water quality, degree of automation, and soil characteristics. Subsurface irrigation is suitable for arid and semi-arid areas having limited water sources, hot and windy climate, and sandy type soils. This method of water application eliminates the surface irrigation affects such as crusting, ponding, surface runoff, and soil erosion (Reich et al. 2014).

17.3.2 Soil Moisture Conservation Practices

Soil moisture management involves monitoring and maintaining optimal soil moisture levels to support crop growth and resilience against climate extremes.

Mulching Applying organic or inorganic materials on the soil surface reduces evaporation, suppresses weed growth, and moderates soil temperature. Any material except for soil or plants that is spread over the ground for prolonged protection is considered mulch (Jordán et al. 2011). This method of moisture conservation enabled enhanced soil moisture retention, improved soil structure and fertility, and reduced soil erosion. The use of mulch boosts soil quality by improving water retention, lowering evaporation rates, maintaining temperature levels, enhancing nutrient uptake by plant roots, decreasing salinity, and managing weed growth, pests, and crop diseases (Han et al. 2020a, b). Mulches are divided into two categories: organic or biodegradable and non-organic, which are often made of plastic. Plastic material, crop residues, livestock manure, sands, rocks, and cement are utilized for mulching. Even though experts are still unsure about which option is best suited for varying agricultural scenarios and locations. Plastic mulches have become popular among farmers due to their user-friendly nature and widespread availability. In general, these mulches are utilized in either clear or block colors, with each option presenting its own advantages and disadvantages. For example, block is better for weed control, and clear is suitable for solarization. The combined use of these two has brought satisfactory results. However, there are concerns about the environmental issues caused by the use of plastic mulches. Using its biodegradable type can reduce these concerns.

If the purpose of using mulch is to reduce evaporation and thus reduce the need for irrigation, reflective mulches can be used. When mulch is used to raise temperature for growth, it should be noted that higher temperatures lead to more evaporation. In addition, because of raising more under it, black plastic mulches have been found to potentially increase evaporation more than organic mulches (Merwin et al. 1995). Overall, mulching plays a crucial role in maintaining the health and sustainability of water and soil resources. By creating a protective barrier on the soil surface, mulch helps to regulate water flow, reduce erosion, and promote better infiltration. Additionally, the presence of organic matter in mulch enhances soil fertility, encourages beneficial soil organisms like earthworms, and improves the overall plant's root growth and consequently, water and nutrients uptake (Zhang et al. 2021).

Despite that, the significant advantage that we are focusing on is the retention of soil moisture and increased efficiency in water utilization. That being said, it is crucial to understand that by preserving moisture in this specific case, there is a boost in the microbial population within the soil and therefore an improvement in the nutrient cycle. Måge (1982) observed that soil temperature beneath a layer of black PE mulch was 3 °C higher at a depth of 30 cm and the soil moisture tension was reduced by 13–29 kPa when compared to uncovered soil. Although the biggest benefit of employing mulch in combating climate change is the consistency, it brings to soil moisture levels despite fluctuations in vapor pressure deficiency. On the other hand, the existence of mulch can boost transpiration levels and expedite the plants' development (Hou et al. 2010). Mulch usage can have an impact on water use efficiency due to its effectiveness on transpiration.

Cover Cropping Growing cover crops during off-seasons protects the soil from erosion, improves soil structure, and enhances moisture retention. Cover crops are close growing crops that are grown between normal crop production periods to protect the soil, seed and improve the physical, chemical and biological properties of the soil (SSSA 2008). Cover cropping is added to fields to increase sustainability in cultivation. Cover crops improve the physical (permeability and hydraulic conductivity), chemical (carbon sequestration and nitrogen fixation), and biological properties of the soil. Also, cover crops have positive environmental effects by absorbing excess fertilizers used and reducing the need for fertilizers (such as nitrogen fixation by legumes). In new agricultural methods like no-till farming, which does not incorporate plant residues into the soil and minimizes the addition of organic matter to the soil, cover crops are crucial for providing organic matter to the soil. The utilization of cover crops on farms leads to a noticeable decline in penetration resistance (Ruis et al. 2020), attributed to the reduced traffic from agricultural machinery. Evidence from multiple studies has shown that cover crops have lasting effects on soil properties like porosity, infiltration rates, and water holding capacity (e.g., Leuthold et al. 2021; Schomberg et al. 2023). In past studies, various reasons have been mentioned for the effect of cover crops on water infiltration. Blanco-Canqui (2018) highlights the role of canopy protection in protecting soil from water droplets, thus preventing pore closure. Blanco-Canqui et al. (2011) point to an abundance of earthworms in CC during the summer months as the cause for the rise in cumulative penetration.

Cover crops can act as a key in climate change conditions. Li et al. (2021a, b) discovered that the inclusion of wheat cover crops led to a reduction in the correlations of temperature or precipitation extremes with grain yields and major water losses. They stated that cultivation of cover crops can help lessen the impact of climate extremes on the WUE in a corn–soybean cropping system. However, the success of utilizing cover crops may fluctuate based on variables like the particular cover crop selected, soil properties, tillage methods, crop rotation strategies, management practices employed in the past, and the overall climate in the area (Blanco-Canqui et al. 2011). Although many experiments on cover crops have been conducted, most of these have focused in on short-term effects (Blanco-Canqui et al. 2011). Long-term studies are required to fully comprehend the diverse effects of cover crops on agriculture as well as the environment.

17.3.3 Rainwater Harvesting and Storage

Techniques such as building check dams, percolation tanks, and farm ponds help capture and store rainwater for agricultural use. This practice helps to mitigate water scarcity during dry periods, reduce dependence on groundwater, and enhance groundwater recharge. In semi-arid regions of Africa, rainwater harvesting systems were implemented for wheat cultivation. These systems captured and stored rainwater, reducing the impact of droughts and ensuring a stable water supply for the crops. As a result, wheat yield increased by 30%. Rainwater harvesting can be done with various techniques such as: Micro-catchment systems collect and store rainwater in small basins, which can be used during a dry spell, and reduce runoff and soil erosion. Farm ponds' construction improves water availability for irrigation. It provides a reliable water source during drought periods and enhances groundwater recharge. Rooftop rainwater harvesting involves collecting rainwater from rooftops and storing it for agricultural use.

In arid and semi-arid regions, precipitations are not distributed throughout the year, and most of them occur in a limited period of time. The concentration of precipitations in a certain period, which is often short, creates a flood season and a dry period in the rest of the year. The rainy season may not coincide with the growing season of many crops. It affects yield and farmers' livelihood, seriously. The usual solution in these cases is irrigation in dry seasons. However, this is certainly not the most sustainable solution (because of some activities such as pumping or construction of big water reservoirs). Consequently, irrigation using conventional methods (even advanced methods) has conflicts with environmental sustainability and requires economic investment. Rainwater harvesting is a smart method in which storing water in the soil or reservoirs in the rainy season to meet the plant water requirement in drought periods. The advantage of rainwater harvesting is not only providing water for plants or domestic use, but also preventing erosion, and avoiding excessive evaporation. One way to use harvested rainwater is to store it in underground sources, which faces problems such as high groundwater levels, high excavation and construction costs, and the need for labor. In rainwater harvesting, every factor that enhances the penetration of water into the soil and then holding it should be taken into consideration: adjusting the slope of the land, using plastic sheets, using absorbents, adding clay materials to the soil, tillage, and counter farming.

Rainwater harvesting is also used in urban agriculture. In addition to providing a reliable source of irrigation water, rainwater harvesting can also help mitigate flooding in urban areas by reducing stormwater runoff. The technique works particularly well in urban areas lacking natural water sources or the proper network of pipes to provide irrigation water. In urban environments, the main challenge with using this method lies in the shortage of knowledge, limited design methods, and the necessity for implementing innovative projects to accommodate the diverse nature of urban settings (Amos et al. 2018).

The main issue when employing this technique is the risk of excessive water extraction, which has the potential to disrupt the basin's water distribution. It is crucial to carefully manage water harvesting practices to ensure sustainable use of water resources and prevent detrimental effects on the environment and society. Implementing proper regulations, monitoring systems, and conservation measures can help mitigate the risks associated with excessive water harvesting and maintain a healthy water balance in the basin. Collaboration between stakeholders, including government agencies, local communities, and water resource management organizations, is essential to address this issue effectively and safeguard the long-term sustainability of water resources.

17.3.4 Integrated Water Resource Management (IWRM)

IWRM is a comprehensive approach that coordinates the management of water, land, and related resources to maximize economic and social welfare without compromising the sustainability of vital ecosystems. Key components include:

Watershed Management Watershed management is based on the concept of sustainability and meeting the needs of the present population without compromising the interests of future generations for the efficient utilization of water and other natural resources. Ensuring sustainable water use at the watershed level involves maintaining the carrying capacity, promoting afforestation, and implementing soil and water conservation measures. Watershed management is important for conservation of soil and water and to manage and utilize the runoff water for a useful purpose, and control soil erosion. Watershed management practices include agronomical measures such as strip and contour cropping, pasture cropping, grassland farming and engineering measures (structural practices) such as contour bunding, terracing, construction of earthen embankment, construction of check dams, construction of farm ponds, and gully controlling structures and providing vegetative and stone barriers. The concept of watershed management may be expressed in symbolic form by "POWER" Singh (2000).

P	= Production of food-fiber-fodder-fuel combine on a sustained basis
	= Pollution control, and
	= Prevention of floods
O	= Over exploitation of resources to be minimized like over grazing, and
	= Operational practicability of all on-farm operations
W	= Water storage at convenient location for different purposes, and
	= Wildlife and indigenous plant life conservation at selected places
E	= Erosion control ecosystems,
	= Ecosystem safety,
	= Economic stability, and
	= Employment generation.
R	= Recharge of groundwater
	= Reduction of drought hazards
	= Recreation, and
	= Reduction of siltation in multi-purpose reservoirs

The concept of watershed management recognizes interrelationships among land use, soil, and water and the linkage between uplands and downstream areas.

Groundwater Management Efficient use of groundwater resources involves practices like regulated pumping, artificial recharge, and monitoring groundwater levels to prevent over-extraction and ensure long-term availability. IWRM should be applied according to the scale of river basin to make sure the water access for drinking, industrial, agriculture, and environmental purposes. For this approach, the Integrated Quantity-Quality Model (IQQM) can assist in the planning, management, and evaluating the water resources polices under a decision support system. Similarly, the concept of Water Banking is helpful for IWRM to store water for the time of drought, and requirement like Managed Aquifer Recharge (MAR) technique can maximize natural storage and increase water supply system resilience during periods of low flows and high seasonal variability. This sustainable groundwater management technique enhances groundwater quality and quantity. The water managers, experts, and policymakers should adopt such IWRM techniques at basin and farm levels to sustain the available water resources for future generations.

17.4 Tools and Technologies for Climate-Smart Water Management

There is a need to put great emphasis on the use of modern tools and techniques to address the challenges being faced by our farmers related to water resources management. The enhancement of water productivity pertaining to smart and precision irrigation is the key for sustainable development. The land and water conservation for water management under the Internet of Things and Geographic Information Systems (GIS) provide early weather warnings and crop-related instructions to the farmers. Water profession is considered as the sun and moon revolve around the water—in all the countries around the world. We have major floods, while the other side has very important droughts. Water is very important in the political framework. When the flood is over, the political interest will be gone until the next drought or flood. Water should be on the top of political agenda for economic and environmental security. For any policy such as industrial and environmental, first must to see water as a criterion. Every policy should bend on the knee to water. Everything revolves around water, but think how water can be used as a catalyst for economic development. Urbanization is increasing, the population is increasing, per capita water is increasing, and water use in the agriculture sector is increasing. Industrial water use is increasing hyperbolically. So IWRM should be adopted for small-scale basins where indigenous people/farmers, stakeholders can potentially take care of own available water resources. Innovative tools and technologies play a critical role in enhancing water management: remote sensing and GIS technologies to monitor soil moisture, crop health, and water usage on a large scale. Soil moisture sensors and devices provide real-time data to optimize irrigation practices. Decision support systems software and tools help to make informed decisions on water management. Mobile applications platforms offer accessible information and guidance to farmers, helping them adopt best practices. Effective water management should be integrated with other CSA practices to achieve maximum benefits. This includes:

- Pest and disease management in water-scarce conditions.
- Soil fertility management through the use of organic and inorganic fertilizers.
- Crop diversification to reduce the risk of crop failure due to water stress and to enhance water efficiency.
- Agroforestry practices to improve water infiltration and reduce soil erosion.

There are different challenges and opportunities related to climate-smart IWRM techniques in the context of CSA, such as high initial costs for implementing advanced water management technologies, limited access to technology and knowledge among smallholder farmers, and institutional and policy barriers. Government policies and institutional support are crucial for the successful implementation of climate-smart water management practices. Key areas of focus include developing and disseminating water-efficient technologies, capacity building and training for farmers on water management practices, and providing financial incentives and subsidies for adopting efficient irrigation systems. The advancement in technology, such as precision agriculture and remote sensing, provides new opportunities for improving water use efficiency. This increased awareness of climate change impacts and the need for sustainable water management practices.

17.5 Conclusion

Water management is integral to climate-smart agriculture, offering numerous benefits for crop production, water conservation, and climate resilience. By adopting efficient irrigation systems, soil moisture management techniques, rainwater harvesting, and integrated water resource management, agricultural systems can become more resilient and productive, and farmers can enhance their productivity and sustainability. Policymakers, researchers, and farmers must collaborate to implement these strategies effectively, ensuring food security and environmental sustainability for future generations. Integrating these practices with supportive policies and institutional frameworks will ensure the successful implementation of climate-smart water management strategies.

References

Ahmed I, Rahman MH, Ahmed S, Hussain J, Ullah A, Judge J (2018) Assessing the impact of climate variability on maize using simulation modeling under semi-arid environment of Punjab, Pakistan. Environ Sci Pollut Res 25:28413–28430

Abera K, Crespo O, Seid J et al. (2018) Simulating the impact of climate change on maize production in Ethiopia, East Africa. Environ Syst Res 7:4. https://doi.org/10.1186/s40068-018-0107-z

Adhikari P, Ale S, Bordovsky JP, Thorp KR, Modala NR, Rajan N, Barnes EM (2016) Simulating future climate change impacts on seed cotton yield in the Texas High Plains using the CSM-CROPGRO-cotton model. Agric Water Manag 164:317–330

Alexandratos N, Bruinsma J (2012) World agriculture towards 2030/2050: The 2012 Revision. ESA Working Paper No. 12–03. FAO Agricultural Development Economics Division, Rome, Italy

Allara M, Kugbei S, Dusunceli F, Gbehounou G (2012) Coping with changes in cropping systems: plant pests and seeds. In FAO/OECD Workshop on Building Resilience for Adaptation to Climate Change in the Agriculture Sector (23–24 April 2012)

Amos CC, Rahman A, Karim F, Gathenya JM (2018) A scoping review of roof harvested rainwater usage in urban agriculture: Australia and Kenya in focus. J Clean Prod 202:174–190

Amouzou KA, Naab JB, Lamers JPA et al (2018) CROPGRO-Cotton model for determining climate change impacts on yield, water-and N-use efficiencies of cotton in the Dry Savanna of West Africa. Agric Syst 165:85–96

Andersen JK, Boldrin A, Samuelsson J, Christensen TH, Scheutz C (2010) Quantification of greenhouse gas emissions from windrow composting of garden waste. J Environ Qual 39(2):713–724

Anjum MN, Cheema MJM, Hussain F, Wu RS (2023) Precision irrigation: challenges and opportunities. Precision Agric:85–101

Asseng S, Foster IA, Turner NC (2011) The impact of temperature variability on wheat yields. Glob Chang Biol 17:997–1012

Asseng S, Cammarano D, Basso B, Chung U, Alderman PD, Sonder K et al (2017) Hot spots of wheat yield decline with rising temperatures. Glob Chang Biol 23:2464–2472

Bregaglio S, Hossard L, Cappelli G, Resmond R, Bocchi S, Barbier J-M, Ruget F, Delmotte S (2017) Identifying trends and associated uncertainties in potential rice production under climate change in Mediterranean areas. Agricultural and Forest Meteorology (237–238):219–232. https://doi.org/10.1016/j.agrformet.2017.02.015

Blanco-Canqui H (2018) Cover crops and water quality. Agron J 110(5):1633–1647

Blanco-Canqui H, Mikha MM, Presley DR, Claassen MM (2011) Addition of cover crops enhances no-till potential for improving soil physical properties. Soil Sci Soc Am J 75(4):1471–1482

Chaudhary QU (2017) Climate change profile of Pakistan Asian Development Bank. https://doi.org/10.22617/TCS178761

Chataut G, Bhatta B, Joshi D, Subedi K, Kafle K (2023) Greenhouse gases emission from agricultural soil: A review. J Agric Food Res 11:100533
Chen C, Pang Y, Pan X, Zhang L (2015) Impacts of climate change on cotton yield in China from 1961 to 2010 based on provincial data. J Meteorol Res 29(4):515–524
Chum H, Faaij A, Moreira J et al (2011) Renewable energy sources and climate change mitigation: special report of the intergovernmental panel on climate change. Cambridge University Press, pp 209–332
Diarra A, Barbier B, Zongo B, Yacouba H (2017) Impact of climate change on cotton production in Burkina Faso. African J Agric Res 12(7):494–501
Fangueiro D, Becerra D, Albarran A et al (2017) Effect of tillage and water management on GHG emissions from Mediterranean rice growing ecosystems. Atmos Environ 150:303–312
FAO (2013) Climate smart agriculture-sourcebook. Food and Agriculture Organization of the United Nations, Rome
FAO (2016) The state of food and agriculture 2016 (SOFA): climate change, agriculture and food security. Food and Agriculture Organization, Rome, Italy
FAO (2018) Ministry of Agriculture, Livestock and Fisheries: Climate Smart Agriculture - Training Manual for Extension Agents in Kenya
Groenigen JWV (2010) Towards an agronomic assessment of N2O emissions: a case study for arable crops. J Sci Food Agric 90(5):903–913
Gupta K, Kumar R, Baruah KK, Hazarika S, Karmakar S, Bordoloi N (2021) Greenhouse gas emission from rice fields: a review from Indian context. Environ Sci Pollut Res 28(24):30551–30572
Han S, Tian F, Gao L (2020) Current status and recent trend of irrigation water use in China. Irrig and Drain 1–11. https://doi.org/10.1002/ird.2441
Han Q, Harayama H, Uemura A et al (2020a) High biomass productivity of short-rotation willow plantation in boreal Hokkaido achieved by mulching and cutback. Forests 11(5):505
Han S, Tian F, Gao L (2020b) Current status and recent trend of irrigation water use in China. Irrig Drain 69:25–35
Haq SU, Boz I, Shahbaz P (2021) Adoption of climate-smart agriculture practices and differentiated nutritional outcomes among rural households: a case of Punjab province Pakistan. Food Secur. https://doi.org/10.1007/s12571-021-01161-z
Haynes RJ, Murtaza G, Naidu R (2009) Inorganic and organic constituents and contaminants of biosolids: implications for land application. Adv Agron 104:165–267
Hochman Z, Gobbett DL, Horan H (2017) Climate trends account for stalled wheat yields in Australia since 1990. Glob Chang Biol 23:2071–2081
Hou XY, Wang FX, Han JJ, Kang SZ, Feng SY (2010) Duration of plastic mulch for potato growth under drip irrigation in an arid region of Northwest China. Agric For Meteorol 150(1):115–121
Intergovernmental Panel on Climate Change (IPCC) (2007) Technical Summary. In: Climate change 2007: mitigation. Contribution of Working Group III to the Fourth Assessment Report of the IPCC. Cambridge University Press, Cambridge, UK, and New York, USA
Izaurralde RC, McGil WB, Robertson JA, Juma NG, Thurston JJ (2001) Carbon balance of the Breton classical plots over half a century. Soil Sci Soc Am J 65(2):431–441
Javadi A, Ghahremanzadeh M, Sassi M et al (2024) Impact of climate variables change on the yield of wheat and rice crops in Iran (application of stochastic model based on Monte Carlo simulation). Comput Econ 63:983–1000. https://doi.org/10.1007/s10614-023-10389-0
Jensen ME (2007) Beyond irrigation efficiency. Irrig Sci 25(3):233–245
Jiang P, Xu F, Zhang L, Liu M, Xiong H, Guo X (2021) Grain yield and N use efficiency of direct-seeded rice under different N management practices aimed to reduce N input. Arch Agron Soil Sci 67:1578–1590. https://doi.org/10.1080/03650340.2020.1800643
Jordán A, Zavala LM, Muñoz-Rojas M (2011) Mulching, effects on soil physical properties. In: Gliński J, Horabik J, Lipiec J (eds) Encyclopedia of agrophysics. Springer, Dordrecht, pp 492–496
Kabange NR, Kwon Y, Lee SM, Kang JW, Cha JK, Park H et al (2023) Mitigating greenhouse gas emissions from crop production and management practices, and livestock: A review. Sustain For 15:15889. https://doi.org/10.3390/su152215889

Khan MA, Sultan MS, Khan H (2022) Climate smart agriculture training module: a reference manual for agriculture extension officers, researchers, students and policy makers. Available at: https://www.apn-gcr.org/wp-content/uploads/2022/04/CSA-module-book-_-08-Apr-2022-2.pdf

Khanal U, Wilson C, Hoang V-N, Lee B (2018) Farmer's adaptation to climate change, Its determinants and impacts on rice yield in nepal. Ecological Economics 144(C):139–147

Leuthold SJ, Salmeron M, Wendroth O, Poffenbarger H (2021) Cover crops decrease maize yield variability in sloping landscapes through increased water during reproductive stages. Field Crop Res 265:108111

Li LC, Wang B, Feng PY, Wang HH, He QS, Wang YK et al (2021a) Crop yield forecasting and associated optimum lead time analysis based on multi-source environmental data across China. Agric For Meteorol 308-309:108558

Li Y, Tian D, Feng G, Yang W, Feng L (2021b) Climate change and cover crop effects on water use efficiency of a corn-soybean rotation system. Agric Water Manag 255:107042

Linquist B, Van Groenigen KJ, Adviento-Borbe MA, Pittelkow C, Van Kessel C (2012) An agronomic assessment of greenhouse gas emissions from major cereal crops. Glob Chang Biol 18(1):194–209

Liu Y, Villalba G, Ayres RU, Schroder H (2008) Global phosphorus flows and environmental impacts from a consumption perspective. J Ind Ecol 12(2):229–247

Liu B, Asseng S, Müller C, Ewert F, Elliott J, Lobell DB et al (2016) Similar estimates of temperature impacts on global wheat yield by three independent methods. Nat Clim Chang 6:1130–1136

Lipper L, McCarthy N, Zilberman D, Asfaw S, Branca G (eds) (2018) Climate Smart Agriculture: Building Resilience to Climate Change. Springer, Cham. ISBN 978-3-319-61194-5

Lobell DB, Burke MB, Tebaldi C, Mastrandrea MD, Falcon WP, Naylor RL (2008) Prioritizing climate change adaptation needs for food security in 2030. Science 319:607–610

Iizumi T, Yokozawa M, Nishimori M (2011) Probabilistic evaluation of climate change impacts on paddy rice productivity in Japan. Climatic Change 107:391–415. https://doi.org/10.1007/s10584-010-9990-7

Ma W, Meng L, Wei F, Opp C, Yang D (2021) Spatiotemporal variations of agricultural water footprint and socioeconomic matching evaluation from the perspective of ecological function zone. Agric Water Manag 249:106803

Måge F (1982) Black plastic mulching compared to other orchard soil management methods. Sci Hortic 16(2):131–136

Merwin IA, Rosenberger DA, Engle CA, Rist DL, Fargione M (1995) Comparing mulches, herbicides, and cultivation as orchard groundcover management systems. HortTechnology 5(2):151–158

Moin K, Kamil A (2018) Drip irrigation in India: prospects, potential and challenges. J Geogr Environ Earth Sci Int 17(4):1–14

Özdoğan M (2011) Modeling the impacts of climate change on wheat yields in Northwestern Turkey, Agriculture, Ecosystems & Environment, 141(1–2):1–12. https://doi.org/10.1016/j.agee.2011.02.001

Pereira LS, Gonçalves JM (2018) Surface irrigation. In Oxford Research Encyclopedia of Environmental Science

Prakash S, Verma JP (2016) Global perspective of potash for fertilizer production. In: Potassium solubilizing microorganisms for sustainable agriculture. Springer, New Delhi, pp 327–331

Rahman MH, Ahmad A, Wang X et al (2018) Multi-model projections of future climate and climate change impacts uncertainty assessment for cotton production in Pakistan. Agric For Meteorol 253:94–113

Rao BB, Chowdary PS, Sandeep VM, Pramod VP, Rao VUM (2015) Spatial analysis of the sensitivity of wheat yields to temperature in India. Agric For Meteorol 200:192–202. https://doi.org/10.1016/j.agrformet.2014.09.023

Reich D, Godin R, Chávez JL, Broner I (2014) Subsurface drip irrigation (SDI), Fact Sheet No. 4.716, Crop Series/Irrigation, Colorado State University Extension. Available at: https://extension.colostate.edu/docs/pubs/crops/04716.pdf

Ruis SJ, Blanco-Canqui H, Elmore RW et al (2020) Impacts of cover crop planting dates on soils after four years. Agron J 112(3):1649–1665

Schomberg HH, White KE, Thompson AI et al (2023) Interseeded cover crop mixtures influence soil water storage during the corn phase of corn-soybean-wheat no-till cropping systems. Agric Water Manag 278:108167

Shakoor A, Shahbaz M, Hassan T et al (2021a) A global meta-analysis of greenhouse gases emission and crop yield under no-tillage as compared to conventional tillage. Sci Total Environ 750:142299

Shakoor A, Shakoor S, Rehman A et al (2021b) Effect of animal manure, crop type, climate zone, and soil attributes on greenhouse gas emissions from agricultural soils: A global meta-analysis. J Clean Prod 278:124019

Singh R (2000) Watershed Planning and Management. Yash Publishing House, Bikaner, pp. 470

Soil Science Society of America (SSSA) (2008) Glossary of soil science terms 2008. ASA-CSSA-SSSA

Tabari H (2020) Climate change impact on flood and extreme precipitation increases with water availability. Sci Rep 10(1):13768

Thangarajan R, Bolan NS, Tian G, Naidu R, Kunhikrishnan A (2013) Role of organic amendment application on greenhouse gas emission from soil. Sci Total Environ 465:72–96

Tongwane M, Mdlambuzi T, Moeletsi M, Tsubo M, Mliswa V, Grootboom L (2016) Greenhouse gas emissions from different crop production and management practices in South Africa. Environ Dev 19:23–35

Vadez V, Pilloni R, Grondin A et al (2023) Water use efficiency across scales: from genes to landscapes. J Exp Bot 74(16):4770–4788

Valizadeh J, Ziaei SM, Mazloumzadeh SM (2014) Assessing climate change impacts on wheat production (a case study). J Saudi Soc Agric Sci 13:107–115. https://doi.org/10.1016/j.jssas.2013.02.002

Wang F, Wang X, Sayre KD (2004) Comparison of conventional, flood irrigated, flat planting with furrow irrigated, raised bed planting for winter wheat in China. Field Crops Research 87(1):35–42. https://doi.org/10.1016/j.fcr.2003.09.003

Wang X et al (2020) Emergent constraint on crop yield response to warmer temperature from field experiments. Nat Sustain 3:908–916

Williams A, White N, Mushtaq S et al (2015) Quantifying the response of cotton production in eastern Australia to climate change. Clim Chang 129(2):183–196

Wu RS, Hussain F (2024) Managing the impacts of climate change, droughts, and water scarcity on agriculture. In: Knox JW (ed) Improving Water Management in Agriculture: Irrigation and Food Production. Burleigh Dodds Series in Agricultural Science

Yang C, Fraga H, Ieperen WV, Santos JA (2017) Assessment of irrigated maize yield response to climate change scenarios in Portugal. Agricultural Water Management 184:178–190. https://doi.org/10.1016/j.agwat.2017.02.004

Ye X, Liu H, Zhang X et al (2020) Impacts of irrigation methods on greenhouse gas emissions/absorptions from vegetable soils. J Soils Sediments 20(2):723–733

Yusuf RO, Noor ZZ, Abba AH et al (2012) Methane emission by sectors: a comprehensive review of emission sources and mitigation methods. Renew Sust Energ Rev 16(7):5059–5070

Zafar MM, Jia X, Shakeel A, Sarfraz Z, Manan A, Imran A et al (2022) Unraveling heat tolerance in upland cotton (Gossypium hirsutum L.) using univariate and multivariate analysis. Front Plant Sci 12:727835. https://doi.org/10.3389/fpls.2021.727835

Zhang H, Miles C, Gerdeman B, LaHue DG, DeVetter L (2021) Plastic mulch use in perennial fruit cropping systems—a review. Sci Hortic 281:109975

Zhang T, He Y, DePauw R et al (2022) Climate change may outpace current wheat breeding yield improvements in North America. Nat Commun 13:5591. https://doi.org/10.1038/s41467-022-33265-1

Zhao C, Liu B, Piao S, Wang X, Lobell DB, Huang Y et al (2017) Temperature increase reduces global yields of major crops in four independent estimates. Proc Natl Acad Sci USA 114:9326–9331. https://doi.org/10.1073/pnas.1701762114

Forward Genetic Approaches for Gene Identification Controlling Nutrition in Staple Crops

18

Aiman Hina, Raheela Amin, Tahira Kabeer, Asim Abbasi, Inzamam Ul Haq, Muhammad Farhan Sarwar, Nader R. Abdelsalam, and Hayssam M. Ali

Abstract

Nutritional deficiencies in staple crops like rice, wheat, and maize pose critical challenges to global food security and public health. Addressing these issues requires a deep understanding of the genetic factors influencing nutrient composition and bioavailability. This chapter delves into forward genetic approaches aimed at identifying genes that control nutrition in staple crops. Traditionally, forward genetics has relied on mutagenesis and phenotypic screening, but recent advancements in molecular biology have enabled more precise and efficient gene

A. Hina (✉)
Ministry of Agriculture (MOA) Key Laboratory of Biology and Genetic Improvement of Soybean (General), National Centre for Soybean Improvement, State Key Laboratory for Crop Genetics and Germplasm Enhancement, Nanjing Agricultural University, Nanjing, China

R. Amin · T. Kabeer · M. F. Sarwar
College of Agricultural Biochemistry and Biotechnology, University of Agriculture, Faisalabad, Pakistan

A. Abbasi
Department of Entomology, University of Agriculture, Faisalabad, Pakistan

I. U. Haq
State Key Laboratory of Ecological Pest Control for Fujian and Taiwan Crops, Key Laboratory of Biopesticide and Chemical Biology, MOE, College of Plant Protection, Fujian Agriculture and Forestry University, Fuzhou, China

N. R. Abdelsalam
Agricultural Botany Department, Faculty of Agriculture (Saba Basha), Alexandria University, Alexandria, Egypt

H. M. Ali
Department of Botany and Microbiology, College of Science, King Saud University, Riyadh, Saudi Arabia

© The Author(s), under exclusive license to Springer Nature Singapore Pte Ltd. 2025

S. Faiz et al. (eds.), *Climate Smart Agriculture for Future Food Security*,
https://doi.org/10.1007/978-981-96-4499-5_18

identification. We discuss a range of strategies, including mapping populations, quantitative trait locus (QTL) analysis, and mutant libraries, that have been effective in uncovering genes associated with nutritional traits. Furthermore, the chapter explores the integration of high-throughput sequencing, CRISPR-Cas9, and other emerging technologies, which are revolutionizing gene discovery and functional validation in plants. These cutting-edge innovations offer new opportunities for accelerating the breeding of crops with improved nutritional profiles. By leveraging forward genetic approaches, researchers can make substantial progress in enhancing the nutritional quality of staple crops, contributing to better health outcomes and greater food security worldwide. This chapter provides a comprehensive overview of these methods, highlighting their potential to transform staple crop nutrition and address pressing global challenges.

Keywords

Forward genetics · Nutritional traits · Staple crops · Gene identification · QTL analysis · Mutagenesis · CRISPR-Cas9

18.1 Introduction

The enhancement of nutritional quality in staple crops is a critical goal in agricultural research, particularly in the context of global food security and the increasing demand for nutritionally rich diets. As the global population continues to rise, the pressure to develop crops that can meet both caloric and nutritional needs intensifies. One promising avenue in this pursuit is the application of forward genetic approaches to identify and manipulate genes that control key nutritional traits in staple crops. Forward genetics, which involves the identification of genes responsible for particular phenotypes by screening and analysing mutant populations, offers a robust strategy for enhancing crop nutrition. The process begins by identifying mutants with intriguing traits, aiming to uncover the functions of the genes responsible for the observed mutations. This method follows a "phenotype-to-gene" approach (Alonso and Ecke 2006). This approach has been instrumental in unravelling the complex genetic networks that govern traits such as nutrient content, bioavailability, and resistance to environmental stressors, which in turn influence the nutritional quality of crops (Das et al. 2024; Satyavathi et al. 2024).

The development of staple crops with improved nutritional profiles is particularly vital for addressing micronutrient deficiencies, commonly referred to as "hidden hunger," which affects millions of people worldwide. For instance, deficiencies in essential nutrients like iron, zinc, and vitamin A are prevalent in many developing countries, where diets are heavily reliant on staple crops such as rice, wheat, and maize (Reddy et al. 2024). Traditional breeding methods have had limited success in enhancing the nutritional content of these crops due to the complexity of the traits involved, which are often controlled by multiple genes and influenced by environmental factors. Forward genetic approaches, however, have enabled researchers to overcome these challenges by providing a more targeted and efficient means of identifying and manipulating the genetic determinants of nutrition (Bollam et al. 2024).

One of the most significant advancements in this field has been the integration of CRISPR/Cas genome editing technology, which has revolutionized the ability to precisely modify genes of interest in crops (Chaudhry et al. 2023; Abbasi et al. 2024a). This technology has been particularly effective in improving the nutritional quality of crops by enabling the fine-tuning of metabolic pathways involved in the synthesis and accumulation of essential nutrients (Freitas-Alves et al. 2024). For example, recent studies have demonstrated the successful application of CRISPR/Cas in soybean to enhance grain nutritional quality and improve resistance to biotic and abiotic stresses (Razzaq et al. 2023; Freitas-Alves et al. 2024; Hina et al. 2023). In addition to CRISPR/Cas, other forward genetic techniques, such as TILLING (Targeting Induced Local Lesions in Genomes) and EMS (Ethyl Methanesulfonate) mutagenesis, have also been employed to identify genes controlling nutritional traits in crops. These methods have been particularly useful in crops with complex genomes, where traditional breeding methods face significant challenges. For example, the application of TILLING in lentils has led to the discovery of novel alleles associated with enhanced nutritional content and improved resistance to biotic stresses (Das et al. 2024). Furthermore, the integration of forward genetics with advanced genomic tools, such as high-throughput sequencing and genome-wide association studies (GWAS), has accelerated the pace of gene discovery in crop research. These tools allow for the rapid identification of genetic loci associated with key nutritional traits, facilitating the development of biofortified crops through marker-assisted selection and genomic selection (Abbasi et al. 2024b; Habib et al. 2024; Weiland et al. 2024; Hina et al. 2023). For instance, the sequencing of the pearl millet genome has provided valuable insights into the genetic architecture of traits related to nutrient content, enabling the development of improved cultivars with higher nutritional value (Bollam et al. 2024). The combination of forward genetics and genomics has thus opened new avenues for crop improvement, paving the way for the creation of staple crops that are not only high-yielding but also nutritionally enhanced. This chapter explores the application of forward genetics in staple crops, with an emphasis on improving nutrition. It outlines various approaches and discusses emerging methods and technologies that hold potential for advancing both basic and applied plant sciences, ultimately enhancing the nutritional value of staple crops through forward genetics.

18.2 Forward Genetics

Natural mutations occur spontaneously and are the driving force behind genetic variation, which is crucial for biological diversity. Early humans unknowingly harnessed this genetic diversity when they began domesticating crops. However, it was Gregor Mendel in 1865 who laid the groundwork for genetics as a scientific discipline. His experiments with garden peas revealed heritable traits such as seed shape and colour, showing that these traits followed predictable inheritance patterns, even though the molecular mechanisms were not yet understood. Over time, scientists have delved deeper into the genetic basis of Mendel's observations. For example, the difference between round and wrinkled pea seeds was later found to result from a transposon insertion in a gene responsible for starch branching (Bhattacharyya et al. 1990). The difference between tall and short pea plants is believed to stem from a single nucleotide mutation that leads to an amino acid substitution (Ellis et al. 2011). Following the rediscovery of Mendel's laws in the early twentieth century, scientists quickly recognized the limitations imposed by the naturally low rate of spontaneous mutations. To address this, Hermann Muller made a ground-breaking discovery in 1927: by exposing fruit flies (*Drosophila melanogaster*) to ionizing radiation, he could significantly increase the occurrence of abnormal phenotypes (Muller 1927). This finding opened up new possibilities for genetic research by enhancing mutation rates artificially. At the same time, Stadler's experiments with inducing mutations in cereal crops (Stadler 1928) paved the way for mutation induction and breeding in agriculture. While Muller saw great potential in using induced mutations for plant breeding, Stadler was more cautious. He expressed concerns about the high occurrence of mutations that resulted in negative phenotypic traits, which made him more sceptical of the method's immediate value for improving crops (Sigurbjoernsson and Micke 1974). The introduction of methods to induce genetic variation opened up new possibilities for research. Stadler's findings, for instance, demonstrated that polyploid organisms, which have multiple sets of homologous genes, were more resistant to the damaging effects of ionizing radiation. This discovery highlighted the protective role of gene redundancy in polyploids, offering new insights into the impact of mutations on genetic stability (Stadler 1929). This phenomenon was later confirmed at the DNA level in chemically mutagenized wheat, where TILLING (Targeted Induced Local Lesions IN Genomes) assays revealed that polyploids are capable of accumulating a greater number of induced point mutations. This finding further underscored the ability of polyploids to tolerate genetic changes, providing a clearer understanding of mutation dynamics in polyploid species (Slade et al. 2005). The development of recombinant DNA and sequencing technologies in the 1970s allowed scientists to pinpoint induced genetic lesions responsible for heritable phenotypic variations (Sanger et al. 1977). These advancements provided all the tools necessary for what is now recognized as the classical forward genetics approach. As described by Zakhrabekovaa et al. (2013), forward genetics involves three key steps: obtaining mutants (either through natural variation or induced mutations), selecting the desired phenotype, and then identifying the gene responsible for that trait and determining its function (Fig. 18.1).

18.2.1 Comparison with Reverse Genetics

Forward genetics begins by generating mutations, either through natural processes or by using mutagens, and then selecting a phenotype for further study. The gene

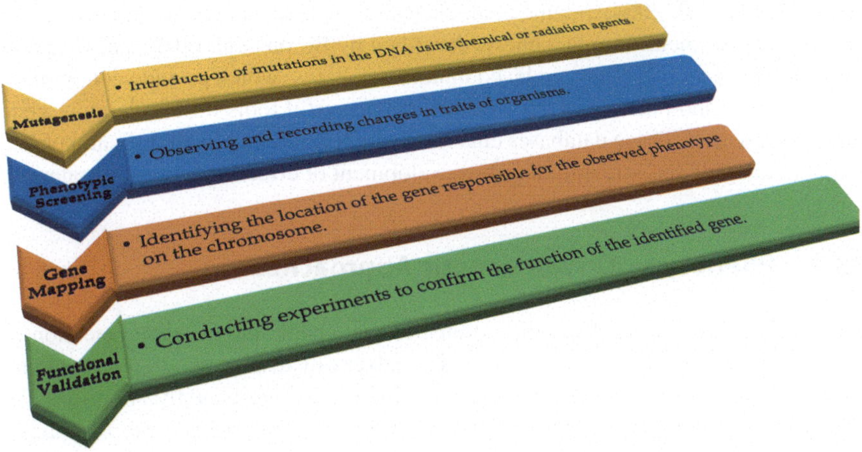

Fig. 18.1 Key Steps of Forward Genetics: Generate or identify mutants; Select and analyse phenotypes; Map and identify the responsible genes; and Determine gene functions

responsible is subsequently identified, and its function is clarified. In reverse genetics, however, the process starts with a known gene sequence, but its function remains unknown. To determine its role, researchers induce mutations or modify the gene's expression, observing the resulting phenotypic changes to uncover its function (Zakhrabekovaa et al. 2013). While the reverse genetics approach can provide valuable insights into the function of specific genes, it is constrained by assumptions about the expected phenotypic outcomes of the targeted genetic changes. In contrast, forward genetics starts with a biological trait or phenomenon of interest and seeks to identify the genes responsible for that trait. This method involves generating mutant lines that exhibit variations in phenotype, which are then analysed to uncover the underlying genetic causes. The underlying mutational causes can be identified through techniques such as mapping and positional cloning. Variations in phenotype expression resulting from mutations in different genes can reveal the significance of specific genes in the overall process of phenotypic expression (Lukowitz et al. 2000; Janny et al. 2003).

Combining forward and reverse genetics can leverage the strengths of both approaches while mitigating their individual limitations. A notable example of this integrated strategy is seen in TILLING projects. Perry and his colleagues at the John Innes Centre, for instance, utilized both approaches in Lotus japonicus, creating a phenotype-indexed population to enhance reverse genetic studies (Perry et al. 2003). In reverse genetic screens targeting genes involved in root symbiosis, researchers managed to identify crucial genic lesions from a focused subset of plants displaying specific root phenotypes. Rather than screening the entire mutant collection of 3697 plants, they concentrated on a smaller group of 275, resulting in a more than tenfold reduction in the effort required for mutation discovery. This efficient approach has also proven effective in other crops, such as barley, tomato, pea, Brachypodium, and flax, demonstrating its broad applicability in plant genetic research (Dalmais et al.

2008; Menda et al. 2004; Talame et al. 2008). Other reverse genetic initiatives, like the soybean fast neutron mutagenesis described by Bolon et al. (2011), also benefit from similar resources. Fully indexed mutant databases, which provide comprehensive genotype and phenotype information, are powerful tools for accelerating gene function research. These databases enable researchers to quickly identify and utilize advantageous alleles, streamlining the development of effective breeding strategies.

18.3 Phenotype-Driven Genetic Approaches

There are various forward genetics strategies that enable the efficient detection of multiple genes or alleles (Table 18.1). The advent of next-generation sequencing (NGS) along with molecular marker technologies has significantly simplified the localization of causative mutations. Forward genetics approaches can be categorized based on the type of individuals used in genome sequencing projects.

18.3.1 Map-Based/Positional Cloning

Initially, genes linked to mutational traits were uncovered through the map-based cloning method. To begin with forward genetics, researchers explore natural or insertional mutations present in inbred lines to identify traits of interest. After determining a desired trait, they cross these inbred lines with another line that showcases a contrasting trait, enabling a comprehensive analysis of genetic differences and their implications. To pinpoint the causal gene (mutant gene), genetic linkage analysis was employed, using numerous DNA markers and segregating populations. Despite its utility, this method has challenges, such as the need for an extensive count of DNA markers to refine the genomic region of interest, along with the requirement for a sizable mapping population (Jander et al. 2002). Using chromosome walking proved to be a powerful technique for navigating the areas surrounding a known target sequence. This method facilitated the identification, isolation, and cloning of sequences in proximity to the mapped gene, enhancing our understanding of genetic architecture (Bender et al. 1983). Nonetheless, its utility was hindered by the limited number of available markers, which regulated its broader use.

18.3.2 Sequencing-Based Mapping

With advancements in genomic sequencing methods, the development of NGS platforms enabled the generation of vast amounts of genomic data, facilitating the identification of causal mutations. As a result, the traditional map-based gene cloning method, which was slow and had limited throughput, was gradually replaced by the more efficient and scalable NGS technique (Schneeberger and Weigel 2011). The

Table 18.1 Enlisting different forward genetic approaches along with their principles, advantages, and limitations

Genetic Approach	Principle	Advantages	Limitations
Mutagenesis	Inducing random mutations using chemicals (e.g. EMS) or radiation (e.g. X-rays) to generate genetic diversity.	Creates a wide range of mutations; can be applied to any crop species.	Mutations are random; may require screening of large populations to identify desired traits.
TILLING (targeting induced local lesions IN genomes)	Combines mutagenesis with high-throughput screening to identify specific point mutations.	Enables identification of mutations in specific genes; non-GMO technique.	Requires prior knowledge of target gene sequences; labor-intensive screening process.
QTL mapping	Identifies genomic regions (quantitative trait loci) associated with specific traits by analysing the offspring of crosses between different parent lines.	Useful for studying complex traits controlled by multiple genes; can identify multiple QTLs.	Resolution is often low; requires large populations and detailed phenotyping.
Association mapping (GWAS)	Correlates genetic variation with phenotypic traits across a natural population to identify candidate genes.	High resolution; can identify genes associated with traits in diverse germplasm.	Requires large and diverse populations; can be influenced by population structure.
Insertional mutagenesis	Introduces mutations by inserting a known DNA sequence (e.g. T-DNA, transposons) into the genome, disrupting gene function.	Allows for easy identification of mutated genes; insertion sites are often known.	Limited to species where transformation is feasible; may cause unintended mutations elsewhere in the genome.
Bulked Segregant analysis (BSA)	Combines the DNA of individuals with extreme phenotypes to identify genetic markers linked to traits.	Quick and cost-effective; useful for traits with simple inheritance.	Less effective for traits controlled by multiple genes; requires clear phenotypic segregation.

introduction of NGS greatly accelerated the process of identifying and cloning Quantitative Trait Loci (QTL), eliminating the necessity for large pedigree crosses. Additionally, this strategy created an opportunity for isolating genes in non-model crops that lack extensive genetic data, which is why it has been referred to as next-generation mapping. This approach relies on aggregating DNA or RNA from individuals who exhibit extreme phenotypes. The bulked samples are then sequenced to uncover Single Nucleotide Polymorphisms (SNPs) associated with specific characteristics (Austin et al. 2011). Several sequence-based mapping techniques are outlined in the following section.

18.3.3 Bulk-Segregant Analysis (BSA)

Bulked Segregant Analysis (BSA), introduced by Michelmore in 1991 (Michelmore et al. 1991), is an efficient method for gene mapping. In BSA, two genetically distinct parental lines, such as one resistant to a disease and one susceptible, are crossed to produce a mapping population, such as an F2 generation. Plants within this population are phenotyped, and those exhibiting extreme traits such as high resistance or susceptibility are selected for further analysis (Majeed et al. 2022). To conduct the analysis, DNA is aggregated from the ten highest tolerant and ten highest vulnerable plants to establish two separate bulks. These bulks, along with DNA from both parent lines, are analysed using a wide array of genomic markers. Markers that show polymorphisms between the parent lines and the bulks are identified as being associated with the putative gene, such as disease resistance. Gene mapping initially utilized RFLP markers, but the introduction of NGS technologies greatly expedited the identification of target genes through the BSA method. Later, sequence-based mapping techniques expanded on the bulk sequencing concept, incorporating various modifications to improve the efficiency and accuracy of gene discovery (Schneeberger and Weigel 2011).

18.3.4 SHOREmap

SHOREmap (SHOrtREad map), introduced by Schneeberger et al. (2009), employs a combination of traditional genetic mapping and high-throughput sequencing to identify candidate genes in Arabidopsis thaliana. The process involves crossing a mutant line with a wild-type parent to create a mapping population, followed by deep sequencing of recombinants to detect the gene associated with the trait. To create an Illumina library, DNA from homozygous mutant lines and wild-type plants within the segregating F2 population is aggregated. Analysing the differences in allele frequencies between these pools helps identify genetic linkages related to specific traits. The identified mutations in targeted genomic regions are then compared with similar regions in a reference genome. This data is subsequently utilized with the SHOREmap 'annotate' function for further interpretation and analysis. This function aids in predicting the effects of substitution mutations by ranking these alterations based on their proximity to the peak of allele distribution. The gene associated with the phenotypic change is recognized as the most significant one arising from the divergent parent. However, sequencing of mutant genomes can be constrained by numerous background mutations, necessitating utilizing previous mapping intervals to refine the search for potential gene of interest (Schneeberger et al. 2009).

18.3.5 MutMap

MutMap, a modified version of SHOREmap introduced by Abe et al. (2012), is used to pinpoint causal genomic regions and SNPs in mutagenized rice that exhibit pastel green leaves and semi-dwarf characteristics. Using this technique, mutations are introduced into an inbred line through the use of chemical or physical mutagens to generate the necessary genetic variation. The mutant line is crossed with its parental counterpart to produce F1 offspring, which is then self-fertilized to create an F2 generation. Mutant plants are selected, and DNA is pooled in equal amounts from each plant. The pooled DNA is sequenced using NGS technologies with adequate coverage, and the resulting short reads are aligned to the reference parental genome to identify the locations of SNPs. To determine the SNP index, researchers count the short reads with parental and mutant SNP alleles. To determine the SNP index, the ratio of reads with the mutant allele to the total number of reads covering the SNP site is calculated. A mutation that is closely linked to the gene of interest is expected to show an SNP ratio approaching 1, whereas unlinked mutations are generally characterized by a ratio indicative of weaker association. An advanced iteration of MutMap, known as MutMap+, identifies causal mutations by analysing SNP frequencies in pooled DNA samples from both mutant and wild-type progeny of the M3 generation, derived from the self-fertilization of an M2 heterozygous plant. This method eliminates the need for crossing between mutant and wild-type lines, making it particularly useful for detecting mutations that cause early developmental lethality, sterility, or other issues that complicate crossing, as well as for crops that are difficult to cross artificially (Fekih et al. 2013). An enhanced variant of MutMap, MutMap-Gap employs whole-genome sequencing to identify causal mutations in genomic regions that are missing or exhibit gaps in the reference genome. By focusing on these underrepresented or missing areas, this method helps uncover genetic variations that may be overlooked by standard reference-based approaches, thus providing a more comprehensive understanding of genomic regions that are crucial for trait development (Takagi et al. 2013).

18.3.6 Bulk-Segregant RNA-Sequencing (BSR-Seq)

A variant of BSA, known as Bulk-Segregant RNA-Sequencing (BSR-Seq), was developed to map the glossy3 (gl3) gene, which controls the collection of epicuticular wax in maize seedlings. This method leverages RNA sequencing (RNA-Seq) data from segregating maize populations to identify novel SNPs and measure gene expression. BSR-Seq is particularly beneficial when polymorphic markers are absent in the target populations. Additionally, it allows for the examination of how the mutation influences overall gene expression patterns, providing a broader understanding of the mutation's impact (Liu et al. 2012). By analysing expression profiles within a specific genomic interval, researchers can identify potential candidate genes, particularly since mutant pools often display reduced expression levels. Furthermore, identifying polymorphic SNPs associated with the gene of interest aids in streamlining the gene cloning process. As with other methods, the size of the mutant and non-mutant pools, sequencing depth, along with polymorphism level within the population impact the extent of the mapping interval.

18.3.7 QTL-Sequencing (QTL-Seq)

QTL-sequencing (QTL-Seq), an extension of Bulked Segregant Analysis (BSA), leverages high-throughput sequencing technologies (Takagi et al. 2013). This method involves QTL mapping through whole-genome sequencing or skimGBS of pooled DNA samples from the most extreme 5–10% of progenies, which display contrasting traits for the characteristic of interest (Hina et al. 2020). The resulting sequence reads are aligned to the reference genome of the parents, and SNP-index plots for both the high and low bulks are compared. SNPs that exhibit contrasting patterns between these bulks are identified as the causal variants responsible for the traits. QTL-Seq was first applied in rice to map genes related to partial resistance against fungal rice blast disease and traits associated with the growth vigor during seedling stage. QTL-Seq offers an advantage compared to mutant-derived populations by containing a greater quantity of beneficial alleles, which have been acquired through evolutionary change, particularly for agronomically significant QTLs. In contrast, mutants typically possess a more limited set of beneficial alleles. Moreover, this method enables the study of evolutionary processes and selection sweeps, whether natural or artificial, by comparing bulked samples from regions with extreme climates or diverse geographical areas. In a similar way, QTL mapping is essential for pinpointing genes related to nutrition in staple crops. It identifies genetic regions linked to differences in nutrient content, aiding in the creation of crop varieties with improved and enhanced nutritional qualities (Table 18.2).

Table 18.2 Summarizing key QTL mapping studies related to nutritional traits in staple crops

Crop Species	Nutritional Trait	Number of QTLs Identified	Genetic Markers Used	References
Rice (Oryza sativa)	Iron and zinc content	12	SSR, SNP	Stangoulis et al. (2007)
Maize (Zea mays)	Provitamin A content	6	SSR, SNP	Yan et al. (2010)
Wheat (Triticum aestivum)	Grain protein content	9	RFLP, SSR	Prasad et al. (2003)
Barley (Hordeum vulgare)	Grain zinc content	5	AFLP, SSR	Lonergan et al. (2009)
Sorghum (Sorghum bicolor)	Grain iron and zinc content	8	SSR, RFLP	Rami et al. (1998)
Cassava (Manihot esculenta)	Carotenoid content	7	SSR	Esuma et al. (2016)
Pearl millet (Pennisetum glaucum)	Grain iron content	4	SSR, DArT	Kumar et al. (2016)

Innovative genome editing tools such as zinc finger nucleases, TALENs, and CRISPR/Cas9 can also be utilized as forward genetics tools to construct genome-wide libraries with precisely edited target sites. These libraries can be screened to discover and analyse new genetic variations related to traits of interest, including those linked to disease resistance, without relying solely on natural mutations (Mojica et al. 2000; Bibikova et al. 2002; DeFrancesco 2011; Shan et al. 2013). Among genome editing tools, CRISPR/Cas9 stands out as the simplest and most efficient. Its key advantage lies in its precision, whereas other forward genetics methods, such as those involving chemical or physical mutagens, are constrained by the randomness of induced mutations (Shalem et al. 2014; Chaudhry et al. 2023).

18.4 Nutritional Gene Profiling in Key Staple Crops

Nutritional gene profiling involves studying the genes of key staple crops to enhance their nutritional value (Table 18.3). By examining these crops' genetic information, scientists can identify genes that affect levels of essential nutrients like vitamins, minerals, and proteins. This process helps in developing crop varieties with improved nutritional content, addressing deficiencies and supporting better health. For example, profiling can increase the iron content in rice, boost protein levels in wheat, or enhance the vitamin content in maize (Table 18.4). These improvements contribute to better public health and food security. Examples of staple crops include wheat, rice, maize, legumes, and barley. Gene identification in these major staples is briefly discussed below.

18.4.1 Wheat (*Triticum aestivum*)

Wheat is a fundamental staple crop that accounts for approximately 20% of global dietary calories and protein intake, making it a cornerstone of food security worldwide (Nazir et al. 2021; Ahmad et al. 2023; Shaukat et al. 2024). Its importance is

Table 18.3 Summary of identified genes controlling nutritional traits in staple crops

Gene Name	Crop Species	Method used for Identification	Nutritional Trait Affected	References
OsNAS2	Rice (*Oryza sativa*)	QTL mapping	Iron and zinc content	Lee et al. (2009)
crtRB1	Maize (*Zea mays*)	Association mapping	Provitamin A content	Yan et al. (2010)
GPC-B1	Wheat (*Triticum aestivum*)	Positional cloning	Grain protein content	Uauy et al. (2006)
OsPT8	Rice (*Oryza sativa*)	TILLING	Phosphate and iron content	Wang et al. (2016)
HvNAM-1	Barley (*Hordeum vulgare*)	Gene expression analysis	Grain protein and micronutrient content	Distelfeld et al. (2008)

Table 18.4 Variation in specific nutritional traits among different mutant or natural populations of staple crops

Crop Species	Population Type	Nutritional Trait	Trait Variation	References
Rice (Oryza sativa)	Mutant (Golden Rice)	Vitamin A (β-carotene)	0.0–37 μg/g (β-carotene)	Paine et al. (2005)
Maize (Zea mays)	Natural (high Provitamin A)	Vitamin A (Provitamin A)	0.3–15 μg/g (Provitamin A)	Harjes et al. (2008)
Wheat (Triticum aestivum)	Mutant (high zinc)	Zinc content	20–65 ppm	Velu et al. (2012)
Cassava (Manihot esculenta)	Natural (high cyanogenic glycosides)	Iron bioavailability	Low (due to high phytate and cyanide levels)	Sayre et al. (2011)
Sorghum (Sorghum bicolor)	Mutant (low Phytate)	Iron bioavailability	10–50% increase compared to wild type	Krishnan et al. (2014)
Potato (Solanum tuberosum)	Natural (Andean varieties)	Vitamin C content	10–30 mg/100 g	Burgos et al. (2009)
Pearl millet (Pennisetum glaucum)	Natural (biofortified)	Iron content	47–77 ppm	Gupta et al. (2015)

particularly pronounced in regions such as North Africa, the Middle East, and Central Asia, where it serves as the primary food source for over 2.5 billion people (Curtis et al. 2002; FAO 2021). Given its critical role in human nutrition, enhancing the nutritional quality of wheat through genetic improvements is crucial for addressing malnutrition and improving public health globally.

Forward genetics has proven to be a powerful tool in identifying and manipulating genes that enhance wheat's nutritional profile. One significant breakthrough was the identification of the GPC-B1 gene by Uauy et al. (2006). This gene is linked to increased grain protein content by accelerating leaf senescence, which facilitates the remobilization of nitrogen from the leaves to the grain. The increased protein content is vital in regions where wheat is a primary source of protein, offering a promising avenue for developing wheat varieties with enhanced protein levels. Another important advancement came from the identification of NAM (NAC) genes by Distelfeld et al. (2008). These genes are crucial for nutrient remobilization during grain filling, specifically for zinc and iron. Enhancing the micronutrient content in wheat through the manipulation of NAM genes addresses deficiencies in populations that rely heavily on wheat as a staple food. This biofortification effort is significant for improving public health, particularly in developing countries where micronutrient deficiencies are prevalent. Research by Bhati et al. (2016) also contributed to wheat's nutritional improvement by focusing on reducing phytic acid content. Phytic acid is an anti-nutritional factor that binds essential minerals, reducing their bioavailability. By identifying mutants with lower phytic acid levels, this study has paved the way for developing wheat varieties that offer better mineral bioavailability, thereby improving the overall nutritional profile of wheat.

Additionally, forward genetics has been employed to increase amylose content in wheat, which enhances dietary fibre intake. Researchers targeted the GBSSI (Granule-Bound Starch Synthase I) gene using gamma radiation mutagenesis to generate wheat mutants with higher amylose content. These mutants demonstrated increased amylose levels and improved indices of large-bowel health in animal studies (Shimbata et al. 2005; Regina et al. 2006). Varieties such as "Amilo" wheat, developed with high-amylose traits, have shown promise in increasing dietary fibre and potentially reducing the risk of colon cancer (Chia and Morell 2009). However, high-amylose wheat varieties often face challenges in maintaining desirable bread-making quality. The increased amylose content can affect dough properties and bread texture, making it essential to balance nutritional improvements with other agronomic traits such as yield and disease resistance. Furthermore, advancements in increasing protein content have been achieved by identifying and manipulating genes involved in protein synthesis, such as the Gliadin and Glutenin gene families. These genes have led to the development of wheat varieties with improved protein quality and quantity, addressing the need for higher protein content in wheat (Shewry et al. 2003). The application of forward genetics in wheat breeding has led to significant advancements in improving the nutritional quality of this essential staple crop. By identifying and manipulating key genes related to protein content, micronutrient levels, and anti-nutritional factors, researchers are making strides towards better global health and food security. Continued research and development are crucial for overcoming challenges and ensuring that genetic improvements do not compromise other important agronomic traits.

18.4.2 Rice (*Oryza sativa*)

Rice, as a fundamental staple food for one-half of the total population, is integral to food security, particularly in Asia and Africa where it provides more than 20% of daily caloric intake (FAO 2019; IRRI 2020). Enhancing rice's nutritional profile is crucial for addressing micronutrient deficiencies and improving public health. Forward genetics has emerged as a powerful approach in this regard, enabling the development of rice varieties with improved nutrient content through the identification and manipulation of key genes.

One of the significant achievements in this field is the identification of the OsNAS3 (*Nicotianamine Synthase* 3) gene, which enhances the accumulation of iron and zinc in rice grains. Zheng et al. (2010) discovered that OsNAS3 plays a pivotal role in synthesizing nicotianamine, a chelator that facilitates the transport and accumulation of these essential micronutrients. Through forward genetic screens of mutagenized rice populations, researchers identified mutants with higher levels of iron and zinc, demonstrating the gene's potential in biofortification. Another notable advancement is the discovery of the OsNAS2 gene, also associated with increased iron and zinc content in rice. Lee et al. (2009) found that overexpressing OsNAS2 significantly boosts these micronutrients in polished rice grains without negatively impacting plant growth or yield. This finding holds promise for addressing iron and zinc deficiencies, which affect billions globally and contribute to health issues such as anaemia and weakened immune function. Forward genetics has also contributed to improving the protein content of rice. The OsGpc1 gene, a homolog of Gpc-B1 found in rice, was identified by Wei et al. (2013) through mutational analysis and mapping. Variations in OsGpc1 are associated with increased grain protein content, which is vital for populations dependent on rice as their primary protein source. Enhancing protein content in rice can support better health outcomes and development, especially in children.

Furthermore, advancements in increasing the vitamin E content of rice have been achieved through forward genetics. Chun et al. (2017) identified mutations in the OsVTE1 gene that led to elevated tocopherol (vitamin E) levels in rice grains. Vitamin E is an important antioxidant that helps protect cells from oxidative stress, offering additional health benefits. Despite these advancements, challenges persist, such as maintaining high yields while enhancing nutrient content and addressing consumer acceptance of genetically modified organisms (GMOs). Genetic modifications aimed at improving nutrient profiles which can sometimes result in yield penalties due to complex interactions between nutritional and agronomic traits (Trijatmiko et al. 2016). Additionally, consumer perceptions and preferences regarding GMOs can affect the adoption of biofortified rice varieties. Addressing these challenges through continued research and transparent communication is crucial for the successful integration of biofortified rice into global food systems.

18.4.3 Maize (*Zea mays*)

Maize, also known as corn, is a critical staple crop globally (Ahmed et al. 2022; Zia et al. 2023), providing approximately 19% of the world's daily calorie intake and serving as a primary source of food, feed, and industrial products in many countries, particularly in Sub-Saharan Africa and Latin America (FAO 2020; Shiferaw et al. 2011). Its significance in food security is immense, given its adaptability to diverse climates and its role as a major source of carbohydrates, vitamins, and minerals (Gillani et al. 2022; Haq et al. 2022). Improving the nutritional quality of maize through forward genetics is a key research focus, especially in addressing deficiencies in essential nutrients like vitamin A, protein, and essential amino acids, which are critical for populations relying heavily on maize as a dietary staple. One of the most significant achievements in this area is the development of Quality Protein Maize (QPM), which was accomplished through the identification and manipulation of the Opaque2 (O2) gene. This gene regulates the expression of lysine and tryptophan, two essential amino acids that are deficient in traditional maize varieties. By using forward genetics, researchers identified mutants with a non-functional O2 gene that led to an increase in lysine and tryptophan content, significantly improving the protein quality of maize. The QPM varieties have been widely adopted in regions where protein malnutrition is prevalent, providing a sustainable solution to enhance the nutritional value of maize (Prasanna et al. 2001).

Another important study identified the crtRB1 gene, which is associated with the biosynthesis of provitamin A carotenoids in maize. Using forward genetics and association mapping, Yan et al. (2010) identified favourable alleles of crtRB1 that significantly increase the levels of β-carotene, a precursor of vitamin A, in maize kernels. This breakthrough is particularly important in combating vitamin A deficiency, which is a leading cause of preventable blindness and immune deficiencies in children in developing countries. The biofortified maize varieties developed from this research, often referred to as "orange maize," have been successfully deployed in several African countries, improving the dietary intake of vitamin A. Moreover, research on the ZmMADS47 gene, which plays a role in regulating ear development, has shown potential for increasing the kernel number and size, thereby enhancing the overall yield and nutritional output of maize. Through forward genetics, Guo et al. (2018) identified natural variants of ZmMADS47 that could be used in breeding programs to develop maize varieties with improved yield and nutritional content, addressing both food security and nutritional needs. These advancements in forward genetics highlight the potential to improve the nutritional quality of maize, making it a more effective tool in the fight against malnutrition and food insecurity. By targeting key genes that influence essential nutrients and yield, researchers and breeders can develop maize varieties that better meet the dietary needs of populations worldwide.

18.4.4 Sorghum (*Sorghum bicolor*)

Sorghum (*Sorghum bicolor*) is a crucial cereal crop, particularly in arid and semi-arid regions of Africa and Asia, where its resilience to drought and heat makes it a significant source of calories, protein, and essential nutrients (Rooney and Serna-Saldivar 2000). Given its vital role in food security, enhancing sorghum's nutritional quality through genetic improvements is essential for addressing the dietary needs of populations that rely heavily on this crop. Forward genetics has played a pivotal role in improving sorghum's nutritional profile. One notable advancement is the identification of the SbNAC1 gene, which regulates drought tolerance. Using forward genetics, researchers screened mutant populations for traits associated with drought resistance and identified SbNAC1 as a key regulator of stress responses, significantly enhancing the plant's ability to withstand water deficits (Kadier et al. 2017). This improvement is critical for maintaining yield stability under challenging environmental conditions.

In terms of nutritional enhancement, significant progress has been made in increasing sorghum's iron and zinc content. The SbMT2 gene, a metallothionein gene, has been identified as a key player in biofortifying sorghum grains with these essential micronutrients. Kannan et al. (2020) demonstrated that overexpressing SbMT2 substantially improves iron and zinc accumulation, addressing deficiencies that are prevalent in sorghum-consuming regions. Protein quality and content in sorghum have also been targeted for improvement through forward genetics. The SbPPR gene, involved in protein synthesis and processing, was identified as a candidate for increasing protein content in sorghum grains. Reddy et al. (2019) used forward genetics to explore variations in SbPPR, leading to enhanced protein levels. This development is particularly important for populations that rely on sorghum as a primary protein source, thereby improving nutritional outcomes. Addressing anti-nutritional factors has been another focus of genetic research. The SbTLP gene, which influences tannin synthesis, has been modified to reduce tannin levels in sorghum. Tannins can inhibit the absorption of essential nutrients, and research by Boudet et al. (2015) showed that reducing tannin content through genetic modification improves nutrient bioavailability and overall nutritional quality.

These advancements highlight the potential of forward genetics to enhance the nutritional profile of sorghum, addressing key deficiencies in iron, zinc, protein, and reducing anti-nutritional factors. Such improvements are essential for bolstering food security and nutritional health in regions where sorghum is a staple crop.

18.4.5 Pearl Millet (*Pennisetum glaucum*)

Pearl millet (*Pennisetum glaucum*) is a crucial staple crop in arid and semi-arid regions of Africa and Asia, known for its resilience to harsh climates and its nutritional benefits. It provides an important source of dietary energy, protein, and essential minerals such as iron and zinc. Pearl millet is particularly valued for its drought tolerance, making it a key crop for ensuring food security in regions prone to water scarcity (Bidinger et al. 2009). Given its significance, improving the nutritional quality of pearl millet through forward genetics has become an important area of research, particularly to address micronutrient deficiencies and enhance its overall nutritional profile. One major advancement in this field is the identification of genes associated with iron and zinc biofortification. The PpMT2a gene, a metallothionein gene, was identified through forward genetics as playing a critical role in enhancing iron and zinc accumulation in pearl millet grains. Research by Pucella et al. (2019) demonstrated that overexpression of PpMT2a significantly increases the levels of these essential micronutrients in the grains, contributing to improved nutritional quality. This advancement is particularly valuable in regions where deficiencies in iron and zinc are prevalent, impacting millions of people with conditions such as anaemia and impaired immune function.

Another significant breakthrough is the identification of genes involved in enhancing protein content and quality. The PpDREB1 gene, which is involved in stress responses, has been linked to improved protein content in pearl millet under drought conditions. This gene was identified through forward genetics approaches that targeted stress-responsive pathways, demonstrating its role in increasing protein accumulation in response to environmental stressors (Kumar et al. 2017). Enhancing protein content is crucial for improving the nutritional value of pearl millet, especially in regions where it forms a major part of the diet.

Additionally, research has focused on identifying genes involved in increasing the digestibility and nutritional value of pearl millet. The PpR1 gene, associated with reducing anti-nutritional factors such as phytates, was identified as a key target for improving mineral bioavailability. Phytates can bind essential minerals, reducing their absorption in the human digestive system. Studies have shown that modifying the expression of PpR1 can lead to lower phytate levels and higher mineral availability in pearl millet grains (Singh et al. 2020). This improvement is crucial for addressing mineral deficiencies and enhancing the overall nutritional quality of pearl millet. These advances in forward genetics demonstrate the potential for developing pearl millet varieties with enhanced nutritional profiles, addressing key micronutrient deficiencies, and improving food security in regions where pearl millet is a staple crop.

18.4.6 Barley (*Hordeum vulgare*)

Barley is one of the oldest cultivated grains, playing a crucial role in food security and nutrition, particularly in arid and semi-arid regions. It is a staple crop in several countries and serves multiple purposes, including animal feed, malting for beer production, and direct human consumption. Barley provides around 10% of the world's cereal production and is especially important in regions where harsh climatic conditions limit the cultivation of other crops (FAO 2019). Given its adaptability and nutritional value, improving the nutritional quality of barley through forward genetics has been a significant focus in recent agricultural research. One major success in this area is the identification and utilization of the Hordein gene, which is responsible for the storage proteins in barley. These proteins are rich in essential amino acids such as lysine, which is typically low in cereal grains. Through forward genetics, researchers have identified mutations in the Hordein gene that result in increased lysine content, enhancing the nutritional profile of barley. The development of high-lysine barley varieties is particularly beneficial in regions where barley is a major dietary component, contributing to improved protein intake and overall nutrition (Shewry and Morell 2001). Another critical advancement is the identification of genes involved in beta-glucan synthesis, particularly the HvCslF gene family. Beta-glucans are soluble fibre that have been linked to various health benefits, including lowering blood cholesterol levels and improving glycaemic control. A study by Burton et al. (2006) utilized forward genetics to identify mutations in the HvCslF6 gene that significantly increase beta-glucan content in barley grains. This discovery has led to the development of barley varieties with enhanced health benefits, making barley an even more valuable crop for human consumption and addressing dietary fibre deficiencies. Moreover, forward genetics has facilitated the discovery of the Nud gene, which controls the hulled or hulless nature of barley. The hulless (or "naked") barley, identified through a mutation in the Nud gene, offers improved digestibility and nutritional value, particularly for human consumption. Hulless barley varieties have higher protein and starch content compared to hulled varieties, making them more suitable for direct human consumption and food processing (Taketa et al. 2008). These varieties are increasingly being adopted in regions where barley is consumed as a staple food, providing a more nutritious and versatile cereal option. These case studies illustrate the potential of forward genetics in improving the nutritional quality of barley. By targeting key genes associated with protein content, fibre composition, and grain structure, researchers can develop barley varieties that not only thrive in challenging environments but also offer enhanced nutritional benefits, contributing to better food security and public health.

18.4.7 Cassava (*Manihot esculenta*)

Cassava is a major source of carbohydrates in tropical regions, but its low nutritional density necessitates improvements. Forward genetics have been employed to enhance its nutritional profile. Research efforts have focused on increasing the content of essential vitamins and minerals, such as vitamin A and iron. For instance, the development of provitamin A-enriched cassava varieties involved forward genetic screens to identify genes affecting carotenoid biosynthesis. The incorporation of these genes into cassava varieties has resulted in improved vitamin A content (Gibson et al. 2016). Forward genetics has also been used to develop cassava varieties resistant to major pests and diseases, such as cassava mosaic disease. Identifying and introgressing resistance genes from wild relatives have been successful in creating resilient cassava varieties (Fregene et al. 2019). Enhancing nutritional content without negatively impacting storage root quality, such as texture and taste, is challenging. Combining improved nutrition with disease resistance is complex, as different traits may be influenced by various genetic factors.

18.5 Conclusion

In conclusion, forward genetic approaches have become a cornerstone in the field of crop genetics, offering transformative solutions for enhancing the nutritional quality of staple crops. These methods enable the precise identification and manipulation of genes that govern critical nutrients, such as proteins, vitamins, and minerals, leading to the development of crops with significantly improved nutritional profiles. This advancement is particularly important for addressing global malnutrition, as it enhances the quality and quantity of essential nutrients in staple foods like wheat, rice, maize, pearl millet, and sorghum. The integration of forward genetics with advanced genomic technologies, including high-throughput sequencing and CRISPR/Cas9 gene-editing, is set to further revolutionize crop improvement. These technologies provide more precise and efficient tools for targeting genes and understanding complex trait interactions, thus accelerating the development of crops with optimized nutritional attributes. Future research will likely focus on several key areas: enhancing the bioavailability of nutrients, developing crops that are resilient to climate change, and improving multifactorial traits that involve complex gene-environment interactions. Additionally, addressing challenges related to consumer acceptance and regulatory approval will be crucial for the widespread adoption of these genetically modified crops. As these advancements continue, forward genetics will play a pivotal role in developing nutritionally enriched crops that not only ensure food security but also contribute to better health outcomes for populations around the world.

Author Contribution Statement Aiman Hina led the conceptualization, coordination, and writing of the chapter. Raheela Amin, Tahira Kabeer, Asim Abbasi, Inzamam Ul Haq, Muhammad Farhan Sarwar, Nader R. Abdelsalam, and Hayssam M. Ali contributed to data collection, analysis, writing, and revisions, ensuring the chapter's overall quality and coherence.

References

Abbasi A, Hina A, Chaudhry A, Ishfaq T, Abbasi NA, Aslam A (2024a) CRISPR/Cas-mediated biotic stress resistance in cereals for achieving zero hunger. In: Fiaz S, Prakash CS (eds) OMICs based interventions for global food security. Wiley. https://doi.org/10.1002/9781394209156.ch10

Abbasi A, Waheeb SA, Dawood AI, Zafar S, Hina A, Zaman Q, Ali HM, Jaremko M, Abdelsalam NR (2024b) Biofortification of leafy vegetables: maximizing nutritional potential for Well-being of humans. In: Tiwari S, Singh B (eds) Harnessing crop biofortification for sustainable agriculture. Springer, Singapore. https://doi.org/10.1007/978-981-97-3438-2_14

Abe A, Kosugi S, Yoshida K, Natsume S, Takagi H, Kanzaki H, Matsumura H, Yoshida K, Mitsuoka C, Tamiru M, Innan H (2012) Genome sequencing reveals agronomically important loci in rice using MutMap. Nat Biotechnol 30:174–178

Ahmad Z, Waraich EA, Barutçular C, Hina A, Abbasi A, Mohiuddin M, Haq IU, Ayub MA, Sana S (2023) Climate resilience Technologies for Wheat Production. In: Hasanuzzaman M (ed) Climate-Resilient Agriculture, vol 2. Springer, Cham. https://doi.org/10.1007/978-3-031-37428-9_9

Ahmed W, Zaman Q, Mehmood F, Abbasi A, Ashraf K, Fatima K, Nazir S, Komal N, Waqas M (2022) The potential of commercial compost in reducing heavy metals and metalloids phytoaccumulation in maize. Pak J Bot 55:10.30848/PJB2023-2(25)

Alonso JM, Ecke JR (2006) Moving forward in reverse: genetic technologies to enable genome-wide phenomic screens in Arabidopsis. Nat Rev Genetics 9:38–45

Austin RS, Vidaurre D, Stamatiou G, Breit R, Provart NJ, Bonetta D, McCourt P (2011) Next-generation mapping of Arabidopsis genes. The Plant J 67:715–725

Bender W, Spierer P, Hogness DS, Chambon P (1983) Chromosomal walking and jumping to isolate DNA from the ace and rosy loci and the bithorax complex in Drosophila melanogaster. J Mol Biol 168:17–33

Bhati J, Sharma S, Rawat V, Meher PK, Prasad R (2016) Identification of mutants affecting Phytic acid content in wheat (Triticum aestivum L.) grains. Front. Plant Sci 7:1576

Bhattacharyya MK, Smith AM, Ellis TH (1990) The wrinkled-seed character of pea described by Mendel is caused by a transposon-like insertion in a gene encoding starch-branching enzyme. Cell 60(1):115–122

Bibikova M, Golic M, Golic KG, Carroll D (2002) Targeted chromosomal cleavage and mutagenesis in drosophila using zinc-finger nucleases. Genetics 161:1169–1175

Bidinger FR, French RJ, Crouch LR (2009) Pearl millet: its potential as a grain crop in the semi-arid regions of Africa. Field Crop Res 112:281–289

Bollam S, Deva PA, Sai Bindu KC, Varshney RK, Srivastava RK (2024) Pearl millet genome sequencing: utilization of sequencing information for pearl millet improvement. In: The pearl millet genome. Springer International Publishing, Cham, pp 57–68

Bolon YT, Haun WJ, Xu WW et al (2011) Phenotypic and genomic analyses of a fast neutron mutant population resource in soybean. Plant Physiol 156:240–253

Boudet AM, Penel C, Thouvenot C (2015) Role of SbTLP gene in reducing tannin content in sorghum. J Cereal Sci 65:13–21

Burgos G, Amoros W, Morote M, Stangoulis J, Bonierbale M (2009) Iron and zinc concentration of native Andean potato varieties from a human nutrition perspective. J Food Sci Technol 44:1231–1238

Burton RA, Jobling SA, Harvey AJ, Shirley NJ, Mather DE, Bacic A, Fincher GB (2006) The genetics and transcriptional profiles of the cellulose synthase-like HvCslF gene family in barley. Plant Physiol 140:599–610

Chaudhry A, Hassan AU, Khan SH, Abbasi A, Hina A, Mt K, Abdelsalam NR (2023) The changing landscape of agriculture: role of precision breeding in developing smart crops. Funct Integr Genom 23:167

Chia T, Morell M (2009) High-amylose wheat: its development and impact on nutrition. J Cereal Sci 49:341–348

Chun L, Kim Y, Lee H (2017) Identification of novel alleles enhancing vitamin E content in Rice (Oryza sativa L.) grains through forward genetics. J Agri Food Chem 65:2506–2512

Curtis BC, Rajaram S, Gómez Macpherson H (2002) Bread wheat: improvement and production. Food and Agriculture Organization of the United Nations (FAO). Retrieved from http://www.fao.org/3/y4011e/y4011e00.htm

Dalmais M, Schmidt J, Le Signor C et al (2008) UTILLdb, a Pisum sativum in silico forward and reverse genetics tool. Genome Biol 9:R43

Das A, Murmu M, Barman M, Roy S, Dash SSS (2024) Genomics-enabled breeding for Manoeuvring biotic stresses in lentil. Springer

DeFrancesco L (2011) Move over ZFNs: a new technology for genome editing may put the zinc finger nuclease franchise out of business, some believe. Not so fast, say the finger people. Nat Biotechnol 29:681–685

Distelfeld A, Korol A, Dubcovsky J, Uauy C, Blake T, Fahima T (2008) Collinearity between the barley grain protein content (GPC) QTL on chromosome arm 6HS and the wheat Gpc-B1 region. Mol Breed 22:25–38

Ellis TH, Hofer JM, Timmerman-Vaughan GM (2011) Mendel, 150 years on. Trends Plant Sci 16(11):590–596

Esuma W, Kawuki RS, Herselman L, Labuschagne M, Rabbi I (2016) Quantitative trait loci mapping of provitamin a carotenoid content in cassava roots. BMC Plant Biol 16:219

FAO (2019) Barley production worldwide 2018. Food and Agriculture Organization of the United Nations. Retrieved from http://www.fao.org/faostat/en/#data/QC

FAO (2020) Maize in human nutrition. Food and Agriculture Organization of the United Nations. Retrieved from http://www.fao.org/3/T0395E/T0395E00.htm

FAO (2021) The state of food security and nutrition in the World 2021. Food and Agriculture Organization of the United Nations. Retrieved from https://www.fao.org/publications

Fekih R, Takagi H, Tamiru M, Abe A, Natsume S, Yaegashi H, Sharma S, Sharma S, Kanzaki H, Matsumura H, Saitoh H (2013) MutMap+: genetic mapping and mutant identification without crossing in rice. PLoS One 8:7

Fregene M, Ospina Y, El-Sharkawy M (2019) Development of cassava with improved resistance to cassava mosaic disease. Plant Dis 103:1224–1232

Freitas-Alves NS, Moreira-Pinto CE, Távora FTKP, Paes-de-Melo B, Arraes FBM, Lourenço-Tessutti IT, Moura SM, Oliveira AC, Morgante CV, Qi Y, Grossi-de-Sa MF (2024) CRISPR/Cas genome editing in soybean: challenges and new insights to overcome existing bottlenecks. J Adv Res

Gibson RS, Kesson R, Hotz C (2016) Biofortification of cassava for improved micronutrient nutrition. J Nutr 146:358–365

Gillani SFA, Rasheed A, Abbasi A, Majeed Y, Abbas M, Hassan MU, Qari SH, Amin NBN, Kashgry TA, Tahir MM, Yunling P (2022) Comparative gene enrichment analysis for drought tolerance in contrasting maize genotype. Genes 14:31

Guo T, Mu Q, Wang J, Vanous AE, Onogi A, Iwata H, Holland JB (2018) Natural variation at ZmMADS47 regulates ear development in maize. Nat Genetics 50:1279–1284

Gupta SK, Velu G, Rai KN, Sumalini K (2015) Biofortification of pearl millet for increasing levels of grain iron and zinc. Cereal Chem 92:343–348

Habib M, Hina A, Sajjad M, Ahmad MQ, Azhar MT, Qayyum A (2024) Marker assisted selection in wheat. In: Zencirci N, Altay F, Baloch FS, Nadeem MA, Ludidi N (eds) Advances in wheat breeding. Springer, Singapore. https://doi.org/10.1007/978-981-999478-6_15

Haq IU, Idrees A, Abbasi A, Ali S, Asad M, Li C, Liu CZ, Zhang KX, Yasin M, Asghar MA, Iqbal J (2022) Silicon accumulation in maize and its effects on demographical traits of fall armyworm, [*Spodoptera frugiperda* (J. E. Smith)]. Silicon. https://doi.org/10.1007/s12633-022-02250-0

Harjes CE, Rocheford TR, Bai L, Brutnell TP, Kandianis CB, Sowinski SG, Buckler ES (2008) Natural genetic variation in lycopene epsilon cyclase tapped for maize biofortification. Science 319:330–333

Hina A, Cao Y, Song S, Li S, Sharmin RA, Elattar MA, Bhat JA, Zhao T (2020) High-resolution mapping in two RIL populations refines major "QTL hotspot" regions for seed size and shape in soybean (Glycine max L.). Int J Mol Sci 21(3):1040

Hina A, Abbasi A, Imtiaz S, Ul Haq I, Ul Hassan A, Chaudhry A (2023) General aspects of genetic improvement (traditional and transgenic methods) aiming to food biofortification. Nadeem, M.A., et al. Legumes Biofortification. Springer, Cham, In. https://doi.org/10.1007/978-3-031-33957-8_4

International Rice Research Institute (IRRI) (2020) Rice and the global food security. Retrieved from https://www.irri.org/rice-and-global-food-security

Jander G, Norris SR, Rounsley SD, Bush DF, Levin IM, Last RL (2002) Arabidopsis map-based cloning in the post-genome era. Plant Physiol 129:440–450

Janny L, Peters CF, Gerats T (2003) Forward genetics and map-based cloning approaches. Trends Plant Sci 8:484–491

Kadier Y, Zu YY, Dai QM, Song G, Lin SW, Sun QP, Pan JB, Lu M (2017) Genome-wide identification, classification and expression analysis of NAC family of genes in sorghum [Sorghum bicolor (L.) Moench]. Plant Growth Regul 83:301–312

Kannan MS, Ram M, Smith R (2020) Identification and functional characterization of the SbMT2 gene for enhanced iron and zinc biofortification in sorghum. Plant Sci 292:110384

Krishnan HB, Kim WS, Oehrle NW, Baxter I (2014) Iron bioavailability in sorghum is impacted by the phytate content of the grain and is affected by biofortification. J Agri Food Chem 62:6017–6023

Kumar S, Hash CT, Thirunavukkarasu N, Singh G, Rajaram V, Rathore A, Yadav RS (2016) Mapping QTLs for heat tolerance in pearl millet. Theor Appl Gene 129:647–659

Kumar S, Kumar V, Rai P (2017) Identification and functional characterization of PpDREB1 gene for enhancing protein content in pearl millet under drought stress. J Plant Physiol 217:1–11

Lee S, Jeon US, Lee SJ, Kim YK, Persson DP, Husted S, Schjoerring JK, Kakei Y, Masuda H, Nishizawa NK, An G (2009) Iron fortification of rice seeds through activation of the nicotianamine synthase gene. Proc Nat Acad Sci 106:22014–22019

Liu S, Yeh CT, Tang HM, Nettleton D, Schnable PS (2012) Gene mapping via bulked segregant RNA-Seq (BSR-Seq). PLoS One 7:e36406

Lonergan PF, Pallotta MA, Lorimer M, Paull JG, Barker SJ, Graham RD, Rathjen AJ (2009) Mapping quantitative trait loci associated with barley zinc accumulation. Theor Appl Gen 118:401–410

Lukowitz W, Gillmor CS, Scheible W (2000) Positional cloning in Arabidopsis. Why it feels good to have a genome initiative working for you. Plant Physiol 123:795–805

Majeed A, Johar P, Raina A, Salgotra RK, Feng X, Bhat JA (2022) Harnessing the potential of bulk segregant analysis sequencing and its related approaches in crop breeding. Front Genet 13:944501

Menda N, Semel Y, Peled D (2004) In silico screening of a saturated mutation library of tomato. Plant J 38:861–872

Michelmore RW, Paran I, Kesseli RV (1991) Identification of markers linked to diseaseresistance genes by bulked segregant analysis: a rapid method to detect markers in specific genomic regions by using segregating populations. Proc Nat Acad Sci USA 88:9828–9832

Mojica FJ, Díez-Villaseñor C, Soria E, Juez G (2000) Biological significance of a family of regularly spaced repeats in the genomes of Archaea, Bacteria and mitochondria. Mol Microbiol 36:244–246

Muller HJ (1927) Artificial transmutation of the gene. Science 66(1699):84–87. https://doi.org/10.1126/science.66.1699.84

Nazir S, Zaman Q, Abbasi A, Komal N, Riaz U, Ashraf K, Ahmad N, Agarwal S, Nasir R, Chen Y (2021) Bioresource nutrient recycling in the rice–wheat cropping system: cornerstone of organic agriculture. Plan Theory 10:2323

Paine JA, Shipton CA, Chaggar S, Howells RM, Kennedy MJ, Vernon G, Drake R (2005) Improving the nutritional value of Golden Rice through increased pro-vitamin a content. Nat Biotechnol 23:482–487

Perry JA, Wang TL, Welham TJ et al (2003) A TILLING reverse genetics tool and a web-accessible collection of mutants of the legume Lotus japonicus. Plant Physiol 131:866–871

Prasad M, Varshney RK, Roy JK, Balyan HS, Gupta PK (2003) The use of microsatellites for detecting DNA polymorphism, genotype identification and genetic diversity in wheat. Theor Appl Gene 106:669–677

Prasanna BM, Vasal SK, Kassahun B, Singh NN (2001) Quality protein maize. Curr Sci 81:1308–1319. Retrieved from https://www.jstor.org/stable/24106290

Pucella J, Kuang L, Li H (2019) Identification of PpMT2a gene for iron and zinc biofortification in pearl millet. Plant Sci 279:1–8

Rami JF, Dufour P, Trouche G, Fliedel G, Mestres C, Davrieux F, Glaszmann JC (1998) Quantitative trait loci for grain quality in sorghum (Sorghum bicolor L. Moench): towards the identification of the underlying genes. Theor Appl Gene 96:833–840

Razzaq MK, Hina A, Abbasi A, Karikari B, Ashraf HJ, Mohiuddin M, Maqsood S, Maqsood A, Haq IU, Xing G, Raza G, Bhat JA (2023) Molecular and genetic insights into secondary metabolic regulation underlying insect-pest resistance in legumes. Funct Integr Genom 23:217

Reddy BK, Vijayreddy D et al (2024) A comprehensive review on biofortification in vegetable crops. J Adv Biotechnol 27(8):1448–1458

Reddy BV, Nagaraj K, Srinivas T (2019) Genetic improvement of protein content in sorghum through forward genetics targeting the SbPPR gene. Mol Breed 39:1–14

Regina A, Bird A, Topping D, Bowden S, Freeman J, Barsby T, Morell M (2006) High-amylose wheat generated by RNA interference improves indices of large-bowel health in rats. Proc Nat Acad Sci 103:3546–3551

Rooney LW, Serna-Saldivar SO (2000) Sorghum and Millets. In: Cereal grain quality. CRC Press, pp 181–220. Retrieved from https://www.crcpress.com/Sorghum-and-Millets/Rooney-Serna-Saldivar/p/book/9780849333687

Sanger F, Nicklen S, Coulson AR (1977) DNA sequencing with chain-terminating inhibitors. Proc Natl Acad Sci 74(12):5463–5467

Satyavathi CT, Mukesh Sankar S, Singh SP (2024) Breeding climate resilient pearl millet cultivars for India. Springer

Sayre R, Beeching JR, Cahoon EB, Egesi C, Fauquet C, Fellman J, Zhang P (2011) The BioCassava plus program: biofortification of cassava for sub-Saharan Africa. Ann Rev Plant Biol 62:251–272

Schneeberger K, Weigel D (2011) Fast-forward genetics enabled by new sequencing technologies. Tre Plant Sci 16:282–288

Schneeberger K, Ossowski S, Lanz C, Juul T, Petersen AH, Nielsen KL, Jørgensen JE, Weigel D, Andersen SU (2009) SHOREmap: simultaneous mapping and mutation identification by deep sequencing. Nat Methods 6:550–551

Shalem O, Sanjana NE, Hartenian E, Shi X, Scott DA, Mikkelsen TS, Heckl D, Ebert BL, Root DE, Doench JG, Zhang F (2014) Genome-scale CRISPR-Cas9 knockout screening in human cells. Science 343:84–87

Shan Q, Wang Y, Li J, Zhang Y, Chen K, Liang Z (2013) Targeted genome modification of crop plants using a CRISPR-Cas system. Nat Biotechnol 31:686–688

Shaukat M, Abbasi A, Ramzan K, Hina A, Memon SQ, Maqsood Z, Gaafar AZ, Hodhod MS, Lamlom SF (2024) Ameliorating heat stressed conditions in wheat by altering its physiological and phenotypic traits associated with varying nitrogen levels. Not Bot Horti Agrobot Cluj-Napoca 52:13471

Shewry PR, Morell MK (2001) Manipulating the structure and function of cereal seed storage proteins. J Cereal Sci 34:139–146

Shewry PR, Halford NG, Tatham AS (2003) Wheat gluten. J Exp Bot 54:1237–1257

Shiferaw B, Prasanna BM, Hellin J, Bänziger M (2011) Crops that feed the world 6. Past successes and future challenges to the role played by maize in global food security. Food Secur 3:307–327

Shimbata T, Ohm JB, Yoshida M (2005) Effects of high-amylose wheat on bread quality and digestibility. Cereal Chem 82:552–557

Sigurbjoernsson B, Micke A (1974) Philosophy and accomplishments of mutation breeding. In: Polyploidy and induced mutations in plant breeding. Proceedings of two meetings joint FAO/IAEA Division and EUCARPIA, Bari 1972. International Atomic Energy Agency, Vienna, pp 303–343

Singh S, Dubey RS, Lal SK (2020) Genetic improvement of pearl millet for reduced phytate content and improved mineral bioavailability. Front Plant Sci 11:123

Slade AJ, Fuerstenberg SI, Loeffler D (2005) A reverse genetic, nontransgenic approach to wheat crop improvement by TILLING. Nat Biotechnol 23(1):75–81

Stadler LJ (1928) Mutations in barley induced by x-rays and radium. Science 68(1756):186–187. https://doi.org/10.1126/science.68.1756.186

Stadler LJ (1929) Chromosome number and the mutation rate in *Avena* and *Triticum*. Proc Natl Acad Sci 15(12):876–881

Stangoulis JCR, Huynh BL, Welch RM, Choi EY, Graham RD (2007) Quantitative trait loci for phytate in rice grain and their relationship with grain micronutrient content. Euphytica 154:289–294

Takagi H, Uemura A, Yaegashi H, Tamiru M, Abe A, Mitsuoka C, Utsushi H, Natsume S, Kanzaki H, Matsumura H, Saitoh H (2013) Mut Map-Gap: whole-genome resequencing of mutant F 2 progeny bulk combined with de novo assembly of gap regions identifies the rice blast resistance gene Pii. New Phytol 200:276–283

Taketa S, Kikuchi S, Awayama T, Yamamoto S, Ichii M, Kawasaki S (2008) Nud encodes a barley ER transmembrane protein that is necessary for lipid biosynthesis and transfer from the testa to the pericarp. Plant Cell 20:1833–1848

Talame V, Bovina R, Sanguineti MC et al (2008) TILLMore, a resource for the discovery of chemically induced mutants in barley. Plant Biotechnol J 6(5):477–485

Trijatmiko KR, Dueñas C, Tsakirpaloglou N, Torrizo L, Arines FM, Adeva C, Slamet-Loedin IH (2016) Biofortified indica rice attains iron and zinc nutrition dietary targets in the field. Nat Commun 7:1–12

Uauy C, Distelfeld A, Fahima T, Blechl A, Dubcovsky J (2006) A NAC gene regulating senescence improves grain protein, zinc, and iron content in wheat. Science 314:1298–1301

Velu G, Ortiz-Monasterio I, Cakmak I, Hao Y, Singh RP (2012) Biofortification strategies to increase grain zinc and iron concentrations in wheat. J Cereal Sci 56:127–135

Wang F, Deng M, Xu J, Zhu X, Zheng T, Zhang W (2016) TILLING by sequencing: a successful approach for identifying mutant alleles in rice. Plant Biotechnol J 14:1865–1875

Wei Y, Shohag MJI, Yang X, Yibin Z (2013) Genetic mapping and characterization of a major gene for high grain protein content in Rice. J Integ Agri 12:964–971

Weiland C, Grieb J, Bauer D, Chala D (2024) Dataspace integration for agrobiodiversity digital twins with RO-crate. Biodiv Info Sci Stand 8:e134479. https://doi.org/10.3897/biss.8.134479

Yan J, Kandianis CB, Harjes CE, Bai L, Kim EH, Yang X, Skinner DJ, Fu Z, Mitchell S, Li Q, Fernandez MG (2010) Rare genetic variation at *Zea mays* crtRB1 increases provitamin A and is under selection in Africa. Nat Gene 42:322–327

Zakhrabekovaa SM, Gougha S, Lundhb L, Hansson M (2013) Functional genomics and forward and reverse genetics approaches for identification of important QTLs in plants. Proc Azerbaijan Natl Acad Sci 68:23–28

Zheng L, Cheng Z, Ai C, Jiang X, Bei X, Zheng Y, Glahn RP, Welch RM, Miller DD, Lei XG, Shou H (2010) Nicotianamine, a novel enhancer of rice iron bioavailability to humans. PLoS One 5:e10190

Zia UU, Niazi AR, Ahmad Z, Alharby HF, Waraich EA, Abbasi A, Iqbal MA, Ahmed S, Hina S (2023) Dose optimization of silicon for boosting Arbuscular Mycorrhizal fungi colonization and cadmium stress mitigation in maize (*Zea mays* L.). Environ Sci Pollut Res 30:67071. https://doi.org/10.1007/s11356-023-26902-9

Genetically Modified Crops Combating Climate Change and Environmental Protection

19

Yunus Emre Arvas

Abstract

Genetically modified crops play a significant role in combating climate change by providing resistance to biotic and abiotic stress factors. These crops possess traits such as herbicide tolerance, antibiotic resistance, insect resistance, and drought stress tolerance. These features have increased the productivity of many agricultural products, such as alfalfa, canola, cotton, corn, potatoes, rice, and soybeans, which are commercially used. Climate change, caused by the increase in greenhouse gases, leads to temperature changes worldwide and negatively affects agricultural production. Food security is threatened by increased bacterial production at high temperatures and irregular rainfall patterns due to climate change. In this situation, alternative solutions in agricultural activities should be sought. Genetically modified crops stand out as an effective tool for adapting to climate change and protecting the environment due to their resistance to environmental stress factors. Many countries, such as the United States, Argentina, Brazil, and Canada, permit the commercial cultivation of genetically modified crop varieties and emphasize that the approval for the release of these crops should not be delayed. Although some countries, like those in the European Union, still follow these concerns seriously and have banned the cultivation and human consumption of these products, they allow limited access for use as animal feed. Undoubtedly, with the use of these technologies, agricultural productivity and food security can be enhanced. However, socioeconomic impacts and issues such as food supply security should also be given necessary importance,

Y. E. Arvas (✉)
Faculty of Science, Department of Molecular Biology and Genetics, Van Yüzüncü Yıl University, Tusba, Van, Turkey

Faculty of Science, Department of Biology, Karadeniz Technical University, Ortahisar-Trabzon, Turkey
e-mail: yunusemrearvas@yyu.edu.tr

© The Author(s), under exclusive license to Springer Nature Singapore Pte Ltd. 2025

S. Faiz et al. (eds.), *Climate Smart Agriculture for Future Food Security*,
https://doi.org/10.1007/978-981-96-4499-5_19

and legal guarantees should be provided. The acceptance and widespread adoption of genetically modified crops, with their benefits and potential risks, is now an inevitable fact as they are the strongest candidates to address the global food problems caused by climate change. Moreover, as the impact of climate change becomes more pronounced with each passing day, the adoption of new technologies in agricultural production and the promotion of environmentally friendly, economical, accessible, and soil-appropriate alternative solutions should be encouraged.

Keywords

Food security · Global warming · Plant production · Transgenic crops

19.1 Introduction

Global warming is the increase in Earth's surface temperature due to the greenhouse effect caused by gases released into the atmosphere. This phenomenon has become more detectable and significant over the last 50 years (Singh 2024). The excessive exploitation of natural resources, leading to the disruption of natural balances, has caused major problems. Processes such as hunger, thirst, extinction of species, destruction of vegetation and soil, climate change, thinning and depletion of the ozone layer, and environmental pollution are the main issues triggering global warming (Ruiz Serrano et al. 2024).

Crop production in the world and in Turkey is directly or indirectly affected by global warming, and it cannot be independent of climate events such as temperature, drought, increasing salinity, rainfall, sunlight, frost, wind, and humidity (Hasan et al. 2010; Chege et al. 2024). As a result, it is thought that erosion, diseases, and pests will increase, agricultural difficulties will arise, and consequently, there will be declines in crop yield and quality. In addition, with the continuous increase in the world population, efforts are ongoing to ensure that agricultural activities reach a level capable of feeding the population to meet the growing food demand (Hekimoğlu and ALTINDEĞER 2008; Popp et al. 2013; Tıraşcı and Erdoğan 2021).

By 2050, UN predicts that the global population will increase from 7 billion to >9 billion (Division 2015; Varzakas and Smaoui 2024). According to these data, a rapid population growth will be observed globally. Rapid population growth will bring about several problems. While it is known that the most important of these problems is providing sufficient, balanced, and healthy nutrition for the increasing population, another issue is the damage to the natural environment resulting from increased economic activities (Junaid and Gokce 2024). To address the food needs of the ever-growing world population, it is crucial to increase the amount of product obtained per unit area rather than expanding the already maximized cultivable land. Therefore, significant successes have been achieved in the quality and quantity of plant products through classical breeding methods applied since the Green Revolution. However, results with these breeding methods are obtained after many years of research. To improve and increase the effectiveness of classical breeding methods used in the production of high-value agricultural products, studies in the fields of molecular biology and genetics have given rise to the concept of "Plant Biotechnology" worldwide (Aziz and Masmoudi 2024).

The increase in studies in this field and the integration of some biotechnological methods with breeding studies have enabled the completion of research in a much shorter time. However, in the following years, the long-term persistence of chemical pesticides used against plant diseases and pests in the soil and the food chain has become an increasing concern for animal and human health as well as the environment (Zhou et al. 2024). Therefore, studies on gene transfer to plants using modern biotechnology methods began, making it possible to transfer genes that could not be transferred using classical breeding methods. Agricultural biotechnology and genetically modified organisms (GMOs) have provided significant contributions to solving problems in agricultural products that are economically important but cannot be resolved using classical breeding methods (Singh et al. 2006; Estrada et al. 2017; Bakhsh et al. 2021). The benefits of GMOs include the enrichment of food components, reduction of adverse effects such as allergies caused by foods, reduction in the use of herbicides and pesticides, increased yield per unit area, and extended shelf life (Kramkowska et al. 2013; Gaharwar et al. 2021; Ghimire et al. 2023). Potential harms of GMOs include the possibility of antibiotic resistance genes spreading to nature, increased risk of allergic reactions, loss of genetic diversity in natural species, disruption of the natural balance due to the release of genetically modified microorganisms into the environment, and the risk of undesirable resistance in virus populations resulting from the transfer of resistance genes derived from viruses to other viruses (Hug 2008; Azadi and Ho 2010; Tsatsakis et al. 2017).

The risks mentioned above have raised concerns about the gradual loss of existing biological diversity. To prevent the negative impacts of biotechnological applications and to ensure the continuous preservation of biological diversity, the "United Nations Cartagena Protocol on Biosafety" was established (Caradus 2023). This protocol has been in effect since 2003 and serves as a binding legal document. It was prepared to prevent potential risks to the environment and human health caused by GMOs, covering a broad range from the research stage to the release and transit of genetically modified organisms. This protocol aims to ensure the permanent protection of biological diversity, taking into account the possible adverse effects on human health, and it applies to the import, export, and use of all genetically modified living organisms (Kıvılcım 2012).

In light of this information, this section emphasizes the global status of genetically modified crops, the impacts of climate change on plant production and consumption, and the potential effects of genetically modified crops in combating climate change and protecting the environment.

19.2 Climate Change

Climate is defined as the average state of all weather conditions experienced or observed in any part of the Earth's surface over long periods, or more systematically, as the synthesis defined by the long-term statistics of the variability and average values of weather conditions and atmospheric elements in a specific area (Demir et al. 2008). Human activities, such as the use of fossil fuels, changes in land use, deforestation, and industrial processes, have led to a rapid increase in the accumulation of greenhouse gases in the atmosphere, strengthening the natural greenhouse effect and causing changes in the Earth's average surface temperatures, which in turn lead to climatic changes (Singh and Kumar 2023; Bajoria et al. 2024). Therefore, according to the United Nations Framework Convention on Climate Change, climate change is defined as a change resulting from human activities that directly or indirectly alter the composition of the global atmosphere, in addition to the natural climate variability observed over comparable time periods (United Nations 2023; Usman et al. 2024). Global warming, on the other hand, is defined as the increase in Earth's surface temperature due to the greenhouse effect caused by gases. It is thus understood that global warming and climate change do not refer to the same phenomenon (Kweku et al. 2018).

Global warming and climate change are different but related concepts. Global warming describes the recent increase in the Earth's temperature, which is only one aspect of climate change. Global warming is a cause of climate change. Climate change refers to the many different impacts of global warming on the Earth's climate system. These impacts include rising sea levels, melting glaciers, changing precipitation patterns, changing frequencies of extreme weather events (such as flash floods and heatwaves), changing lengths of seasons, and changing crop yields. The increase in average temperature due to climate change will cause increased drought and reduced agricultural production in some regions (Soutter and Mõttus 2020; Gulev et al. 2021; Ahmed et al. 2023).

The food sector, considered one of the strategic sectors of the twenty-first century, faces a significant threat in feeding the world population, which is projected to reach 10 billion by 2050 (Aryal et al. 2022). Research indicates that agricultural and food production must increase by at least 50% from current levels to feed the global population in 2050 (Ritchie and Roser 2024). The Intergovernmental Panel on Climate Change (IPCC) Assessment Reports predict that, due to the impacts of climate change, water shortages and droughts will increase worldwide, agricultural productivity will decline, and global food prices could rise by up to 85% (Kogan 2023). It is known that some agricultural areas are becoming arid due to changing precipitation patterns, the maturation periods of agricultural products are changing, some agricultural lands are becoming unusable due to flooding, or becoming salinized due to rising sea levels, resulting in decreased yields (Guimond and Michael 2021). Bacterial production in food will increase at high temperatures. Climate change also threatens food security. Therefore, there must be ongoing efforts to find alternative solutions in agricultural activities to meet the nutritional needs of the population (Kogan 2023; Mottaleb et al. 2023).

One of these efforts is undoubtedly genetically modified crops. Today, genetically modified products can be developed with various traits to combat biotic and abiotic stress factors such as 2,4-D herbicide tolerance, antibiotic resistance, Lepidopteran, Hemipteran, and Coleopteran insect resistance, glyphosate, glufosinate, dicamba herbicide tolerance, drought stress tolerance, improved photosynthesis/yield, enhanced provitamin A content, and resistance to late blight disease. Using these traits, genetically modified varieties of many crops such as alfalfa, canola, cotton, cowpea, eggplant, corn, papaya, potato, rice, soybean, sugarcane, sweet pepper, tobacco, tomato, and wheat have been developed and are commercially available (Platani et al. 2023; Rout et al. 2023; ISAAA 2024; Xoconostle-Cázares et al. 2024).

19.3 Genetically Modified Crops in Adapting to Climate Change

Genetically modified (GM) crops are expected to enhance the sustainability and productivity of agricultural production by providing various benefits such as higher yields, resistance to drought and salinity, reduced pesticide use, reduced fertilizer use, reduced carbon footprint, improved food security, and increased nutritional value, such as higher trace element content (Lönnerdal 2003). These advantages make GM crops a significant tool in combating climate change and protecting the environment. The adoption and widespread implementation of these technologies are crucial for agricultural sustainability and food security (National Research Council et al. 2010; Kumar et al. 2020).

From the dawn of humanity to the present, humans have utilized, influenced, and been influenced by the natural resources around them to meet their needs. Over the past century, human population growth has accelerated rapidly, along with urbanization and industrialization (Davis 2015). Concurrently, the Earth has undergone climatic changes over centuries during its formation. These climatic changes have directly impacted living beings on Earth, either positively or negatively. Some species have become dominant, while others have gone extinct. Atmospheric temperatures have continually fluctuated, and climate change has occurred over long periods. The Ice Age lasted for approximately a thousand years, followed by a warming period characterized by milder and warmer seasons. The climate continues to undergo changes over long intervals (Wilson et al. 2000; Sánchez Goñi et al. 2018).

In the past century, especially due to the burning of fossil fuels, the industrial revolution, urbanization, and similar human activities, the concentrations of carbon dioxide, methane, chlorofluorocarbons, and other greenhouse gases in the atmosphere have caused the planet to warm up further. The causes of this global warming can be both human-induced and natural, such as changes in Earth's orbit, the El Niño effect, and solar influences. The increase in greenhouse gases in the atmosphere enhances the natural greenhouse effect, leading to a rise in temperatures in the lower layers of the atmosphere and on Earth's surface, a phenomenon known as global warming (Karl and Trenberth 2003; Nda et al. 2018).

The effects of global warming are felt differently across regions globally. For instance, while the southeastern parts of the Americas have experienced a drop in temperature, the northern parts have seen an increase. Therefore, a homogeneous distribution is not applicable to global warming. In Southeast Asian countries such as Indonesia and the Philippines, soil erosion has begun to be seen as an effect of global warming. Some areas in the Pacific Ocean, like the Kiribati islands, have started to become submerged, forcing people to migrate to higher ground (Campbell 2014). Due to its complex climate structure, Turkey is expected to feel the effects of global warming significantly (Özşahin 2023). Turkey is among the countries at risk. Surrounded by seas on three sides and located in different climate zones, global warming will have varying impacts and manifestations across different regions of Turkey (Turkes 2020).

It is expected that global warming will have vital effects on agricultural production in various regions of the world. The main effects include problems such as irrigation issues in agricultural production, chemical rainfalls like acid rain, decrease in agricultural and forest products, proliferation of harmful microorganisms, changes in plant habitats, reduction or extinction of certain plant species, or invasion of some species becoming invasive, and significant negative impacts on agricultural crop production such as early emergence of insects and birds (Mariappan et al. 2023; Subedi et al. 2023; Janni et al. 2024).

19.4 Genetically Modified Organisms

Throughout history, humanity has utilized organisms that developed through natural selection according to their needs, without human intervention. Like in ancient Egypt and Rome, people engaged in agriculture preserved the seeds of the best-quality plants to sow the following year. For centuries, humans have cultivated and improved plants and animals that benefit them. Later, classical breeding methods were used to cultivate and enhance certain plants, resulting in increased yields. Concurrently, some animals such as cattle and goats were domesticated for greater utility (Atsan and Kaya 2008).

After World War II, the world population began to increase rapidly. During the years known as the Green Revolution (1965–1985), significantly higher yields were achieved even from limited areas. This increase in yield was due to advancements in classical breeding methods, the use of machinery, and the introduction of fertilizers, especially synthetic ones, into agriculture, which proved effective. During this period, not only did yield increase, but the quality of products also improved. Plant and animal breeding methods were also advanced to unprecedented levels (Aydın 2012).

However, as the world population continued to grow and agricultural needs could not be fully met, the Green Revolution became insufficient. This led to searches for alternatives due to increasing demands. Disadvantages such as the time-consuming nature and intensive labor costs associated with traditional breeding methods, as well as the limited number of species suitable for hybridization, rendered classical breeding methods inadequate. Thus, genetically modified organisms began to enter our lives (Nayeri et al. 2023; Yaseen et al. 2024).

Genetically Modified Organisms (GMOs) refer to organisms created by transferring a new gene to the organism from another organism or from the same species, using recombinant DNA technologies. GMOs can carry genes not only from their own species but also from different species and genera. For example, using recombinant DNA technologies, genetically modified tobacco can be obtained by transferring genes from bacteria, known as genetically modified tobacco (Kaya et al. 2013; Saha et al. 2023). Similarly, a genetically modified banana can be produced by transferring a gene from fish living in polar regions to the banana plant. Lastly, a genetically modified tomato can be obtained by transferring a gene to the tomato plant using recombinant technologies (Narayanan and Glick 2023; Tariq et al. 2024).

Genetically modified organisms, whose initial studies began in the USA, continued to develop after the first genetically modified organism was obtained in the laboratory in 1973. In 1983, the world's first genetically modified tobacco plant was obtained. In 1995, the first genetically modified maize plant containing the gene from the "*Bacillus thuringiensis*" bacterium was cultivated in open fields. The first commercially traded genetically modified plant is the tomato. The first genetically modified tomato, called "*Flavr Savr*," had a longer shelf life compared to other varieties. In subsequent years, the number of products continued to increase (Arvas and Kaya 2019; Arvas and Kocacaliskan 2020).

In the past 20 years, the cultivation of genetically modified crops has been increasing. Globally, while 1.7 million hectares of GMO crops were cultivated in 1996, this figure reached 206.3 million hectares in 2023 (Fig. 19.1), marking a 1.9% increase from the previous year and setting a new record. GM crops are grown in 27 countries with 11 different GM crops. Soybean, with 100.9 million hectares, is the most widely planted GM crop, surpassing 100 million hectares for the first time. It

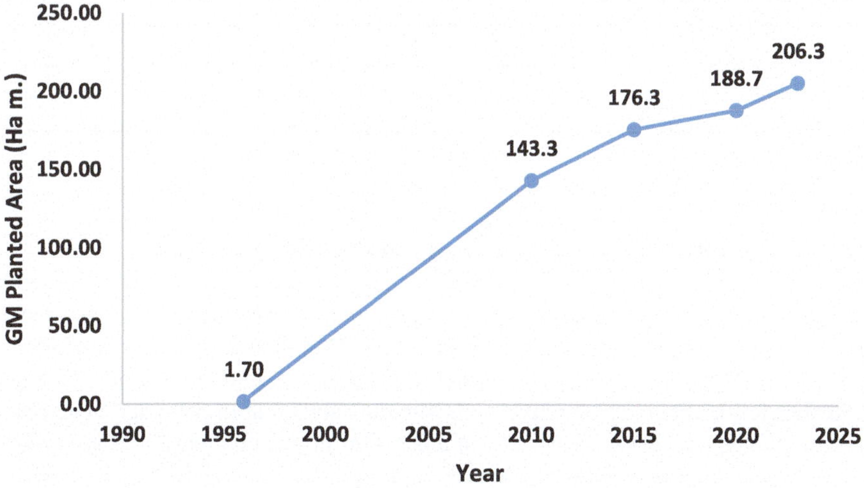

Fig. 19.1 Changes in GM products in the world over the years (AgbioInvestor 2023)

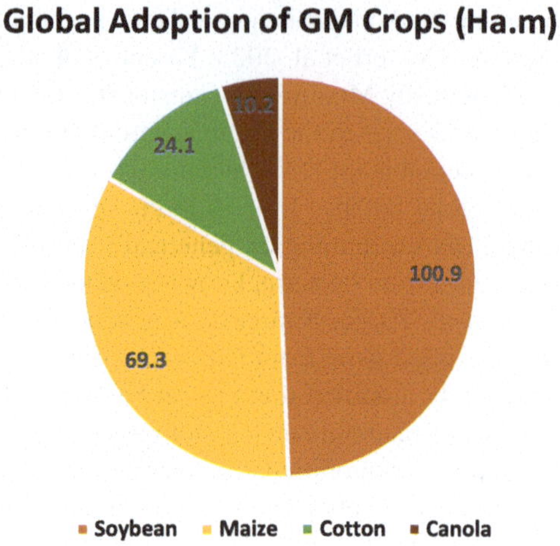

Fig. 19.2 Distribution graph of the top 4 most cultivated varieties (AgbioInvestor 2023)

is followed by maize with 69.3 million hectares and cotton with 24.1 million hectares. Since 2020, 27 countries have been cultivating GM crops, a number that was also reached in 2015 (AgbioInvestor 2023).

As of 2023, the most widely produced genetically modified agricultural products on Earth, from least to most, are economically valuable crops such as canola, cotton, maize, and soybeans (Fig. 19.2). Cotton holds the highest adoption rate of GM technology, covering 76.0% of the global cotton cultivation area. Soybeans account for 72.4%, maize for 34.0%, and canola for 24.0% of respective cultivation areas. Adoption rates for other crops decrease rapidly (AgbioInvestor 2023).

One of the potential harms caused by genetically modified organisms is genetic pollution. Genetic pollution refers to the ease with which foreign genes, particularly from genetically modified plants, can spread to many different ecosystems through factors such as bees and wind. This can lead to contamination in the DNA of plants in fields where organic and traditional farming is practiced. An example of the damage it can cause to beneficial organisms is the report that the pollen from transgenic maize plants, which contain Bt genes claimed to only resist corn borers, also caused the death of various beneficial butterflies. This poses a significant threat to natural ecosystems (Quan and Wu 2023; Sabat and Tripathy 2024).

19.5 Genetically Modified Crops in Plant Production

The development and commercialization of genetically modified (GM) products involve a multistep process spanning many years. Maintaining the integrity of product quality throughout the development of a GM product, its identification, labeling, and the development of effective tools for traceability at every stage constitute essential components for the success of a new GM product (Gupta and Ram 2004). The commercial tomato variety released under the trade name Flavr Savr™ used an antisense RNA to regulate the expression of polygalacturonase (PG) enzyme genes

Table 19.1 Countries producing the most GM crops in the world (AgbioInvestor 2023)

GM Crop Area by Leading Country 2023 (million hectar)		
1.	USA	74.4
2.	Brazil	66.9
3.	Argentina	23,10
4.	India	12,10
5.	Canada	11,50
6.	Paraguay	4,30
7.	South Africa	3,30
8.	China	2,80
9.	Pakistan	2,30
10.	Bolivia	1,50
11.	Others	4,10

(Irfan and Kumar 2024). The PG enzyme is crucial in pectin metabolism during fruit ripening. It is the most abundant protein in ripe tomatoes and is responsible for fruit softening (Prasanna et al. 2007; Uluisik and Seymour 2020). By reducing the expression of the PG gene using antisense RNA, the soluble pectin content decreases in ripened tomatoes, allowing them to stay fresh for longer periods.

Following the introduction of the first GM product to the market, a new and rapidly growing sector emerged globally, and the cultivation of GM seeds has steadily increased in agricultural areas. Between 1966 and 2023, the production of GM products has increased by more than 100-fold, reaching a total production area of 206.3 million hectares. Therefore, GMOs are currently seen as the technological products that have most rapidly adapted to agricultural areas in recent times (AgbioInvestor 2023).

As of 2023, the GM cultivation area exceeded 1 million hectares in the top 10 countries: USA (74.4), Brazil (66.9), Argentina (23.10), India (12.10), Canada (11.50), Paraguay (4.30), South Africa (3.30), China (2.80), Pakistan (2.30), and Bolivia (1.50) (Table 19.1). According to the report by AgbioInvestor (2023), soybean has been the most produced and fastest-growing GM crop in the past decade. GM crops of 11 different varieties were grown in 27 countries, with soybeans having the largest cultivation area of 100.9 million hectares, surpassing 100 million hectares for the first time. This was followed by maize with 69.3 million hectares and cotton with 24.1 million hectares (AgbioInvestor 2023).

19.5.1 First Generation GMO Traits Introduced

In the early years of commercializing GM products, efforts were focused on traits directly benefiting farmers, such as herbicide tolerance, insect resistance, and pathogen resistance—features that assist in the cultivation of agricultural crops. The primary objective in developing first-generation GMOs during production stages has been to increase farmers' profitability and crop yields (Korth 2008).

Herbicide tolerance, widely introduced, significantly reduces product costs for farmers. Additionally, crops engineered with genes from Bacillus thuringiensis (Bt), particularly effective against damaging caterpillars in maize and cotton cultivation areas, have led to decreased pesticide use during production. This economically viable practice helps mitigate the adverse environmental and human health impacts of applied chemical treatments (James 2016a).

Products in this category continue to be commercially cultivated, with global cultivation spanning millions of hectares. While countries like the USA, Brazil, and Argentina boast the largest cultivation areas, Turkey and the European Union have restrictions on the cultivation and human consumption of these crops but permit their use as animal feed. As of 2024, Turkey has approved forty-eight varieties of transgenic products. Since 2022, twenty-four transgenic products have been authorized for cultivation (Table 19.2) (TAGEM 2024).

19.5.2 Herbicide Resistance

Losses in yield due to weeds, which are unwanted plants in agricultural areas, are estimated globally at an average of 10–15% (Junaid and Gokce 2024). Therefore, herbicides are used to reduce yield losses and control weeds. Herbicides act selectively, targeting unwanted weeds with specific morphological and physiological characteristics. Herbicide resistance conferred through genetic transfer can generally be approached in two different mechanisms (Vats 2015; Kumar et al. 2024). First, it provides tolerance to the plant by modifying the area affected by the herbicide so that it does not harm the plant. The other mechanism involves transferring resistance genes that encode proteins that inactivate the active ingredient in herbicides, thus providing resistance. These resistance genes are isolated either from microorganisms or naturally resistant plants that possess these genes. As seen in Fig. 19.3, resistance to herbicides has increased over time (Öktem 2004a).

Herbicide tolerance has been an important agricultural trait used effectively for thirty years to control weeds. Many genes, including the dehE gene and EPSPS gene, are used to develop herbicide-tolerant plants. The *dehE* gene, isolated from Pseudomonas spp., detoxifies the herbicide dalapon by producing dehalogenase E enzyme, which converts it to pyruvic acid. After isolation from bacteria, this gene was transferred to Agrobacterium tumefaciense using recombinant DNA methods and then transferred to *Nicotiana tabacum* TAPM24. The resulting genetically modified tobacco plants acquired resistance to dalapon herbicide (Kaya et al. 2013).

In America, cultivation of genetically engineered (GE) crops resistant to potent herbicides such as glyphosate, glufosinate, and dicamba is underway. According to USDA 2023 survey data, the percentage of domestic soybean acres planted with HT seeds was reported to be 94% in 2014, 95% in 2021, and remained at the same level in 2023. HT cotton planting area was noted at 95% in 2019 and decreased to 94% in 2023. Although adoption rates for HT corn grew relatively slowly following the commercialization of GE seeds, by 2023, 91% of domestic corn acres were reportedly planted with HT seeds (USDA 2024).

Table 19.2 Transgenic products approved by official institutions in Turkey (2022–2024) (TAGEM 2024)

No	GM Plant	Variety	ID	Approval Year
1.	GDMO	Proteaz/Mukorpepsin (EC 3.4.23.23) enzimi	LIVZ-105	2022
2.	Corn	4114	DP-ØØ4114–3	
3.	Soybean	SYHT0H2	SYN-ØØØH2–5	
4.	Soybean	MON87751	MON-87751-7	
5.	Soybean	DAS-81419-2	DAS-81419-2	
6.	Corn	MON87411	MON-87411-9	
7.	Corn	MZIR098	SYN-ØØØ98–3	
8.	GDMO	Fitaz	LIVZ-104	
9.	Corn	5307	SYN-Ø53Ø7–1	
10.	GDMO	Lipaz/Triasilgliserol Lipaz(EC 3.1.1.3)	LIVZ-106	
11.	GDMO	Glikoz oksidaz (EC 1.1.3.4)	LIVZ-107	
12.	GDMO	İnaktive edilmiş mikrobiyal biyokütle	BT-101	
13.	GDMO	Endo-1,4-beta- glukanaz (E.C. 3.2.1.4) ve Endo-1,4- beta-ksilanaz (EC 3.2.1.8) içeren karışımı	LIVZ-108	
14.	Corn	NK603xMON810	MON-ØØ6Ø3-6xMON-ØØ81Ø-6	
15.	Corn	Bt11	SYN-BTØ11-1	2024
16.	Corn	DAS1507	DAS-Ø15Ø7-1	
17.	Corn	DAS59122	DAS-59122-7	
18.	Corn	NK603	MON-ØØ6Ø3-6	
19.	Corn	GA21	MON-ØØØ21-9	
20.	Corn	MON89034	MON-89Ø34-3	
21.	Corn	MON88017	MON-88Ø17–3	
22.	GDMO	Mikrobiyal Kolajenaz (EC 3.4.24.3)	LIVZ-109	
23.	GDMO	Fosfolipaz A1 (EC 3.1.1.32)	LIVZ-110	
24.	GDMO	Proteaz Termomikolin (EC 3.4.21.65)	LIVZ-111	

19.5.3 Stress Tolerance

The discovery of new genes, determining their expression patterns in response to abiotic stress, and understanding their roles in stress adaptation provide the foundation for effective DNA technologies that will lead to greater stress tolerance (Cushman and Bohnert 2000). By incorporating traits such as water deficiency tolerance, high salinity, temperature extremes, and heavy metal resistance, crops can be cultivated under extreme conditions. The development of plants that adapt to various stress areas using gene transfer is still in its early stages. For example, the

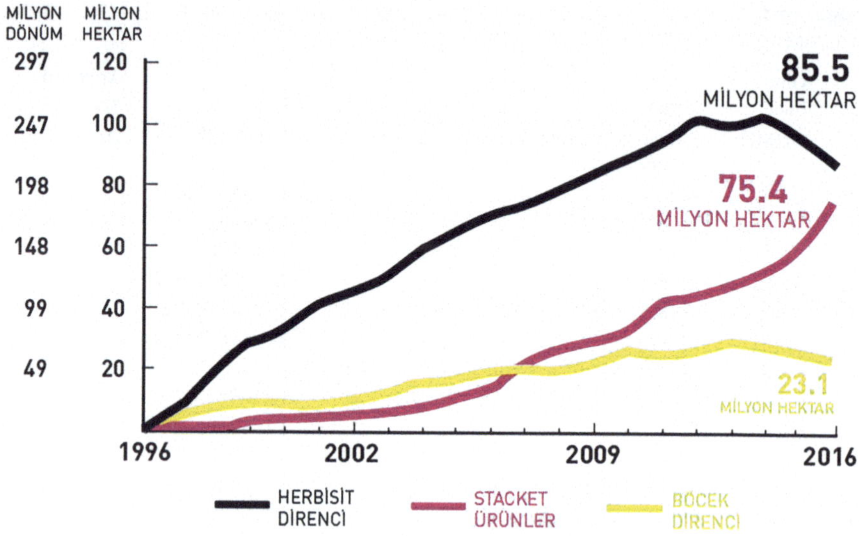

Fig. 19.3 Change in tolerance to herbicides over the years (James 2016b)

transfer of the Mn-SOD gene into tobacco (*Nicotiana tabacum*) has enabled the detoxification of superoxide anions, a common form of reactive oxygen species, by converting them into hydrogen peroxide and subsequently into water via superoxide dismutase (SOD) and catalases. This process has conferred active oxygen tolerance in tobacco plants (Greenberg and Glick 1993; Sharma et al. 2012).

Similarly, the introduction of the trehalose synthase gene (TPS), which regulates trehalose expression, into maize and tobacco has produced stress-resistant lines (Zhang et al. 2005). Genes that enhance the saturation of chloroplast membranes with lipids are selected to increase cold stress tolerance. An example is the transfer of a cDNA responsible for glycerol-3-phosphate acyltransferase from squash to tobacco, which confers cold tolerance (Sanghera et al. 2011).

To produce transgenic plants resistant to frost, the AFP gene encoding the anti-freeze protein from the fish Pseudopleuronectes americanus has been transferred into tobacco and tomato plants (Duman and Wisniewski 2014). Methods have been developed to create resistance to salt stress as well. For instance, in genetically modified rice, overexpression of the glutamine synthetase enzyme enhances resistance to salt stress (James et al. 2018; Yin et al. 2022). Additionally, transferring the choline oxidase gene from *Arthrobacter globiformis* into Arabidopsis has led to the accumulation of glycine-betaine in the plants, thereby increasing their salt stress tolerance (Malick et al. 2021).

19.5.4 Disease Tolerance

Plants possess defense systems that can exhibit various effects to limit disease infections. Examples include lignification of the cell wall, proliferation of reactive oxygen species, production of small molecules used as antibiotics, and the death of host cells in the damaged area. Resistance acquired systematically against viruses, bacteria, fungi, and nematodes occurs through the concerted action of numerous resistance genes (de Ronde et al. 2014). Although traditional plant breeding has made significant progress in developing resistant seeds against important diseases, the time-consuming processes of hybridization and backcrossing and the difficulties in selecting the desired resistant offspring make traditional plant breeding less effective against newly developed harmful fungal species (Mohanan 2024).

The increasing concern about environmental pollution has led to research aimed at reducing chemical pesticide usage, thereby lowering production costs. The development of molecular techniques and plant transformation to enhance resistance against fungal diseases have played a crucial role in the genetic improvement of seeds (Azizi-Dargahlou and Pouresmaeil 2024).

19.5.5 Pesticide Resistance

In addition to causing mechanical damage to plant tissues, insects can also lead to infections by pathogens such as fungi, bacteria, and viruses (Nazarov et al. 2020). Advances in agricultural biotechnology have enabled the use of effective genetic technology methods to combat pests. According to 2016 data, 23.1 million hectares were planted with insect-resistant transgenic cereals. Herbicide-resistant plants were grown on 47% of the total area where transgenic crops were cultivated, while insect-resistant plants were grown on 12%. The share of stacked transgenic crops in the total cultivation area reached 41% (ISAAA 2016).

The first successful gene transfer to insects in agricultural warfare was achieved in 1982 with the P element in *Drosophila melanogaster* Meig (Handler 2001). Gene transfer to other insects was accomplished 13 years later, using the Minos vector system in the Mediterranean fruit fly *Ceratitis capitata* Wied (Loukeris et al. 1995). Following these successful studies, researchers began exploring ways to use gene transfer techniques in some vector insects and agricultural pest control. The gene transfer method generally comprises four essential elements: the vector (carrier), a gene for making suitable changes in the species, a helper and dispersal mechanism for spreading the transferred gene within the population, and marker genes. The marker gene used in insects that undergo gene transfer during the egg stage is critical. Transgenic insects can be easily distinguished by marker genes (green fluorescent eyes) (Wimmer 2003; Yorulmaz and Ay 2006).

cry genes are used to impart this resistance to insects. *cry* genes are natural and synthetic genes responsible for the synthesis of *Bacillus thuringiensis* (Bt) delta-endotoxin proteins. *Bacillus thuringiensis* produces a substance toxic to certain insect species. Because this substance does not harm animals and humans, it has been used in various plant species to create resistance to harmful insects (Korth 2008).

For instance, a cholesterol-oxidase gene as an insect resistance gene from Streptomyces species and a lectin gene from pea plants have been successfully isolated. The lectin gene disrupts normal digestion by binding to nutrient compounds in the insect gut, protecting plants containing this protein from insect damage. By transferring these genes to plants and transgenic plants, the aim is to impart multiple resistance traits, thereby slowing or preventing the rapid development of pest insect populations on plants (Kumari et al. 2022).

Due to the high costs and labor involved in chemical interventions (insecticide use) against pests, which often fail to prevent losses, different control methods using gene technology are preferred. The primary focus of these methods is developing transgenic plants by transferring insecticidal proteins with toxic effects on insects (Öktem 2004b).

Today, the most commonly used genes for these studies are *cry* genes responsible for the synthesis of *Bacillus thuringiensis* delta-endotoxin proteins (bt), a gram-positive soil bacterium. The natural insecticidal activity of Bt endotoxin genes offers an attractive alternative to synthetic chemical insecticides, which often have nonselective toxic effects on beneficial insects, birds, fish, and mammals. *Cry* genes encoding toxic proteins like Bt are named after the crystal inclusions that form within the bacterium during the sporulation phase. These crystals typically contain multiple specific cry gene products. Before becoming toxic, the Bt proteins encoded by the cry gene exist as protoxins and need to be activated in the insect's digestive system. When they enter the digestive system of a susceptible insect, the crystals dissolve in the alkaline environment of the midgut, where the pH is usually above eight. At this stage, the ends of the Bt protoxin protein are cleaved by specific protease enzymes in the gut, releasing the toxic protein. The active protein then binds to specific protein receptors on the microvilli membrane in the insect midgut. When the active Bt toxin binds to the receptor, it oligomerizes with multiple proteins to form a pore as it penetrates the insect cell membrane (Korth 2008). The formation of pores causes ion leakage from the membrane, leading to osmotic lysis and membrane disruption (Cai et al. 2024).

Insect-resistant crops, which contain genes from the soil bacterium Bt (*Bacillus thuringiensis*) and produce insecticidal proteins, have been available for corn and cotton since 1996. In the United States, the acreage of domestic Bt corn was 19% in 2000 and increased to 85% in 2023. Bt cotton acreage in the U.S. was 37% in 2001, and by 2023, 89% of U.S. cotton acres were planted with insect-resistant transgenic seeds. According to the USDA 2023 report, approximately 86% of cotton acres and 82% of corn acres were planted with stacked seeds in 2023.

Of course, these rates may change depending on the development of new commercial varieties resistant to corn-specific pests such as the corn rootworm and the corn earworm. Similarly, the adoption of Bt cotton will vary based on the prevalence of pests like the tobacco budworm, bollworm, and pink bollworm (USDA 2024).

19.6 Effects of Genetically Modified Plants

Biotechnology has played a very significant role in preserving biodiversity from the past to the present. When examining plant genetic resources, it is evident that biotechnology assists in the tissue culture of stubborn seed species like oak and chestnut, species problematic to reproduce by seeds like orchids and ornamental plants, and plants that propagate vegetatively like fruit trees, where classical conservation methods cannot be used. Additionally, biotechnology methods are employed in applications such as preservation under extreme cold conditions, preservation of DNA, and preservation of pollen (Arvas and Kaya 2019; Arvas and Kocacaliskan 2020).

In hybridization studies conducted with the help of biotechnological methods, it has been observed that biodiversity is increased and the genetic diversity of the same species is better understood. Furthermore, the use of molecular markers in selection studies and the transfer of desired genes found in wild species to cultivated plants are made possible. With these aspects mentioned above, biotechnology is seen as an indispensable tool for preserving plant genetic resources and ensuring sustainable agriculture (Karagöz 2010). Given that our country is quite rich in endemic species (İpek and Gürbüz 2010), comprehensive measures should be taken and enhanced to minimize the negative impacts of genetically modified plants (Arvas and Kaya 2019; Arvas and Kocacaliskan 2020).

19.6.1 Effects on Environment and Biodiversity

Biodiversity establishes a connection with the past of a society. Therefore, the ecology of a country and its cultural and spiritual riches have gained significant importance. However, with the increasing population and the consequent rise in the consumption rate of natural resources, the reduction in genetic diversity has led to concerns in recent years about the potential adverse effects that genetically modified organisms (GMOs) might cause (Yılmaz 2014). The negative impacts of the release of GMOs into the environment are mainly expressed as risks arising from uncontrolled pollination, gene escape, and wild hybridization (Fig. 19.4). These negative outcomes could lead to the transfer of the "foreign gene" modified for a particular trait to other species (Özdemir 2007). Should such a situation occur, it would pose a significant problem not only for human and animal health but also for biodiversity and the entire ecosystem (Yılmaz 2014).

Recent studies have demonstrated that the release and use of GMOs in the environment have led to some negative impacts. Research indicates that the production of GM crops has increased the use of chemical pesticides. In 1999, there was an approximately 17% increase in soybean acreage in Argentina, accompanied by a doubling of the amount of pesticides used and a slight decrease in yield during the same period (Olhan 2010). Thus, the increase in pesticide usage could not be attributed solely to the increase in soybean acreage. Another concern associated with the expansion of GMO cultivation worldwide is the pollution of groundwater and

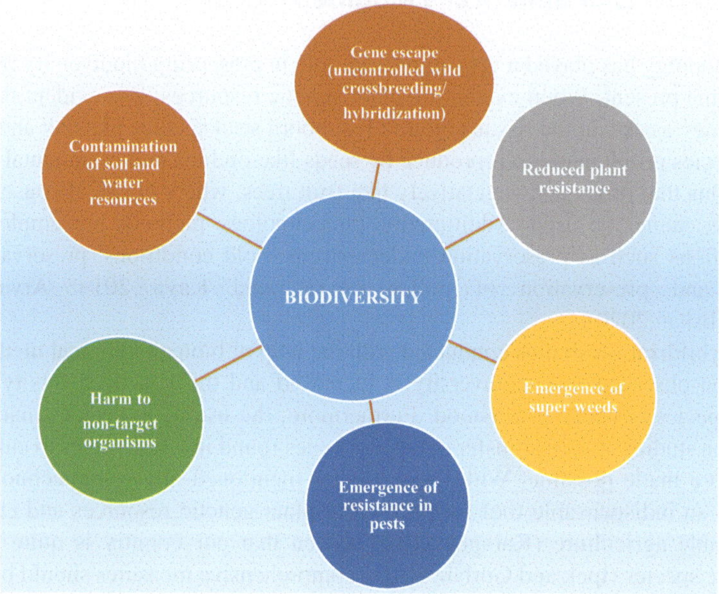

Fig. 19.4 Effects of GMOs on biodiversity (Özdemir 2007)

rivers, along with soil degradation (Abbasi et al. 2014). Another critical issue regarding the effects of genetically modified plants on biodiversity is the "terminator" gene technology. In this case, there is concern about the possibility of new resistant bacteria emerging as a result of the transfer of new genes to soil bacteria or due to chemicals used to activate the terminator gene (Korkut and Soysal 2013).

19.7 Recent Developments on Genetically Modified Organisms in the European Union

In countries like the United States, Argentina, Brazil, Canada, Japan, and India, genetically modified crops are cultivated extensively, but in the European Union (EU), regulations that came into effect before the advent of new technologies impose extremely restrictive barriers to the cultivation of genetically modified organisms (GMOs). Even if these products are proven safe in EU countries, their use is prohibited. The laws prohibiting the cultivation of GMOs in the EU support activities aimed at agricultural ecology and organic farming that respect animal welfare, support biological diversity, and enable farmers to produce healthy and nutritious food for everyone. Despite the EU's ban on cultivating genetically modified crops across the EU under Council Directive 90/220/EEC from 2001, large quantities are imported for animal feed. Proponents argue that banning the cultivation of genetically modified crops across the EU provides solutions to plant diseases, reduces pesticide and fertilizer use, and promotes the cultivation of more sustainable crops that are drought- and flood-resistant.

Scientists also anticipate finding ways to increase yield and produce more nutritious crops. Nevertheless, the discovery of CRISPR and the resulting surge in genetic editing research in Europe have prompted scientists and agricultural enterprises across the continent to demand revisions to GMO regulations. Recently, the European Union has signaled its intention to relax regulations on genetically modified organisms in legislative revisions (Genetic Literacy Project 2023; European Union 2023, 2024).

19.8 Results and Recommendations

Undoubtedly, genetically modified crops hold great potential in addressing global hunger and malnutrition issues. To realize this potential, it is crucial to facilitate necessary accommodations while considering scientific research alongside ethical and environmental impacts. Vigilant monitoring of advancements in genetic engineering technology should encourage research into new technologies that benefit humanity and the environment.

Over the past century, technological advancements have led to the emission of harmful gases and climate change due to natural causes, a phenomenon that continues to persist. Global warming affects the entire world, both positively and negatively. To mitigate human-induced global warming, environmentally friendly policies must minimize its causes.

Agricultural research should focus on crop production, an area directly impacted by global warming. Despite efforts to reduce global warming, its dangers will persist. Therefore, crop production remains essential for human survival. Particularly over the past 20 years, the development of recombinant DNA technologies has facilitated genetic transformation in crops. For instance, rice cultivation, currently flourishing, may face reduced yield and quality due to future water shortages caused by global warming. Therefore, manipulating rice genomes to introduce genes from wheat or other organisms for drought resistance is imperative.

Similarly, transferring genes to wheat to enhance heat tolerance exemplifies the potential of recombinant DNA technologies. Genetically modified crops developed through these technologies should assess environmental stress conditions such as extreme temperatures, droughts, excessive rain, or harmful chemicals caused by global warming to obtain stress-resistant GMO crops.

Given the escalating impact of climate change, agricultural production should adopt diverse technologies. These solutions should be eco-friendly, economical, accessible to all, and compatible with soil structures.

References

Abbasi A, Sajid A, Haq N, Rahman S, Misbah Z-T, Sanober G et al (2014) Agricultural pollution: an emerging issue. In: Improvement of crops in the era of climatic changes, vol 1. Springer, New York, pp 347–387

AgbioInvestor (2023) AgbioInvestor GM monitor [Online]. Available: https://gm.agbioinvestor.com/ (Accessed: 04.09.2024) [Accessed]

Ahmed M, Ahmad S, Kheir AM (2023) Climate change: an overview. In: Global agricultural production: resilience to climate change. Springer, Cham, pp 1–30

Arvas YE, Kaya Y (2019) Genetiği Değiştirilmiş Bitkilerin Biyolojik Çeşitliliğe Potansiyel Etkileri. Yüzüncü Yıl Üniversitesi Tarım Bilimleri Dergisi 29(1):168–177

Arvas YE, Kocacaliskan İ (2020) Genetiği değiştirilmiş bitkilerin biyogüvenlik riskleri. Türk doğa ve fen dergisi 9(2):201–210

Aryal JP, Manchanda N, Sonobe T (2022) Expectations for household food security in the coming decades: a global scenario. In: Future foods. Elsevier, pp 107–131

Atsan T, Kaya TE (2008) Genetiği değiştirilmiş organizmaların (GDO) tarım ve insan sağlığı üzerine etkileri. Uludağ Üniversitesi Ziraat Fakültesi Dergisi 22(2):1–6

Aydın H (2012) Sağlıklı Nesiller için Önce Sağlıklı Tohum. İstanbul Ticaret Odası Yayınları, İstanbul

Azadi H, Ho P (2010) Genetically modified and organic crops in developing countries: a review of options for food security. Biotechnol Adv 28(1):160–168

Aziz MA, Masmoudi K (2024) Molecular breakthroughs in modern plant breeding techniques. Horticult Plant J

Azizi-Dargahlou S, Pouresmaeil M (2024) Agrobacterium tumefaciens-mediated plant Transformation: a review. Mol Biotechnol 66(7):1563–1580

Bajoria A, Kanpariya J, Bera A (2024) Greenhouse gases and global warming. In: Advances and technology development in greenhouse gases: emission, capture and conversion. Elsevier, pp 121–135

Bakhsh A, Sırel IA, Kaya RB, Ataman IH, Tillaboeva S, Dönmez BA et al (2021) Contribution of genetically modified crops in agricultural production: success stories. In: Policy issues in genetically modified crops. Elsevier, pp 111–142

Cai Y, Hou B, Fabrick JA, Yang Y, Wu Y (2024) The role of aquaporins in osmotic cell lysis induced by Bacillus thuringiensis Cry1Ac toxin in Helicoverpa armigera. Pestic Biochem Physiol 106068:106068

Campbell JR (2014) Climate-change migration in the Pacific. Contemp Pac 26:1–28

Caradus JR (2023) Processes for regulating genetically modified and gene edited plants. GM Crop Food 14(1):1–41

Chege C et al (2024) Adapting green innovation centers to climate change: reducing post-harvest loss and improving processing for target groups

Cushman JC, Bohnert HJ (2000) Genomic approaches to plant stress tolerance. Curr Opin Plant Biol 3(2):117–124

Davis K (2015) The urbanization of the human population. In: The city reader. Routledge, pp 43–53

de Ronde D, Butterbach P, Kormelink R (2014) Dominant resistance against plant viruses. Front Plant Sci 5:307

Demir İ, Kılıç G, Coşkun M (2008) Türkiye ve bölgesi için PRECIS bölgesel iklim modeli çalışmaları. İklim Değişikliği ve Çevre 1(1):11–17

Division U.D.s.P. (2015) World population projected to reach 9.7 billion by 2050 [Online]. Available: https://www.un.org/en/development/desa/news/population/2015-report.html (Accessed: 04.09.2024) [Accessed 2024]

Duman JG, Wisniewski MJ (2014) The use of antifreeze proteins for frost protection in sensitive crop plants. Environ Exp Bot 106:60–69

Estrada AC, Díaz DV, Hernández CAM (2017) The role of biotechnology in agricultural production and food supply. Ciencia e investigación agraria: Revista Latinoamericana de Ciencias de la Agricultura 44(1):1–11
European Union (2023) Directive 2001/18/EC of the European Parliament and of the Council of 12 March 2001 on the deliberate release into the environment of genetically modified organisms and repealing Council Directive 90/220/EEC - Commission Declaration [Online]. Available: https://eur-lex.europa.eu/legal-content/EN/TXT/?uri=CELEX:32001L0018 (Accessed: 04.09.2024) [Accessed]
European Union (2024) European Union: crops/food. A public resource compiled by the genetic literacy project [Online]. Available: https://crispr-gene-editing-regs-tracker.geneticliteracyproject.org/european-union-crops-food/ (Accessed: 04.09.2024) [Accessed]
Gaharwar US, Verma A, Singh R, Prasad T, Rajamani P (2021) Health risks and environmental concerns of GM crop adoption. In: Policy issues in genetically modified crops. Elsevier, pp 371–400
Genetic Literacy Project (2023) Why the European Union needs to grow genetically-engineered crops [Online]. Available: https://geneticliteracyproject.org/2023/04/04/why-the-european-union-needs-to-grow-genetically-engineered-crops/ (Accessed: 04.09.2024) [Accessed]
Ghimire BK, Yu CY, Kim W-R, Moon H-S, Lee J, Kim SH et al (2023) Assessment of benefits and risk of genetically modified plants and products: current controversies and perspective. Sustain For 15(2):1722
Greenberg B, Glick B (1993) The use of recombinant DNA technology to produce genetically modified plants. In: Methods in plant molecular biology and biotechnology. Routledge, pp 1–10
Guimond JA, Michael HA (2021) Effects of marsh migration on flooding, saltwater intrusion, and crop yield in coastal agricultural land subject to storm surge inundation. Water Resour Res 57(2):e2020WR028326
Gulev SK et al. (2021) Changing state of the climate system
Gupta M, Ram R (2004) Development of genetically modified agronomic crops. In: The GMO handbook. Springer, pp 219–241
Handler AM (2001) Genetic transformation of insects for pest management. Annual Review of Entomology 46:667–692
Hasan K, Şahin MD, Şenay A (2010) İklim değişikliğinin Uşak'ta tarım ürünlerine etkisi. Biyoloji Bilimleri Araştırma Dergisi 3(1):39–46
Hekimoğlu B, Altindeğer M (2008) Küresel Isinma Ve İklim Değişikliği. Samsun Valiliği İl Tarım Müdürlüğü, pp 1–79
Hug K (2008) Genetically modified organisms: do the benefits outweigh the risks? Medicina 44(2):87–99
İpek A, Gürbüz B (2010) Türkiye florasında bulunan Salvia türleri ve tehlike durumları. Tarla Bitkileri Merkez Araştırma Enstitüsü Dergisi 19(1–2)
Irfan M, Kumar P (2024) Bioengineered fruit and vegetables: trends in postharvest quality improvement. CRC Press
ISAAA (2016) Gobal status of commercialized biotech/GM crops
ISAAA (2024) Biotech/genetically modified approval database [Online]. Available: https://www.isaaa.org/gmapprovaldatabase/ (Accessed: 04.09.2024) [Accessed]
James C (2016a) 20th Anniversary of the global commercialization of biotech crops (1996 to 2016) and biotech crop highlights in 2016 [Online]. Available: http://www.isaaa.org/resources/publications/briefs/52/default.asp (Accessed 19.06.2017)
James C (2016b) Biotech crop highlights in 2016 [Online]. Available: https://www.google.com/url?q=http://www.isaaa.org/resources/publications/pocketk/16/&sa=U&ved=0ahUKEwjWr8Kh_snUAhVE2hoKHavZAIcQFggEMAA&client=internal-uds-cse&usg=AFQjCNHUVB5M_hWtiEJPBpToCgBEnS0Wpw (Accessed 19.06.2017)
James D, Borphukan B, Fartyal D, Achary V, Reddy M (2018) Transgenic manipulation of glutamine synthetase: a target with untapped potential in various aspects of crop improvement. In: Biotechnologies of crop improvement, Volume 2: Transgenic approaches. Springer, Cham, pp 367–416

Janni M, Maestri E, Gullì M, Marmiroli M, Marmiroli N (2024) Plant responses to climate change, how global warming may impact on food security: a critical review. Front Plant Sci 14:1297569
Junaid M, Gokce A (2024) Global agricultural losses and their causes. Bull Biol All Sci Res 2024(1):66–66
Karagöz A (2010) Genetiği Değiştirilmiş Organizmaların Bitkisel Biyolojik Çeşitlilik Üzerine Olası Etkileri. Farklı BoyutlarıylaGenetiği Değiştirilmiş Organizmalar ankara tabip odası ankara, pp 15–21
Karl TR, Trenberth KE (2003) Modern global climate change. Science 302(5651):1719–1723
Kaya Y, Yilmaz S, Marakli S, Gozukirmizi N, Huyop F (2013) Transformation of Nicotiana tabacum with dehE gene. J Food Agri Environ 11(3–4)
Kıvılcım Z (2012) Cartagena Protokolü ve Türkiye Biyogüvenlik Mevzuatı. Marmara Avrupa Araştırmaları Dergisi 1(20):99–121
Kogan F (2023) The IPCC reports on global warming and land changes. In: Remote sensing land surface changes: the 1981–2020 intensive global warming. Springer, pp 67–79
Korkut D, Soysal A (2013) Genetiği Değiştirilmiş Organizmalar. Halk Sağlığı Uzmanları Derneği (HASUDER), Ankara
Korth KL (2008) Transgenik Bitkilerle İlgili Genler ve Özellikler, Bitki Biyoteknolojisi ve Genetik
Kramkowska M, Grzelak T, Czyzewska K (2013) Benefits and risks associated with genetically modified food products. Ann Agric Environ Med 20(3):413
Kumar K, Gambhir G, Dass A, Tripathi AK, Singh A, Jha AK et al (2020) Genetically modified crops: current status and future prospects. Planta 251(4):91
Kumar T, Wang J-G, Xu C-H, Lu X, Mao J, Lin X-Q et al (2024) Genetic Engineering for Enhancing Sugarcane Tolerance to Biotic and Abiotic Stresses. Plan Theory 13(13):1739
Kumari P, Jasrotia P, Kumar D, Kashyap PL, Kumar S, Mishra CN et al (2022) Biotechnological approaches for host plant resistance to insect pests. Front Genet 13:914029
Kweku DW, Bismark O, Maxwell A, Desmond KA, Danso KB, Oti-Mensah EA et al (2018) Greenhouse effect: greenhouse gases and their impact on global warming. J Sci Res Rep 17(6):1–9
Lönnerdal B (2003) Genetically modified plants for improved trace element. Nutrition (133):1430–1433
Loukeris T, Livadaras I, Arca B, Zabalou S, Savakis C (1995) Gene transfer into the medfly, Ceratitis capitata, with a Drosophila hydei transposable element. Science 270(5244):1941–1942
Malick CEH, Ling-Feng M, Da-Dong L, Fan Y (2021) Insights into metabolic engineering of the biosynthesis of glycine betaine and melatonin to improve plant abiotic stress tolerance. In: Botany-recent advances and applications. IntechOpen
Mariappan S, David Raj A, Kumar S, Chatterjee U (2023) Global warming impacts on the environment in the last century. In: Ecological footprints of climate change: adaptive approaches and sustainability. Springer, pp 63–93
Mohanan K (2024) Essentials of plant breeding. PHI Learning Pvt. Ltd
Mottaleb KA, Kruseman G, Frija A, Sonder K, Lopez-Ridaura S (2023) Projecting wheat demand in China and India for 2030 and 2050: implications for food security. Front Nutr 9:1077443
Narayanan Z, Glick BR (2023) Biotechnologically engineered plants. Biology 12(4):601
National Research Council; Division on Earth and Life Studies; Board on Agriculture and Natural Resources; Committee on the Impact of Biotechnology on Farm-Level Economics and Sustainability (2010) The impact of genetically engineered crops on farm sustainability in the United States. National Academies Press
Nayeri S, Dehghanian Z, Lajayer BA, Thomson A, Astatkie T, Price G (2023) CRISPR/Cas9-Mediated genetically edited ornamental and aromatic plants: A promising technology in phytoremediation of heavy metals. J Clean Prod 428:139512
Nazarov PA, Baleev DN, Ivanova MI, Sokolova LM, Karakozova MV (2020) Infectious plant diseases: etiology, current status, problems and prospects in plant protection. Acta Nat 12(3):46
Nda M, Adnan MS, Ahmad KA, Usman N, Razi MAM, Daud Z (2018) A review on the causes, effects and mitigation of climate changes on the environmental aspects. Int J Integ Eng 10(4)

Öktem A (2004a) Herbisitlere Dayanıklı Transgenik Bitkilerin Yetiştirilmesi. Bitki Biyoteknolojisi II–Doku Kültürü ve Uygulamaları
Öktem H (2004b) Böceklere Dayanıklı Transgenik Bitkilerin Geliştirilmesi. Bitki Biyoteknolojisi II, Genetik Mühendisliği ve Uygulamaları. Selçuk Üniversitesi Vakfı Yayınları, Konya, pp 208–238
Olhan E (2010) Modern Biyoteknolojinin Tarımda Kullanımının Politik Ve Ekonomik Yönden Değerlendirilmesi. Farklı Boyutlarıyla Genetiği Değiştirilmiş Organizmalar Ankara Tabip Odası, pp 9–14
Özdemir O (2007) Gen Kaynaklarının Sürdürülebilirliği Açısından GDO'ların Sosyo-Ekonomik Etkileri. Ankara Biyoteknoloji Günleri: Biyoteknoloji, Biyogüvenlik ve Sosyo-Ekonomik Yaklaşımlar, Ankara Üniversitesi Biyoteknoloji Enstitüsü, Ankara
Özşahin E (2023) Climate change effect on soil erosion using different erosion models: a case study in the Naip Dam basin, Türkiye. Comp Electron Agri 207:107711
Platani M, Sokefun O, Bassil E, Apidianakis Y (2023) Genetic engineering and genome editing in plants, animals and humans: Facts and myths. Gene 856:147141
Popp J, Pető K, Nagy J (2013) Pesticide productivity and food security. A review. Agron S ustain Dev 33:243–255
Prasanna V, Prabha TN, Tharanathan RN (2007) Fruit ripening phenomena–an overview. Crit Rev Food Sci Nutr 47(1):1–19
Quan Y, Wu K (2023) Managing practical resistance of Lepidopteran pests to Bt cotton in China. Insects 14(2):179
Ritchie H, Roser M (2024) How much of the world's land would we need in order to feed the global population with the average diet of a given country? Our World in Data
Rout GR, Swain R, Sahoo DP (2023) Journey of genetically modified crops: status and prospects. Magna Sci Adv Res Rev 7(1):103–128
Ruiz Serrano A, Musumeci A, Li JJ, Ruiz Serrano M, Serrano Barquin C (2024) Rationality and the exploitation of natural resources: a psychobiological conceptual model for sustainability. Environ Dev Sustain. https://doi.org/10.1007/s10668-024-04470-3
Sabat M, Tripathy A (2024) Genetically modified and gene-edited food crops: recent status and future prospects. In: Food production, diversity, and safety under climate change. Springer, Cham, pp 211–222
Saha PU, Ali A, Khan J (2023) Genetically engineered microorganisms. In: Genetic engineering. Apple Academic Press, pp 125–157
Sánchez Goñi MF, Desprat S, Fletcher WJ, Morales-Molino C, Naughton F, Oliveira D et al (2018) Pollen from the deep-sea: A breakthrough in the mystery of the Ice Ages. Front Plant Sci 9:38
Sanghera GS, Wani SH, Hussain W, Singh N (2011) Engineering cold stress tolerance in crop plants. Curr Genomics 12(1):30
Sharma P, Jha AB, Dubey RS, Pessarakli M (2012) Reactive oxygen species, oxidative damage, and antioxidative defense mechanism in plants under stressful conditions. J Bot 2012(1):217037
Singh V (2024) Global warming and climate change. In: Textbook of environment and ecology. Springer, pp 283–295
Singh U, Kumar K (2023) Exploring the interconnection between anthropogenic activities and greenhouse gas emissions: an empirical study. J Energ Environ Policy Opt 6(4):43–53
Singh OV, Ghai S, Paul D, Jain RK (2006) Genetically modified crops: success, safety assessment, and public concern. Appl Microbiol Biotechnol 71:598–607
Soutter ARB, Mõttus R (2020) "Global warming" versus "climate change": A replication on the association between political self-identification, question wording, and environmental beliefs. J Environ Psychol 69:101413
Subedi B, Poudel A, Aryal S (2023) The impact of climate change on insect pest biology and ecology: implications for pest management strategies, crop production, and food security. J Agri Food Res 14:100733
TAGEM (2024) Approved GMO list
Tariq S, Gul A, Negri S, Mujtaba T (2024) Genetic engineering in tomato. In: Targeted genome engineering via CRISPR/Cas9 in plants. Elsevier, pp 101–133

Tıraşcı S, Erdoğan Ü (2021) Küresel Isınmanın Tarıma Etkisi. J Agri Food Environ Anim Sci 2(1):16–33

Tsatsakis AM, Nawaz MA, Kouretas D, Balias G, Savolainen K, Tutelyan VA et al (2017) Environmental impacts of genetically modified plants: a review. Environ Res 156:818–833

Turkes M (2020) Climate and drought in Turkey. In: Water resources of Turkey. Springer, Cham, pp 85–125

Uluisik S, Seymour GB (2020) Pectate lyases: Their role in plants and importance in fruit ripening. Food Chem 309:125559

United Nations (2023) What is climate change? United Nations Available as retrieved on 5th December, United Nations; 2023. [Online]. Available: https://www.un.org/en/climatechange/what-is-climate-change (Accessed: 04.09.2024) [Accessed]

USDA (2024) Recent trends in GE adoption [Online]. Available: https://www.ers.usda.gov/data-products/adoption-of-genetically-engineered-crops-in-the-u-s/recent-trends-in-ge-adoption/ (Accessed: 10.07.2024) [Accessed]

Usman S, Jayeoba JO, Kundiri A (2024) Climate change at a global concept: impacts and adaptation measures. Int J Environ Clim Chang 14(6):445–459

Varzakas T, Smaoui S (2024) Global food security and sustainability issues: the road to 2030 from nutrition and sustainable healthy diets to food systems change. Food Secur 13(2):306

Vats S (2015) Herbicides: history, classification and genetic manipulation of plants for herbicide resistance. Sustain Agri Rev 15:153–192

Wilson RCL, Drury SA, Chapman JL (2000) The great ice age: climate change and life. Psychology Press

Wimmer E (2003) Applications of insect transgenesis. Nat Rev 4:225–232

Xoconostle-Cázares B, Vidal LCT, Domínguez-Fernández YG, Obando-González R, Padilla-Viveros A, Ruiz-Medrano R (2024) Harvesting in progress: the crucial role of genetically improved crops in Latin America. In: Genetically modified organisms. Intechopen

Yaseen N, Mumtaz A, Mazhar MW, Umair M, Naseer S (2024) A Comprehensive Review on Crop Modification Techniques Using Traditional and Modern Biotechnological Approaches. Trad Med 5(1):1–12

Yılmaz F (2014) Bitkisel Üretimde Genetiği Değiştirilmiş Organizmalar ve Ürünleri ile Biyogüvenlik

Yin H, Yang F, He X, Du X, Mu P, Ma W (2022) Advances in the functional study of glutamine synthetase in plant abiotic stress tolerance response. Crop J 10(4):917–923

Yorulmaz S, Ay R (2006) Genetigi Degistirilmis Organizmaların (GDO) Entomoloji Alanındaki Uygulama Olanakları. Süleyman Demirel Üniversitesi Ziraat Fakültesi Dergisi 1(2):53–59

Zhang SZ, Yang BP, Feng CL, Tang HL (2005) Genetic transformation of tobacco with the trehalose synthase gene from Grifola frondosa Fr. enhances the resistance to drought and salt in tobacco. J Integr Plant Biol 47(5):579–587

Zhou W, Li M, Achal V (2024) A Comprehensive Review on Environmental and Human Health Impacts of Chemical Pesticide Usage. Emerg Contam 11:100410

Rural Development and Role of Extension Workers for Climate-Smart Agriculture

20

Muhammad Abdul Rahman and Zain ul Abideen

Abstract

Climate-smart agriculture (CSA) is a holistic approach to addressing the challenges posed by climate change on food security, rural development, and agricultural sustainability. CSA integrates practices and technologies that enhance productivity, build resilience, and reduce greenhouse gas emissions, while ensuring the long-term sustainability of agri-food systems. Its relevance to food security lies in its ability to increase crop yields, improve the resilience of farming systems to climate-induced disruptions, and promote environmentally sustainable practices, thereby securing food supply for growing populations. Additionally, CSA practices, such as agroforestry, water management, and climate-resilient crop varieties, are crucial for adapting agricultural systems to changing environmental conditions. Rural development plays a complementary role by improving infrastructure, access to markets, and the socioeconomic conditions of rural communities, making them more resilient to food insecurity. Extension workers are pivotal in facilitating the adoption of CSA practices at the grassroots level. By disseminating knowledge, promoting sustainable farming techniques, and providing farmers with practical solutions, extension workers act as change agents in improving agricultural productivity and rural livelihoods. This interconnected approach, combining CSA, rural development, and extension services, offers a pathway to achieve sustainable food security in the face of climate change and economic challenges.

M. A. Rahman (✉) · Z. ul Abideen
Department of Economics and Agricultural Economics, PMAS Arid Agriculture University, Rawalpindi, Pakistan
e-mail: arahman@uaar.edu.pk

© The Author(s), under exclusive license to Springer Nature Singapore Pte Ltd. 2025

S. Faiz et al. (eds.), *Climate Smart Agriculture for Future Food Security*,
https://doi.org/10.1007/978-981-96-4499-5_20

Keywords

Greenhouse gas emissions · Agroecological requirements · Agroforestry and livestock management · Climate-smart innovations

20.1 Context and Importance of Climate-Smart Agriculture

The notion of climate-smart agriculture (CSA) refers to sustainable and adaptive techniques against changing climate, which covers the scope of agriculture and livestock combined with forests and fisheries (FAO 2024). CSA practices also include such climate resilient and environment-friendly approaches that highlight concerns for changing climate patterns and necessary adaptation strategies in agri-food systems, which are consistent with food security risks (Lipper et al. 2014). CSA addresses and attempts to tackle major objectives, like increasing agricultural production and incomes for sustainability and adaption to changing climate including excess rainfall and increased greenhouse gas emissions (Barua and Mitra 2024)..

20.1.1 CSA's Relevance to Climate Change and Food Security

How CSA and food security are interlinked? CSA is important for food security because it incorporates agricultural practices and technologies that aid in achieving diverse public policy goals such as feeding growing population by means of increased crop productivity, better resilience, and reduced greenhouse gas emissions (Lipper et al. 2017). Conventional techniques of agriculture together with the principles and concerns of sustainability can deal with global threats of food security (Haldar 2023).

CSA uses a variety of strategies to highlight and address matters related to food security and climate change. It also explains and addresses climate change within the agrifood system, paying attention to complementariness and trade-offs between productivity, adaptation and mitigation (Teklu et al. 2024). CSA also includes practices and technologies which are developed for the masses' specific agroecological requirements, as well within the socioeconomic context (Ahmad et al. 2020). Food security can be ensured with climate-resilient crop varieties, sustainable agriculture techniques, agroforestry, and livestock management leveraging precision farming tools including water advanced management techniques or strategies (Taylor 2018; Wakweya 2023).

20.2 Overview of Rural Development and Role of Extension Workers

20.2.1 Rural Development

In this chapter we will be focusing on importance of rural development in reducing/adapting threats of food insecurity and for achieving sustainable development for rural settlements. Purpose of rural development is to increase the standard and quality of life in rural areas; this includes development in every sector like establishment of infrastructure and provision of basic necessities of life (Belshaw 1977). Rural development is also to create an environment that offers sustainability and promotes climate-smart agricultural practices as this increases the income source of villagers. Less-developed countries mostly rely on agricultural income and fostering rural development in combination with CSA approaches can lead to increase in income level (Tomashuk 2017). In addition to all these, provision of electricity, construction of basic transportation routes, and infrastructures targeting to link rural communities with markets, cities, or industries are key components of rural development (Riva et al. 2018).

In the modern world, the rural development cannot be conceptualized by the integration of CSA practices. CSA helps in attaining its major goals such as increased agricultural produce which leads to several benefits like the creation of job opportunities and improving livelihoods of inhabitants. Also, adopting climate-smart agriculture practices reduces risk of food security (Mccarthy et al. 2018; Chavula 2021).

20.2.2 Role of Extension Workers

Extension workers have a pivotal role in agricultural sector of a county; their primary role includes dissemination of latest research and technology to the stakeholder of agriculture and farmers in particular; extension workers help in effectively implementing the research in practical terms by sharing information and promoting sustainable farming practices as well (David and Samuel 2014). Traditionally, extension workers have been tasked with educating farmers about modern agricultural technologies and promoting implementation of novel practices to increase farm output (Suvedi and Kaplowitz 2016). This stimulates and encourages agriculturists to increase their agricultural outputs; extension department offers assistance to farmers that is tested and validated by extensive research; they also conduct outreach programs where the farmers' grievances are heard and resolved (Okoronkwo et al. 2024).

Extension workers are sometimes referred to as agents of change. Indeed, the extension agents function as consultants and motivators for innovation. They perform several duties to improve farmer capacities introducing best-adapted and climate-smart innovations (Abegaz and Wims 2015). Conservation of biodiversities and the sustainable farming practices are areas in which extension workers involve actively. Other roles are to advocate for the rights of agriculturist, documentation of local agricultural varieties, to assist in establishment of seed banks, or serving as support base to the government in community engaging projects (Adeleye and Atere 2022).

20.3 Rural Development and Climate-Smart Agriculture

20.3.1 Integrating Climate-Smart Agriculture Practices while Strategizing Rural Development

In order to shrink risk of food security and increase in rural livelihood under changing climatic patterns, it is necessary for rural development policymakers must integrate practices which fall within a framework CSA. Chikaire et al. (2015) revealed by his research conducted in Southeast Nigeria that underscored the importance of conventional adaptation strategies, i.e., use of organic manure and afforestation. Similarly, the Asia-Pacific Network is making efforts jointly in order to deal with climate modification difficulties underlining on greener skills comparable micro-irrigation & zero-tillage farming (Kaushik et al. 2023). So far, CSA projects executed in emergent economies of South and Southeast Asia and Sub-Saharan Africa highlight the necessity for adopting institutional, technological and regulatory responses to address increasing attempts at managing climate change impacts on crop production systems (Abegunde et al. 2019). The installment of climate-smart villages (CSVs) in various locations, including India, demonstrated a comprehensive approach integrating weather, water and market smart technologies, ensuring sustainable agricultural production under volatile climates (Aggarwal et al. 2013; Aggarwal et al. 2018).

Climate-smart agricultural techniques strengthen the rural development through climate-smarter agriculture technology and strategies, resiliency, sustainability and higher life paying crops can be achieved in local communities. Climate-smart agriculture (CSA) enables and advocates such land use system which provides multifaceted benefits to rural communities in terms of increase agricultural production, reducing emissions and improving climate resilient practices (Pokorny et al. 2013). The benefits of the implementation of CSA strategies which might include conservation farming, soil fertility management, diversification and sector-based irrigation could lead to increased crop yields which result in more farm revenue as well as raising the living standards of rural farmers (Ariani et al. 2018; Taylor 2018).

Consequently, according to the Sustainable Development Goals of UN climate-smart agriculture (CSA) also fosters food security and helps in poverty reduction in agricultural areas worldwide while enhanced their abilities to adapt or be resilient against climatic changes (Haldar 2023). Climate-smart agriculture (CSA) technologies that are integrated with conventional agricultural techniques can improve the economy and productivity of agriculture as well show features as environment sustainability, positive externalities in health benefit (Abhilash et al. 2021; Takacs-Gyorgy and Takács 2022). Furthermore, initiatives like climate-smart villages and organizations can support to promote the adoption of climate-smart agriculture approaches through improved credit availability, push for global collaboration in technology transfer for mainstreaming climate-smart agriculture practices (Nagar et al. 2023).

20.3.2 Capacity Building for Climate Resilient Agriculture in Rural Areas

A climate-smart agriculture is only possible through strengthening the skills, knowledge and resources of farmers as well as other stakeholders to better manage and adapt to changing patterns of climate change. At the heart of this process, there are wide-ranging education and training programs on sustainable farming practices as well as through compelling farmers for effective management of resources through climate-smart technologies (Kombat et al. 2021). Capacity building for climate resilient agriculture in rural areas can be ensured by the timely communication of useful knowledge and skill development of farmers through workshops, field demonstrations and on-field research engagement programs that is done to give agriculturist the chance to see themselves the practical demonstration of new approved practices (Kurgat et al. 2020). These initiatives provide practically implementation and hands-on training about cropping patterns and introduction with revamped varieties to boost annual yields and soil erosion.

Investing in capacity building is also crucial to improve the potential of agricultural institutions like extension department and other community-based organizations that are spreading awareness about climate adaptation subject (Amadu et al. 2020). This comprise of equipping these institutions with the tools, resources as well as training disposal to provide uninterrupted mentoring and support to farming community which can also be made available or ensured through the use of ICT approaches (Cynthia and Nwabugwu 2016). Building networks and links with stakeholders such as government, nongovernmental organizations (NGO), universities, research centers and businesses to market sustainable agriculture practices and climate resilient approaches (Dinesh et al. 2017). Secondly, the combination of conventional agro-practices with modern scientific approaches will result in more geocentric solutions which means that practicing such types should make sense not only scientifically but also practically (Singh and Singh 2017). Empowering rural communities through the capacity building programs will result in more resilience to the adversities employed by a changing climate.

20.4 Contribution of Extension Workers in Climate-Smart Agriculture

20.4.1 Adaptation of Roles to Promote Climate-Smart Agriculture Practices

To promote climate-smart agriculture (CSA) practices, various roles of extension workers need adaptation. First, the implementation of CSA requires the realignment of current agricultural practices and the introduction of smart techniques to minimize negative environmental impacts and enhance resilience to climate change (Morecroft et al. 2012). Additionally, the adoption of CSA practices necessitates the use of precision agriculture tools like remote sensing and data analytics for site-specific management decisions. Furthermore, the incorporation of advanced technologies such as Internet of Things (IoT), artificial intelligence (AI), and blockchain can significantly contribute to CSA by monitoring climate and fertility status, enhancing crop management, and ensuring transparency and traceability in the agricultural supply chain (Roy and George 2020; Gupta et al. 2022). Moreover, educating farmers about CSA, providing workshops, and enhancing their knowledge through educational programs are crucial roles that institutions like the department of agriculture, land reforms authorities, and rural development authorities can play in promoting CSA adoption (Mizik 2021). By synergizing traditional practices with modern innovations and technologies, conventional farming practices can be effectively modernized to meet the challenges of climate change and ensure agricultural sustainability (Singh and Singh 2017).

20.4.2 Dissemination of Climate-Smart Agriculture Knowledge and Techniques

Adoption of Climate-Smart Agriculture (CSA) technologies is paramount in enhancing the adaptation and mitigation capacity against climate change-related impacts on agricultural productivity. The process consists of the methodological and periodic transfer through information and innovations for improving production while helping to mitigate climate change, making it more resilient among environmental changes (Ogunyiola et al. 2022). Dissemination strategies include extension services, Farmer Field Schools (FFS), demonstration plots, and digital platforms. These are routes through which specific and practical information is given to farmers within their contexts or problems timely.

Senyolo et al. (2021) highlighted that partnerships with research institutions, nongovernmental organizations, and governmental agencies involved in dissemination are highly relevant in this case. Collaborative approaches can serve as the link between scientific research and successful practical applications, providing farmers not only with access to climate-smart agriculture techniques but also enabling them how to implement it in real-time scenarios. In addition, adoption strategies that include traditional knowledge and practice can increase the relevance of climate-smart agriculture techniques and influence acceptance among farmers for establishing a sustainable agriculture (Rodríguez-Barillas et al. 2024).

20.4.3 Outreach Programs in Affected Areas

Community outreach in climate-impacted areas is important for creating awareness about resiliency and helping communities pivot to new environmental-tweezed norms. This includes programs that are intended to provide short- and long-term relief by providing resources, training, and support to people who may be at the greatest risk from climate impacts (Franzel et al. 2019). A great asset in effective outreach is working with community leaders, local organizations, and residents to learn through designing solutions that are practical as well culturally competent. This helps reinforce that the direction of implementation towards concrete results is not limited in scope, providing for sustainable solutions.

In vulnerable regions, awareness raising campaigns should emphasize educational and capacity building activities that allow communities to adapt climate-smart mitigation measures leading towards sustainable development. This might involve conducting training, demonstration, and publication and dissemination of materials and relevant literature on best practices in agriculture (Dinesh et al. 2018). Improved self-reliance and climate resilience drives and programmes like outreach can enable those exposed to the risks of climate change to better withstand and cope with the prospective threats of climate change.

20.5 Capacity Building for Extension Workers

20.5.1 Essential Skills and Knowledge for Climate-Smart Agriculture

To implement climate-smart agriculture, extension workers should possess skills and knowledge about sustainable farming practices, usage of advanced agricultural technologies, and integrating ecological principles into farmlands. To develop the awareness about climate variability and food security, it is necessary for extension workers to be competent in crop diversification techniques, practising soil management, and farming practices like rainwater harvesting or drip irrigation. They should be able to ensure effective crop health and resource management by using advanced technology process such as GPS-based geolocation, remote sensing which helps in monitoring remotely (Turyasingura and Chavula 2022). Insights into agroforestry and integrated pest management can also lend a helping hand to curb down the environmental burden while enhancing farm productivity. Extension workers also need capacity to analyze and interpret data in order to make predictions on the weather (Al-Zahrani et al. 2018). Continuous learning and mastering of innovative practices are crucial in spreading awareness among agriculturists.

20.5.2 Training Programs, Workshop, and Value Addition

Capacity building is important as it strengthens the feedback pathways from field to extension staff and researchers, increasing knowledge and skills required for keeping up with rapidly changing technological innovations (Ogunbameru et al. 2013).

Capacity building of extension workers not only helps in training farmers and strengthening the innovation process, but also creates links around researchers and feedback of farmers (Photakoun and Millar 2009). Improved capacity building of agriculture extension relevant to climate-smart agriculture resource management, poverty alleviation, and adaptive cropping strategies is imperative. The implementation of climate-smart agriculture projects will never boost the required agricultural growth unless farmers, groups, and organizations have been capacitated (Iqbal 2014). We have no other choice but to increase the size of the extension department if we are to solve farmer problems and meet climate-smart agricultural policy objectives.

20.6 Implementation Strategies for Climate-Smart Agriculture

In order to support sustainable agricultural systems, implementation of CSA through a participatory framework is necessary that brings farmers, researchers, policymakers, and financial institutions together in an environment to work collaboratively; this can contribute towards building resilient agricultural systems against climate change impacts.

20.6.1 Promoting Sustainable Farming Practices

Promoting sustainable farming methods is a solid education and outreach strategy. Information on agroecological techniques such as crop rotation, intercropping, and conservation tillage can help to improve soil health. Digital platforms may help in wide dispersion of such initiatives as training and dissemination. In addition, partnering with agricultural schools and cooperatives can help promote the transition to practices reducing harmful chemical-based inputs and increasing ecological harmony (Vishnoi and Goel 2024). While these practices are context-specific, the success of management depends on their adaptation to local conditions and active participation by farmers in experiential learning.

20.6.2 Initiation of Pilot Projects

CSA techniques can be implemented to pilot projects for testing and then implementation at a smaller scale before being rolled out more broadly. Such projects should be farmer-led to guarantee local input and make them truly effective. Pilot projects offer useful insights and proof of concepts. The lessons learned from these projects can help to scale up efficiency of work thus providing a ripple effect that accelerates the overall uptake (Raj and Garlapati 2020). Additionally, successful pilot projects illustrate the value of ongoing feedback and incremental advancement in fostering adaptive agricultural systems.

20.6.3 Subsidized Projects for Climate-Smart Agriculture

One of the most important ways to encourage CSA practices is by offering financial subsidies and this strategy applied in all cases, mostly where smallholder farmers are producing without having enough technical resources (Tankha et al. 2020). The subsidies fund the related expenditure of buying climate-resilient varieties or hybrids, installing water-efficient methods as well using renewable energy solutions. And here, subsidized training programs can help the farmers to understand and learn these innovations in a better way (Westermann et al. 2018). In addition to these benefits, financial incentives help disperse the economic risk of changing farming practices and provide credibility that a focus on long-term resilience really is being supported.

20.7 Lessons Learned from the Developed World

The developed countries have lessons to offer for climate-smart agriculture adoption and adaption as well as fostering food security via proper rural development practices such as agricultural extension services. The synthesis of all these case studies shows how integrated solutions that involve multiple areas such as technical innovation, financial support, and local community-friendly strategies can help create some change at our traditional agriculture.

A good case in point is the Netherlands where farmers are taken through extensive government programs on sustainable practices. The programs involved have offered subsidies, technical training, and market access on a participatory basis between farmers, researchers, and policymakers. Altogether, this combined support system worked well to strengthen the resilience of agriculture against climate change (Long et al. 2016).

In Australia, extension services have been key to enable climate adaptation strategies in the agriculture sector. Through the Australian Climate Change Research Program, farmers have access to leading research on farm-scale adaptation and better management practices. Extension workers translate complicated scientific info into easy to understand data, and supported by this, farmers will become a lot of conversant in climate-resilient practices (Palutikof et al. 2019; Cobon et al. 2021).

Another interesting case is Precision Agriculture, where the US stands out. Today, farming has become more technologically driven thanks to equipment guided by GPS, remote sensing devices, and data analytics, making this possible as we strive towards optimizing resource use and maximizing crop yield. The extension services help in training and operationalization of these technologies, thereby showing the promising impacts that technology-driven interventions can have over sustainable agricultural practices (Mcfadden et al. 2023).

20.8 Challenges Faced in CSA Techniques Implementation

Implementing climate-smart agriculture in the rural areas is a challenging task which often comes with obstacles that can prevent progress. These include everything from financial and logistical problems to social, political, and cultural issues that must be solved in order for a conducive environment for sustainable farming practices.

20.8.1 Resource and Financial Constraints in Rural Development

In many cases, the lack of resources and funding is a major obstacle for rural communities to implement CSA approaches. Smallholder farmers have limited access to credit and financial services which hampers their abilities to invest in the technologies or infrastructure they need. These are exacerbated by poor road networks, unreliable electricity supply, and in consequence limited market access for farmers as well as long egg journeys (Nasir 2024). These constraints can be overcome through investments in specific types of rural infrastructure and with the development of innovative finance mechanisms designed for small-scale farmers.

20.8.2 Financial and Logistical Challenges for Extension Workers

Extension workers face many financial and logistical constraints that hamper their effectiveness in assisting farmers. Limited resources restrict the range of extension work that can be conducted and low transportation and communication facilities in remote areas reduce accessibility to services. Better funding to support extension services, leveraging mobile technologies for providing remote assistance, and partnerships with local organizations are key in enhancing the efficiency of existing infrastructure (Msuya et al. 2017).

20.8.3 Social and Cultural Barriers and Instinct Behavior of Opposing Change

Adoption of CSA practices may be negatively affected by social and cultural barriers, so this should represent a major concern. Hence, there will be considerable resistance to new ways of doing things if they are seen as risky or if traditional farming practices have been inculcated over generations (Zakaria et al. 2020). Meeting these challenges requires listening to and involving local communities in order to take their opinions into account as project designs are developed, creating demonstration projects that can prove the benefits of a more sustainable approach.

20.8.4 Lack of Interest and Ignorance

Other challenges include a low level of CSA awareness and interest among some farmers. Farmers may be reluctant to change without realising the significance and benefits of sustainability. Concerted campaigns and educational efforts, held alongside with community meetings, are essential for ensuring both awareness raising and interest generation. Global Community Dialogues (GCD) are suitable tools to illustrate the positive effects of CSA by utilizing the services of local media (Moutouama et al. 2022).

20.8.5 Fixed Costs of Changing Cropping Patterns

Farmers may want to use CSA practices, but are stopped because the amount of fixed costs that come with changing cropping patterns will always be too high. For small-sized landholder and farmers, investing in new seeds, equipment, and infrastructure can cost too much. That kind of interventions would likely include financial incentives, subsidies, and technical assistance to overcome those barriers and make the transition to climate-resilient cropping systems more economically attractive for farmers (Mutenje et al. 2019).

20.9 Innovative Approaches for Future Directions

Novel techniques are vital to progress climate insightful cultivating and cultivate as a predominant practice. They must be technology-enabled, community-owned, and financially viable pathways to a sustainable and climate-resilient agricultural future.

20.9.1 Use of Digital Marketing and Artificial Intelligence

Artificial intelligence is capable of evaluating enormous amounts of data on meteorological conditions, the quality and fertility of the soil, and the performance of crops for customised recommendations in planting, irrigation, or even preventing pests. Artificial intelligence can be utilized for digitization and at platforms of larger scale, empowering a farmer with resources and the right support system connected or supported via AI, which can detect whether a farmland has adopted or is following sustainable and climate-friendly agricultural practices (Singh et al. 2020; Ahmed et al. 2022).

20.9.2 Community-Based Approaches and Participatory Methods

It is also important to rely on community-based approaches and participatory methods for the implementation of CSA initiatives that can be advertised. These approaches engage farmers in the design, delivery, and assessment of interventions to guarantee their relevance to local needs and circumstances. The participatory model through interventions such as farmer field schools and community-based workshops not only incorporates but enhances learning among communities to engender a culture of knowledge sharing, ownership, adaptive capacity for uptake of climate-smart practices (Lee 2017).

20.9.3 Fund Raising and Investment Ventures on Green Agriculture

It is essential to raise the resources and investment in green agriculture as a precursor for scaling-up of CSA practices. By mobilizing fiscal resources from the public as well as private sector, it will help in nurturing sustainable agriculture projects. New cost-effective financing models such as green bonds, impact investing, and climate finance can help to raise the capital needed for investments in renewable energy projects, sustainable agriculture methods, or more eco-friendly infrastructure. Apart from contributing to environmental sustainability, these investments ensure a financial return turning green agriculture into an attractive venture (Arif et al. 2020; Casey et al. 2021).

20.10 Conclusion

This discourse has established that rural development plays a critical role in promoting climate-smart agriculture, enhancing food security, and ensuring climatic resilience. By delivering the infrastructure, resources and support systems that farmers need to put sustainable practices into action. Agricultural extension workers connect researchers and practitioners, providing advice that responds to the specific needs of local areas while also facilitating innovation and community involvement. Their role is key to support sustainable agriculture and productivity, as well as build resilience against climate impacts.

Future usage of CSA should revolve around digital technologies to scale up community-based approaches in implementation and financial readjustment for sustainability. They enable rural communities to better cope with climate change, securing sustainable food for the long term. Rural development policies and extension workers working together in a complementary manner are crucial for building climate-resilient agriculture, resilient to the impacts of climate change.

References

Abegaz DM, Wims P (2015) Extension agents' awareness of climate change in Ethiopia. J Agric Educ Ext 21(5):479–495

Abegunde VO, Sibanda M, Obi A (2019) The dynamics of climate change adaptation in sub-Saharan Africa: a review of climate-smart agriculture among small-scale farmers. Climate 7(11):132

Abhilash, Rani A, Kumari A, Singh RN, Kumari K (2021) Climate-smart agriculture: an integrated approach for attaining agricultural sustainability. Climate Change and Resilient Food Systems: Issues, Challenges, and Way Forward: 141–189

Adeleye IA, Atere OB (2022) Perception of extension workers on climate-smart agricultural practices in Ekiti state. GSJ 10:7

Aggarwal PK, Zougmoré RB, Kinyangi J (2013) Climate-smart villages: a community approach to sustainable agricultural development

Aggarwal, P. K., Jarvis A. Campbell BM, Zougmoré RB, Khatri-Chhetri A, Vermeulen SJ, Loboguerrero A, Sebastian LS, Kinyangi J, Bonilla-Findji A,Radeny M, Recha J, Martinez-Baron D, Ramirez-Villegas J, Huyer S, Thornton P, Wollenberg E, Hansen J, Alvarez-Toro P, Aguilar-Ariza A, Arango-Londoño D, Patiño-Bravo V, Rivera O, Ouedraogo M, Tan Yen B (2018). The climate-smart village approach: framework of an integrative strategy for scaling up adaptation options in agriculture. Ecology and Society 23(1):14. https://doi.org/10.5751/ES-09844-230114

Ahmad F, Talukdar NR, Uddin M, Goparaju L (2020) Climate smart agriculture, need for 21st century to achieve socioeconomic and climate resilience agriculture in India: a geospatial perspective. Ecol Quest 31(1):78–100

Ahmed RA, Hemdan EED, El-Shafai W, Ahmed ZA, El-Rabaie ESM, Abd El-Samie FE (2022) Climate-smart agriculture using intelligent techniques, blockchain and internet of things: concepts, challenges, and opportunities. Trans Emerg Telecommun Technol 33(11):e4607

Al-Zahrani K, Aldosari F, Baig M, Shalaby M, Straquadine G (2018). Assessing the competencies and training needs of agricultural extension workers in Saudi Arabia. J. Agr. Sci. Tech. (2017) Vol. 19: 33-46.

Amadu FO, Mcnamara PE, Miller DC (2020) Yield effects of climate-smart agriculture aid investment in southern Malawi. Food Policy 92:101869

Ariani M, Hervani A, Setyanto P (2018) Climate smart agriculture to increase productivity and reduce greenhouse gas emission–a preliminary study. IOP Publishing, p 012024

Arif M, Jan T, Munir H, Rasul F, Riaz M, Fahad S, Adnan M, Mian IA, Amanullah (2020) Climate-smart agriculture: assessment and adaptation strategies in changing climate. Global Climate Change and Environmental Policy: Agriculture Perspectives: 351–377

Barua P, Mitra A (2024) Review on climate smart agriculture practice: a global perspective. J Clim Change 10(1):11–20

Belshaw DG (1977) Rural development planning: concepts and techniques. J Agric Econ 28(3):279–292

Casey J, Bisaro A, Valverde A, Martinez M, Rokitzki M (2021) Private finance investment opportunities in climate-smart agriculture technologies. Retrieved on May, 25: 2023

Chavula P (2021) A review between climate smart agriculture technology objectives' synergies and tradeoffs. future 17:18

Chikaire J, Ani A, Atoma C, Tijjani A (2015) Capacity building: key to agricultural extension survival. Sch J Agric Vet Sci 2(1A):13–21

Cobon D, Jarvis C, Reardon-Smith K, Guillory L, Pudmenzky C, Nguyen-Huy T, Mushtaq S, Stone R (2021) Northern Australia Climate Program: supporting adaptation in rangeland grazing systems through more targeted climate forecasts, improved drought information and an innovative extension program. Rangeland J 43(3):87–100

Cynthia EN, Nwabugwu EN (2016) Challenges to adoption of ICT tools by agricultural extension workers in Anambra State, Nigeria. Asian J Agric Extension Econ Sociol 10(4):1–6

David MM, Samuel HS (2014) The role of agriculture extension in the 21 century: reflections from Africa. Int J Agric Ext 2(1):89–93

Dinesh D, Aggarwal P, Khatri-Chhetri A, Rodríguez AML, Mungai C, Sebastian L, Zougmore R (2017) The rise in climate-smart agriculture strategies, policies, partnerships and investments across the globe. Agric Dev 30:4–9

Dinesh D, Zougmore RB, Vervoort J, Totin E, Thornton PK, Solomon D, Shirsath PB, Pede VO, Lopez Noriega I, Läderach P (2018) Facilitating change for climate-smart agriculture through science-policy engagement. Sustain For 10(8):2616

FAO (2024) Climate-smart agriculture

Franzel S, Kiptot E, Degrande A (2019) Farmer-to-farmer extension: a low-cost approach for promoting climate-smart agriculture. The climate-smart agriculture papers: Investigating the Business of a productive, resilient and Low emission future: 277–288

Gupta D, Gujre N, Singha S, Mitra S (2022) Role of existing and emerging technologies in advancing climate-smart agriculture through modeling: a review. Eco Inform 71:101805

Haldar N (2023) Sustainable and climate smart agriculture for food security: a review. J Exp Agric Int 45(11):229–239

Iqbal M (2014) Capacity building through extension education. Int J Agric Ext:93–100

Kaushik N, Siddique K, Ahmed MS, Hakimi S (2023) Climate-smart agriculture strategies for South Asia to address the challenges of climate change: identification of climate-resilient agriculture practices for India, Bangladesh, and Afghanistan. APN Sci Bull 13(1):241–262

Kombat R, Sarfatti P, Fatunbi OA (2021) A review of climate-smart agriculture technology adoption by farming households in sub-saharan africa. Sustain For 13(21):12130

Kurgat BK, Lamanna C, Kimaro A, Namoi N, Manda L, Rosenstock TS (2020) Adoption of climate-smart agriculture technologies in Tanzania. Front Sustain Food Syst 4:55

Lee J (2017) Farmer participation in a climate-smart future: evidence from the Kenya Agricultural Carbon Project. Land Use Policy 68:72–79

Lipper L, Thornton P, Campbell BM, Baedeker T, Braimoh A, Bwalya M, Caron P, Cattaneo A, Garrity D, Henry K (2014) Climate-smart agriculture for food security. Nat Clim Chang 4(12):1068–1072

Lipper L, Mccarthy N, Zilberman D, Asfaw S, Branca G (2017) Climate smart agriculture: building resilience to climate change. Springer Nature

Long TB, Blok V, Coninx I (2016) Barriers to the adoption and diffusion of technological innovations for climate-smart agriculture in Europe: evidence from The Netherlands, France, Switzerland and Italy. J Clean Prod 112:9–21

Mccarthy N, Lipper L, Zilberman D (2018) Economics of climate smart agriculture: an overview. Climate smart agriculture: building resilience to climate change: 31–47

Mcfadden J, Njuki E, Griffin T (2023). Precision Agriculture in the Digital Era: Recent Adoption on U.S. Farms, EIB-248, U.S. Department of Agriculture, Economic Research Service.

Mizik T (2021) Climate-smart agriculture on small-scale farms: a systematic literature review. Agronomy 11(6):1096

Morecroft MD, Crick HQ, Duffield SJ, Macgregor NA (2012) Resilience to climate change: translating principles into practice. J Appl Ecol 49(3):547–551

Moutouama FT, Tepa-Yotto GT, Agboton C, Gbaguidi B, Sekabira H, Tamò M (2022) Farmers' perception of climate change and climate-smart agriculture in Northern Benin, West Africa. Agronomy 12(6):1348

Msuya CP, Annor-Frempong FK, Magheni MN, Agunga R, Igodan C, Ladele AA, Huhela K, Tselaesele NM, Msatilomo H, Chowa C (2017) The role of agricultural extension in Africa's development, the importance of extension workers and the need for change. Int J Agric Ext 5(1):51–58

Mutenje MJ, Farnworth CR, Stirling C, Thierfelder C, Mupangwa W, Nyagumbo I (2019) A cost-benefit analysis of climate-smart agriculture options in Southern Africa: balancing gender and technology. Ecol Econ 163:126–137

Nagar AK, Namdeo S, Verma B, Shrivastava S, Sarkar R (2023) Role of climate-smart agriculture towards sustainable food production. Int J Environ Clim Change 13(12):465–472

Nasir J (2024) Development challenges of Pakistan: constraints and choices. Springer, pp 235–256

Ogunbameru B, Mustapha S, Idrisa Y (2013) Capacity building for climate change adaptation: modules for agricultural extension curriculum development. Russ J Agric Soc-Econ Sci 14(2):61–66

Ogunyiola A, Gardezi M, Vij S (2022) Smallholder farmers' engagement with climate smart agriculture in Africa: role of local knowledge and upscaling. Clim Pol 22(4):411–426

Okoronkwo DJ, Ozioko RI, Ugwoke RU, Nwagbo UV, Nwobodo C, Ugwu CH, Okoro GG, Mbah EC (2024) Climate smart agriculture? Adaptation strategies of traditional agriculture to climate change in sub-Saharan Africa. Front Clim 6:1272320

Palutikof JP, Boulter SL, Stadler F, Vidaurre ACP (2019) Tracking the progress of climate change adaptation: an Australian case study. Environ Sci Pol 101:126–135

Photakoun V, Millar J (2009) Evaluating capacity building methods to strengthen livestock extension outcomes in Laos. Ext Farm Syst J 5(2):91–100

Pokorny B, De Jong W, Godar J, Pacheco P, Johnson J (2013) From large to small: reorienting rural development policies in response to climate change, food security and poverty. Forest Policy Econ 36:52–59
Raj S, Garlapati S (2020) Extension and advisory services for climate-smart agriculture. Global climate change: Resilient and smart agriculture: 273–299
Riva F, Ahlborg H, Hartvigsson E, Pachauri S, Colombo E (2018) Electricity access and rural development: review of complex socio-economic dynamics and causal diagrams for more appropriate energy modelling. Energy Sustain Dev 43:203–223
Rodríguez-Barillas M, Klerkx L, Poortvliet PM (2024) What determines the acceptance of climate smart technologies? The influence of farmers' behavioral drivers in connection with the policy environment. Agric Syst 213:103803
Roy T, George KJ (2020). Precision farming: a step towards sustainable, climate-smart agriculture. Global climate change: resilient and smart agriculture: 199–220
Senyolo MP, Long TB, Omta O (2021) Enhancing the adoption of climate-smart technologies using public-private partnerships: lessons from the WEMA case in South Africa. Int Food Agribus Manag Rev 24(5):755–776
Singh R, Singh G (2017) Traditional agriculture: a climate-smart approach for sustainable food production. Energy Ecol Environ 2:296–316
Singh V, Huseynov R, Jhajj S (2020) Role of digital marketing for smart agriculture: their implications, practices and future direction. Economic Agents:167–202
Suvedi M, Kaplowitz MD (2016) What every extension worker should know: core competency handbook. Michigan State University, Department of Community Sustainability
Takacs-Gyorgy K, Takács I (2022) Towards climate smart agriculture: how does innovation meet sustainability? Ecocycles 8(1):61–72
Tankha S, Fernandes D, Narayanan N (2020) Overcoming barriers to climate smart agriculture in India. Int J Clim Change Strategies Manag 12(1):108–127
Taylor M (2018) Climate-smart agriculture: what is it good for? J Peasant Stud 45(1):89–107
Teklu A, Simane B, Bezabih M (2024) Climate smart agriculture impact on food and nutrition security in Ethiopia. Front Sustain Food Syst 7:1079426
Tomashuk I (2017) Problems and prospects of management of rural development. Balt J Econ Stud 3(5):214–220
Turyasingura B, Chavula P (2022) Climate-smart agricultural extension service innovation approaches in Uganda: Review Paper. International Journal of Food Science and Agriculture, 6(1). DOI:10.26855/ijfsa.2022.03.006
Vishnoi S, Goel RK (2024) Climate smart agriculture for sustainable productivity and healthy landscapes. Environ Sci Pol 151:103600
Wakweya RB (2023) Challenges and prospects of adopting climate-smart agricultural practices and technologies: implications for food security. J Agric Food Res 100698
Westermann O, Förch W, Thornton P, Körner J, Cramer L, Campbell B (2018) Scaling up agricultural interventions: case studies of climate-smart agriculture. Agric Syst 165:283–293
Zakaria A, Azumah SB, Appiah-Twumasi M, Dagunga G (2020) Adoption of climate-smart agricultural practices among farm households in Ghana: the role of farmer participation in training programmes. Technol Soc 63:101338

Climate Change and Technological Innovations: Regulatory Concerns

21

Syeda Anber Zahra, Youzhi Wu, Javed Iqbal, Banzeer Ahsan Abbasi, Faiza Munir, Shoaib Khan, Rashid Iqbal, Ghulam Murtaza, and Majid Hussain

Abstract

Climate change poses an unprecedented threat to humanity, necessitating immediate action to mitigate its far-reaching consequences. This chapter contends that technological innovations are crucial in addressing climate change, and that a multifaceted approach is required to reduce carbon emissions and promote sustainability. The transition to renewable energy sources, adoption of smart agriculture practices, and development of carbon capture technologies are vital components of our efforts to combat climate change. Sustainable practices and low-carbon technology adoption can only be accelerated with strong policies and regulations. Businesses and individuals can be encouraged to adopt sustainable practices by governments and international organizations that establish frameworks that encourage investment in low-carbon technologies. Taking a collab-

Syeda Anber Zahra and Javed Iqbal contributed equally with all other contributors.

S. A. Zahra (✉) · Y. Wu
School of Food and Pharmacy, Shanghai Zhongqiao Vocational and Technical University, Shanghai, China

J. Iqbal (✉)
Department of Botany, Bacha Khan University, Charsadda, Khyber Pakhtunkhwa, Pakistan

B. A. Abbasi
Department of Botany, Rawalpindi Women University, Rawalpindi, Pakistan

F. Munir
Department of Agricultural Sciences and Technology, Atta-ur-Rahman School of Applied Biosciences, National University of Sciences and Technology, Islamabad, Pakistan
e-mail: faiza.munir@asab.nust.edu.pk

S. Khan
Department of Chemistry, Abbottabad University of Science and Technology (AUST), Abbottabad, Pakistan

© The Author(s), under exclusive license to Springer Nature Singapore Pte Ltd. 2025

S. Faiz et al. (eds.), *Climate Smart Agriculture for Future Food Security*,
https://doi.org/10.1007/978-981-96-4499-5_21

orative approach is crucial to tackle climate change, and this chapter stresses the significance of technological innovation in this regard. Sustainable land use practices, lower carbon emissions, and a prosperous planet for future generations can all be achieved through the application of technological innovation with a focus on sustainability. Expanding the reach of these technologies, removing obstacles to their widespread adoption, and discovering uncharted territories in the fight against and adaptation to climate change should all be the priorities of future studies.

Keywords

Technological innovation · Climate change · Renewable energy · Smart agriculture · Carbon emissions · Carbon capture and storage

21.1 Introduction

Climate change is one of the "greatest challenges" humanity is facing in the twenty-first century affecting economy, environment, society, affecting our agriculture, threatening food security, and even our moral compass. For millennia, a delicate equilibrium existed between life on earth and the climate. However, this balance began to shift following the Industrial Revolution (around 1850), with the effects accelerating in the mid-twentieth century. Climate change, driven by human activities, has become a paramount threat to human well-being and biodiversity (Shivanna 2022). Human activities have caused a steady rise in greenhouse gas (GHG) emissions, driving climate change (Fig. 21.1). The primary culprits are carbon dioxide (76%), methane (16%), and nitrous oxide (2%) (Chen et al. 2023). The steadiness of sunshine redirected back into space is crucial, as it balances the temperature immersed by the world, keeping a stable universal heat. However, the accumulation of greenhouse gases, primarily alongside water vapor, acts like a transparent blanket in the atmosphere (Viebahn et al. 2007). These gases trap infrared radiation (heat energy) emitted from Earth's surface, reradiating it back and causing a gradual rise in global temperature, known as the greenhouse effect. Early human societies

R. Iqbal
Department of Agronomy, Faculty of Agriculture and Environment,
The Islamia University of Bahawalpur, Bahawalpur, Pakistan

Department of Life Sciences, Western Caspian University, Baku, Azerbaijan
e-mail: rashid.iqbal@iub.edu.pk

G. Murtaza
School of Agriculture, Yunnan University, Kunming, Yunnan, China

M. Hussain
Department of Forestry and Wildlife Management, University of Haripur,
Haripur City, KP, Pakistan
e-mail: majid@uoh.edu.pk

Fig. 21.1 Different factors affecting climate change

caused limited deforestation through subsistence farming, clearing land for crops to feed themselves and their communities. During this preindustrial era, a natural balance existed between the amount of carbon dioxide released and the amount absorbed by plants, with forests acting as the primary carbon sinks. However, the Industrial Revolution marked a turning point. Deforestation rates have skyrocketed due to the demands of urbanization, expanding industries, and large-scale agriculture (Mac Kinnon et al. 2018).

Satellite data reveals a staggering increase in agricultural land use. Over the past two decades, an additional 1 million square kilometers of land has been converted for crops, with roughly half replacing forests and other ecosystems (Potapov et al. 2022). Furthermore, recent decades have seen a surge in deforestation to accommodate commercial plantations (oil palm, coffee, etc.), cattle ranching, and mining activities, leading to a significant decline in global forest cover. The World Wildlife Fund (WWF) estimates a loss of over 43 million hectares of forest between 2004 and 2017, highlighting the alarming rate of deforestation worldwide (Pacheco et al. 2021).

Human impact on the environment has intensified over the past three centuries. Anthropogenic carbon dioxide emissions threaten to significantly alter global climate for millennia. This growing influence was recognized as early as 1873 by geologist Antonio Stoppani, who likened human activity to a powerful geological force (Crutzen and Brauch 2016). Population growth and resource exploitation have accelerated this impact. The human population has ballooned tenfold in the last three centuries, and deforestation for agriculture and development is rampant (Zahra et al. 2021). Tropical rainforests, crucial carbon sinks, are disappearing at an alarming rate, further accelerating climate change and biodiversity loss (Boota et al. 2024).

Fossil fuel burning and industrial activities have exacerbated the problem. Energy use has soared, leading to substantial increases in greenhouse gases like carbon dioxide and methane. These concentrations now exceed levels seen in the past 400,000 years, with more to come. The consequences, including acid rain, smog, and climate warming, are already being felt. The ocean acts as a giant heat sink for Earth's climate system, with a heat capacity roughly 1000 times greater than the atmosphere (Soomro et al. 2024). This translates to a dominant role in absorbing excess heat. Scientific consensus overwhelmingly confirms a substantial warming of the global ocean since 1955. This warming accounts for over 80% of the changes in Earth's climate system's energy content during this period (Cochrane et al. 2009).

Observations reveal widespread warming in the upper 700 m of the global ocean. However, the Atlantic Ocean exhibits deeper warming (up to 3000 m) compared to the Pacific, Indian, and Southern Oceans (Shi et al. 2024). This phenomenon is attributed to the North Atlantic's deep overturning circulation. Furthermore, surface water temperatures in lakes and rivers across Europe, North America, and Asia have increased by 0.2–2 °C since the 1960s. To enable this long-term decarbonization, a significant shift in the technologies used for energy production and consumption is imperative, requiring substantial investments in innovation (Han et al. 2024).

The Intergovernmental Panel on Climate Change (IPCC) emphasizes that the cost of emission reduction policies will heavily depend on future investments in research, development, and demonstration (RD&D) activities. Achieving progress in combating climate change and other environmental challenges entails innovation across various low-carbon technologies, covering all stages of technological advancement, from idea creation to widespread adoption. Lowering the costs of existing low-carbon technologies like offshore wind turbines and solar panels is crucial to enable their extensive deployment, while advancing fundamental research is essential for technologies such as smart grids and energy storage systems (Dechezleprêtre et al. 2019).

Climate change policies can significantly influence investment in low-carbon technologies, leading to a shift in innovation efforts from polluting technologies to low-carbon ones. Empirical evidence suggests that climate change regulations encourage the diffusion of environmentally friendly technologies, driving innovation activities in low-carbon technologies, thereby favoring R&D in this sector. Higher energy prices resulting from climate change policies spur innovation, with studies showing a positive correlation between increased energy prices and energy patenting in low-carbon technologies (Holtz et al. 2018). This demonstrates the influence of policy-induced price changes on innovative activity.

The innovative response to climate change policy measures like the EU Emission Trading System is swift, with increased innovation activity observed in low-carbon technologies among participating companies (Saraji and Streimikiene 2023). This highlights the effectiveness of policy mechanisms in stimulating innovation towards environmentally sustainable solutions. Climate change regulations play a vital role in steering economies away from polluting economic trajectories towards low-carbon pathways, which can contribute to enhanced profits, competitiveness, and growth through the development and adoption of innovative low-carbon technologies (Shen et al. 2023).

21.2 Technological Innovation

Technological innovation is a pivotal driver of economic growth, societal progress, and sustainable development. The relentless pursuit of novel solutions and iterative refinements has yielded remarkable advancements in various domains (Garud et al. 1997). As we navigate the complexities of the digital age, technological innovation continues to redefine the contours of industries, governments, and individuals (Fig. 21.2). Through the lens of disruptive technologies, we witness the democratization of access, the acceleration of processes, and the emergence of new paradigms (Garud et al. 2013). The confluence of artificial intelligence, blockchain, and the Internet of Things (IoT) is crafting a future where efficiency, transparency, and adaptability are paramount (Kumaraswamy et al. 2018) Fig. 21.3.

Technological innovation is a major strategy for energy conservation and emission reduction. Energy efficiency, emissions, and consumption can all be improved, and the expansion of renewable energy sources can be accelerated. A high level of renewable energy technology innovation (RETI) can transform the energy structure and expand renewable energy supply capacity at a reduced cost (Xin et al. 2022). Due to its zero carbon emissions, renewable energy is acknowledged as the energy of the future, therefore utilizing renewable energy on a wide scale can reduce climate change and provide energy security (Liu et al. 2022). For this reason, RETI is frequently regarded as an affordable means of achieving a low-carbon society. Renewable energy sources are generating more power than ever before, while low-carbon living and production methods are being developed. These are clear signs of the positive impact of RETI on reducing CO_2 emissions in society (Turkenburg and

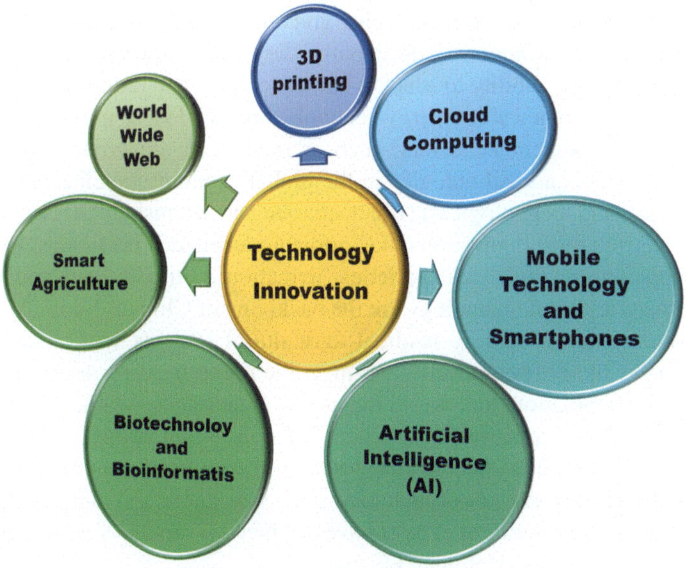

Fig. 21.2 Different forms of technological innovation

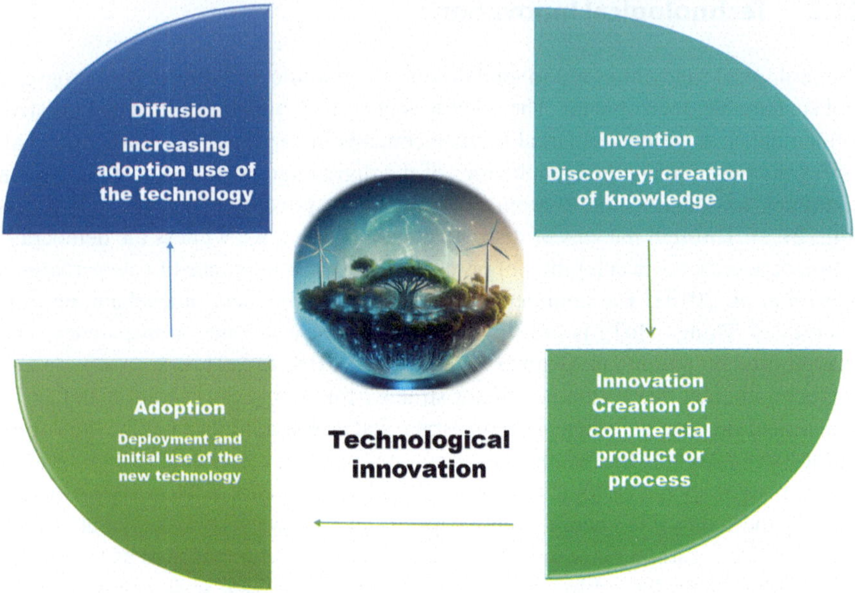

Fig. 21.3 Phases of technological innovation

Faaij 2000). The growing power generated by renewable energy sources and the advancement of low-carbon living and production methods are clear signs of the positive impact of RETI on reducing CO_2 emissions in society (Si et al. 2024).

As Renewable Energy Technologies (RETs) reduce air pollution, they offer a cleaner alternative to fossil fuels, reducing greenhouse gas emissions and contributing to a healthier environment. Additionally, they aid in lowering greenhouse gas emissions. A country's ability to adapt to climate change and reduce greenhouse gas emissions through renewable energy technologies (RETs) is influenced by both endogenous and exogenous factors, such as inadequate governance and limited access to resources and technologies in the global economy (socioeconomic standard, financial, and technological) (Bhattacharjee and Jahanshahi 2024).

Modern agricultural innovations like inorganic fertilizers, pesticides, foodstuffs, dietary supplements, high-yield varieties, irrigation methods, land management, and modern agricultural technology are the backbone of Climate-Smart Agriculture (CSA), which aims to achieve resilient agricultural systems with abundant crop yields. In order to alleviate poverty and meet the rising need for food caused by a fast increasing population, this was of the utmost importance. Mathews et al. (2018) found that in order to improve the well-being of small-scale farmers in the short and long term, it is essential to understand the costs and benefits of possible CSA choices (Mathews et al. 2018). As part of their endeavors to adapt to and mitigate the effects of climate change, agricultural experts have produced a variety of agricultural technologies and approaches (Hellin and Fisher 2019).

21.3 Renewable Energy Technology

Renewable energy technology (RETs) sources have such a profound impact on monetary output and social progress that they are intrinsically linked to sustainable development. There is hope that renewable energy sources can help stabilize energy prices, boost economic and social growth, make energy more accessible, slow the rate of climate change, and lessen negative effects on human and environmental health (Gielen et al. 2019). The most well-known and significant RETs are enhanced water mills, solar, wind, and micro and macro hydropower (Fig. 21.4).

One sustainable and endless source of energy is solar power. Energy that is directly harvested from the Sun is known as solar energy. Solar energy's versatility has led to its increased use across the world. It may be used for a variety of purposes, including producing heat, desalinating water, and generating electricity. Photovoltaics (PVs) and Concentrated Solar Power (CSP) are the two primary taxonomies of solar energy applications (Shahsavari and Akbari 2018). Producing energy does not affect the environment by releasing gases like CO_2, SO_2, and NOx, or greenhouse gases (GHGs). One square meter of sunshine may produce 1700 kilowatt-hours of power annually on a worldwide scale. The exact locations with the maximum sunlight exposure include desert regions and tropical regions close to equator (fur Sonnenenergie 2009).

To use solar power, you can do one of two things: Photovoltaic (PV) electricity is great for homes and businesses alike, whereas concentrated solar power (CSP) is better suited to industrial and commercial uses on a larger scale. The precise locations are from 3 GW in 2003 to 219 GW in 2015 and even more to 430 GW in 2018; the world's use of photovoltaic (PV) solar electricity has grown substantially. The first sun-powered mirrors were used in the seventh century, marking the beginning of the historical chronology of solar energy. Scientists have been able to convert sunlight into electricity for several decades after its discovery in 1893 (the photovoltaic effect). Scientists all over the globe have put in a lot of time and effort to create

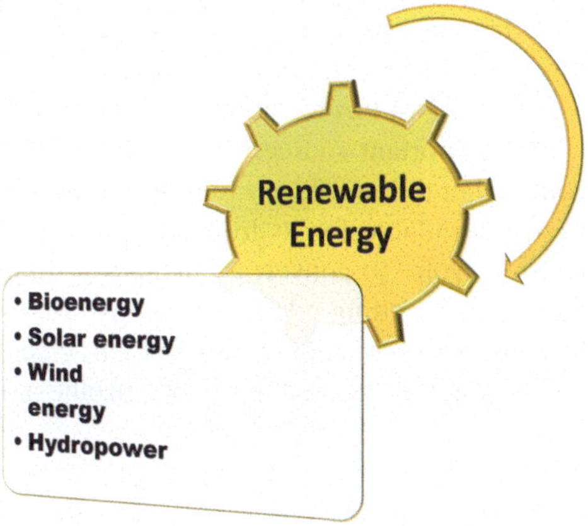

Fig. 21.4 Types of renewable energy

solar thermal and solar PV, the two main uses of solar energy technology (Aman et al. 2015).

Solar panels use photovoltaic (PV) technology to collect sunlight and transform it into electricity. These photovoltaic (PV) devices have surpassed all others in terms of affordability as a means of producing power in many parts of the world, thanks to their widespread use. Electricity generation costs from solar photovoltaic facilities fell 77% between 2010 and 2018. But the quantity of solar PV electricity that was installed increased 100-fold between 2005 and 2018. In order to meet the aims of the 2030 Sustainable Development Goals (SDGs) and the Paris Climate Agreement, solar photovoltaics (PV) have become an integral part of the low-carbon sustainable energy system that is necessary to supply cheap and dependable power (International Energy Agency 2009).

Solar cells, or photovoltaic (PV) devices, are electronic components that convert sunlight into electricity. Among today's renewable energy sources, photovoltaics (PVs) are among the fastest-growing. As a result, its future contribution to global power generation is expected to be enormous. One of the most common ways to desalinate saltwater or other saline water using solar PV is electrodialysis, or ED. Consequently, solar photovoltaic (PV) modules are closely linked to the desalination process. This process uses DC electricity to draw salt out of saltwater or water that has already been salted (Suman 2021).

In photovoltaic thermal (PV-T) technology, a thermal collector is attached to the back of a standard solar photovoltaic (PV) module. With this setup, domestic hot water may be preheated. More of the incident solar energy can be transformed into usable thermal and electrical energy on the collector. If a building can balance its energy consumption and production, it will have zero net energy emissions and will not emit any carbon dioxide. This type of construction is known as a zero-energy building (Edenhofer et al. 2011). For the goal of supplying energy, buildings can make use of building-integrated photovoltaic (BIPV) technology, which combines solar energy with equipment already installed within the building. So, building-integrated photovoltaics that use thermal energy (BIPV/T) use state-of-the-art technology like solar cooling. Concentrated Photovoltaic (CPV) technology enhances sunshine coverage on PV cells by using sparkly or refractive concentrators, thus make best use of energy output (Yumkella 2012). In order to achieve their great efficiency, solar cells typically use a stack of semiconductor materials. The efficiency rating of this technique is more than 47%.

One important source of energy is hydropower, which uses the flow of water from higher to lower levels to turn turbines that generate electricity. There is a range of sizes available for hydropower projects, including run-of-river, in-stream, and dam/reservoir. To convert a fluctuating supply into usable electricity, hydropower plants need cutting-edge technology. Reducing the impact of both floods and droughts is a common function of hydroelectric reservoirs. Gravity and the height at which water is poured into the turbine are the principal factors that generate electricity. The quantity of energy generated is affected by the little drop in reservoir level caused by water outflow. A turbine's design takes into account the possibility of an optional water flow. They can store energy for a long time, improve quickly, and release very little emissions (Førsund 2015).

One kind of sustainable energy is bioenergy, which is derived from naturally occurring biological sources. Cooking, heating, generating power, and transporting products using biodiesel are all possible with bioenergy, making it a flexible energy supply. Bioenergy creates power from a variety of sources, including leftovers of agriculture (such as sugar cane waste), forestry (such as wood wastes), and animal husbandry (such as cow manure). By-products, residues, or waste products are common names for the fuel that comes from the sources listed above. One major advantage of using biomass to power generators is this. Most importantly, it keeps arable land and fuel land from competing for resources (Ajanovic 2011).

Although biofuel production is still relatively modest on a worldwide scale, it is steadily increasing. Fifteen billion liters of biodiesel were used in the US in 2006. A volume of 30 billion liters is targeted to be achieved by the end of 2012, with an annual growth rate ranging from 30% to 50%. The Earth's interior provides a natural source of thermal energy, which is known as geothermal energy. There is a tight relationship between the planet's interior structure and the physical processes that occur there and the heat source. Despite being abundant in the earth's crust, even in the lowest regions, heat is often dispersed, seldom concentrated, and situated at depths too great to be physically recovered (Barbier et al. 2008). Approximately 30 °C per km is the mean geothermal gradient. Drilling allows access to parts of the Earth's interior with far steeper gradients than the norm. To harvest heat from geothermal reservoirs, a variety of technologies and wells are used. Naturally occurring hot and permeable reservoirs are known as hydrothermal reservoirs. The opposite is true with enhanced geothermal systems (ESGs), which are reservoirs that have had their heat and permeability increased via hydraulic stimulation. Thermoelectricity and other types of thermal energy may be generated from fluids of varying temperatures once they reach the surface (Owusu and Asumadu-Sarkodie 2016).

An increasingly important renewable energy source on a worldwide scale is wind power. Wind exists on every continent, and in certain areas it packs a powerful punch. Using the kinetic energy of wind, we can generate electricity. The primary goal of reducing greenhouse gas emissions is to generate power from massive turbines installed either on land (onshore) or in water (offshore), whether saltwater or freshwater (Bhargava et al. 2021).

The combination of ocean heat, waves, tides, and currents has the potential to double the world's energy consumption. In 2008, the first commercial ocean energy devices were introduced. SeaGen in the UK and Pelamis in Portugal were the lucky recipients of these units. Wind, waves, tides, and the temperature differential between shallow and deep water may now be used to generate energy from coastal regions (Esteban and Leary 2012).

21.4 Combating Climate Change Through Technological Innovations and SDG 7

Goal 7 of the Sustainable Development Agenda focuses on Clean and cheap Energy, which means providing energy services that are accessible, cheap, dependable, and ecologically sustainable (Khanal et al. 2019). Achieving carbon neutrality, reducing emissions of greenhouse gases, and stopping the increase in global temperatures are all worldwide initiatives and objectives in response to the climate change threat. Heat waves, heavy precipitation, droughts, and typhoons are becoming more common, and the acceleration of global warming due to increases in carbon emissions is threatening both human and ecological survival, as stated in the IPCC's sixth assessment report, "Climate Change 2021: Basis of Natural Science." (Fig. 21.5). The warmest decade in recent Earth history, lasting from 2011 to 2020, saw a worldwide surface temperature increase of 1.09 °C compared to the temperature during the Industrial Revolution. In its fifth assessment report, the United Nations Intergovernmental Panel on Climate Change laid out the scientific justification for the idea that emissions of greenhouse gases, the most abundant of which is carbon dioxide, are the primary cause of global warming. Cutting down on carbon dioxide emissions is the most effective way to slow down global warming. In order to mitigate the global warming crisis, "carbon control" is a must-do for all countries. People and cities must confront the possibilities and threats posed by climate

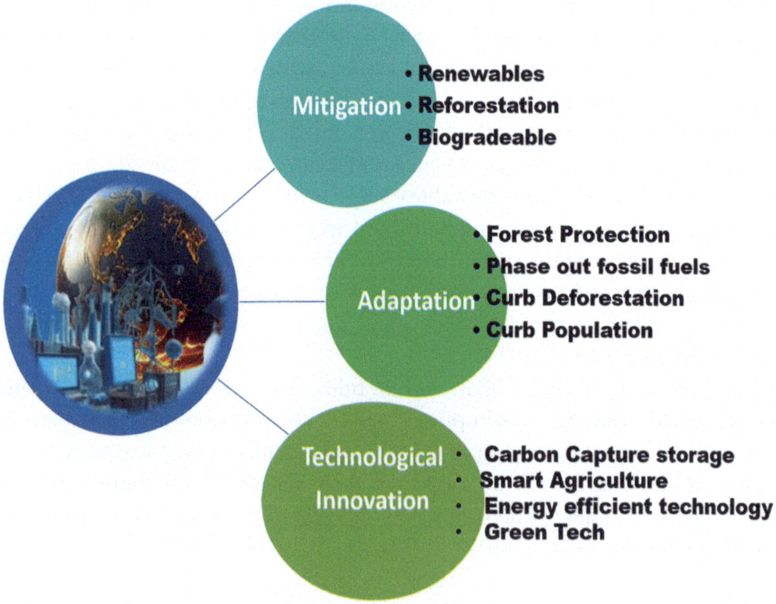

Fig. 21.5 Approaches and Solution for Climate Change

change, and they must take decisive action to transition to a low-carbon future (Elavarasan et al. 2021).

The Climate Change International Coordinating Mechanism, comprised of the Paris Agreement, the Kyoto Protocol, and the United Nations Framework Convention on Climate Change, is based on the principle of "common but differentiated responsibilities" for developed and developing nations. Each country's emission responsibilities and measures to reduce emissions are defined under this process. As the low-carbon shift in economic growth becomes increasingly obvious, the international community is reaching a consensus on how to combat climate change. Sixty nations have pledged to achieve zero net carbon emissions by the year 2050, according to the UNFCCC report from September 2019. Carbon emissions and digital technologies promoting the development of digital technologies form a crucial initial step in combating the climate change danger and achieving carbon peaking and carbon neutrality objectives. One of the primary principles of sustainable development is the ability to keep ecosystems healthy and biodiversity (Owusu and Asumadu-Sarkodie 2016). The carbon capture and storage (CCS) technology has a substantial environmental impact. Frequent monitoring of the atmosphere, terrain, bodies of water, and natural resources is commonly employed to ascertain this. CCS projects can potentially affect biodiversity, ecosystems, and cropland during their implementation.

The construction of transport routes through agricultural areas can lead to a reduction in food production. Relocation becomes imperative in specific projects to facilitate the establishment of robust and secure infrastructure. In addition, the decimation of marine life could render the environment highly vulnerable. Once carbon capture and storage (CCS) is implemented, the majority of projects related to fossil fuels will necessitate an increased number of skilled workers for maintenance purposes (Fan et al. 2011). Consequently, an increase in employment opportunities will result in an improvement in individuals' quality of life. Additional personnel will be required for the regular functioning of the facilities, and continuous examination and supervision of the storage site will also be imperative.

This will serve as a sustainable and enduring means of generating employment opportunities. Despite their limited duration, infrastructure construction projects will also generate employment opportunities. Furthermore, it will facilitate the training of a greater number of individuals in the surrounding areas. The global economy will achieve stability through the creation of additional employment opportunities. Employees will be guaranteed secure and stable employment, leading to an improvement in their overall quality of life. Establishing a project in a region characterized by a significant unemployment rate will effectively enhance the quality of life in that community and serve as a program to mitigate poverty (Viebahn et al. 2007). With the increase in job opportunities, there will be a large reduction in poverty levels and participants agree to adhere to health and safety protocols; the implementation of the technology will lead to a safer work environment.

The extraction of natural resources may improve energy infrastructure and stimulate economic growth in many countries, particularly developing ones (Chen et al. 2016). The nation's climate change outcomes may be greatly improved by incorporating carbon capture and storage (CCS) into fossil fuel projects. The energy produced by these sophisticated linked processes is oftentimes ecologically benign; Carbon Capture and Storage (CCS) energy systems may provide more attractive incentives than non-CCS facilities. At the same time as this would strengthen the national economy, it will also help CCS-related initiatives move forward (Keith et al. 2006). There is evidence in the literature that the energy market will benefit from CCS integration into energy-related initiatives. As a result, there is a greater amount of capital being allocated to the industry. Integrating CCS technology into the system reduces the carbon emissions associated with the use of fossil fuels. This is a compelling method through which the utilization of CCS convinces investors to finance the advancement of this state-of-the-art technology. CCS technology is essential for addressing climate change and serves as a strategic solution for meeting global energy demands (Seevam 2010).

Most policies regarding the implementation of CCS in projects related to fossil fuels primarily concentrate on the transportation and containment of CO_2. According to the International Energy Agency (IEA) in 2008, the European Union has had the most significant involvement with CCS (Carbon Capture and Storage) technology up to this point. CCS technology conventions are categorized into two primary groups.

Carbon Capture and Storage (CCS) is recognized as an attractive solution for decreasing greenhouse gas emissionsits. Additionally, the United Nations Convention on the Law of the Sea and the London Convention govern the process of oceanic sequestration for captured CO_2.

By 2050, China hopes to have cut its carbon emissions by 40–45%. To do this, we need to make some serious structural changes to the economy and put some effective plans in place to reduce emissions, such as cutting back on coal use. Total emissions from the energy sector can be decreased by integrating carbon capture and storage (CCS) technology into the main sources of energy production (Socolow et al. 2004).

The main goal of carbon capture and storage technology is to collect emissions that are produced by projects that use fossil fuels. Approximately 85–95% of CO_2 emissions may be absorbed by it. It shows that by using carbon capture and storage (CCS) technology in more than half of its thermal production processes, the power industry could cut CO_2 emissions by over 1 billion tons (Shen et al. 2023).

The practice of carbon capture and storage, or CCS, offers hope for a future free of fossil fuels and their negative environmental impacts. On the other hand, more energy is needed to capture and compress CO_2, which might reduce the energy conversion efficiency from 48% to 36%. Coal use may rise if carbon capture and storage is integrated into half of the thermal electricity used in homes.

While carbon capture and storage (CCS) might lessen our reliance on fossil fuels, it could also boost energy efficiency and consumption, which could necessitate the use of more coal (Baratti 2022).

21.5 Global Resilience Funding Programs for Climate Regulations and SDG 13

Sustainable development, industrial transformation, and economic growth are all goals that "science and technology for goodness" aspires to assist mankind achieve. Goal 13 emphasizes the need for immediate measures to address climate change and its consequences. It is inherently interconnected with all 16 of the other Goals of the 2030 (Singh et al. 2023) Agenda for Sustainable Development. A global climate compensations program cannot be launched without first establishing a global system. A long-standing organization, the Green Climate Fund (GCF) seeks to alleviate the financial burden of climate change on low-income nations. The United Nations Framework Convention on Climate Change's Conference of the Parties approved the GCF in 2010. The business finally got going a few years later, with headquarters in Incheon, South Korea (Baratti 2022).

As a result of the disproportionate impact on low-income communities and other disadvantaged groups, the global phenomena of climate change poses a significant human rights challenge. High Commissioner for Human Rights Michele Bachelet warned that human rights will be severely compromised by the projected levels of global warming in her 2019 opening remarks to the 42nd session of the Human Rights Council. The United Nations High Commissioner for Human Rights, Michele Bachelet, issued a dire warning about the human rights implications of the projected levels of global warming in her opening remarks to the 42nd session of the Human Rights Council in 2019 (Fig. 21.6).

Global warming is already having a negative impact on efforts to improve people's lives and the economy by enforcing economic and social rights, and it's just going to get worse. As a result of the climate catastrophe, it is becoming more difficult to safeguard and advance some human rights. These include basic human rights like the right to live and the right to be healthy, as well as a host of socioeconomic elements including housing, availability of clean water and sanitation, and access to healthy food. An increase in water- and vector-borne infections, asthma, allergies, heat-related ailments, violent incidents, and mental health issues are all ways in which climate change is negatively impacting public health (Hamann et al. 2015). Rising sea levels, extreme weather events, more frequent and intense rainfall in certain areas, and droughts in others are some of the environmental impacts. Reduced agricultural productivity, water purity, and worker productivity are all negative outcomes of these occurrences. In addition, many people may be forced to flee their homes since the land would be too flooded or unusable for farming. The negative effects of climate change on human health include heat-related diseases, infections spread by vectors, water- and food-borne illnesses, respiratory and allergy disorders, starvation, social instability, and mental health issues. One billion children are likely to experience negative consequences due to the climate catastrophe, according to a newly released UNICEF research (Mitchell et al. 2018).

Cooperation among states to promote human rights on a global scale is a matter of legal need. According to the High Commissioner for Human Rights, this partnership should involve the allocation of sufficient financial resources to help with

Fig. 21.6 Global regulatory bodies for climate change

climate change adaptation, mitigation, and remediation, especially from the well-off (Blanco 2021). Furthermore, in order for the global response to prioritize equity, the nations that caused the environmental disaster must make greater efforts to fix the harm they've caused, especially to the most vulnerable nations. Compensation becomes a concern in this context.

The disparity in the financial resources of different stakeholders is at the heart of Shue's "second principle" of equality. When several people or groups are working together towards a common objective, the reasoning goes, those with more resources should chip in more. This is called a collective effort.

It is the biggest worldwide fund devoted to combating climate change at the moment (Mehling et al. 2021). Supporting developing nations in their efforts to adapt to and mitigate the effects of climate change is central to this organization's mission. There is a balanced representation of high- and low-income nations on the 24-member board. International and national banks, private equity firms, and multilateral, regional, and national development organizations are among the public and private institutions that contribute to this organization's funding (Lucchetta 2023). With the Resilience Funding Programmes for Loss and Damage, appeals following significant environmental and climate-related disasters are no longer associated with the "charity" designation. Therefore, the foundation for constructing community and broader socio-ecological resilience is solidarity. Future developments and incremental impacts are expected to increasing because of global action is taken to lessen greenhouse gas emissions. This will be focused on longer-term initiatives that assist locally driven and democratically run programmes to lessen the effects of climate change, such as planting new mangroves or initiating coastal restoration projects in proportion to what has been destroyed or lost (Perry 2020).

The terms "loss and damage" allude to a certain class of costly damages (such as ruined infrastructure) and irreversible losses (such as the extinction of species, the loss of life, or the destruction of land) connected to the negative effects of climate change that can be seen as an adjunct to adaptation. Therefore, loss and damage take into account not just the current needs for climate adaptation, which are terribly financed, but also the connection between existing trends and the repercussions of climate change that are being felt today (Khan et al. 2020). In the Paris Agreement of 2015, it was first excluded, but as a result of pressure from vulnerable nations in the forum, they have drawn more attention to the issue in the climate talks at COP25 in Madrid, which did not result in a suitable settlement. In the Bahamas, for example, the assessed damage was US $2.5 billion, which included estimated losses of up to $717.3 million and additional costs of $220.9 million (Bailey et al. 2022).

21.6 Conclusion and Future Perspectives

In conclusion, the fight against climate change demands a comprehensive strategy that harnesses the power of technological innovation to slash carbon emissions and foster sustainability. A swift transition to renewable energy sources, such as solar and wind power, is vital to wean ourselves off fossil fuels and decrease greenhouse gas emissions. Additionally, embracing smart agriculture practices like precision farming and vertical farming can optimize agricultural productivity while minimizing environmental degradation (Potapov et al. 2022). The development and deployment of carbon capture technologies are also crucial in reducing industrial carbon footprints. By leveraging carbon capture and storage (CCS) technologies, we can significantly decrease emissions from power plants and industrial processes, making them a vital component of our climate change mitigation efforts (Bailey et al. 2022).

Technological innovation plays a pivotal role in promoting sustainable land use practices. Advanced technologies like satellite imaging and remote sensing enable us to monitor deforestation and land degradation, empowering policymakers to make informed decisions about land use management. Moreover, precision agriculture and vertical farming can optimize crop yields while conserving water and fertilizer usage (Shivanna 2022).

Effective policy and regulation are also essential in driving the adoption of low-carbon technologies and sustainable practices. Governments and international organizations must create frameworks that incentivize investment in low-carbon technologies and encourage businesses and individuals to adopt sustainable practices. This can be achieved through carbon pricing mechanisms, tax incentives, and subsidies for renewable energy. Ultimately, our ability to combat climate change hinges on our capacity to leverage technological innovation and prioritize sustainability in all aspects of our lives. By collaborating and embracing innovation, we can create a sustainable future and ensure a thriving planet for generations to come (Smith et al. 2023).

Established in 2006, the Global Facility for Disaster Risk and Reduction (GFDRR) is a global grant-funding mechanism for disaster risk projects, involving the World Bank, United Nations agencies, and bilateral contributors. GFDRR focuses on initiatives that aim to strengthen hydrometeorological services and early-warning systems as part of its broader mission to reduce vulnerability to natural hazards and climate change impacts (Clarvis, n.d.). The World Bank typically encourages and places conditions on initiating reforms to boost capital markets or private sector engagement and collaboration, even if the funds are typically provided on a grant or concessional basis and do not particularly seek to achieve the requirements of the relevant populations, but they also cover other interests. As a result, rather than being used as recompense or as a means of fostering greater unity and ultimately achieving climate justice, these payments are viewed as charitable contributions and are actively marketed as "pledges" by nations.

A number of players, nations, institutions, and regions would need to come together in order to create a global climate stabilization fund and resilience funding scheme for loss and damage (Khan and Roberts 2013). They would also need to understand that compensating for Donor countries' "charity" or "goodwill" does not justify the loss and suffering caused by climate change in disadvantaged and previously colonial countries. However, it also serves as a call to action, acknowledging that the historical, economic, and structural conditions of these countries have generated unique vulnerabilities that are reflected in the significant impacts of climate change on their political, cultural, and economic well-being. Thus, persuading nations—particularly those that were once colonial powers—that climate reparations are necessary in response to this exploitative connection will necessitate a multifaceted political strategy. Discussions, lobbying, and pertinent scholarly and policy-focused research amply demonstrate this connection. Working at the levels of regional organizations, civil society, and the international arena is necessary.

Author Contribution Statement Syeda Anber Zahra led the chapter's conceptualization, coordination, and writing, while Youzhi Wu, Javed Iqbal, Banzeer Ahsan Abbasi, Faiza Munir, Shoaib Khan, Rashid Iqbal, Ghulam Murtaza, and Majid Hussain contributed to data collection, analysis, and manuscript revisions, ensuring the chapter's quality and coherence.

References

Ajanovic A (2011) Biofuels versus food production: does biofuels production increase food prices? Energy 36(4):2070–2076

Aman MM, Solangi KH, Hossain MS, Badarudin A, Jasmon GB, Mokhlis H et al (2015) A review of Safety, Health and Environmental (SHE) issues of solar energy system. Renew Sust Energ Rev 41:1190–1204

Bailey J, Becker-Birck C, Bissoon D, Fox A, Gischler C, Hampton D, Sloan W (2022) Building a more resilient and low-carbon Caribbean: report 4: infrastructure resilience in the Caribbean through nature based solutions

Baratti RM (2022) The green climate fund, climate change and corporate due diligence: what role for the private facility sector? In: European yearbook of international economic law 2022. Springer, Cham, pp 63–81

Barbier EB, Koch EW, Silliman BR, Hacker SD, Wolanski E, Primavera J et al (2008) Coastal ecosystem-based management with nonlinear ecological functions and values. Science 319(5861):321–323

Bhargava V, Kasuba S, Maddula SP, Jagadish D, Khan MA, Padhy CP et al (2021) A case study of wind turbine loads and performance using steady-state analysis of BEM. Int J Sustain Energ 40(1):22–40

Bhattacharjee A, Jahanshahi AA (2024) Simulation of green supply chain design with renewable energy and green technology for intensifying sustainability after COVID-19. In: Green supply chain management practice and principles. IGI Global, pp 219–248

Blanco AM (2021) From stakeholders to rightsholders: assessing public participation in the international climate regime. CCLR 15:282

Boota MW, Zwain HM, Rasta M, Hu C, Liu C, Li Y, Soomro MHAA (2024) From lake to fisheries: interactive effect of climate and landuse changes hit on lake fish catch? Environ Res 258:119397

Chen HQ, Wang X, He L, Chen P, Wan Y, Yang L, Jiang S (2016) Chinese energy and fuels research priorities and trend: a bibliometric analysis. Renew Sust Energ Rev 58:966–975

Chen L, Chen Z, Zhang Y, Liu Y, Osman AI, Farghali M et al (2023) Artificial intelligence-based solutions for climate change: a review. Environ Chem Lett 21(5):2525–2557

Clarvis MH (n.d.) Financing of emergency preparedness and the Global Facility for Disaster Reduction and Recovery (GFDRR)

Cochrane K, De Young C, Soto D Bahri T (2009) Climate change implications for fisheries and aquaculture. FAO Fisheries and aquaculture technical paper, 530, 212

Crutzen PJ, Brauch HG (eds) (2016) Paul J. Crutzen: a pioneer on atmospheric chemistry and climate change in the Anthropocene, vol 50. Springer

Dechezleprêtre A, Martin R, Bassi S (2019) Climate change policy, innovation and growth. In: Handbook on green growth. Edward Elgar Publishing, pp 217–239

Edenhofer O, Pichs-Madruga R, Sokona Y, Seyboth K, Kadner S, Zwickel T et al (eds) (2011) Renewable energy sources and climate change mitigation: special report of the intergovernmental panel on climate change. Cambridge University Press

Elavarasan RM, Pugazhendhi R, Jamal T, Dyduch J, Arif MT, Kumar NM et al (2021) Envisioning the UN Sustainable Development Goals (SDGs) through the lens of energy sustainability (SDG 7) in the post-COVID-19 world. Appl Energy 292:116665

Esteban M, Leary D (2012) Current developments and future prospects of offshore wind and ocean energy. Appl Energy 90(1):128–136

Fan Y, Zhu L, Zhang X (2011) Analysis of global CCS technology, regulations and its potential for emission reduction with focus on China. Adv Clim Chang Res 2(2):57–66

Førsund FR (2015) Hydropower economics, vol 217. Springer

fur Sonnenenergie DG (2009) Planning and installing photovoltaic systems: a guide for installers, architecs and engineers. Earthscan

Garud R, Nayyar PR, Shapira Z, Shapira ZB (eds) (1997) Technological innovation: oversights and foresights. Cambridge University Press

Garud R, Tuertscher P, Van de Ven AH (2013) Perspectives on innovation processes. Acad Manag Ann 7(1):775–819

Gielen D, Boshell F, Saygin D, Bazilian MD, Wagner N, Gorini R (2019) The role of renewable energy in the global energy transformation. Energ Strat Rev 24:38–50

Hamann A, Roberts DR, Barber QE, Carroll C, Nielsen SE (2015) Velocity of climate change algorithms for guiding conservation and management. Glob Chang Biol 21(2):997–1004

Han X, Boota MW, Soomro SEH, Ali S, Soomro SGH, Soomro NEH et al (2024) Water strategies and management: current paths to sustainable water use. Appl Water Sci 14(7):154

Hellin J, Fisher E (2019) Climate-smart agriculture and non-agricultural livelihood transformation. Climate 7(4):48

Holtz G, Xia-Bauer C, Roelfes M, Schüle R, Vallentin D, Martens L (2018) Competences of local and regional urban governance actors to support low-carbon transitions: development of a framework and its application to a case-study. J Clean Prod 177:846–856

International Energy Agency (2009) World energy outlook. OECD/IEA, Paris, p 17

Keith DW, Ha-Duong M, Stolaroff JK (2006) Climate strategy with CO_2 capture from the air. Clim Chang 74(1):17–45

Khan MR, Roberts JT (2013) Adaptation and international climate policy. Wiley Interdiscip Rev Clim Chang 4(3):171–189

Khan M, Robinson SA, Weikmans R, Ciplet D, Roberts JT (2020) Twenty-five years of adaptation finance through a climate justice lens. Clim Chang 161(2):251–269

Khanal RC, Shakya SR, Bajracharya TR (2019) Contribution of renewable energy technologies (RETs) in climate resilient approach and SDG 7. J Inst Eng 15(3):393

Kumaraswamy A, Garud R, Ansari S (2018) Perspectives on disruptive innovations. J Manag Stud 55(7):1025–1042

Liu C, Xin L, Li J, Sun H (2022) The impact of renewable energy technology innovation on industrial green transformation and upgrading: beggar thy neighbor or benefiting thy neighbor. Sustain For 14(18):11198

Lucchetta M (2023) Climate bonds: are they invested efficiently? J Environ Manag 345:118864

Mac Kinnon MA, Brouwer J, Samuelsen S (2018) The role of natural gas and its infrastructure in mitigating greenhouse gas emissions, improving regional air quality, and renewable resource integration. Prog Energy Combust Sci 64:62–92

Mathews JA, Wentink GJ, Kruger L (2018) Climate-smart agriculture for sustainable agricultural sectors: the case of Mooifontein. Jàmbá J Disas Risk Stud 10(1):1–10

Mehling M, Obergassel W, Arens C, Beuermann C, Brandemann V, Hermwille L, Spitzner M (2021) From stakeholders to rightsholders: assessing public participation in the 282 international climate regime. Carbon Clim Law Rev 15(4):282–301

Mitchell LE, Lin JC, Bowling DR, Pataki DE, Strong C, Schauer AJ et al (2018) Long-term urban carbon dioxide observations reveal spatial and temporal dynamics related to urban characteristics and growth. Proc Natl Acad Sci 115(12):2912–2917

Owusu PA, Asumadu-Sarkodie S (2016) A review of renewable energy sources, sustainability issues and climate change mitigation. Cogent Engin 3(1):1167990

Pacheco P, Mo K, Dudley N, Shapiro A, Aguilar-Amuchastegui N, Ling PY et al (2021) Deforestation fronts: drivers and responses in a changing world. WWF, Gland, p 125

Perry KK (2020) Realising climate reparations: towards a global climate stabilization fund and resilience fund programme for loss and damage in marginalised and former colonised societies. Available at SSRN 3561121

Potapov P, Turubanova S, Hansen MC, Tyukavina A, Zalles V, Khan A et al (2022) Global maps of cropland extent and change show accelerated cropland expansion in the twenty-first century. Nat Food 3(1):19–28

Saraji MK, Streimikiene D (2023) Challenges to the low carbon energy transition: a systematic literature review and research agenda. Energ Strat Rev 49:101163

Seevam P (2010) Transporting the next generation of CO_2 for carbon capture and storage. Doctoral dissertation, University of Newcastle Upon Tyne

Shahsavari A, Akbari M (2018) Potential of solar energy in developing countries for reducing energy-related emissions. Renew Sust Energ Rev 90:275–291

Shen Y, Yang Z, Zhang X (2023) Impact of digital technology on carbon emissions: evidence from Chinese cities. Front Ecol Evol 11:1166376

Shi X, Guo J, Jalbani S, Asad M, Anwar MI, Hu C et al (2024) Effects of seasonal temperature regimes: does Cyprinus carpio act as a health hazard during the construction of Suki Kinari hydropower project on Kunhar River in Pakistan? Sci Total Environ 907:168023

Shivanna KR (2022) Climate change and its impact on biodiversity and human welfare. Proc Indian Natl Sci Acad 88(2):160–171

Si R, Wang Y, Cao M, Wen H (2024) Does green technology innovation promote green economic growth?–examining regional heterogeneity between resource-based and non-resource-based cities. Int Rev Econ Finan 94:103406

Singh P, Kaur S, Baabdullah AM, Dwivedi YK, Sharma S, Sawhney RS, Das R (2023) Is# SDG13 trending online? Insights from climate change discussions on Twitter. Inf Syst Front 25(1):199–219

Smith C, Baker JCA, Spracklen DV (2023) Tropical deforestation causes large reductions in observed precipitation. Nature 615(7951):270–275

Socolow R, Hotinski R, Greenblatt JB, Pacala S (2004) Solving the climate problem: technologies available to curb CO_2 emissions. Environ Sci Policy Sustain Dev 46(10):8–19

Soomro SEH, Soomro AR, Batool S, Guo J, Li Y, Bai Y, Wang Y (2024) How does the climate change effect on hydropower potential, freshwater fisheries, and hydrological response of snow on water availability? Appl Water Sci 14(4):65

Suman A (2021) Role of renewable energy technologies in climate change adaptation and mitigation: a brief review from Nepal. Renew Sust Energ Rev 151:111524

Turkenburg WC, Faaij A (2000) Renewable energy technologies. In: UNDP/UNDESA/WEC: energy and the challenge of sustainability. World energy assessment. UNDP, New York, pp 219–272

Viebahn P, Nitsch J, Fischedick M, Esken A, Schüwer D, Supersberger N et al (2007) Comparison of carbon capture and storage with renewable energy technologies regarding structural, economic, and ecological aspects in Germany. Int J Greenhouse Gas Contr 1(1):121–133

Xin L, Sun H, Xia X, Wang H, Xiao H, Yan X (2022) How does renewable energy technology innovation affect manufacturing carbon intensity in China? Environ Sci Pollut Res 29(39):59784–59801

Yumkella KK (2012) Multilateralism and energy for development. In: Energy for development: resources, technologies, environment. Springer, Dordrecht, pp 45–56

Zahra SA, Iqbal J, Abbasi BA, Shahbaz A, Kanwal S, Shah SL, Mahmood T (2021) Antimicrobial, cytotoxic, antioxidants, enzyme inhibition activities, and scanning electron microscopy of Lactuca orientalis (Boiss.) Boiss. seeds. Microscop Res Tech 84(6):1284–1295

22 Legume Production Under the Changing Climate

Amna Kiyani, Mingguang Lei, Ghanwa Rabbani, M. Muddassir Sardar, M. Kashif Ilyas, Umar Iqbal, and Ayesha T. Tahir

Abstract

Food security is a major concern globally both in terms of quality and quantity, with one in every three individuals being affected by malnutrition. Rapidly growing world population and climate change further amplify the problem. Legumes are often referred to as "superfoods" due to their high protein content and the presence of nutritive bioactive elements like vitamins, minerals, dietary fiber, and low glycemic index carbohydrates. Moreover, they also improve soil fertility through their symbiotic relationship with beneficial bacteria, thus promoting sustainable agricultural practices. Climate-induced stresses like high temperature, drought, and salinity pose a significant threat to legumes along with other food crops with a predicted decrease of up to 2.3% in global food production by the 2050s. These stresses alter the morphophysiological, biochemical, and molecular mechanisms in plants, resulting in changes in the oil and nutritional content, reduced nitrogen fixation, and loss of yield. The understanding of how these stresses affect legume plants, and the morphological, physiological, and biochemical responses to cope with these stresses, is crucial for developing climate-resilient crops. This chapter discusses the negative impact of these stresses on legume production and provides a comprehensive account of modern and conventional approaches that have been utilized to harness the potential of existing genotypes to develop stress-resilient legume varieties. An integrated strategy that

A. Kiyani · G. Rabbani · M. M. Sardar · A. T. Tahir (✉)
Department of Biosciences, COMSATS University Islamabad, Islamabad, Pakistan
e-mail: ayesha.tahir@comsats.edu.pk

M. Lei
College of Life and Environmental Sciences, Hangzhou Normal University, Hangzhou, China

M. K. Ilyas · U. Iqbal
National Agriculture Research Centre, Islamabad, Pakistan

© The Author(s), under exclusive license to Springer Nature Singapore Pte Ltd. 2025

S. Faiz et al. (eds.), *Climate Smart Agriculture for Future Food Security*,
https://doi.org/10.1007/978-981-96-4499-5_22

combines various management options and simultaneously addresses all abiotic stresses could be the answer to sustainable food production and security.

22.1 Introduction

Health and nutrition remain significant challenges, especially in low- and middle-income countries (LMICs). Globally, one in every three individuals is affected by at least one form of malnutrition such as hunger, wasting, stunting, micronutrient deficiencies (MND), or obesity, which contributes to the rise of diet-related noncommunicable diseases (NCDs) across populations (International Food Policy Research Institute [IFPRI] 2024). Studies have suggested that following food-based dietary guidelines (FBDGs) leading to a dietary shift with a 5% increase in legume consumption along with other food groups could reduce premature deaths from NCDs by 15% and lower greenhouse gas emissions by 13%. Hence, legumes are often referred to as superfoods and are key to food security and sustainable agriculture. It is the second largest family of seed plants comprising 600 genera and approximately 13,000 species, of which 20 are generally preferred by humans for dietary consumption (Kumar and Pandey 2020). In most cultures, legumes are a part of traditional diet with common beans, chickpeas, cowpeas, groundnuts, lentil, pigeon peas, and soybean being highly preferable (Nigam et al. 2021). Despite the broad-spectrum advantages of legumes and their promotion through FBDGs, research shows significant global variation in intake patterns, partly due to the absence of a unified daily target for legume consumption. While studies recommend intake of 50 g/day (Willett et al. 2019), global consumption remains well below half this target, averaging only 21 g/day (Rawal and Navarro 2019).

Nutritional significance of the legumes is owing to higher protein content, threefold higher than cereals (Abobatta et al. 2021; Semba et al. 2021). They also contain other valuable and nutritive bioactive molecules, including low glycemic index carbohydrates, dietary fiber, minerals such as iron, zinc, magnesium, calcium, potassium, sodium, and phosphorus, as well as vitamins A, B-6, C, D, E, K, folate, niacin, riboflavin, and thiamin, making them indispensable for human diet. Many studies have reported the beneficial effects of these bioactive molecules on human health. They help in the prevention and treatment of various diseases, including diabetes, cardiovascular diseases, and obesity (Cakir et al. 2019; Grdeń and Jakubczyk 2023; Torheim and Fadnes 2024). In addition to their nutritional significance for human health, legumes are also beneficial for the environment as they improve soil fertility through their symbiotic relationship with beneficial bacteria. The bacteria form nodules on their roots and fix atmospheric nitrogen, enriching the soil and reducing the need for nitrogen fertilizers, thus promoting sustainable agricultural practices (Kebede 2021).

Legumes are cultivated in over 120 countries across the continents. According to FAOSTAT (2021), Brazil is the largest producer of legumes, followed by the USA, India, China, Argentina, Albania, Angola, Algeria, Canada, and Russia. Other notable contributors include Turkey, Bangladesh, Nepal, Iran, Mexico, Myanmar,

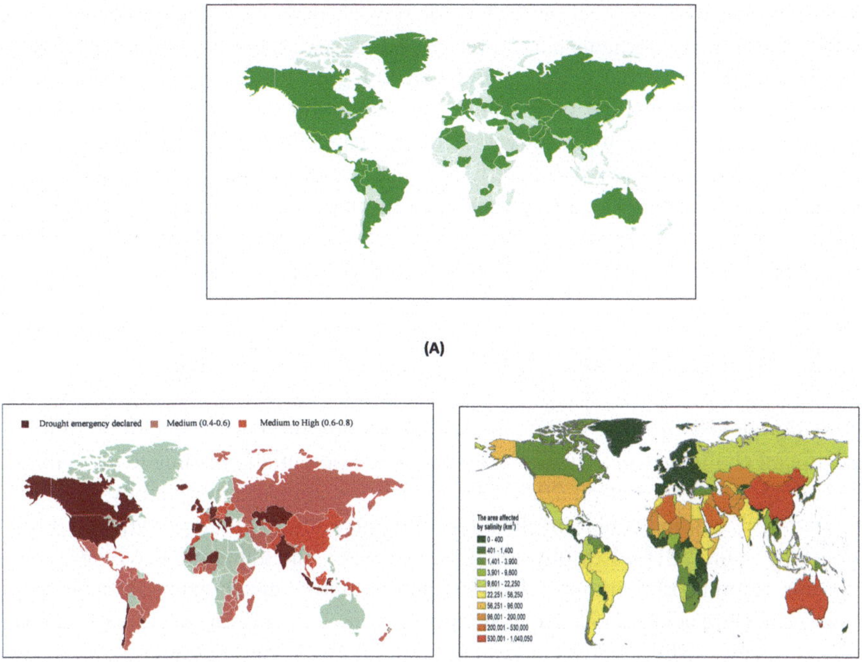

(A)

(B) (C)

Fig. 22.1 (**a**) Legume production worldwide (Mapped from data of Dell'Olmo et al. (2023)). (**b**) Drought stress (https://www.statista.com/chart/25101/countries-by-drought-risk/). (**c**) Salinity stress (https://www.eurekalert.org/multimedia/707020)

Ethiopia, Australia, Spain, Pakistan, Syria, Morocco, and Egypt. Climate change poses significant challenges to legume production worldwide (Al-Tawaha et al. 2022). Although plants face a combination of biotic and abiotic stresses, this chapter will focus on three main abiotic stresses, including heat stress, drought, and salinity. Figure 22.1 highlights the legume-producing regions around the globe along with drought- and salinity-affected areas. The maps clearly show that legume-producing areas are significantly affected by these environmental stresses that are tremendously increasing with climate change over the years.

Climate change, combined with a growing global population, slows down the economy and intensifies the adverse impacts on agricultural systems, thereby increasing food insecurity by affecting livelihoods, incomes, and food systems at local, national, and global levels (Toromade et al. 2024). Calzadilla et al. (2013) assessed the potential impacts of climate change on global agriculture using GTAP-W model. The results predicted a reduction in global food production by 0.5% in the 2020s and up to 2.3% in the 2050s, which could lead to a rise in food prices by 39–43% for all crops.

Given the numerous benefits of legume crops and the need to meet the nutritional requirements of a growing population, it is essential to recognize the climatic constraints in their production and work toward managing and resolving these issues. This chapter presents current research and case studies that provide valuable insights into the complex relationship between legumes and a changing climate. It examines both threats and opportunities by identifying challenges and exploring potential solutions to mitigate adverse effects. This chapter discusses the negative impacts of climate change particularly abiotic stresses, such as heat stress, drought, and salinity on global legume production. Furthermore, it explores strategies to enhance these crops' tolerance and resilience to environmental stresses, enabling them to fulfill their vital role in global food production amidst climate change.

22.2 Abiotic Stresses Affecting Legume Production

Climate-induced stresses, such as high temperature, drought, and salinity, present significant challenges to agricultural productivity, potentially leading to yield losses exceeding 50% (Sher et al. 2024). These stresses alter the morphophysiological, biochemical, and molecular mechanisms of plants, limiting their growth and development (Pereira 2016). Key physiological processes, including photosynthesis, respiration, reproduction, nitrogen fixation, and oxidative metabolism, are negatively influenced (Sita et al. 2017). These stresses influence the overproduction of reactive oxygen species (ROS) which can damage cellular DNA, cause protein denaturation, and induce lipid peroxidation, thereby disturbing cellular homeostasis and resulting in significant yield losses (Sultana et al. 2014). Their impact on legumes is particularly complex due to their symbiotic relationship with Rhizobia, a defining feature of these crops (Dave et al. 2024). Understanding how these stresses affect legume plants is crucial for mitigation and ensuring sustainable crop production. A detailed discussion of the impacts of these stresses on legumes is presented in the later sections. Figure 22.2 illustrates how these stresses affect key morphophysiological features, influencing growth, yield, and seed quality.

22.2.1 Heat Stress

Temperature is among the most critical factors influencing plant growth, determining where crops can be grown based on their sensitivities and directly affecting crop phenology and yield. Among environmental stresses, temperature stress has the broadest and most significant impact on crops' growth and metabolism, often resulting in a substantial reduction in yield potential (Bhandari et al. 2017).

Legume production and quality are particularly sensitive to temperature changes. The optimum temperature range for grain legume crops is between 10 and 36 °C (Farooq et al. 2017) with specific thresholds as follows: chickpea 15–30 °C, common bean 20–24 °C, cowpea 18–28 °C, faba bean 25 °C, groundnut 30–35 °C, lentil 15–30 °C, lupins 20–30 °C, mung bean 28–35 °C, pea 15–25 °C, pigeon pea 18–30 °C, soyabean 23–26 °C, and urd bean 25–35 °C (Dave et al. 2024). An increase in temperature beyond the threshold, even by 1°, can induce heat stress,

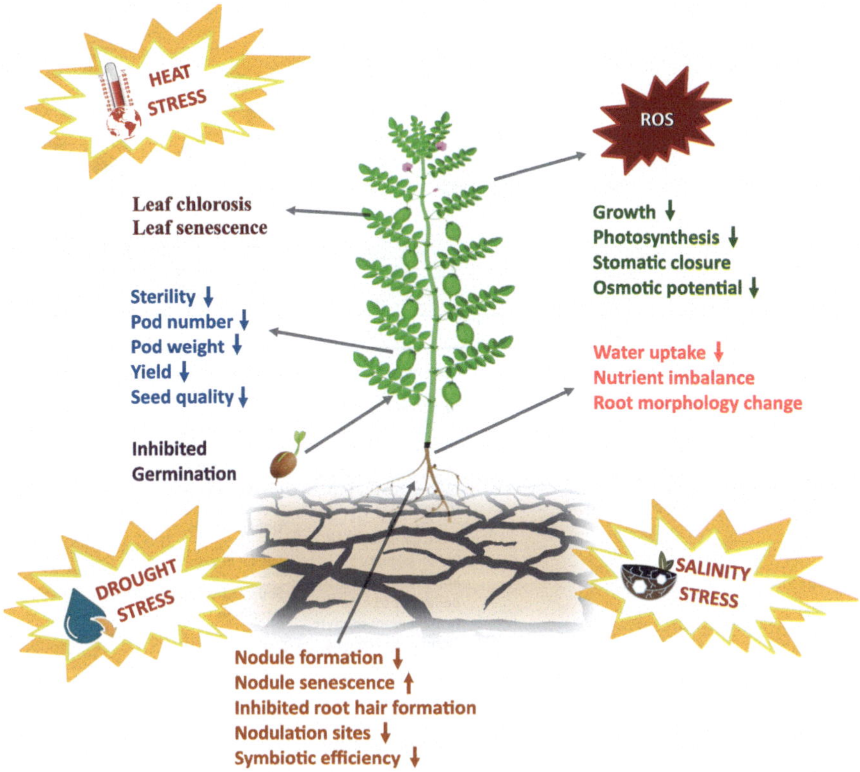

Fig. 22.2 Overview of the impact of abiotic stresses on the morphophysiological features of legumes

leading to significant productivity losses (Bhandari et al. 2017; Basu et al. 2023). Projections suggest that if global warming remains at 1.5 °C, legume crop yields may decline by about 10%. However, if temperature rises by 2.5–4.8 °C by 2100, yields of beans, chickpea, cowpea, soybean, and lentil could decrease by 20–49% (Dave et al. 2024). Cool-season legumes like chickpeas and lentil, which are adapted to cooler environments, are particularly susceptible to heat stress (Sarkar et al. 2021).

Heat stress disrupts the entire plant life cycle, affecting morphological, reproductive, and developmental processes. Heat stress affects legumes at various organizational levels, impacting both vegetative and reproductive stages (Fig. 22.3). The severity of these effects largely depends on the crop growth stage and the duration of heat exposure. Heat stress hinders crop growth by affecting soil moisture content, nutrient availability, and gas exchange, while indirectly influencing plant–water relations (Farooq et al. 2017). This stress compromises membrane stability, reduces leaf water content, and decreases chlorophyll levels and photosynthetic efficiency, impairing overall cellular functions (Devi et al. 2023). Morphological symptoms of heat stress include sunburn and scorching of leaves, twigs, and stems; leaf curling and yellowing; withering; fruit discoloration; reductions in shoot growth, root

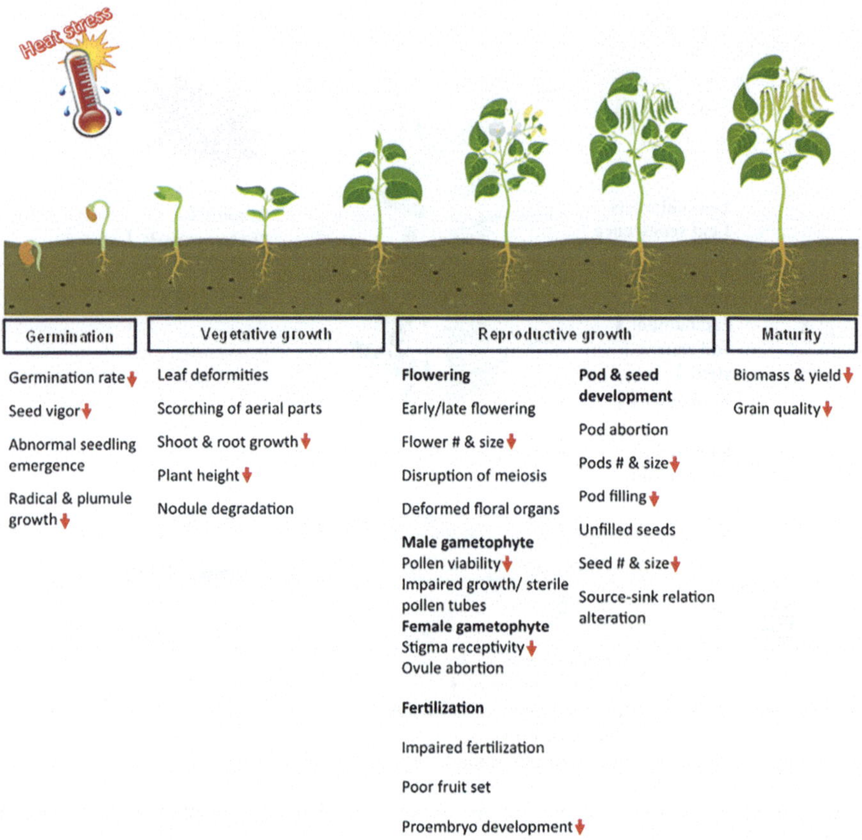

Fig. 22.3 Impact of heat stress on legume development across growth stages

growth, number of roots, root diameter; and senescence and abscission (Kumari et al. 2020, 2021).

Generation of reactive oxygen species (ROS) by heat stress impairs cell function by damaging membranes, lipids, and proteins leading to membrane instability (Sher et al. 2024). Plasma membranes are particularly heat-sensitive, serving as primary sites of injury. Elevated temperatures compromise membrane structure and function, increasing fluidity due to protein denaturation and high levels of unsaturated fatty acids. This loss of membrane integrity results in increased electrolyte leakage and reduced thermostability (Sita et al. 2017). Consequently, this instability disrupts chloroplast function, affecting photosynthesis and assimilate transport, ultimately reducing yield and quality (Kumari et al. 2021).

22.2.2 Drought

Drought is characterized as a prolonged period of dry weather during which soil moisture becomes insufficient to meet the demands of potential evapotranspiration (Sultana et al. 2014). High air temperatures combined with low soil and atmospheric humidity create an imbalance between transpiration and water uptake. This disproportion results in water stress, making it difficult for plants to maintain proper hydration for growth. This phenomenon is further exacerbated by climate change, which increases both its severity and frequency (Schwalm et al. 2017). Moreover, global warming contributes to soil moisture deficits through increased evaporation from the land surface. Reduced annual rainfall and erratic monsoon patterns further heighten the risk of more frequent droughts worldwide (Kumari et al. 2021). By the end of this century, over 25% of the world's population may face severe water shortages if temperatures rise by 2 °C (Seleiman et al. 2021).

Drought poses a significant challenge to agriculture, limiting plant yield and productivity globally (Lesk et al. 2022). This is particularly concerning for legumes, which are often rain-fed crops and thus highly vulnerable to drought (Álvarez-Aragón et al. 2023). The effects of drought stress are generally unpredictable, as various factors such as rainfall patterns, soil water holding capacity, and evapotranspiration influence its impact (Dave et al. 2024). Water shortages negatively impact the morphological, physiological, phenological, and biochemical processes in legumes (Rane et al. 2021). The timing and duration of drought stress significantly influence plant development, affecting various growth stages from germination to reproduction (Rane et al. 2021). Key symptoms of drought stress in these plants include leaf twisting, wilting, burning, and premature shedding (Petrushin et al. 2023). Drought-induced cell dehydration disrupts key growth processes such as cell expansion, division, and maintaining turgidity, resulting in smaller leaves, a reduced photosynthetic area, limited stem elongation, and poor root proliferation, all of which contribute to lower crop yield (Dollete et al. 2024).

22.2.3 Salinity

Soil salinity is an escalating global threat that negatively impacts soil health, as well as the growth, development, and yield of plants, leading to annual losses exceeding USD$ 10 billion (Cabusora 2024). It is characterized by excessive concentrations of soluble salts in the soil posing a serious risk to agricultural productivity worldwide, particularly in irrigated, arid, and semi-arid regions (Corwin 2021). Globally, 10% of the total arable land is affected by saline or sodic soils (Mishra et al. 2023), and about 25–30% of irrigated lands are salinized (Zaman et al. 2018). According to the FAO and the Intergovernmental Technical Panel on Soils (ITPS), the extent of salinity varies among continents with Asia having the largest affected area (193.8 Mha), followed by Africa (122.9 Mha), Central and South America (71.5 Mha), Australia (17.6 Mha), Europe (6.7 Mha), and North America (6.2 Mha) (Hossain et al. 2023).

All soils naturally contain salts that support normal growth and development of plants. However, when the salt concentration exceeds a certain threshold, it becomes harmful. Saline soils have an electrical conductivity of 4 dSm^{-1} or greater at 25 °C (Mary et al. 2020) and contain salts such as chloride (Cl^-), sulfate (SO_4^{2-}), sodium (Na^+), potassium (K^+), calcium (Ca^{2+}), and magnesium (Mg^{2+}) (Provin and Pitt 2024). These soils negatively affect agriculture by degrading soil structure, making it more susceptible to water and wind erosion, compaction, and surface crust formation. These changes in soil physiochemical properties reduce soil fertility, productivity, and nutrient uptake, impair soil microbial diversity, and degrade groundwater quality (Shaaban 2024).

In arid and semi-arid regions, which comprise approximately 40% of the Earth's total land area (Sarkar et al. 2021), salinity issues arise when the rate of evapotranspiration exceeds annual precipitation, causing salts to accumulate in the soil. Climate change has a significant role in increasing soil salinity as higher temperature and drought accentuate soil salinity via numerous mechanisms, including evapotranspiration. Coastal regions also face salinity due to the intrusion of salt water (Zaman et al. 2018). Global warming intensifies this problem by increasing evaporation rates and rising sea levels (Mary et al. 2020). Over the past 25 years, soil salinity in coastal agricultural lands has risen from 1% to 33% due to sea-level rise, causing widespread concern (Ullah et al. 2021). Coastal soil salinity ranges from 0.5 dSm^{-1} during the rainy season to as high as 50 dSm^{-1} in the summer season, predominately composed of NaCl and Na_2SO_4. These soils, often referred to as "white alkali" or "solonchak" soils due to the white crust of salts on the surface, remain permeable to water and air, but their high salt content hampers crop growth (Chaudhary 2023). The ongoing impact of climate change on coastal agricultural lands is alarming, with projections that if current trend persists, up to 50% of arable land could be salinized by 2050, posing a significant threat to global food security (Rahman et al. 2018; Ullah et al. 2021).

Plants vary in their response to salinity, with salt-tolerant plants adjusting more effectively to soil salinity than salt-sensitive ones. The extent of the salt injury depends on the crop species, cultivar, growth stage, ecological factors, and the type of salts present in the soil (Rane et al. 2021). For instance, salt injury is less severe in cool weather due to lower transpiration compared to warm and dry conditions (Chaudhary 2023). Most legume plants are sensitive to high salt concentrations (El Moukhtari et al. 2024); however, tolerance to salinity varies widely among species and cultivars. In chickpea, for example, salinity of 3 dSm^{-1} in field soils is the threshold for reduced shoot growth and yield (Rao et al. 2002), with reproductive stages being more sensitive than vegetative stages (Sultana et al. 2014). Some chickpea varieties can tolerate salinity levels ranging from 6 to 9 dSm^{-1} (Sarita et al. 2023). In contrast, pulses like cowpea, soybean, and pigeon pea have much lower salinity thresholds, as low as 1.3 dSm^{-1} (Sultana et al. 2014).

Symptoms of salt stress resemble those of drought inducing wilting, yellowing (chlorosis), stunted growth, leaf tip burning, necrosis, and scorching (Negrão et al.

2017). Salinity disrupts various morphological, physiological, and biochemical processes, including seed germination, plant growth, and the uptake of water and nutrients (Shrivastava and Kumar 2015), affecting all stages of development in legumes (Bouzroud et al. 2023) and reducing legume production by 12–100% (El Moukhtari et al. 2024). The effects of salinity and other abiotic stresses on different physiological aspects and growth stages in legumes are discussed in the following sections.

22.2.4 Impact of Abiotic Stresses on Seed Germination and Vegetative Growth

Several legumes, including chickpea, mung bean, pea, lentil, and soybean, exhibit reduced seed germination at higher temperatures (Sita et al. 2017). A study revealed that soybean seeds produced at 27/22 °C (day/night) showed 84% germination strength. In contrast, seeds exposed to heat stress of 32/28 °C (day/night) for 35 and 60 days had 73% and 56% germination percentages, respectively (Staniak et al. 2023). The germination rate of pea seeds collected from heat-stressed plants was lower by 4–8% compared to those harvested under normal conditions (Devi et al. 2023). During the vegetative growth, heat stress can reduce germination rates and seed vigor, increase the emergence of abnormal seedlings, and reduce overall biomass (Bhandari et al. 2017; Sita et al. 2017).

Like heat stress, high salinity levels can negatively impact seed germination and vegetative growth in legumes. Seed germination is influenced by both internal and external stimuli. High concentrations of salts in the soil can significantly delay the germination process, as well as reduce the rate of germination and seedling growth (Zhao et al. 2020). This is attributed to the high osmotic potential outside the seed, which inhibits water uptake (Rane et al. 2021) and decreases the activity of α-amylase, an enzyme essential for starch breakdown (Atta et al. 2023). Salinity adversely affects seed germination in legumes, including chickpea, faba beans, and mung beans (Rane et al. 2021). At a salt concentration of 7.47 g L^{-1}, the germination rate of chickpea decreases by approximately 87%. Similarly, at a 300 mM salt concentration, the soybean germination decreases by 70–80% (Bouzroud et al. 2023). The common bean experiences a 50% decrease in seed germination due to increased salinity, with salinity levels ranging from 0 to 180 mM NaCl (Nadeem et al. 2019b).

Food legumes are highly sensitive to salinity stress at the seedling stage (Sarkar et al. 2021). Mi et al. (2023) conducted a study to observe changes in the physiological and biochemical characteristics of legume seedlings under NaCl stress. According to their findings, salinity adversely affects seedling growth in both vetch and alfalfa, each exhibiting distinct physiological responses to stress. Elevated NaCl levels in common beans result in a reduction in leaf area, leaf number, and plant height (Torche et al. 2018). Similarly, in soybeans, a 7-day salt stress treatment with NaCl has been reported to inhibit overall plant growth (Ning et al. 2018).

22.2.5 Impact of Abiotic Stresses on Reproductive Development

Reproductive development is more sensitive to heat stress than the vegetative stage (Sher et al. 2024). Heat stress affects the entire seed development process, including the formation of male and female gametophytes, fertilization, and seed maturation (Sarkar et al. 2021). Legumes are particularly sensitive to heat stress during the flowering, pod set, and seed-filling stages. Heat stress can severely affect crop yield and quality by reducing pollen viability, impairing fertilization, causing flower abortion, and leading to pod set failure, which ultimately results in smaller seeds and yield losses (Basu et al. 2023). Heat stress disrupts grain development by disintegrating the tapetum layer, which reduces nutrient supply to microspores and causes premature anther dehiscence. This hampers carbohydrate synthesis and distribution to the grain, shortening the grain-filling duration, and resulting in low grain weight, poor pod development, and fractured embryos (Sarkar et al. 2021).

Several studies have been conducted to investigate the impact of heat stress on legume reproduction. Pollen germination in chickpea genotypes decreases to 58–61%, if the temperature is raised to 40/30 °C. In contrast, lentil genotypes experience 11–12.5% decline in pollen germination at the same temperature. Furthermore, tube growth in chickpeas decreases by 62–64% compared to the control at 40/30 °C, while lentil genotypes exhibit a more pronounced decrease of 85–90% (Bhandari et al. 2020). All legume species experience 22–45% increase in flower or pod abortion during brief periods of severe temperatures of 35 °C or above. For instance, chickpeas exhibit nearly 100% reduction in reproductive ability at temperatures exceeding 36 °C, even under sufficient moisture (Dave et al. 2024). High temperatures in mung bean can lead to the shedding of up to 79% of flowers (Bhardwaj et al. 2023). Chickpea pod numbers drop by 18–25% at 38/28 °C, while lentil pod numbers decrease by 38–44% at the same temperature. At 38/28 °C, chickpea seed yield reduced to 4.98–5.76 g per plant (19–23% of control), whereas lentil genotypes yield 1.87–1.98 g per plant (33–51% of control) (Bhandari et al. 2020).

The seed-filling stage is crucial in the life cycle of leguminous crops, as it significantly impacts the quality and nutritional value of legumes (Farooq et al. 2017; Choukri et al. 2022). This stage involves the synthesis and mobilization of essential elements, such as carbohydrates, fatty acids, and proteins along with the accumulation of various components in developing seeds (Farooq et al. 2018). Heat stress can accelerate the seed-filling rate while simultaneously reducing seed size in peas, soybean, white lupin, cowpea, chickpea, and lentil. Heat stress disrupts leaf chlorophyll content, photosynthetic efficiency, and carbon fixation, leading to reduced leaf water content and impaired seed filling.

High temperature also affects the visual quality of seeds by impairing the mechanisms involved in seed filling. Heat stress induces browning, a phenomenon of discoloration of seed coats which reduces both the quality and visual appearance of seeds, ultimately decreasing their market value. Browning due to heat stress negatively impacts the seed coats of soybean, cowpea, lentil and faba bean. Additionally, exposure of cowpea to high temperatures damages the seeds, resulting in asymmetrically twisted cotyledons and ruptured structures (Farooq et al. 2017).

Like heat stress, drought stress also significantly impacts legume reproduction, with flowering and reproductive stages being particularly sensitive to drought stress (Sehgal et al. 2018). Drought stress causes significant yield losses in various legumes, including beans, black gram, chickpea, cowpea, faba bean, lentil, mung bean, pigeon pea, and soybean (Nadeem et al. 2019a; Rane et al. 2021; Dave et al. 2024). It reduces flower size, causes floral abortion, and impairs embryo development, resulting in smaller grains and lower yields (Sehgal et al. 2018). Post-anthesis drought stress is especially harmful to grain production (Nadeem et al. 2019a). For instance, drought stress in soybean reduces pollen germination by 17% (Poudel et al. 2023), while in chickpea, it reduces biomass, plant yield, and pod filling by 29–33% (Sachdeva et al. 2022). Annually, up to 40% of global chickpea yield losses are attributed to drought stress (Khatun et al. 2021). In lentil, a 12.1% reduction in seed weight and a 38.9% reduction in yield have been reported (Bansal et al. 2023). Additionally, drought conditions during reproductive stages predispose grain legumes to fungal infections, such as dry root rot and black root rot caused by *Rhizoctonia bataticola* and *Fusarium solani*, further compounding yield losses (Sinha et al. 2019).

In addition to heat and drought stress, salinity is another major environmental factor that adversely affects legume reproduction, particularly during the reproductive stage. Salinity reduces pollen viability, leading to floret sterility, flower abortion, slow development, hampered leaf expansion, and decreased plant dry weight. These effects result in poor pod and seed development, increased rates of plant growth reduction, elevated senescence, defoliation, and ultimately contribute to reproductive failure and plant death (Das and Biswas 2022; Sarita et al. 2023). Salt stress reduces the grain yield of food legumes. For instance, increasing NaCl concentrations in faba beans negatively impacts yield components, such as the number of pods per plant, the number of seeds per pod, and seed weight. Plants grown with 200 mM NaCl exhibited a 54%, 45%, and 46% reduction in the number of pods, seeds, and seed weight, respectively (Bouzroud et al. 2023). In chickpeas, a yield decrease of 69% was observed at a salinity level of 8.3 dSm^{-1}. Furthermore, when 8.3 and 3.3 dSm^{-1} salt were added, the yield loss for faba beans and lentil was 53% and 100%, respectively (El Moukhtari et al. 2024).

22.2.6 Impact of Abiotic Stresses on Nutritional Components

Heat stress also causes denaturation and aggregation of grain storage proteins by disrupting covalent and noncovalent interactions, adversely affecting protein structure and metabolism. This reduces the activity of enzymes involved in protein synthesis, ultimately decreasing protein content (Farooq et al. 2017). A temperature rise to 32/26 °C reduces protein content in peanuts by 19.6%, while soybean experiences a sharp decrease above 40/30 °C (Farooq et al. 2018). In chickpeas, heat stress results in a 13.7% reduction in protein content, with iron and zinc concentrations decreasing by up to 39% and 31%, respectively (Samineni et al. 2022). Similarly, high temperatures reduce protein (14.3%), iron (18.4%), and zinc (22%)

concentrations in lentil (Choukri et al. 2022). Additionally, heat stress leads to smaller seed sizes and inhibits the accumulation of carbohydrates, fats, and essential minerals in the seeds (Devi et al. 2023). It impedes the enzymatic processes necessary for starch synthesis, particularly affecting their accumulation (Sarkar et al. 2021). Legume seeds typically contain 25–65% carbohydrates, with starch being the primary carbohydrate in most legumes. Elevated temperatures inhibit starch synthesis and accumulation by reducing the activity of enzymes such as sucrose synthase, starch synthase, and invertase. For example, temperatures exceeding 32/20 °C and 35 °C significantly reduce starch levels in chickpeas and peas, respectively (Farooq et al. 2017).

Drought significantly influences the nutritional composition of legumes by altering their mineral, protein, fat, and carbohydrate contents, as well as antioxidants. It also has a pronounced effect on the mineral profile of legumes. For instance, in soybeans, drought can enhance the concentrations of calcium, copper, manganese, molybdenum, phosphorus, and zinc. Conversely, drought reduces calcium, potassium, and sodium levels in chickpea but increases proline content, which aids in stress response (Farooq et al. 2018). In lentil, it decreases iron content by 24.2%, zinc by 31.1%, and protein content by 25.3% (Bansal et al. 2023). Interestingly, mild water deficits during flowering may enhance protein assimilation, resulting in an increase in grain protein content by 6–21% in legumes, such as chickpeas and mung bean. However, severe drought conditions can significantly reduce protein levels in legumes like lupins by 19–35% (Ullah and Farooq 2022).

Salinity also affects grain composition and quality in food legumes by reducing carbohydrate, polysaccharides, total amino acid, protein, and oil content. Excessive concentrations of Na^+ and Cl^- in the rhizosphere interfere with the uptake of essential nutrients, affecting the absorption of N, P, K, Ca, Zn, B, Mg, Cu, and Fe (Sarkar et al. 2021), while leading to increased Ca, Na, Mg, and Cl contents in food grains (Farooq et al. 2018). For example, in mung bean, total carbohydrate, polysaccharides, amino acid, and protein content decrease with increasing salt stress. Additionally, K and P concentrations also decline under salt stress, while the concentrations of Na, Ca, Mg, and Cl increase (Sarkar et al. 2021). However, in some cases, protein content may increase when N uptake and movement to non-root plant parts are not affected by salt stress (Farooq et al. 2018).

22.2.7 Impact of Abiotic Stresses on Oil Content

Lipids are the primary stores of carbon and energy for germination in oilseeds. Heat stress can significantly alter the oil content in oilseed crops (Devi et al. 2023), with effects varying across species. It may positively affect grain oil content; for instance, a 20% increase in the oil content in peanuts was reported. However, kidney beans experienced a 23% decline in the oil content under heat stress (Sarkar et al. 2021). A 40/30 °C heatwave resulted in a 5% reduction in soybean oil and protein content, particularly during the critical seed-filling phase (Djanaguiraman et al. 2011). In another study, a 6 °C increase in temperature during seed-filling stage resulted in a 2.6% decrease in oil content in soybean (Kumar et al. 2023).

Drought conditions can reduce oil content and quality of these seeds (Sehgal et al. 2018) by altering the fatty acid composition of soybean grains, affecting oil level, stability, and overall composition during grain filling (Sarkar et al. 2021).

22.2.8 Impact of Abiotic Stresses on Nodulation and N-Fixation

The nitrogen fixation ability of legumes declines drastically beyond optimal temperatures, highlighting the sensitivity of this vital process to heat stress (Hungria and Kaschuk 2014). Nitrogen fixation occurs within the root nodules, where bacteria are isolated from host legume by the symbiosome membrane, which regulates the exchange between the two. The plant provides bacteria with energy and carbon skeletons from photosynthesis for nitrogen fixation, and in return, they supply ammonium, the biologically active form of nitrogen, essential for plant growth (Jumrani et al. 2024). Increased oxygen diffusion resistance in root Bacteroides hampers nitrogenase function, reducing nitrogen fixation, which negatively affects seed quality and yield (Sehgal et al. 2018). High temperatures above 30–35 °C exacerbate this by slowing nitrogenase enzyme kinetics and causing amino acid deformities in nodule tissues, further reducing the efficiency of this symbiosis (van der Pol et al. 2022). Other direct effects of heat stress on nitrogen fixation include reduced nodule formation, accelerated nodule senescence, and impaired functionality. Indirectly, heat stress inhibits root hair formation, reduces nodulation sites, and affects bacterial adherence to root hairs, further limiting nitrogen fixation (Aranjuelo et al. 2014). The negative effects of high temperature on nitrogen fixation have been demonstrated in common bean, peanut, and soybean (Hungria and Kaschuk 2014; Bhandari et al. 2017). Common bean plants exposed to heat stress (35–38 °C) form nodules that are inefficient in nitrogen fixation (Hungria and Kaschuk 2014). In peanut and soybean, heat stress prevents nodule formation, with no nodules formation at 40 °C in peanut and 37 °C in soybean (Bhandari et al. 2017).

In addition to heat stress, drought stress also severely impacts legume nodulation, nitrogen fixation, and soil nitrogen availability, especially during the flowering stage. Water deficit inhibits the exchange of signaling molecules needed for nodulation, resulting in fewer nodules and reduced nitrogen fixation. Drought diminishes rhizobia populations and their effectiveness, limiting root infection and nodulation (Dollete et al. 2024). It hinders nitrogen fixation by causing carbon shortage for rhizobia. Additionally, drought stress reduces nitrogenase activity (measured through acetylene reduction), ultimately lowering biological nitrogen fixation in legumes (McCulloch et al. 2021).

The effects of heat and drought are particularly severe when these stresses are combined together, creating an even greater challenge for legumes. High temperature coupled with water stress reduces carbon dioxide availability in soybean that negatively affects nitrogen fixation resulting in a decline in nodule activity with increasing temperatures (Jumrani et al. 2024). Similar results have been observed in faba bean, chickpea, and lentil where increased temperatures (40/30 °C) combined with water stress during the seed-filling stage have been reported to cause a 29–76% reduction in nitrogen fixation rates as compared to ambient conditions (Sehgal et al. 2017). Combined drought and heat stress during critical phases like seed-filling can reduce nitrogen fixation rates in legumes, such as soybean, pea, faba bean, and chickpea, by up to 76% (Sehgal et al. 2018; Dave et al. 2024).

Similar to heat and drought stresses, salinity has long been known to adversely affect legume-rhizobia symbiosis. It disrupts nodulation by distorting root hairs, inhibiting the infection of rhizobia, and preventing the release of rhizobia from infection threads. As a result, only a few, ineffective nodules form, which exhibit decreased nitrogenase and leghemoglobin activity (Kirova and Kocheva 2021; Chakraborty and Harris 2022). It has been observed that irrigating faba beans with 35% sea water suppresses nodule production (Bouzroud et al. 2023). Additionally, changes in nitrate uptake and nitrogen metabolism under high salinity conditions result in reduced oil, protein, and starch content (Das and Biswas 2022).

22.3 Management Approaches

As discussed, the challenges posed by abiotic stressors to legumes threaten global food security and socioeconomic well-being, underscoring the need for climate-adaptive varieties. The changing climate scenario is complex, exacerbated by overlapping stressors. Abiotic stressors are often interconnected; for example, heat stress can lead to drought which in turn may increase soil salinity and alkalinity, ultimately depleting soil nutrients (Sindhu et al. 2020). Developing climate-resilient crops requires understanding of plant coping mechanisms under stress. In response to stress, plants adopt various strategies: *avoidance*, involving morphological and anatomical changes to reduce exposure; *escape*, through traits like seed dormancy, early maturity, and a shorter life cycle to evade stress; and *tolerance*, by making physiological, biochemical, and molecular adjustments to endure challenging conditions (Cabusora 2024). To get climate-resilient legume varieties that can thrive under environmental challenges, it is critical to understand the mechanisms the plants can adopt under abiotic stress(es), which helps identify traits that contribute to resilience, thus laying foundation for developing climate-resilient crops (Rane et al. 2021).

Various approaches can harness the potential of existing genotypes to develop climate-resilient legume varieties that combine high yield, adaptability, and stress resistance.

22.3.1 Agronomic Approaches

Agronomic practices such as water conservation, nutrient application, the use of plant growth-promoting rhizobacteria (PGPR), adjusting planting and harvesting times, and employing stress-tolerant varieties are effective in mitigating adverse effects of heat and drought stress (Bhandari et al. 2017; Kumari et al. 2021; Chaudhry and Sidhu 2022). Applying macro- and micro-nutrients like boron, calcium, iron, manganese, potassium, selenium, and zinc alleviates stress while improving the nutritional quality of legumes (Kumari et al. 2021). Techniques like conservation tillage and soil cover help to retain water, while adjusting sowing times to protect heat-sensitive crop stages can improve legume yields under heat stress (Farooq et al. 2017).

Studies show that adjusting plant geometry during sowing, along with proper fertilizer management, as well as practices like mulching, zero tillage, and deep plowing during the rainy season, helps alleviate drought stress in legumes. Optimizing plant density and row spacing has enhanced drought tolerance by improving water use efficiency and yield in soybean and chickpea (Nadeem et al. 2019a). In addition, biochar application, along with exogenous application of osmoprotectants, PGPRs, and arbuscular mycorrhizal fungi (AMF), has been reported to improve yield and enhance tolerance against drought stress (Khatun et al. 2021). Both endogenous modulation and exogenous application of phytohormones, along with phytohormone-producing microbes, enhance stress tolerance and yield under drought conditions in soybean (Shaffique et al. 2023). Recent studies have explored microbial communities to enhance sustainable food production, showing that plant-associated microbes can help alleviate drought stress (Petrushin et al. 2023; Dollete et al. 2024). Álvarez-Aragón et al. (2023) revealed that rhizobial symbiosis in pea and vetch plants promotes drought tolerance, providing an effective mechanism for plants to cope with water deficit conditions (Bhandari et al. 2017). Kumari and colleagues reported that seed priming with potassium dihydrogen phosphate (KH_2PO_4), disodium hydrogen phosphate (Na_2HPO_4), or water has been found to develop drought resistance in crops (Kumari et al. 2021).

Various strategies have been effective in mitigating salinity stress in legumes. Seed priming using natural agents such as sugars, amino acids and their derivatives, vitamins, along with synthetic agents like potassium nitrate (KNO_3), silica (SiO_2), hydrogen peroxide (H_2O_2), and group of chemicals known as reactive oxygen, nitrogen, and sulfur species (RONSS) helps alleviate salt stress in legumes (Bouzroud et al. 2023). Additionally, the exogenous application of PGPR and AMF can also mitigate the effects of salinity (Bouzroud et al. 2023). Furthermore, the application of biochar and nanoparticles has been shown to reduce the harmful effects of salinity on legume crops (Li et al. 2022; Tahir et al. 2022; Mishra et al. 2023).

22.3.2 Exploring Genetic Diversity

Genetic diversity serves as the foundation for selection response and is a prerequisite for success of any crops' breeding efforts (Singh et al. 2018). The variability in stress responses among genotypes of the same plant, due to specificity of tolerance traits, underpins breeding strategies aimed at enhancing crop performance under abiotic stresses (Kumari et al. 2020). Over time, crops have lost important genes that help them cope with environmental stress owing to breeding for other traits, such as yield. However, genetic resources like wild or older crop varieties still possess these essential genes, making their exploration advantageous (Tahir et al. 2020; Hussain et al. 2022; Sardar et al. 2025). Scientists can reintroduce these lost traits through advances in genomics and breeding techniques. Information on genetic resources of crops, including legumes, conserved in national and international gene banks, is available online at global portals such as Genesys and Germplasm Resource Information Network (GRIN). The focused identification of germplasm strategy (FIGS) has proven effective for various adaptive traits, including heat, drought, and salinity tolerance, and can be utilized to identify and apply valuable genes in preferred agronomic environments (Ali et al. 2022).

Although a wide genetic variation is necessary for successful breeding, the natural gene pool often lacks the desired variation. Therefore, the process of spontaneous mutation, which broadens the genetic variation but occurs infrequently, can be induced using physical or chemical mutagens to enhance genetic variability (Cabusora 2024).

22.3.3 Conventional Phenotype-Assisted Breeding

Breeding approaches manipulate the genetic makeup of plants to enhance tolerance and edit their genetic architecture for permanent solution (Kumari et al. 2020). Conventional breeding involves selecting plants with a history of desirable traits, such as the ability to withstand abiotic stress, followed by crossing these plants to create desired offspring through gene exchange. Producing abiotic stress-tolerant plants often requires several attempts. In legume crops, breeding methods comprise four stages: hybridization and mutation, selection from the initial population, assessment of selected lines, and the release of cultivar. This process enables the development of crops that are both highly productive and stress resistant. Many heat-tolerant legumes, including chickpea, common bean, mung bean, snap bean, pea, groundnut, and cowpea, have been developed using conventional breeding methods (Sita et al. 2017). Chickpea with desirable traits like stress tolerance can be identified for breeding using an ecogeographic strategy, where plants growing in different regions may exhibit unique traits (Farooq et al. 2017).

Khatun et al. (2021) reported certain promising traits for drought tolerance in grain legumes that could be useful for screening genotypes. Breeders have leveraged genetic variation within and between species (intraspecific, interspecific, and intergeneric) to create abiotic stress-resistant varieties (Cabusora 2024). Wide hybridization has been employed to obtain specific traits within or between species. Many grain legumes have undergone interspecific crosses with varying degrees of success (Khatun et al. 2021). Pedigree backcrossing has also been used in legumes to induce heat tolerance (Farooq et al. 2017). However, conventional breeding approaches have limitations, as they often randomly exploit the tolerance mechanisms of plants and are constrained by the complexity of stress-tolerant traits, which typically involve multiple genes and pathways. Additionally, domestication bottlenecks have resulted in reduced genetic variation (Cabusora 2024).

22.3.4 Quantitative Trait Loci (QTLs) and DNA Marker-Aided Breeding

Phenotype-assisted selection and breeding of plants are increasingly being complemented by genomic-assisted selection and breeding. The advent of genomic research has enabled the genetic improvement of complex traits controlled by multiple genes, such as stress tolerance by using molecular markers to identify and map QTLs associated with these traits (Palit et al. 2020). This facilitates efficient marker-assisted selection (MAS) for breeding crop varieties with enhanced stress tolerance (Gonzalez Guzman et al. 2022).

Genome-based approaches facilitate the identification of desirable alleles in different genomic regions/QTLs that have the potential to influence desired responses (Kuang et al. 2022). Since the morphophysiological traits that impact stress tolerance are quantitatively inherited, marker-assisted selection (MAS) through QTL identification is one of the most promising approaches for breeding climate-resilient crops (Farooq et al. 2014). QTL mapping enables breeders to identify genomic regions associated with complex traits, such as abiotic stress tolerance (Ali et al. 2022), which is essential for improving breeding strategies and advancing genetic research. By dissecting the genetic basis of these complex traits, breeders can develop climate-resilient crops through MAS for traits like heat stress tolerance (Kumari et al. 2020) and drought stress tolerance (Ullah and Farooq 2022), thus saving time and resources compared to conventional breeding methods.

Numerous studies have been conducted to identify cultivated legume accessions with high tolerance to abiotic stresses through biparental linkage mapping or QTL mapping (Narayana and von Wettberg 2020). Identification of QTLs for thermotolerance can be performed using various traits, including thousand grain weight, canopy temperature depression, grain-filling duration, yield, and traits related to senescence (Sita et al. 2017). Heat-tolerant QTLs have been discovered in legumes, including cowpea, chickpea, and adzuki bean (Sita et al. 2017; Kumari et al. 2020). Reduced canopy temperatures indicate efficient water uptake, which is ultimately associated with deep rooting, while increased chlorophyll fluorescence reflects heat-tolerant photosynthesis (Pinto and Reynolds 2015). Traits like leaf morphology, stomatal density, and root characteristics are linked to QTL identification for water stress tolerance (Gonzalez Guzman et al. 2022). Drought stress-tolerant QTLs have been identified in legumes such as chickpea, cowpea, soybean, faba bean, and common bean (Nadeem et al. 2019a; Kumari et al. 2020). Furthermore, QTLs associated with salinity tolerance have also been discovered (Asif and Paull 2021).

One notable trait, "stay-green" encoded by the SGR gene, refers to an important heritable trait that delays senescence and allows plants to retain leaves during stress conditions, thereby supporting active photosynthesis and high yields (Zhang et al. 2022). SGR serves as a crucial regulatory factor in legumes, including pea and soybean (Kachare et al. 2023). In broad bean, the stay-green trait positively affects the uniformity of grain size, while maintaining the stay-green characteristic enhances grain yield in cowpea by increasing photosynthetic activity (Farooq et al. 2017). This trait is essential for improving heat tolerance through QTLs, acting as a marker for selecting genotypes that thrive in warm temperatures. Additionally, it may be employed to design plants tolerant to multiple stressors, including drought (Kumari et al. 2020).

22.3.5 Genome-Wide and Multi-OMICS Approaches

The availability of reference genomes for legumes like mung bean, soybean, groundnut, chickpea, adzuki bean, pigeon pea, lentil, and common bean has opened new opportunities to target key agricultural traits, such as stress tolerance. The mechanisms of stress tolerance are complex, involving multiple genes, post-transcriptional regulation, and other factors. Not all the genes are activated or deactivated simultaneously under stress, and stress can impact at various OMICS levels, leading to phenotypes that cannot be predicted by genotype alone (Ramalingam et al. 2015; Jha et al. 2020; Singh et al. 2020). Thus, combating abiotic stress requires strategies that address transcriptional, translational, and post-translational regulation, along with specific signaling pathways that mediate plant responses to stress. This integration of molecular tools in breeding programs has the potential to facilitate the development of tolerant varieties, thereby sustaining food security under abiotic stress conditions. Scientists are increasingly focusing on understanding the genetic, epigenetic, transcriptomic, proteomic, and metabolic aspects of stress tolerance to overcome the limitations posed by phenotypic-assisted selection and breeding (Tahir et al. 2019; Gao et al. 2022; Sharma and Kumari 2025).

OMICS technologies are essential for advancing our understanding of stress tolerance and improving breeding outcomes. Integrating multiomics—covering genomics (DNA sequence variation), transcriptomics (RNA expression), proteomics (protein expression), metabolomics (cellular metabolites), ionomics (elemental and ion composition), phenomics (large-scale phenotyping), interactomics (proteins and molecular interactions), and epigenomics (DNA and histone modifications)—has facilitated the identification of candidate genes and pathways controlling traits related to abiotic stress responses (Yang et al. 2021; Rasouli et al. 2022; Altaf et al. 2025).

New high-throughput technologies have the potential to bridge the gap between our understanding of genotypes and their respective phenotypes. These advancements assist breeders in identifying promising candidate genes and optimal traits, aiding in marker identification via QTL mapping or GWAS and ultimately increasing the selection efficiency and improving the productivity of legume crops under abiotic stresses (Kumar et al. 2022). Recent research has provided insight into the functions of proteins, metabolites, and many key genes, as well as molecular processes involved in legume responses to heat, drought, and salinity stress (Strozycki et al. 2007; Sita et al. 2017; Khatun et al. 2021; El Moukhtari et al. 2024; Sher et al. 2024).

Protein analysis can be conducted using traditional gel-based techniques, such as 2D-PAGE and 2D difference gel electrophoresis (DIGE), as well as non-gel-based methods, including multidimensional protein identification technology (MudPIT), Isobaric Tag for Relative and Absolute Quantitation (iTRAQ), metabolic labeling, and LC-MS/MS. Proteomic studies in alfalfa, soybean, peanut, chickpea, pigeon pea, groundnut, cowpea, mung bean, lima bean, and faba bean have identified key heat-tolerant proteins, including HSPs, chaperones, and other stress-responsive proteins that play protective roles under high-temperature stress (Sita et al. 2017; Kumari et al. 2020). Similarly, proteomic investigations in drought-stressed soybean roots have discovered proteins that form a defense layer against oxidative stress, which are upregulated during drought. In chickpea, several proteins responsive to dehydration were identified, participating in key processes like nucleocytoplasmic transport, chromatin remodeling, and ROS pathways, which contribute to enhanced drought resilience (Nadeem et al. 2019a).

Metabolomics is a crucial approach in functional genomics that allows the identification and quantification of metabolites within cells, tissues, or organisms, playing a significant role in breeding stress-tolerant legume varieties since metabolites can serve as selection biomarkers. Common metabolite profiling techniques include NMR, gas chromatography (GC)-MS, and liquid chromatography (LC)-MS (Tahir et al. 2019; Narayana and von Wettberg 2020). Metabolomic profiling has shown that antioxidants like flavonoids, tocopherols, phenylpropanoids, and ascorbate contribute to heat tolerance in resilient soybean genotypes (Chebrolu et al. 2016), while sugar, nitrogen, and phytochemical metabolism are critical for adaptation to water deficiency conditions (Das et al. 2017).

Plant phenomics provides quantitative phenotyping using tools, such as drones, sensitive cameras, and scales, to elucidate the genetic bases of agricultural traits and screen germplasm for genetic variation in form, function, and performance. For example, the International Crop Research Institute for the Semi-Arid Tropics employs a custom-built system—a combined lysimeter and scanner—to examine precise phenotypic differences in response to water limitation (Vadez et al. 2017). Phenome–genome association as well as genome-wide association studies (GWAS) offer several advantages over QTL mapping and is increasingly in use for abiotic stress tolerance in legumes (Narayana and von Wettberg 2020).

22.3.6 Post-genomic Era Advancements

In addition to genomic-assisted breeding approaches, the post-genomic era has seen a significant shift in transgenic research due to incorporation of genome editing technologies such as zinc-finger nucleases (ZFNs), transcription activator like effector nucleases (TALENs), and clustered regularly interspaced short palindromic repeats/CRISPR-associated protein 9 (CRISPR/Cas). Genome editing is considered as one of the fastest techniques for crop improvement and the development of climate-smart agriculture systems, with remarkable potential to address food insecurity (Das and Acharjee 2024).

CRISPR/Cas9-mediated genome editing has huge potential for developing elite legume cultivars with enhanced climatic resilience (Khatun et al. 2021). However, only a limited number of studies have reported the application of genome editing, particularly CRISPR/Cas9, for abiotic stress tolerance in legumes. The first successful CRISPR/Cas9-mediated genome editing in legumes was achieved in soybean (Cai et al. 2015). In another study, Cai et al. (2018) reported a link between the CRISPR knockout of the soybean gene (*GmFT2*) associated with flowering time, with GmFT2 mutants exhibiting late flowering under both long-day and short-day conditions. In chickpea, transgenic lines were developed by knocking out drought-tolerant genes *4CL* and *REV7* (Razzaq et al. 2022). Despite the limited studies on genome editing for abiotic stress tolerance in legumes, the potential of these methodologies for future use in legume molecular breeding for complex quantitative traits is immense, on model as well non-model legumes.

22.3.7 Modeling

Modeling innovations are essential for addressing challenges in sustainable food production, nutrition, and natural resource management under a changing climate. Crop models play a crucial role in enabling agriculture to adopt, helping to extrapolate the complexities of climate change and assess its impact (Palit et al. 2020). While genome-based and multi-OMICS approaches have advanced our understanding of the molecular mechanisms behind complex traits, an integrative systems biology approach is now necessary. By combining genomics, systems biology, and phenomics, with suitable crop models, an integrated research framework can provide data-driven insights into optimal G × E × M (genotype × environment × management) strategies for current and projected climate conditions. This approach moves beyond identifying single or a few genes/QTL to understand plant biology at a systems level. By addressing the complexities of climate stressors with larger datasets and multidimensional networks through mathematical modeling, it offers a comprehensive view of plant responses.

Na and Na (2024) utilized advanced artificial intelligence (AI) techniques to develop predictive models for legume crop yields in the context of climate change scenarios. This research encompasses diverse climate change scenarios to assess the resilience of legume crops under varying environmental conditions. By utilizing an extensive dataset comprising legume crop yields, climate change forecasts, and relevant environmental factors, this work not only enhances the precision of agricultural predictive modeling but also underscores the vital role of AI in mitigating the detrimental effects of climate change on food security. Furthermore, developing an integrated approach that includes the recommendations of statutory bodies, policymakers, and stakeholders will help mitigate the deleterious effects of abiotic stressors in the long run (Bolatova et al. 2024).

22.4 Conclusion

Under the changing agroclimatic scenario, the common vision for agriculture is to enhance the resilience of food production systems to meet the ever-increasing demands of the growing population. However, in the face of global climate change, ensuring food security has become an increasingly complex challenge. Rising temperatures, prolonged droughts, soil salinization, and unpredictable climate fluctuations frequently undermine agricultural productivity, particularly affecting legumes, which play a significant role in both nutritional value and sustainable farming.

Plants exposed to unfavorable environmental conditions such as elevated temperature, water deficit, and salinity experience adverse effects on their morphological, developmental, cellular, and molecular processes, ultimately impacting growth and yield. Modern agricultural research has focused on enhancing crop resilience through various agronomic, breeding, and biotechnological interventions. Genomic selection, QTL mapping, GWAS, and CRISPR/Cas9, have emerged as powerful tools for expediting the development of climate-resilient legume varieties. Integrating insights from "omics" studies with breeding practices enables breeders to adopt knowledge-driven strategies rather than relying solely on chance-driven methods, marking a promising advancement in legume improvement programs for the future. By leveraging these techniques, researchers are better equipped to select and cultivate legume crops that can withstand stress conditions without compromising yield, thereby supporting a resilient and sustainable food production system. Continued investment in research and development will be crucial to further enhance the genetic and physiological resilience of legumes.

References

Abobatta WF, El-Hashash EF, Hegab RH (2021) Challenges and opportunities for the global cultivation and adaption of legumes. J Appl Biotechnol Bioeng 8:160–172

Ali A, Altaf MT, Nadeem MA, Karaköy T, Shah AN, Azeem H et al (2022) Recent advancement in OMICS approaches to enhance abiotic stress tolerance in legumes. Front Plant Sci 13:952759

Altaf MT, Liaqat W, Ali A et al (2025) Advancing chickpea breeding: omics insights for targeted abiotic stress mitigation and genetic enhancement. Biochem Genet 63:1063–1115. https://doi.org/10.1007/s10528-024-10954-8

Al-Tawaha ARM, McNeil D, Othman YA, AlRawashdeh IM, Adnan M, Zamin M et al (2022) Legume production and climate change. In: Improvement of plant production in the era of climate change. CRC Press, pp 221–248

Álvarez-Aragón R, Palacios JM, Ramírez-Parra E (2023) Rhizobial symbiosis promotes drought tolerance in Vicia sativa and Pisum sativum. Environ Exp Bot 208:105268

Aranjuelo I, Arrese-Igor C, Molero G (2014) Nodule performance within a changing environmental context. J Plant Physiol 171:1076–1090

Asif MA, Paull JG (2021) An approach to detecting quantitative trait loci and candidate genes associated with salinity tolerance in faba bean (Vicia faba). Plant Breed 140(4):643–653

Atta K, Mondal S, Gorai S, Singh AP, Kumari A, Ghosh T et al (2023) Impacts of salinity stress on crop plants: improving salt tolerance through genetic and molecular dissection. Front Plant Sci 14:1241736

Bansal R, Bana RS, Dikshit HK, Srivastava H, Priya S, Kumar S et al (2023) Seed nutritional quality in lentil (Lens culinaris) under different moisture regimes. Front Nutr 10:1141040

Basu PS, Singh U, Meena SK, Gurumurthy S, Kumar V, Tewari K et al (2023) Implication of climate change on the productivity of legumes. In: Climate change and legumes. CRC Press, pp 207–250

Bhandari K, Sharma KD, Hanumantha Rao B, Siddique KHM, Gaur P, Agrawal SK et al (2017) Temperature sensitivity of food legumes: a physiological insight. Acta Physiol Plant 39:1–22

Bhandari K, Sita K, Sehgal A, Bhardwaj A, Gaur P, Kumar S et al (2020) Differential heat sensitivity of two cool-season legumes, chickpea and lentil, at the reproductive stage, is associated with responses in pollen function, photosynthetic ability and oxidative damage. J Agron Crop Sci 206:734–758

Bhardwaj R, Lone JK, Pandey R, Mondal N, Dhandapani R, Meena SK et al (2023) Insights into morphological and physio-biochemical adaptive responses in mungbean (Vigna radiata L.) under heat stress. Front Genet 14:1206451

Bolatova Z, Bulkhairova Z, Kulshigashova M (2024) Modeling scenarios of climate change impacts on leguminous crop production: a case study in Kazakhstan. Int J Agric Biosci 13(3):367–377

Bouzroud S, Henkrar F, Fahr M, Smouni A (2023) Salt stress responses and alleviation strategies in legumes: a review of the current knowledge. 3 Biotech 13:287

Cabusora CC (2024) Developing climate-resilient crops: adaptation to abiotic stress-affected areas. Technol Agron 4:e005

Cai Y, Chen L, Liu X, Sun S, Wu C, Jiang B et al (2015) CRISPR/Cas9-mediated genome editing in soybean hairy roots. PLoS One 10:e0136064

Cai Y, Chen L, Liu X, Guo C, Sun S, Wu C et al (2018) CRISPR/Cas9-mediated targeted mutagenesis of GmFT2a delays flowering time in soya bean. Plant Biotechnol J 16:176–185

Cakir Ö, Ucarli C, Tarhan Ç, Pekmez M, Turgut-Kara N (2019) Nutritional and health benefits of legumes and their distinctive genomic properties. Food Sci Technol 39:1–12

Calzadilla A, Rehdanz K, Betts R, Falloon P, Wiltshire A, Tol RSJ (2013) Climate change impacts on global agriculture. Clim Chang 120:357–374

Chakraborty S, Harris JM (2022) At the crossroads of salinity and rhizobium-legume symbiosis. Mol Plant-Microbe Interact 35:540–553

Chaudhary DR (2023) Microbial populations and soil fertility in the coastal lands of India. In: Climate change and legumes. CRC Press, pp 153–170

Chaudhry S, Sidhu GPS (2022) Climate change regulated abiotic stress mechanisms in plants: a comprehensive review. Plant Cell Rep 41:1–31

Chebrolu KK, Fritschi FB, Ye S, Krishnan HB, Smith JR, Gillman JD (2016) Impact of heat stress during seed development on soybean seed metabolome. Metabolomics 12:1–14

Choukri H, El Haddad N, Aloui K, Hejjaoui K, El-Baouchi A, Smouni A et al (2022) Effect of high temperature stress during the reproductive stage on grain yield and nutritional quality of lentil (Lens culinaris Medikus). Front Nutr 9:857469

Corwin DL (2021) Climate change impacts on soil salinity in agricultural areas. Eur J Soil Sci 72:842–862

Das D, Acharjee S (2024) Application of CRISPR/Cas-mediated genome editing techniques in leguminous crops. In: Ricroch A, Eriksson D, Miladinović D, Sweet J, Van Laere K, Woźniak-Gientka E (eds) A roadmap for plant genome editing. Springer, Cham. https://doi.org/10.1007/978-3-031-46150-7_15

Das R, Biswas S (2022) Influence of abiotic stresses on seed production and quality. In: Seed biology updates. IntechOpen

Das A, Rushton PJ, Rohila JS (2017) Metabolomic profiling of soybeans (Glycine max L.) reveals the importance of sugar and nitrogen metabolism under drought and heat stress. Plants 6:21

Dave K, Kumar A, Dave N, Jain M, Dhanda PS, Yadav A et al (2024) Climate change impacts on legume physiology and ecosystem dynamics: a multifaceted perspective. Sustainability 16:6026

Dell'Olmo E, Tiberini A, Sigillo L (2023) Leguminous seedborne pathogens: seed health and sustainable crop management. Plants 12. https://doi.org/10.3390/plants12102040

Devi P, Awasthi R, Jha U, Sharma KD, Prasad PVV, Siddique KHM et al (2023) Understanding the effect of heat stress during seed filling on nutritional composition and seed yield in chickpea (Cicer arietinum L.). Sci Rep 13:15450

Djanaguiraman M, Prasad PV, Boyle DL, Schapaugh WT (2011) High-temperature stress and soybean leaves: leaf anatomy and photosynthesis. Crop Sci 51(5):2125–2131

Dollete D, Lumactud RA, Carlyle CN, Szczyglowski K, Hill B, Thilakarathna MS (2024) Effect of drought stress on symbiotic nitrogen fixation, soil nitrogen availability and soil microbial diversity in forage legumes. Plant Soil 495:445–467

El Moukhtari A, Lamsaadi N, Cabassa C, Farissi M, Savouré A (2024) Molecular approaches to improve legume salt stress tolerance. Plant Mol Biol Rep. https://doi.org/10.1007/s11105-024-01432-x

FAO (2021) The impact of disasters and crises on agriculture and food security: 2021. FAO, Rome

Farooq M, Hussain M, Siddique KHM (2014) Drought stress in wheat during flowering and grain-filling periods. CRC Crit Rev Plant Sci 33:331–349

Farooq M, Nadeem F, Gogoi N, Ullah A, Alghamdi SS, Nayyar H et al (2017) Heat stress in grain legumes during reproductive and grain-filling phases. Crop Pasture Sci 68:985–1005

Farooq M, Hussain M, Usman M, Farooq S, Alghamdi SS, Siddique KHM (2018) Impact of abiotic stresses on grain composition and quality in food legumes. J Agric Food Chem 66:8887–8897

Gao X, Dong J, Rasouli F, Pouya AK, Tahir AT, Kang J (2022) Transcriptome analysis provides new insights into plants responses under phosphate starvation in association with chilling stress. BMC Plant Biol 22:26

Gonzalez Guzman M, Cellini F, Fotopoulos V, Balestrini R, Arbona V (2022) New approaches to improve crop tolerance to biotic and abiotic stresses. Physiol Plant 174:e13547

Grdeń P, Jakubczyk A (2023) Health benefits of legume seeds. J Sci Food Agric 103:5213–5220

Hossain MZ, Anawar HM, Chaudhary DR (2023) Legume plants in the context of global climate change: challenges and scopes for environmental sustainability. In: Climate change and legumes. CRC Press, pp 1–22

Hungria M, Kaschuk G (2014) Regulation of N_2 fixation and NO_3^-/NH_4^+ assimilation in nodulated and N-fertilized Phaseolus vulgaris L. exposed to high temperature stress. Environ Exp Bot 98:32–39

Hussain SA, Iqbal MS, Akbar M, Arshad N, Munir S, Ali MA et al (2022) Estimating genetic variability among diverse lentil collections through novel multivariate techniques. PLoS One 17:e0269177. https://doi.org/10.1371/journal.pone.0269177

International Food Policy Research Institute (IFPRI) (2024) Global food policy report 2024. IFPRI, Washington, DC

Jha UC, Bohra A, Nayyar H (2020) Advances in "omics" approaches to tackle drought stress in grain legumes. Plant Breed 139:1–27

Jumrani K, Bhatia VS, Kataria S, Rastogi A (2024) The interactive effect of high temperature and water deficit stress on nitrogen fixation, photosynthesis, chlorophyll fluorescence, seed yield and quality in soybean (Glycine max). Plant Physiol Rep 29:125–140

Kachare S, Tiwari S, Tripathi N (2023) Expression of DREB 1, RBCL, PIP, SGR genes and morpho-physiological changes under water stress in soybean. J Plant Biochem Biotechnol 32(2):338–355

Kebede E (2021) Contribution, utilization, and improvement of legumes-driven biological nitrogen fixation in agricultural systems. Front Sustain Food Syst 5:767998

Khatun M, Sarkar S, Era FM, Islam AKMM, Anwar MP, Fahad S et al (2021) Drought stress in grain legumes: effects, tolerance mechanisms and management. Agronomy 11:2374

Kirova E, Kocheva K (2021) Physiological effects of salinity on nitrogen fixation in legumes—a review. J Plant Nutr 44:2653–2662

Kuang Z, Xiao C, Ilyas MK, Ibrar D, Khan S, Guo L et al (2022) Use of SSR markers for the exploration of genetic diversity and DNA finger-printing in early-maturing upland cotton (Gossypium hirsutum L.) for future breeding program. Agronomy 12:1513

Kumar S, Pandey G (2020) Biofortification of pulses and legumes to enhance nutrition. Heliyon 6:e03682

Kumar P, Choudhary M, Halder T, Prakash NR, Singh V, Vineeth TV et al (2022) Salinity stress tolerance and omics approaches: revisiting the progress and achievements in major cereal crops. Heredity (Edinb) 128:497–518

Kumar S, Bhushan B, Wakchaure GC, Dutta R, Jat BS, Meena KK et al (2023) Unveiling the impact of heat stress on seed biochemical composition of major cereal crops: implications for crop resilience and nutritional value. Plant Stress 9:100183

Kumari P, Rastogi A, Yadav S (2020) Effects of heat stress and molecular mitigation approaches in orphan legume, chickpea. Mol Biol Rep 47:4659–4670

Kumari VV, Roy A, Vijayan R, Banerjee P, Verma VC, Nalia A et al (2021) Drought and heat stress in cool-season food legumes in sub-tropical regions: consequences, adaptation, and mitigation strategies. Plan Theory 10:1038

Lesk C, Anderson W, Rigden A, Coast O, Jägermeyr J, McDermid S et al (2022) Compound heat and moisture extreme impacts on global crop yields under climate change. Nat Rev Earth Environ 3:872–889

Li Z, Zhu L, Zhao F, Li J, Zhang X, Kong X et al (2022) Plant salinity stress response and nano-enabled plant salt tolerance. Front Plant Sci 13:843994

Mary PCN, Murugaragavan R, Ramachandran J, Shanmugasundaram R, Karpagam S, Rakesh SS (2020) Saline soil reclamation. Biot Res Today 2:1070–1072

McCulloch LA, Piotto D, Porder S (2021) Drought and soil nutrients effects on symbiotic nitrogen fixation in seedlings from eight neotropical legume species. Biotropica 53:703–713. https://doi.org/10.1111/btp.12911

Mi J, Ren X, Shi J, Wang F, Wang Q, Pang H et al (2023) An insight into the different responses to salt stress in growth characteristics of two legume species during seedling growth. Front Plant Sci 14. https://doi.org/10.3389/fpls.2023.1342219

Mishra AK, Das R, George Kerry R, Biswal B, Sinha T, Sharma S et al (2023) Promising management strategies to improve crop sustainability and to amend soil salinity. Front Environ Sci 10. https://doi.org/10.3389/fenvs.2022.962581

Na MH, Na IS (2024) AI-powered predictive modelling of legume crop yields in a changing climate. Legum Res 47(8):1390–1395. https://doi.org/10.18805/lrf-790

Nadeem M, Li J, Yahya M, Sher A, Ma C, Wang X et al (2019a) Research progress and perspective on drought stress in legumes: a review. Int J Mol Sci 20:2541

Nadeem M, Li J, Yahya M, Wang M, Ali A, Cheng A et al (2019b) Grain legumes and fear of salt stress: focus on mechanisms and management strategies. Int J Mol Sci 20:799

Narayana PK, von Wettberg EJB (2020) GWAS and genomic approaches in legumes, an expanding toolkit for examining responses to abiotic stresses. In: The plant family fabaceae: biology and physiological responses to environmental stresses, pp 161–180

Negrão S, Schmöckel SM, Tester M (2017) Evaluating physiological responses of plants to salinity stress. Ann Bot 119:1–11

Nigam SN, Chaudhari S, Deevi KC, Saxena KB, Janila P (2021) Trends in legume production and future outlook. In: Genetic enhancement in major food legumes: advances in major food legumes, pp 7–48

Ning L, Kan G, Shao H, Yu D (2018) Physiological and transcriptional responses to salt stress in salt-tolerant and salt-sensitive soybean (Glycine max [L.] Merr.) seedlings. Land Degrad Dev 29:2707–2719

Palit P, Kudapa H, Zougmore R, Kholova J, Whitbread A, Sharma M et al (2020) An integrated research framework combining genomics, systems biology, physiology, modelling and breeding for legume improvement in response to elevated CO_2 under climate change scenario. Curr Plant Biol 22:100149

Pereira A (2016) Plant abiotic stress challenges from the changing environment. Front Plant Sci 7:1123

Petrushin IS, Vasilev IA, Markova YA (2023) Drought tolerance of legumes: physiology and the role of the microbiome. Curr Issues Mol Biol 45:6311–6324

Pinto RS, Reynolds MP (2015) Common genetic basis for canopy temperature depression under heat and drought stress associated with optimized root distribution in bread wheat. Theor Appl Genet 128:575–585

Poudel S, Vennam RR, Shrestha A, Reddy KR, Wijewardane NK, Reddy KN et al (2023) Resilience of soybean cultivars to drought stress during flowering and early-seed setting stages. Sci Rep 13:1277

Provin T, Pitt JL (2024) Managing soil salinity. E-60

Rahman AKMM, Ahmed KM, Butler AP, Hoque MA (2018) Influence of surface geology and micro-scale land use on the shallow subsurface salinity in deltaic coastal areas: a case from southwest Bangladesh. Environ Earth Sci 77:1–8

Ramalingam A, Kudapa H, Pazhamala LT, Weckwerth W, Varshney RK (2015) Proteomics and metabolomics: two emerging areas for legume improvement. Front Plant Sci 6:1116

Rane J, Singh AK, Kumar M, Boraiah KM, Meena KK, Pradhan A et al (2021) The adaptation and tolerance of major cereals and legumes to important abiotic stresses. Int J Mol Sci 22:12970

Rao DLN, Giller KE, Yeo AR, Flowers TJ (2002) The effects of salinity and sodicity upon nodulation and nitrogen fixation in chickpea (Cicer arietinum). Ann Bot 89:563–570

Rasouli F, Kiani-Pouya A, Wang Y, Movahedi A, Yu M, Tahir A et al (2022) Application of omics technologies in single-type guard cell studies for understanding the mechanistic basis of plant adaptation to saline conditions. In: Advances in botanical research. Elsevier, pp 249–270

Rawal V, Navarro DK (2019) The global economy of pulses. FAO

Razzaq MK, Akhter M, Ahmad RM, Cheema KL, Hina A, Karikari B et al (2022) CRISPR-Cas9 based stress tolerance: new hope for abiotic stress tolerance in chickpea (Cicer arietinum). Mol Biol Rep 49:8977–8985

Sachdeva S, Bharadwaj C, Patil BS, Pal M, Roorkiwal M, Varshney RK (2022) Agronomic performance of chickpea affected by drought stress at different growth stages. Agronomy 12:995

Samineni S, Mahendrakar MD, Hotti A, Chand U, Rathore A, Gaur PM (2022) Impact of heat and drought stresses on grain nutrient content in chickpea: genome-wide marker-trait associations for protein, Fe and Zn. Environ Exp Bot 194:104688

Sardar MM, Tahir AT, Ali S, Ayub J, Ali J, Kausar F. Yasmin T, Jabeen Z, Ilyas MK (2025) Insights from Lentil Germplasm resources Leading to crop improvement under changing climatic conditions. Life 15(4):561. https://doi.org/10.3390/life15040561

Sarita, Mehrotra S, Dimkpa CO, Goyal V (2023) Survival mechanisms of chickpea (Cicer arietinum) under saline conditions. Plant Physiol Biochem 205:108168. https://doi.org/10.1016/j.plaphy.2023.108168

Sarkar S, Khatun M, Era FM, Islam AKMM, Anwar MP, Danish S et al (2021) Abiotic stresses: alteration of composition and grain quality in food legumes. Agronomy 11:2238

Schwalm CR, Anderegg WRL, Michalak AM, Fisher JB, Biondi F, Koch G et al (2017) Global patterns of drought recovery. Nature 548:202–205

Sehgal A, Sita K, Kumar J, Kumar S, Singh S, Siddique KH, Nayyar H (2017) Effects of drought, heat and their interaction on the growth, yield and photosynthetic function of lentil (Lens culinaris Medikus) genotypes varying in heat and drought sensitivity. Front Plant Sci 8:1776

Sehgal A, Sita K, Siddique KHM, Kumar R, Bhogireddy S, Varshney RK et al (2018) Drought or/and heat-stress effects on seed filling in food crops: impacts on functional biochemistry, seed yields, and nutritional quality. Front Plant Sci 9:1705

Seleiman MF, Al-Suhaibani N, Ali N, Akmal M, Alotaibi M, Refay Y et al (2021) Drought stress impacts on plants and different approaches to alleviate its adverse effects. Plan Theory 10:259

Semba RD, Ramsing R, Rahman N, Kraemer K, Bloem MW (2021) Legumes as a sustainable source of protein in human diets. Glob Food Sec 28:100520

Shaaban M (2024) Salt-affected soils BT. In: Núñez-Delgado A (ed) Frontier studies in soil science. Springer, Cham, pp 203–215. https://doi.org/10.1007/978-3-031-50503-4_10

Shaffique S, Hussain S, Kang S-M, Imran M, Injamum-Ul-Hoque M, Khan MA et al (2023) Phytohormonal modulation of the drought stress in soybean: outlook, research progress, and cross-talk. Front Plant Sci 14:1237295

Sharma P, Kumari A (2025) Approaches to enhance abiotic and biotic stress tolerance in leguminous crops and microgreens. In: Recent trends and applications of leguminous microgreens as functional foods. Cham, Springer, pp 179–215

Sher A, Noor MA, Li HX, Nasir B, Manzoor MA, Hussain S et al (2024) Heat stress effects on legumes: challenges, management strategies and future insights. Plant Stress 13:100537

Shrivastava P, Kumar R (2015) Soil salinity: a serious environmental issue and plant growth promoting bacteria as one of the tools for its alleviation. Saudi J Biol Sci 22:123–131

Sindhu S, Dahiya A, Gera R, Sindhu SS (2020) Mitigation of abiotic stress in legume-nodulating rhizobia for sustainable crop production. Agric Res 9:444–459

Singh AK, Singh RM, Velmurugan A, Kumar RR, Biswas U (2018) Harnessing genetic resources in field crops for developing resilience to climate change. In: Biodiversity and climate change adaptation in tropical islands. Elsevier, pp 597–621

Singh N, Rai V, Singh NK (2020) Multi-omics strategies and prospects to enhance seed quality and nutritional traits in pigeonpea. Nucleus 63:249–256

Sinha R, Irulappan V, Mohan-Raju B, Suganthi A, Senthil-Kumar M (2019) Impact of drought stress on simultaneously occurring pathogen infection in field-grown chickpea. Sci Rep 9:5577

Sita K, Sehgal A, HanumanthaRao B, Nair RM, Vara Prasad PV, Kumar S et al (2017) Food legumes and rising temperatures: effects, adaptive functional mechanisms specific to reproductive growth stage and strategies to improve heat tolerance. Front Plant Sci 8:1658

Staniak M, Szpunar-Krok E, Kocira A (2023) Responses of soybean to selected abiotic stresses—photoperiod, temperature and water. Agriculture 13:146

Strozycki PM, Szczurek A, Lotocka B, Figlerowicz M, Legocki AB (2007) Ferritins and nodulation in Lupinus luteus: iron management in indeterminate type nodules. J Exp Bot 58:3145–3153

Sultana R, Choudhary AK, Pal AK, Saxena KB, Prasad BD, Singh R (2014) Abiotic stresses in major pulses: current status and strategies. In: Approaches to plant stress and their management, pp 173–190

Tahir AT, Fatmi Q, Nosheen A, Imtiaz M, Khan S (2019) Metabolomic approaches in plant research. In: Essentials bioinformatics, vol III. In silico life sciences: agriculture, pp 109–140

Tahir A, Ilyas MK, Sardar MM, Pouya AK, Rasouli F, Bibi A et al (2020) Selection criteria for yield potential in a large collection of Vigna radiata (L.) accessions. Euphytica 216:1–12

Tahir A, Kang J, Ali J, Bibi A, Abbas S (2022) Nanotechnology for future sustainable plant production under changing environmental conditions. In: Bioresource technology: concept, tools, and experiences, pp 466–492

Torche Y, Blair M, Saida C (2018) Biochemical, physiological and phenological genetic analysis in common bean (Phaseolus vulgaris L.) under salt stress. Ann Agric Sci 63:153–161

Torheim LE, Fadnes LT (2024) Legumes and pulses-a scoping review for Nordic Nutrition Recommendations 2023. Food Nutr Res 68

Toromade AS, Soyombo DA, Kupa E, Ijomah TI (2024) Reviewing the impact of climate change on global food security: challenges and solutions. Int J Appl Res Soc Sci 6:1403–1416

Ullah A, Farooq M (2022) The challenge of drought stress for grain legumes and options for improvement. Arch Agron Soil Sci 68:1601–1618

Ullah A, Bano A, Khan N (2021) Climate change and salinity effects on crops and chemical communication between plants and plant growth-promoting microorganisms under stress. Front Sustain Food Syst 5:618092

Vadez V, Kholova J, Srikanth M, Rekha B, Tharanya M, Sivasakthi K et al (2017) LeasyScan: 3D scanning of crop canopy plus seamless monitoring of water use to harness the genetics of key traits for drought adaptation

van der Pol LK, Robertson A, Schipanski M, Calderon FJ, Wallenstein MD, Cotrufo MF (2022) Addressing the soil carbon dilemma: legumes in intensified rotations regenerate soil carbon while maintaining yields in semi-arid dryland wheat farms. Agric Ecosyst Environ 330:107906

Willett W, Rockström J, Loken B, Springmann M, Lang T, Vermeulen S et al (2019) Food in the Anthropocene: the EAT–Lancet Commission on healthy diets from sustainable food systems. Lancet 393:447–492

Yang Y, Saand MA, Huang L, Abdelaal WB, Zhang J, Wu Y et al (2021) Applications of multi-omics technologies for crop improvement. Front Plant Sci 12:563953
Zaman M, Shahid SA, Heng L, Shahid SA, Zaman M, Heng L (2018) Soil salinity: historical perspectives and a world overview of the problem. In: Guideline for salinity assessment, mitigation and adaptation using nuclear and related techniques, pp 43–53
Zhang J, Li H, Huang X, Xing J, Yao J, Yin T et al (2022) STAYGREEN-mediated chlorophyll a catabolism is critical for photosystem stability during heat-induced leaf senescence in perennial ryegrass. Plant Cell Environ 45:1412–1427
Zhao C, Zhang H, Song C, Zhu J-K, Shabala S (2020) Mechanisms of plant responses and adaptation to soil salinity. Innovation 1:100017

Application of Nanoparticles for Sustainable Crop Production

23

Amarjeet Singh Bhogal, Rahul Gogoi, Namita Borah, Shaik Faheem Akhtar, and Sudipta Sankar Bora

Abstract

Micronutrient deficiencies, including vitamins, iron, zinc, iodine, etc., can seriously affect human health. These deficiencies can lead to stunted development, dementia, neonatal problems, and higher mortality rates among children. Iron deficiency anemia is a significant issue among women in developing nations, often exacerbated by high blood loss during menstruation and parturition. Agronomic biofortification, a method that uses micronutrient-rich fertilizers, can improve the nutritional profile of agricultural products for populations with limited access to a variety of diets or supplements. Studies have shown that nanoparticles can aid agronomic biofortification and improve crop performance and nutrient absorption under various ecological conditions. By delivering agrochemicals in a safe, simple, and effective manner, nanotechnology has the potential to transform farming systems and aid in the biofortification of food crops. As such, by incorporating NPs into crop production, we can biofortify food crops and contribute to better overall nutrition. This book chapter explores the use of nanoparticle strategies to improve the nutritional quality of agricultural crops, highlighting potential benefits for global health and food security.

A. S. Bhogal · N. Borah · S. F. Akhtar
Department of Plant Breeding and Genetics, Assam Agricultural University, Jorhat, Assam, India

R. Gogoi
Department of Agricultural Biotechnology, Assam Agricultural University, Jorhat, Assam, India

S. S. Bora (✉)
Multi-disciplinary Research Unit (MRU), Jorhat Medical College and Hospital, Jorhat, Assam, India

© The Author(s), under exclusive license to Springer Nature Singapore Pte Ltd. 2025

S. Faiz et al. (eds.), *Climate Smart Agriculture for Future Food Security*,
https://doi.org/10.1007/978-981-96-4499-5_23

Keywords

Micronutrient deficiency · Nanoparticle · Agronomic biofortification · Food security · Agriculture

23.1 Introduction

Agronomic biofortification (i.e., the application of micronutrient-rich fertilizers) of crops can significantly enhance the nutritional profile of agricultural products, reducing malnutrition and starvation and ensuring food security at a lower cost (Bhardwaj et al. 2022). Agronomic biofortification, which involves fertilizer application, mineral element solubilization, and source-to-sink mobilization, is an effective strategy for increasing the micronutrient profile of staple food crops (Khush et al. 2012). Agronomic biofortification improves the dietary requirements of primary crops by increasing zinc, iron, and selenium levels in edible parts (Kapoor et al. 2022). It offers a unique advantage over genetic biofortification due to its nonspecific fertilizer forms and application techniques, allowing for quick adaptation to different crops, unlike genetic and transgenic methods, which are time-consuming and resource-intensive (Bhardwaj et al. 2022). Nano-biofortification offers a novel strategy for enriching crops with vital elements to boost an individual's diet, providing nutritious meals loaded with micronutrients to combat malnutrition. Since it enhances farm as well as greenhouse crop nitrogen usage efficacy, it can boost total crop yields and agricultural sustainability. NFs can provide nutrients slowly and steadily when used independently or in combination with chemical or organic fertilizers. Chemical fertilizers take 4–10 days to release their nutrients, while NPs may release them steadily over 40–50 days. Furthermore, NFs can enhance plant resistance to both abiotic and biotic stresses (Seleiman et al. 2021). This chapter explores the overall idea of NP-based agronomic biofortification toward crop health.

23.2 Nanoparticle-Based Nutrient Uptake and Transport by Plants

It is widely accepted that NPs can be absorbed and transported by plants. The conditions under which this occurs are determined by a number of factors, including the size, surface charge, concentration, exposure duration of the NPs, and species of the plant. NPs can enter plant systems through stomata, root hairs, and leaf cracks, as well as propagate through bulk flow, phloem loading, and diffusion.

23.2.1 Uptake of Nanoparticles in Plant Leaves

NPs are sprayed onto leaf surfaces in agricultural applications, where they deposit and are absorbed by plants through stomata or cuticles. The waxy cuticle found on leaf epidermis is composed primarily of wax, cutin, and pectin. It serves as a major natural barrier to keep NPs out of the leaves and shields the leaves from drying out during plant growth. On the waxy stratum corneum's surface, there are two distinct channels: one is lipophilic and the other is hydrophilic (Mittal et al. 2020). Both lipophilic and hydrophilic channels have widths that range from 0.6 to 4.8 nm. Hydrophilic NPs with a diameter of less than 4.8 nm are able to disperse with the aid of the hydrophilic channels. Leaf absorption of lipophilic NPs occurs via infiltration and diffusion through the lipophilic channels in the cuticle surface. Using confocal fluorescence microscopy with excellent spatial and temporal resolution, a recent study has demonstrated that carbon dots smaller than 2 nm could penetrate cotton leaves through the cuticular pathway (Hu et al. 2020b). However, plants' ability to absorb NPs through the epidermis is restricted because of the tiny pore channels in the cuticle. After being applied to the leaf surface, several research findings suggested that NPs might build up in the vascular tissue and epidermis of the leaf.

23.2.2 Absorption of Nanoparticles in Plant Roots

Adsorption on the root surface is how NPs and plant roots first come into contact. The root surface is negatively charged due to the root hairs' ability to discharge chemicals like mucus or organic acids; therefore, positively charged NPs are more likely to aggregate in the root and be readily absorbed on the root surface. NPs may be able to reach the root column by developing lateral roots, which may open up a new adsorption surface for them. Plant roots' epidermis has a similar makeup and purpose to that of plant leaves. On the surface of the plant root tips and the root hairs of the primary and secondary roots, however, the epidermis is not completely grown. The root epidermis is directly touched and crossed by NPs when they are exposed to this region. The epidermal cells in the root cell wall have semipermeable walls (García-Sánchez et al. 2015). Large particles may be impeded from passing through the tiny pores in the root cell wall. When the root lacks an exodermis, NPs may penetrate the root's xylem or central column. Additional research has demonstrated that certain NPs have the ability to rupture the plasma membrane and cause the cell wall to produce new pores, which open up a pathway for large-diameter NPs to enter. Plant tissue can absorb NPs by a variety of mechanisms, including the ion route, endocytosis, interaction with cell membrane proteins, and physical damage (Parkinson et al. 2022).

Research indicates that plant roots absorb NPs, primarily entering cells via the hydrophilic channel, which is only accessible to small NPs due to their tiny channel size. Endocytosis is a significant method by which NPs are absorbed by plant cells. NPs enter cells through the plasma membrane of plant cells through invagination

(Gao et al. 2023). Research shows that plant protoplasts can internalize particles smaller than 1 μm through endocytosis, resulting in NPs lacking particle size selectivity. Carbon-based NPs and carbon nanotubes may be absorbed by *Catharanthus roseus* root cells via endocytosis (Grillo et al. 2021).

Research on the impact of NPs on plant roots is more extensive than the application of various NPs to plant leaves. When the hair cells in the plant roots interact with the root, the NPs are absorbed. They enter the endothelium layer from the epidermis through the symplast or exoplast, preferentially passing through the cell wall. They are subsequently transported to the aboveground portion via the xylem arteries. A pericycle, vascular column, cortex, and epidermis make up the plant root in that order from within to outside. The center of the plant root contains vascular tissue and a pericycle, which make up the vascular column (Rasheed et al. 2022).

23.2.3 Modes of Transport of NPs into Plants

There are two basic ways of classifying NP transport in plants: assisted delivery and passive delivery. When delivering NPs to the plant body, the term "assisted delivery" refers to the application of external force or power to aid in the transfer of NPs into the plant tissues. This external force or power facilitates the delivery process by overcoming obstacles that would otherwise prevent passive transport. A device called a "gene gun" or "biolistic" propels NPs with the use of an external force like pressurized gas or helium. Ag-NPs were used as gene carriers in *Nicotiana tabacum* for biolistic gene delivery, demonstrating significantly higher transformation efficiency compared to Au-microparticles (Rajkumari et al. 2021). Sonoporation, a technique involving ultrasonic vibrations, allows NPs to temporarily penetrate plant cell membranes and enter the cells. Ultrasound's external force facilitates the delivery procedure. Poly-ethyleneimine (PEI)-coated mesoporous silica nanoparticles (MSNs) can be transferred into plant cells economically and simply via ultrasonic treatment. This method eliminates the need for complex equipment and raises no safety concerns (Zolghadrnasab et al. 2021). External magnetic fields are an efficient tool for guiding magnetic NPs to target plant tissues and facilitating their targeted administration. Comprehending the behavior of NPs both in bulk and individual molecule form is essential for characterizing their intrinsic magnetic characteristics.

Assistive delivery methods, due to the external forces involved, can facilitate the transportation of large particles more efficiently. These technologies are useful for bypassing natural barriers and efficiently moving particles. However, several major challenges remain with this mode of transportation, particularly in terms of scaling to greater proportions. As a result, passive delivery continues to outperform. In a controlled laboratory context, it is easier to maintain uniform conditions for assisted delivery, resulting in consistent NP uptake (Mitchell et al. 2021). At larger scales, consistency in a field or region of agriculture becomes increasingly challenging. The efficacy of distribution strategies can be impacted by changes in plant physiology, soil composition, and environmental variables. Furthermore, precise control over parameters such as temperature, humidity, and nutrition levels is made possible via assisted delivery in vitro. Actual agricultural situations differ greatly from one another. It becomes difficult to adapt in vitro techniques to different surroundings, because variables like temperature changes, wind, and precipitation might affect how well NPs are delivered. Large-scale implementation of these assisted delivery techniques is resource-intensive and might not be financially feasible. Regulations pertaining to safety, ethics, and environmental effects may present difficulties for the widespread use of assisted delivery (Milla and Osborne 2021).

Passive delivery refers to the movement of NPs through natural pathways in plants without the aid of proactive mechanisms or external influences. This refers to the range of techniques by which NPs get incorporated into the plant, usually through the roots and leaves. In brief, there are distinctions between assisted delivery and passive delivery. While passive delivery depends solely on natural processes without outside help, assisted delivery uses an intervention to improve NP uptake (Griffin et al. 2017). These methods address a variety of objectives and uses of nanotechnology in plant science research. There are two of these passive delivery methods: foliar and root absorption. Water and other dissolved materials are naturally absorbed by plants, and when they do so, NPs may enter the plant through their roots. This is a typical technique for in vivo NP delivery to plants. Plant leaves can also be sprayed or treated with NPs. The response of *Zea mays* L. to Cd and TiO_2 NPs was strongly influenced by the exposure method (Lian et al. 2020). Plant NPs are carried in the vascular system of a plant, consisting of the xylem and phloem in the stem, after they penetrate the plant's roots or leaves. Water, nutrients, and other materials, including NPs, are distributed to various areas of the plant by the vascular system (Sembada and Wuled Lenggoro 2024).

23.3 Enhancement of Nutrient Bioavailability

Increasing yields is one way that nanotechnology has greatly benefited agriculture (Pandey 2018). Plants acquire nutrition from fertilizers, although most traditional fertilizers are inefficient in terms of nutrient usage and acquisition. Thus, target orientation and resistance to loss are key features of nanofertilizers (Dimkpa and Bindraban 2018). Zinc (Zn) is a significant micronutrient that influences plant growth and production, while deficits may restrict development, reduce the number of tillers, cause chlorosis and smaller leaves, lengthen crop maturity periods, cause spikelet sterility, and result in diminished quality harvested products (Feller et al. 2019). Zinc oxide NPs (ZnO-NPs) are a type of metal oxide nanomaterial (NM) with distinctive physical and chemical properties. They are distinguished by impressive chemical stability, a wide absorbed radiation spectrum, an elevated electromechanical coupling coefficient, and excellent photostability (Rasmidi et al. 2021). ZnO-NPs generated biologically using moringa leaf extract were investigated for their impact on seed germination, growth characteristics, zinc content, and nutrient efficiency of utilization in amaranth crops. Biosynthesized ZnO-NPs showed maximum plant growth and output at 10 ppm concentration, unlike the 50 ppm zinc required for bulk zinc sulphate fertilizer (Reshma and Meenal 2022). ZnO-NPs have been found to promote plant growth and alleviate chromium-induced stress in rice seedlings. Adding ZnO-NPs to the growing medium reduces chromium toxicity while increasing rice seedling growth. ZnO-NPs-mediated alleviation of Cr(VI) stress was related to enhanced photosynthetic efficiency, minimized oxidative markers, and increased regulation of the main antioxidant genes, as well as optimization of AsA-GSH cycle performance (Prakash et al. 2022). A trial in a pot was carried out to investigate the impact of ZnO-NPs under brine stress in a wheat cultivar, "Gemmieza", introduced from Egypt to figure out the salinity issue. A salinity of 10 dS m^{-1} was experimentally generated, and potted wheat received varying dosages of Zn normal fertilizer and ZnO-NPs. ZnO-NPs (0.12 g per pot) considerably improved morphological characteristics when contrasted with the control group undergoing salt stress. The application of ZnO-NPs substantially raised chlorophyll A and B levels by 24.6 and 10%, plant height by vegetative means and maturity stages by 34.6 and 37.4%, shoot and spike lengths by 30.7 and 27.6%, root fresh and dry weights by 74.5 and 63.1%, and wheat grain yield by 42.2% (Adil et al. 2022). ZnO-NPs were used to improve zinc levels in wheat and rice grains grown in saline conditions. The pot trials showed significant improvements in growth, salt tolerance, plant Zn uptake, and grain Zn concentrations, with ZnO-NPs outperforming other bulk sources (Mazhar et al. 2023).

γ-Fe_2O_3-NPs is being increasingly studied for its potential to address iron deficiencies in soils and crops. These NPs stimulate the emergence of seeds, root growth, and chlorophyll water levels. Once γ-Fe_2O_3-NPs are introduced to soil, they liberate iron and move from roots to different connective tissues. *Cucurbita maxima* shoots receiving treatment with Fe^{3+} and γ-Fe_2O_3-NPs had higher iron level compared to the controls or plants treated with Fe (II)-EDTA (Hu et al. 2017). The physiologic efficiency of soybeans treated with foliar or soil supplements comprising Fe_2O_3-NPs, fulvic acid (FA)-coated Fe_2O_3, and Fe-EDTA was assessed in a controlled environment (Yang et al. 2020). The study found no growth problems or toxicological stress in the treatments, with foliar amendments containing nFe_2O_3 and nFe_2O_3-FA significantly improving root developmental indices, plant matter, and chlorophyll content and stimulating biological nitrogen fixation. A study explored the use of iron oxide and EDTA-modified iron oxide nano-micronutrients as a replacement for conventional Fe-fertilizer in mulberry plant growth. The results showed significant morphological and antioxidant effects, with improved sucrose levels and chlorophyll levels, and antioxidant enzyme effectiveness compared to control (Haydar et al. 2022).

Copper oxide NPs are employed as fertilizers as well as plant growth regulators, insecticides, herbicides, and soil remediation compounds. Foliar application of CuO-NPs significantly improved the phytochemical quality, yield, catalase, and peroxidase activity of hydroponic lettuce plants, enhancing leaf breadth, total phenols, flavonoids, and antioxidant capacity (Fortis-Hernández et al. 2024). Folar application of copper nanoparticles significantly enhanced the physical, nutritional, and concentration of copper in melons, enhancing their nutritional value and nutraceutical excellence (Fortis Hernández et al. 2022).

Less than 4% levels of titanium dioxide (TiO_2) NPs were found to improve nitrogen fixation and stimulate photosynthesis in spinach, increasing the plant's overall development (Zheng et al. 2005). The TiO_2-NP treatment of coriander was investigated to assess potential benefits for plant growth and nutritional quality in hydroponic systems. Findings revealed that 50 mg/L TiO_2-NPs enhanced the fresh biomass of the roots and shoots by 13.1% and 4.1%, respectively, when compared to the control (Hu et al. 2020a).

A study examined the impact of cerium oxide nanoparticles ($nCeO_2$) on barley (*Hordeum vulgare* L.) physiology, productivity, and macromolecular composition. Results showed that $nCeO_2$ at 500 mg kg^{-1} soil enhanced plant development, increasing shoot biomass by 331%. $nCeO_2$ at 250 mg kg^{-1} soil significantly increased grain Ce accumulation by 294%, resulting in increased nutrient content in grains (Rico et al. 2015). Rice grains from $nCeO_2$-enriched soils showed reduced Fe, S, prolamin, glutelin, lauric and valeric acids, and starch content. Moreover, antioxidant properties were decreased (Rico et al. 2013).

Noble metal NPs have been extensively investigated in agriculture due to their antibacterial properties and prospective significance for plant disease management. While gold (Au) NPs have been employed sparingly as insecticides and to promote rapid plant growth, silver has been utilized to eradicate undesirable bacteria from soils, plants, and hydroponic systems. It has been revealed that the addition of Au-NPs to *Lactuca sativa* seeds enhances the shoot-to-length ratio without altering the microbial content or producing toxicity (Pestovsky and Martínez-Antonio 2017). Their early flowering, long pods, high levels of sugar and chlorophyll, and improved flavonoid content for free radical scavenging also boost seed germination, growth rate, and yield. Platinum NPs have been experimentally demonstrated to influence plant growth by enhancing the length and weight of the root system (Astafurova et al. 2015).

23.4 Role in Stress Tolerance and Plant Health

23.4.1 Role of NPs in Conferring Stress Tolerance to Plants

The targeted, smart delivery of low doses of NPs improves the plant's response to abiotic stresses, enhancing their resistance to these risks (Lowry et al. 2019). The impact of NPs on plants will depend on various parameters, such as composition, concentration, size, and the physical and chemical properties of the NPs (Shalaby et al. 2016). The agriculture system would become more resilient and less vulnerable to climate change as a consequence of this method, which potentially boosts plants' resilience to biotic and abiotic stresses. Nanotechnology is becoming increasingly popular as an approach to biotic and abiotic stress-related issues. The primary advantage of nanofertilizer is that it helps restore soil health while also providing the optimal amount of micronutrients. The presence of distinct NPs in specific quantities has proven to have significant positive impacts on a wide range of plant species (Zhu et al. 2019).

NPs influence the ability of plants to tolerate salinity stress by changing the levels of hormones, gene expression, defense mechanisms, antioxidant enzyme activity, and ion homeostasis (Zahra et al. 2021). SiNPs, a type of artificial nanosilica, have been found to effectively mitigate the harmful effects of Na^+-derived salt on cucumber seeds and saplings (Alsaeedi et al. 2018). The metabolism of corn seeds primed with nanosilica under salt exposure was investigated. Nano-silica-primed seeds enhanced plant resilience by enhancing antioxidant enzyme activity, reducing reactive oxygen species, and inhibiting lipid peroxidation. It also increased gibberellin levels, enabling energy-releasing chemicals via respiration (Naguib and Abdalla 2019). NPs were reported to cause a rise in the amount of leaf proline and free amino acids in rice, which helps resist NaCl penetration (Abdel-Haliem et al. 2017). Salt stress significantly slows down the developmental processes of tomato plants. Si-NPs significantly enhanced growth attributes, mineral content, photosynthesis, and antioxidative enzyme activity in such plants under salt stress, including height, dry weight, and internal CO_2 concentration, thereby enhancing their growth and activity (Alam et al. 2022).

The application of NPs is an effective method for mitigating the impact of drought on plants by increasing antioxidant enzyme activity, improving phytohormone levels, and affecting physiological properties (El-Saadony et al. 2022). A study investigated the effects of SiO_2-NPs on wheat seeds during severe drought, using individual seeds and NPs application through soil and foliar methods. The findings demonstrated that, in comparison to regular watering, drought stress reduced most features; however, subsequent use of SiO_2-NPs enhanced the relative water content and greenness of the leaves (Behboudi et al. 2018). A study on *Prunus mahaleb* saplings found that pretreatment with SiO_2-NPs significantly reduced the effects of extreme drought on gas exchange, photosynthesis, stomatal conductance, and transpiration rate. Pretreated saplings also maintained root extension and mitigated the effects of extreme drought on root dry mass buildup (Ashkavand et al. 2018). The consequence of SiO_2, Se, and Se/SiO_2-NPs on strawberry growth and output was investigated. Under typical and severe drought conditions, strawberry crops sprayed with treatments containing these NPs showed better growth and yield characteristics. The Se/SiO_2 combinations maintained more photosynthetic pigments and increased concentrations of critical osmolytes. Exogenous spraying of Se/SiO_2 improved drought tolerance and biochemical characteristics of fruits under drought conditions (Zahedi et al. 2020).

Heat stress has become one of the most significant abiotic stresses linked to climate change, and its detrimental effect on crop productivity in all environments has been recognized (Nolan et al. 2020). Stomatal opening in tomato plants induced by TiO_2-NPs considerably decreased heat stress (Qi et al. 2013). A study found that SeNP increased the expression of heat shock factor A4A (HSFA4A) in wheat plants at a concentration of 10 mg/L through specific signaling pathways (Safari et al. 2018). Application of ZnO at a concentration of 30 mgL^{-1} was reported to increase chlorophyll content with upregulated antioxidants and osmoprotectants (LP, GB, and TSP) to prevent oxidative damage under heat stress in *Vigna radiata* L. (Kareem et al. 2022).

A study found that Se-NPs and SeO_3^{2-} significantly improved *Brassica napus* L. growth under Cadmium toxicity conditions by decreasing Cd accumulation, maintaining calcium homeostasis, promoting disulfide bond formation, and restoring the waxy outer layer (Qi et al. 2021). The application of 5–25 mM NPs to rice plants during the flowering stage was found to significantly decrease cadmium levels by up to 64.9%. NPs may reduce aluminum toxicity in *Cicer arietinum* by improving seed germination and plant growth (Chen et al. 2018). NPs ameliorate heavy metal stress in plants by reducing their accumulation and oxidative stress, as well as activating their own defense mechanisms (Chandra et al. 2020). Nanotechnology is crucial for sustainable development, as it can mitigate heavy metals' harmful effects. Its positive results, particularly in increasing plant growth and HM tolerance, suggest potential for remediation in metal-polluted soils (Faizan et al. 2023).

23.4.2 The Potential of NPs to Influence Crop Productivity and Resilience

Plants possess enzyme machinery to counter the oxidative stresses imposed by the environment. The plant defense mechanism brings about serious consequences at physiological and molecular levels. Nanotechnology, along with inadequate water treatment, has been employed to provide a unique strategy for improving medicinal herbs' resistance to moisture stress. Multiple approaches have explored green synthesis of zinc oxide NPs (ZnO-NPs) from *Camellia sinensis* L. leaf extract. Under stressful water conditions, a pot experiment showed that 100 ppm of produced NPs significantly enhanced the antioxidant mechanism by increasing catalases, superoxidases, and APX enzymes. Proline and chlorophyll levels were both increased by applying prepared ZnOx NPs topically (Khan et al. 2021). Improvement in nitrogen assimilation, nitric oxide synthesis, chlorophyll content, and ionic balance under salinity stress was observed in wheat upon application of gold NPs (Au-NP) at the concentration of 300 ppm. The study explored the effects of Au-NPs on wheat crops influenced by salinity. Au-NPs influenced physiological reactions like K^+/Na^+ ratio, chlorophyll level, security mechanisms, nitrogen absorption, stomatal dynamics, and developmental characteristics during salty conditions (Wahid et al. 2022). Under drought conditions, Se-NPs (30 mg/L) applied to wheat (*Triticum aestivum*) enhanced plant growth and reduced ionic leakage and lipid peroxidation, thereby reducing the toxicity of drought-induced cellular damage (Ikram et al. 2020). Se-NP also positively impacts Cadmium stress-induced physiological stress by inhibiting the expression of glycolate oxidases and NADPH oxidases, thereby minimizing membrane lipid damage (Qi et al. 2021). Soil contaminants such as cadmium (Cd) can induce the oxidative degradation of plants, rendering them poisonous. While selenium oxyanions such as selenite and seleniate can mitigate Cd stress, investigation has demonstrated that selenium NPs (Se-NPs) positively affect *Brassica napus* L. growth in Cd stress scenarios. SeNPs reduce reactive oxygen species formation by suppressing NADPH and glycolate oxidases, minimizing oxidative protein and membrane lipid damage. It improves leaf resistance by avoiding Cd buildup, maintaining calcium homeostasis, stimulating disulfide bond formation, and restoring the waxy outer layer (Zhao et al. 2022). The application of nanobiotechnology to transport reactive oxygen species (ROS)-scavenging NMs has the potential to enhance stress tolerance in cereals. ROS accumulates excessively under stressful circumstances, causing biomolecule damage and impeding crop development. ROS-scavenging NMs can reduce environmental stress and enhance stress tolerance in plants. Nanoparticulate silver and copper oxides, which generate ROS, can be successfully administered to plant taxa to boost the immune system, prime stress responses, and improve stress resistance, thereby increasing the productivity of crops.

23.5 Environmental and Health Considerations

Both natural and anthropogenic NMs have positive and negative impacts on biological systems and the environment, which further depend upon the dose, duration, and frequency of exposure (Hochella et al. 2019). In the agricultural sector, NMs are used to enhance crop yield and biofortification and to develop resistance against biotic and abiotic stresses. However, improper and excessive use of nanofertilizers may have a negative impact on crops and, perhaps, the overall ecosystem. NP toxicity can be attributed to environmental variables. The characteristics of NPs that determine their toxicity can be impacted by weather factors such as humidity, temperature, wind speed, latitude, and type of light. Compared to NPs at lower or normal temperatures, NPs can scatter more quickly at higher temperature. NPs exhibit distinct behaviors under various light conditions, including visible and ultraviolet light. The speed of wind can help these minuscule particles penetrate the tissues of plants and animals. Toxic NPs are likely to exhibit similar behavior as certain diseases that have high incidence rates in a particular geographic area.

Due to the environmental fate and ecotoxicological aspects of NPs, the aquatic environment is contaminated with NPs because of the increase in consumer products such as cosmetics and sunscreens, which contain NPs such as TiO_2 (Popov et al. 2005). Aquatic animals such as fish, daphnia, and unicellular organisms, as well as NPs, are toxic (Nowack and Bucheli 2007). The susceptibility of aquatic animals to toxicity varies among species (Griffitt et al. 2008). It is still difficult to pinpoint the precise concentration of NPs in an aquatic environment. The surface charge and chemical reactivity of the particles, in addition to their enormous surface area-to-volume ratio, determine whether or not the NPs are biocompatible with living things (Oberdörster et al. 2004). For aquatic creatures, the size of the particles and their subsequent biodegradation could have dangerous biological consequences. Because of their large surface area, microscopic particles have the potential to produce oxygen-reactive species, which are extremely dangerous and can seriously destroy proteins, DNA, and membranes (Reijnders 2006). NPs may enter aquatic animals through their gills and through ingestion (Oberdörster et al. 2004). The results showed that silver and copper NPs had harmful effects on all investigated organisms when algal species, dolphins, and zebrafish were exposed to the ENPs of nickel, copper, silver, and aluminium (Griffitt et al. 2008). The human genome of zebrafish has a high degree of homology, so zebrafish was chosen for risk assessment of NPs (Bar-Ilan et al. 2009).

It is acknowledged that every toxicity study on the interaction of soil and NPs works with distinct factors, which makes correlation a difficult task. However, recent literature that addresses the fate and effects of ENPs on the soil environment is described here to emphasize their impact on soil. Soil enzyme activity can be quantified to predict human impacts on the soil ecosystem. There is a diversity of bacteria in the soil. At very low concentrations, Ag-NPs are extremely insensitive to harmful effects on enzyme activity (Peyrot et al. 2014). Ag-NPs also had concentration-dependent effects on the population and growth rate of earthworms called *Lumbricus rubellus*. Long-term exposure to high concentrations of Ag-NPs has the potential to completely stop population expansion. Earthworm immune cells are similarly impacted by these particles (van der Ploeg et al. 2014). Ag-NPs were found to be toxic to *Pseudomonas stutzeri* and *Bacillus cereus* in the soil, and after 48 h of exposure, bactericidal effects were observed at a concentration of 5 mg/L (Fajardo et al. 2014). Similarly, ZnO-NPs had a negative impact on earthworm fertility and were more readily absorbed than bulk ZnO and $ZnCl_2$. The denitrification kinetics in soil samples were found to be influenced by cesium NPs, with the process reliant on particle size, concentration, and species (Dahle and Arai 2014). A study investigated the toxicity of CeO-NPs on *Eisenia fetida*, an earthworm, and its long-term exposure has been linked to histopathological alterations (Lahive et al. 2014).

23.6 Effect of NPs on Human Health

The common exposure of the human body to NPs can be described through different mechanisms. Primary exposure for workers, engineers, and scientists occurs during the research-scale and commercial production of NM-based products. Additional sources of exposure may arise from the handling of raw materials during packing and shipping. In the second stage, plants are exposed to such NMs during use and application to crops, which may lead to harmful and toxic effects (Tsuji et al. 2006). The interaction of NPs with the human body can occur through the following processes: (a) penetration through skin nodes, (b) intake by the respiratory system via inhalation, and (c) intake by the digestive system.

The safety and regulatory aspects of NP-fortified crops should be considered, focusing on human health implications and risk assessment. NP emissions from human sources are characterized in two parts. Intentionally produced engineered NPs such as nanotubes, nanowires, nanofibers, and unintentionally produced NPs are produced from combustion sources, vehicle exhaust, and mechanical workshops (Wu et al. 2008). The main source of human exposure to such NPs is polluted air (Kumar et al. 2011). The Clean Air Act, the European Union Directive on ambient air quality (2008/50/EC), and cleaner air for Europe are examples of environmental protection laws and air quality criteria.

NPs, however, are exempt from the present particulate material laws because they are based on mass concentration. The reason for their inapplicability is that non-contrast ultrafine NPs can enter the lungs and deposit (Kumar et al. 2011). Under the Toxic Substances Control Act (TSCA), NPs are regulated by the US Environmental Protection Agency (USEPA). Under the TSCA, the EPA has examined a variety of nanoscale materials, including silica, aluminium NPs for prolonged inhalation, single and multiwall carbon nanotubes, and silica. The "Significant New Use Rule (SNUR)" is another regulation that the USEPA is creating to guarantee the regulatory assessment of products at the nanoscale.

Manufacturers of nanoscale materials are required under the "information gathering rule" to disclose details about the material's safety and health exposure, as well as its production volume and synthesis technique. Testing for environmental health and safety is performed on a specific set of NPs. According to the Canadian Environmental Protection Act (CEPA) (1999), NPs are evaluated by Environment Canada (EC) (http://www.ec.gc.ca/). This act mandates risk assessment for newly synthesized chemicals made in Canada or imported into the country. According to Morimoto and Kobayashi, (2010), the European Commission is regulating engineered NPs under the Registration, Evaluation, Authorization, and Restriction of Chemicals (REACH) program as regular chemical substances (Morimoto et al. 2010). A new law regarding the classification, labeling, and packaging of novel materials has been devised by the European Union (EU). Global institutions such as the Organization for Economic Cooperation and Development (OECD) and the International Organization for Standardization (ISO) are preparing for difficulties related to nanotechnology.

In terms of testing and science-based health and safety procedures involving NMs, the ISO is currently working on creating standards for vocabulary, nomenclature, metrology, and instruments. Similarly, the OECD's Working Party on Manufactured Nanomaterials (WPMN) is working on a number of projects pertaining to exposure measurements, environmental health and safety research databases, collaboration on environmentally sustainable uses of nanotechnology, and testing guidelines for ENPs. The OECD website (http://www.oecd.org/env/ehs/nanosafety/) was used for further information. The US National Institute for Occupational Safety and Health (NIOSH) has released comprehensive guidelines for workers who handle NPs in the workplace. These guidelines cover personal safety, exposure evaluation, possible health problems, and precautionary actions (Hodson et al. 2009).

23.7 Conclusion

There is a dearth of knowledge on the unintended release of nanoparticles into the environment during the manufacturing, application, and disposal of agricultural nanoproducts. Therefore, comprehensive forward testing and future-oriented investigations based on identified knowledge gaps are essential. Understanding the interaction between NPs and soil physicochemical characteristics is crucial for reducing the risk of soil microbiota alteration due to nanotoxicity. The lack of relevant organizations, monitoring bodies, and regulations hinders the commercialization of NP-based agrochemicals. It is essential to combine the framework of nanosafety rules with an emphasis on safety-by-design synthesis methodologies to prevent misuse at a toxic level. Future studies should assess risks and develop a feasible approach for NP-based products in agriculture, focusing on nontoxic, concentration-dependent studies in natural soil systems. Research is also required to optimize soil rotation, convert synthetic to biosynthesized NPs, and investigate long-term physiochemical properties of novel NP-based agrochemicals to reduce toxicity and ensure crop production.

References

Abdel-Haliem MEF, Hegazy HS, Hassan NS, Naguib DM (2017) Effect of silica ions and nano silica on rice plants under salinity stress. Ecol Eng 99:282–289. https://doi.org/10.1016/j.ecoleng.2016.11.060

Adil M, Bashir S, Bashir S, Aslam Z, Ahmad N, Younas T, Asghar RMA, Alkahtani J, Dwiningsih Y, Elshikh MS (2022) Zinc oxide nanoparticles improved chlorophyll contents, physical parameters, and wheat yield under salt stress. Front Plant Sci 13. https://doi.org/10.3389/fpls.2022.932861

Alam P, Arshad M, Al-Kheraif AA, Azzam MA, Al Balawi T (2022) Silicon nanoparticle-induced regulation of carbohydrate metabolism, photosynthesis, and ROS homeostasis in solanum Lycopersicum subjected to salinity stress. ACS Omega 7(36):31834–31844. https://doi.org/10.1021/acsomega.2c02586

Alsaeedi A, El-Ramady H, Alshaal T, El-Garawani M, Elhawat N, Al-Otaibi A (2018) Exogenous Nanosilica improves germination and growth of cucumber by maintaining K+/Na+ ratio under elevated Na+ stress. Plant Physiol Biochem 125:164–171. https://doi.org/10.1016/j.plaphy.2018.02.006

Ashkavand P, Zarafshar M, Tabari M, Mirzaie J, Nikpour A, Bordbar SK, Struve D, Striker GG (2018) Application of SiO2 nanoparticles as pretreatment alleviates the impact of drought on the physiological performance of prunus mahaleb (Rosaceae). Boletín de La Sociedad Argentina de Botánica 53(2):207–219. https://doi.org/10.31055/1851.2372.v53.n2.20578

Astafurova T, Zotikova A, Morgalev Y, Verkhoturova G, Postovalova V, Kulizhskiy S, Mikhailova S (2015) Effect of platinum nanoparticles on morphological parameters of spring wheat seedlings in a substrate-plant system. IOP Conf Series Mater Sci Engin 98:012004. https://doi.org/10.1088/1757-899X/98/1/012004

Bar-Ilan O, Albrecht RM, Fako VE, Furgeson DY (2009) Toxicity assessments of multisized gold and silver nanoparticles in zebrafish embryos. Small 5(16):1897–1910. https://doi.org/10.1002/smll.200801716

Behboudi F, Tahmasebi Sarvestani Z, Kassaee MZ, Modares Sanavi SAM, Sorooshzadeh A (2018) Improving growth and yield of wheat under drought stress via application of SiO2 nanoparticles. J Agri Sci Technol 20(7)

Bhardwaj AK, Chejara S, Malik K, Kumar R, Kumar A, Yadav RK (2022) Agronomic biofortification of food crops: An emerging opportunity for global food and nutritional security. Front Plant Sci 13. https://doi.org/10.3389/fpls.2022.1055278

Chandra J, Chauhan R, Korram J, Satnami ML, Keshavkant S (2020) Silica nanoparticle minimizes Aluminium imposed injuries by impeding cytotoxic agents and over expressing protective genes in Cicer Arietinum. Sci Hortic 260:108885. https://doi.org/10.1016/j.scienta.2019.108885

Chen R, Zhang C, Zhao Y, Huang Y, Liu Z (2018) Foliar application with Nano-silicon reduced cadmium accumulation in grains by inhibiting cadmium translocation in Rice plants. Environ Sci Pollut Res 25(3):2361–2368. https://doi.org/10.1007/s11356-017-0681-z

Dahle JT, Arai Y (2014) Effects of Ce(III) and CeO2 nanoparticles on soil-denitrification kinetics. Arch Environ Contam Toxicol 67(4):474–482. https://doi.org/10.1007/s00244-014-0031-9

Dimkpa CO, Bindraban PS (2018) Nanofertilizers: new products for the industry? J Agric Food Chem 66(26):6462–6473. https://doi.org/10.1021/acs.jafc.7b02150

El-Saadony MT, Saad AM, Soliman SM, Salem HM, Desoky E-SM, Babalghith AO, El-Tahan AM, Ibrahim OM, Ebrahim AAM, Abd El-Mageed TA, Elrys AS, Elbadawi AA, El-Tarabily KA, AbuQamar SF (2022) Role of nanoparticles in enhancing crop tolerance to abiotic stress: a comprehensive review. Front Plant Sci 13. https://doi.org/10.3389/fpls.2022.946717

Faizan M, Alam P, Rajput VD, Faraz A, Shadma Afzal S, Ahmed M, Fang-Yuan Y, Minkina T, Hayat S (2023) Nanoparticle mediated plant tolerance to heavy metal stress: what we know? Sustain For 15(2):1446. https://doi.org/10.3390/su15021446

Fajardo C, Saccà ML, Costa G, Nande M, Martin M (2014) Impact of Ag and Al2O3 nanoparticles on soil organisms: in vitro and soil experiments. Sci Total Environ 473–474:254–261. https://doi.org/10.1016/j.scitotenv.2013.12.043

Feller U, Anders I, Wei S (2019) Distribution and redistribution of 109Cd and 65Zn in the heavy metal Hyperaccumulator solanum Nigrum L.: influence of cadmium and zinc concentrations in the root medium. Plan Theory 8(9):340. https://doi.org/10.3390/plants8090340

Fortis Hernández M, Lopez JO, Rangel PP, Valencia RT, Fortiz EL, Andrade-Sifuentes A, Rueda Puente EO (2022) Biofortification with Copper Nanoparticles (Nps Cu) and its effect on the physical and nutraceutical quality of hydroponic melon fruits. Notulae Botanicae Horti Agrobotanici Cluj-Napoca 50(1):12568. https://doi.org/10.15835/nbha50112568

Fortis-Hernández M, González-Rodríguez T, Espinosa-Palomeque B, Preciado-Rangel P, Gallegos-Robles MA, Rueda-Puente EO (2024) Foliar biofortification with copper nanoparticles and its effect on phytochemical quality and enzymatic activity in lettuce. Hortic Bras 42. https://doi.org/10.1590/s0102-0536-2024-e2617

Gao M, Chang J, Wang Z, Zhang H, Wang T (2023) Advances in transport and toxicity of nanoparticles in plants. J Nanobiotechnol 21(1):75. https://doi.org/10.1186/s12951-023-01830-5

García-Sánchez S, Bernales I, Cristobal S (2015) Early response to nanoparticles in the arabidopsis transcriptome compromises plant defence and root-hair development through salicylic acid signalling. BMC Genomics 16(1):341. https://doi.org/10.1186/s12864-015-1530-4

Griffin S, Masood M, Nasim M, Sarfraz M, Ebokaiwe A, Schäfer K-H, Keck C, Jacob C (2017) Natural nanoparticles: a particular matter inspired by nature. Antioxidants 7(1):3. https://doi.org/10.3390/antiox7010003

Griffitt RJ, Luo J, Gao J, Bonzongo J-C, Barber DS (2008) Effects of particle composition and species on toxicity of metallic nanomaterials in aquatic organisms. Environ Toxicol Chem 27(9):1972–1978. https://doi.org/10.1897/08-002.1

Grillo R, Mattos BD, Antunes DR, Forini MML, Monikh FA, Rojas OJ (2021) Foliage adhesion and interactions with particulate delivery systems for plant nanobionics and intelligent agriculture. Nano Today 37:101078. https://doi.org/10.1016/j.nantod.2021.101078

Haydar, Salman M, Ghosh S, Mandal P (2022) Application of iron oxide nanoparticles as micronutrient fertilizer in mulberry propagation. J Plant Growth Regul 41(4):1726–1746. https://doi.org/10.1007/s00344-021-10413-3

Hochella MF, Mogk DW, Ranville J, Allen IC, Luther GW, Marr LC, Peter McGrail B, Murayama M, Qafoku NP, Rosso KM, Sahai N, Schroeder PA, Vikesland P, Westerhoff P, Yang Y (2019) Natural, incidental, and engineered nanomaterials and their impacts on the earth system. Science 363(6434). https://doi.org/10.1126/science.aau8299

Hodson LL, Methner MM, Zumwalde RD (2009) Approaches to safe nanotechnology; managing the health and safety concerns associated with engineered nanomaterials

Hu J, Guo H, Li J, Wang Y, Xiao L, Xing B (2017) Interaction of γ-Fe2O3 nanoparticles with citrus maxima leaves and the corresponding physiological effects via foliar application. J Nanobiotechnol 15(1):51. https://doi.org/10.1186/s12951-017-0286-1

Hu J, Xinyi W, Fan W, Chen W, White JC, Yang Y, Wang B, Xing B, Tao S, Wang X (2020a) Potential application of titanium dioxide nanoparticles to improve the nutritional quality of coriander (Coriandrum Sativum L.). J Hazard Mater 389:121837. https://doi.org/10.1016/j.jhazmat.2019.121837

Hu P, An J, Faulkner MM, Honghong W, Li Z, Tian X, Giraldo JP (2020b) Nanoparticle charge and size control foliar delivery efficiency to plant cells and organelles. ACS Nano 14(7):7970–7986. https://doi.org/10.1021/acsnano.9b09178

Ikram M, Raja NI, Javed B, Mashwani Z-u-R, Hussain M, Hussain M, Ehsan M, Rafique N, Malik K, Sultana T, Akram A (2020) Foliar applications of bio-fabricated selenium nanoparticles to improve the growth of wheat plants under drought stress. Green Proces Synth 9(1):706–714. https://doi.org/10.1515/gps-2020-0067

Kapoor P, Dhaka RK, Sihag P, Mehla S, Sagwal V, Singh Y, Langaya S, Balyan P, Singh KP, Xing B, White JC, Dhankher OP, Kumar U (2022) Nanotechnology-enabled biofortification strategies for micronutrients enrichment of food crops: current understanding and future scope. NanoImpact 26:100407. https://doi.org/10.1016/j.impact.2022.100407

Kareem HA, Saleem MF, Saleem S, Rather SA, Wani SH, Siddiqui MH, Alamri S, Kumar R, Gaikwad NB, Guo Z, Niu J, Wang Q (2022) Zinc oxide nanoparticles interplay with physiological and biochemical attributes in terminal heat stress alleviation in Mungbean (Vigna Radiata L.). Front Plant Sci 13. https://doi.org/10.3389/fpls.2022.842349

Khan MT, Ahmed S, Shah AA, Shah AN, Tanveer M, El-Sheikh MA, Siddiqui MH (2021) Influence of zinc oxide nanoparticles to regulate the antioxidants enzymes, some Osmolytes and agronomic attributes in Coriandrum Sativum L. grown under water stress. Agronomy 11(10):2004. https://doi.org/10.3390/agronomy11102004

Khush GS, Lee S, Cho J-I, Jeon J-S (2012) Biofortification of crops for reducing malnutrition. Plant Biotechnol Rep 6(3):195–202. https://doi.org/10.1007/s11816-012-0216-5

Kumar P, Robins A, Vardoulakis S, Quincey P (2011) Technical challenges in tackling regulatory concerns for urban atmospheric nanoparticles. Particuology 9(6):566–571. https://doi.org/10.1016/j.partic.2011.06.002

Lahive E, Jurkschat K, Shaw BJ, Handy RD, Spurgeon DJ, Svendsen C (2014) Toxicity of cerium oxide nanoparticles to the earthworm Eisenia Fetida: subtle effects. Environ Chem 11(3):268. https://doi.org/10.1071/EN14028

Lian J, Zhao L, Jiani W, Xiong H, Bao Y, Zeb A, Tang J, Liu W (2020) Foliar spray of TiO2 nanoparticles prevails over root application in reducing cd accumulation and mitigating cd-induced Phytotoxicity in maize (Zea Mays L.). Chemosphere 239:124794. https://doi.org/10.1016/j.chemosphere.2019.124794

Lowry GV, Avellan A, Gilbertson LM (2019) Opportunities and challenges for nanotechnology in the agri-tech revolution. Nat Nanotechnol 14(6):517–522. https://doi.org/10.1038/s41565-019-0461-7

Mazhar Z, Akhtar J, Alhodaib A, Naz T, Zafar MI, Iqbal MM, Fatima H, Naz I (2023) Efficacy of ZnO nanoparticles in Zn fortification and partitioning of wheat and rice grains under salt stress. Sci Rep 13(1):2022. https://doi.org/10.1038/s41598-022-26039-8

Milla R, Osborne CP (2021) Crop origins explain variation in global agricultural relevance. Nat Plant 7(5):598–607. https://doi.org/10.1038/s41477-021-00905-1

Mitchell MJ, Billingsley MM, Haley RM, Wechsler ME, Peppas NA, Langer R (2021) Engineering precision nanoparticles for drug delivery. Nat Rev Drug Discov 20(2):101–124. https://doi.org/10.1038/s41573-020-0090-8

Mittal D, Kaur G, Singh P, Yadav K, Ali SA (2020) Nanoparticle-based sustainable agriculture and food science: recent advances and future outlook. Front Nanotechnol 2. https://doi.org/10.3389/fnano.2020.579954

Morimoto Y, Kobayashi N, Shinohara N, Myojo T, Tanaka I, Nakanishi J (2010) Hazard assessments of manufactured nanomaterials. J Occup Health 52(6):325–334. https://doi.org/10.1539/joh.R10003

Naguib DM, Abdalla H (2019) Metabolic status during germination of nano silica primed Zea Mays seeds under salinity stress. J Crop Sci Biotechnol 22(5):415–423. https://doi.org/10.1007/s12892-019-0168-0

Nolan TM, Vukašinović N, Liu D, Russinova E, Yin Y (2020) Brassinosteroids: multidimensional regulators of plant growth, development, and stress responses. Plant Cell 32(2):295–318. https://doi.org/10.1105/tpc.19.00335

Nowack B, Bucheli TD (2007) Occurrence, behavior and effects of nanoparticles in the environment. Environ Pollut 150(1):5–22. https://doi.org/10.1016/j.envpol.2007.06.006

Oberdörster G, Sharp ZD, Atudorei V, Elder A, Gelein R, Kreyling WG, Cox C (2004) Translocation of inhaled ultrafine particles to the brain. Inhal Toxicol 16:437–445

Pandey G (2018) Challenges and future prospects of agri-nanotechnology for sustainable agriculture in India. Environ Technol Innov 11:299–307. https://doi.org/10.1016/j.eti.2018.06.012

Parkinson SJ, Tungsirisurp S, Joshi C, Richmond BL, Gifford ML, Sikder A, Lynch I, O'Reilly RK, Napier RM (2022) Polymer nanoparticles pass the plant Interface. Nat Commun 13(1):7385. https://doi.org/10.1038/s41467-022-35066-y

Pestovsky YS, Martínez-Antonio A (2017) The use of nanoparticles and Nanoformulations in agriculture. J Nanosci Nanotechnol 17(12):8699–8730. https://doi.org/10.1166/jnn.2017.15041

Peyrot C, Wilkinson KJ, Desrosiers M, Sauvé S (2014) Effects of silver nanoparticles on soil enzyme activities with and without added organic matter. Environ Toxicol Chem 33(1):115–125. https://doi.org/10.1002/etc.2398

Popov AP, Priezzhev AV, Lademann J, Myllylä R (2005) TiO2 nanoparticles as an effective UV-B radiation skin-protective compound in sunscreens. J Phys D Appl Phys 38(15):2564. https://doi.org/10.1088/0022-3727/38/15/006

Prakash V, Rai P, Sharma NC, Singh VP, Tripathi DK, Sharma S, Sahi S (2022) Application of zinc oxide nanoparticles as fertilizer boosts growth in Rice Plant and alleviates chromium stress by regulating genes involved in oxidative stress. Chemosphere 303:134554. https://doi.org/10.1016/j.chemosphere.2022.134554

Qi M, Liu Y, Li T (2013) Nano-TiO2 improve the photosynthesis of tomato leaves under mild heat stress. Biol Trace Elem Res 156(1–3):323–328. https://doi.org/10.1007/s12011-013-9833-2

Qi W-Y, Li Q, Chen H, Liu J, Xing S-F, Meng X, Yan Z, Song C, Wang S-G (2021) Selenium nanoparticles ameliorate brassica Napus L. cadmium toxicity by inhibiting the respiratory burst and scavenging reactive oxygen species. J Hazard Mater 417:125900. https://doi.org/10.1016/j.jhazmat.2021.125900

Rajkumari N, Alex S, Soni KB, Anith KN, Viji MM, Kiran AG (2021) Silver nanoparticles for biolistic transformation in nicotiana tabacum L. 3 Biotech 11(12):497. https://doi.org/10.1007/s13205-021-03043-9

Rasheed A, Li H, Tahir MM, Mahmood A, Nawaz M, Shah AN, Aslam MT, Negm S, Moustafa M, Hassan MU, Ziming W (2022) The role of nanoparticles in plant biochemical, physiological, and molecular responses under drought stress: a review. Front Plant Sci 13. https://doi.org/10.3389/fpls.2022.976179

Rasmidi R, Duinong M, Chee FP (2021) Radiation damage effects on zinc oxide (ZnO) based semiconductor devices– a review. Radiat Phys Chem 184:109455. https://doi.org/10.1016/j.radphyschem.2021.109455

Reijnders L (2006) Cleaner nanotechnology and hazard reduction of manufactured nanoparticles. J Clean Prod 14(2):124–133. https://doi.org/10.1016/j.jclepro.2005.03.018

Reshma Z, Meenal K (2022) Foliar application of biosynthesised zinc nanoparticles as a strategy for ferti-fortification by improving yield, zinc content and zinc use efficiency in Amaranth. Heliyon 8(10):e10912. https://doi.org/10.1016/j.heliyon.2022.e10912

Rico CM, Morales MI, Barrios AC, McCreary R, Hong J, Lee W-Y, Nunez J, Peralta-Videa JR, Gardea-Torresdey JL (2013) Effect of cerium oxide nanoparticles on the quality of rice (Oryza Sativa L.) grains. J Agric Food Chem 61(47):11278–11285. https://doi.org/10.1021/jf404046v

Rico CM, Barrios AC, Tan W, Rubenecia R, Lee SC, Varela-Ramirez A, Peralta-Videa JR, Gardea-Torresdey JL (2015) Physiological and biochemical response of soil-grown barley (Hordeum

Vulgare L.) to cerium oxide nanoparticles. Environ Sci Pollut Res 22(14):10551–10558. https://doi.org/10.1007/s11356-015-4243-y

Safari M, Ardebili ZO, Iranbakhsh A (2018) Selenium Nano-particle induced alterations in expression patterns of heat shock factor A4A (HSFA4A), and high molecular weight Glutenin subunit 1Bx (Glu-1Bx) and enhanced nitrate reductase activity in wheat (Triticum Aestivum L.). Acta Physiol Plant 40(6):117. https://doi.org/10.1007/s11738-018-2694-8

Seleiman MF, Almutairi KF, Alotaibi M, Shami A, Alhammad BA, Battaglia ML (2021) Nano-fertilization as an emerging fertilization technique: why can modern agriculture benefit from its use? Plan Theory 10(1)

Sembada AA, Wuled Lenggoro I (2024) Transport of nanoparticles into plants and their detection methods. Nano 14(2):131. https://doi.org/10.3390/nano14020131

Shalaby SM, Khater MK, Perucho AM, Mohamed SA, Helwa I, Laknaur A, Lebedyeva I, Liu Y, Diamond MP, Al-Hendy AA (2016) Magnetic nanoparticles as a new approach to improve the efficacy of gene therapy against differentiated human uterine fibroid cells and tumor-initiating stem cells. Fertil Steril 105(6):1638–1648.e8. https://doi.org/10.1016/j.fertnstert.2016.03.001

Tsuji JS, Maynard AD, Howard PC, James JT, Lam C-w, Warheit DB, Santamaria AB (2006) Research strategies for safety evaluation of nanomaterials, part IV: risk assessment of nanoparticles. Toxicol Sci 89(1):42–50. https://doi.org/10.1093/toxsci/kfi339

van der Ploeg MJC, Handy RD, Waalewijn-Kool PL, van den Berg JHJ, Herrera ZE, Rivera JB, Molleman B, Baveco JM, Tromp P, Peters RJB, Koopmans GF, Rietjens IMCM, van den Brink NW (2014) Effects of silver nanoparticles (NM-300K) on Lumbricus Rubellus earthworms and particle characterization in relevant test matrices including soil. Environ Toxicol Chem 33(4):743–752. https://doi.org/10.1002/etc.2487

Wahid I, Rani P, Kumari S, Ahmad R, Hussain SJ, Alamri S, Tripathy N, Iqbal M, Khan R (2022) Biosynthesized gold nanoparticles maintained nitrogen metabolism, nitric oxide synthesis, ions balance, and stabilizes the defense systems to improve salt stress tolerance in wheat. Chemosphere 287:132142. https://doi.org/10.1016/j.chemosphere.2021.132142

Wu Z, Min H, Lin P, Liu S, Wehner B, Wiedensohler A (2008) Particle number size distribution in the urban atmosphere of Beijing, China. Atmos Environ 42(34):7967–7980. https://doi.org/10.1016/j.atmosenv.2008.06.022

Yang X, Alidoust D, Wang C (2020) Effects of iron oxide nanoparticles on the mineral composition and growth of soybean (Glycine Max L.) plants. Acta Physiol Plant 42(8):128. https://doi.org/10.1007/s11738-020-03104-1

Zahedi SM, Moharrami F, Sarikhani S, Padervand M (2020) Selenium and silica nanostructure-based recovery of strawberry plants subjected to drought stress. Sci Rep 10(1):17672. https://doi.org/10.1038/s41598-020-74273-9

Zahra N, Wahid A, Shaukat K, Hafeez MB, Batool A, Hasanuzzaman M (2021) Oxidative stress tolerance potential of Milk thistle ecotypes after supplementation of different plant growth-promoting agents under salinity. Plant Physiol Biochem 166:53–65. https://doi.org/10.1016/j.plaphy.2021.05.042

Zhao L, Bai T, Wei H, Gardea-Torresdey JL, Keller A, White JC (2022) Nanobiotechnology-based strategies for enhanced crop stress resilience. Nat Food 3(10):829–836. https://doi.org/10.1038/s43016-022-00596-7

Zheng L, Hong F, Shipeng L, Liu C (2005) Effect of Nano-TiO2 on strength of naturally aged seeds and growth of spinach. Biol Trace Elem Res 104(1):083–092. https://doi.org/10.1385/BTER:104:1:083

Zhu Y, Fang X, Liu Q, Chen M, Liu X, Wang Y, Sun Y, Zhang L (2019) Nanomaterials and plants: positive effects, toxicity and the remediation of metal and metalloid pollution in soil. Sci Total Environ 662:414–421. https://doi.org/10.1016/j.scitotenv.2019.01.234

Zolghadrnasab M, Mousavi A, Farmany A, Arpanaei A (2021) Ultrasound-mediated gene delivery into suspended plant cells using Polyethyleneimine-coated mesoporous silica nanoparticles. Ultrason Sonochem 73:105507. https://doi.org/10.1016/j.ultsonch.2021.105507

The manufacturer's authorised representative in the EU is Springer Nature Customer Service Centre GmbH, Europaplatz 3, 69115 Heidelberg, Germany. If you have any concerns regarding our products, please contact ProductSafety@springernature.com

Printed and bound by CPI Group (UK) Ltd, Croydon, CR0 4YY
01/10/2026
02225929-0018